SOTHIS PRESS

Sothis Press aims to completely reimagine the public domain book with an emphasis on universal access to captivatingly designed print editions. We believe that technology will enable every single book ever published to be perpetually accessible in beautiful print form. Sothis is beauty in ubiquity.

www.sothis.press
1 Waterhouse Square, London, ECIN 2ST

# Elements of the Differential, and Integral Calculus

Revised Edition

William Anthony Granville

SOTHIS PRESS

ISBN 978-1-0365-1817-2

*Elements of the Differential, and Integral Calculus (Revised*
is in the public domain.

This edition is published by
Sothis Press © 2023. The company
is registered in England and Wales.
The Company Number is 14900389.

MATHEMATICAL TEXTS

FOR COLLEGES

EDITED BY

PERCEY F. SMITH, Ph.D.

PROFESSOR OF MATHEMATICS IN THE SHEFFIELD SCIENTIFIC SCHOOL OF YALE UNIVERSITY

Sir Isaac Newton

# ELEMENTS OF THE DIFFERENTIAL AND INTEGRAL CALCULUS

*(REVISED EDITION)*

BY

WILLIAM ANTHONY GRANVILLE, PH.D., LL.D.

FORMERLY PRESIDENT OF GETTYSBURG COLLEGE

WITH THE EDITORIAL COÖPERATION OF

PERCEY F. SMITH, PH.D.

PROFESSOR OF MATHEMATICS IN THE SHEFFIELD SCIENTIFIC SCHOOL
YALE UNIVERSITY

GINN AND COMPANY

BOSTON · NEW YORK · CHICAGO · LONDON
ATLANTA · DALLAS · COLUMBUS · SAN FRANCISCO

PRINTED IN THE UNITED STATES OF AMERICA

627.1

The Athenæum Press
GINN AND COMPANY · PROPRIETORS · BOSTON · U.S.A.

# PREFACE

That teachers and students of the Calculus have shown such a generous appreciation of Granville's "Elements of the Differential and Integral Calculus" has been very gratifying to the author. In the last few years considerable progress has been made in the teaching of the elements of the Calculus, and in this revised edition of Granville's "Calculus" the latest and best methods are exhibited, — methods that have stood the test of actual classroom work. Those features of the first edition which contributed so much to its usefulness and popularity have been retained. The introductory matter has been cut down somewhat in order to get down to the real business of the Calculus sooner. As this is designed essentially for a drill book, the pedagogic principle that each result should be made intuitionally as well as analytically evident to the student has been kept constantly in mind. The object is not to teach the student to rely on his intuition, but, in some cases, to use this faculty in advance of analytical investigation. Graphical illustration has been drawn on very liberally.

This Calculus is based on the method of limits and is divided into two main parts, — Differential Calculus and Integral Calculus. As special features, attention may be called to the effort to make perfectly clear the nature and extent of each new theorem, the large number of carefully graded exercises, and the summarizing into working rules of the methods of solving problems. In the Integral Calculus the notion of integration over a plane area has been much enlarged upon, and integration as the limit of a summation is constantly emphasized. The existence of the limit $e$ has been assumed and its approximate value calculated from its graph. A large number of new examples have been added, both with and without answers. At the end of almost every chapter will be found a collection of miscellaneous examples. Among the new topics added are approximate integration, trapezoidal rule, parabolic rule, orthogonal trajectories, centers of area and volume, pressure of liquids, work done, etc. Simple practical problems have been added throughout; problems that illustrate the theory and at the same time are of interest to the student. These problems do not presuppose an extended knowledge in any particular branch of science, but are based on knowledge that all students of the Calculus are supposed to have in common.

The author has tried to write a textbook that is thoroughly modern and teachable, and the capacity and needs of the student pursuing a first course in the Calculus have been kept constantly in mind. The book contains more material than is necessary for the usual course of one hundred lessons given in our colleges and engineering schools; but this gives teachers an opportunity to choose such subjects as best suit the needs of their classes. It is believed that the volume contains all topics from which a selection naturally would be made in preparing students either for elementary work in applied science or for more advanced work in pure mathematics.

WILLIAM A. GRANVILLE

GETTYSBURG COLLEGE
Gettysburg, Pa.

# CONTENTS

## DIFFERENTIAL CALCULUS

### CHAPTER I

#### COLLECTION OF FORMULAS



### CHAPTER II

#### VARIABLES AND FUNCTIONS



### CHAPTER III

#### THEORY OF LIMITS

## CHAPTER IV

### DIFFERENTIATION



## CHAPTER V

### RULES FOR DIFFERENTIATING STANDARD ELEMENTARY FORMS

## CHAPTER VI

### SIMPLE APPLICATIONS OF THE DERIVATIVE



## CHAPTER VII

### SUCCESSIVE DIFFERENTIATION



## CHAPTER VIII

### MAXIMA AND MINIMA. POINTS OF INFLECTION. CURVE TRACING

## CHAPTER IX

### DIFFERENTIALS



## CHAPTER X

### RATES



## CHAPTER XI

### CHANGE OF VARIABLE



## CHAPTER XII

### CURVATURE. RADIUS OF CURVATURE



## CHAPTER XIII

### THEOREM OF MEAN VALUE. INDETERMINATE FORMS

## CHAPTER XIV

### CIRCLE OF CURVATURE. CENTER OF CURVATURE



## CHAPTER XV

### PARTIAL DIFFERENTIATION



## CHAPTER XVI

### ENVELOPES

## CHAPTER XVII

### SERIES



## CHAPTER XVIII

### EXPANSION OF FUNCTIONS



## CHAPTER XIX

### ASYMPTOTES. SINGULAR POINTS



## CHAPTER XX

### APPLICATIONS TO GEOMETRY OF SPACE

## CHAPTER XXI

### CURVES FOR REFERENCE

# INTEGRAL CALCULUS

## CHAPTER XXII

### INTEGRATION. RULES FOR INTEGRATING STANDARD ELEMENTARY FORMS



## CHAPTER XXIII

### CONSTANT OF INTEGRATION



## CHAPTER XXIV

### THE DEFINITE INTEGRAL

## CHAPTER XXV

### INTEGRATION OF RATIONAL FRACTIONS



## CHAPTER XXVI

### INTEGRATION BY SUBSTITUTION OF A NEW VARIABLE. RATIONALIZATION



## CHAPTER XXVII

### INTEGRATION BY PARTS. REDUCTION FORMULAS



## CHAPTER XXVIII

### INTEGRATION A PROCESS OF SUMMATION

## CHAPTER XXIX

### SUCCESSIVE AND PARTIAL INTEGRATION



## CHAPTER XXX

### ORDINARY DIFFERENTIAL EQUATIONS



## CHAPTER XXXI

### INTEGRAPH. APPROXIMATE INTEGRATION. TABLE OF INTEGRALS

Gottfried Wilhelm Leibnitz

# DIFFERENTIAL CALCULUS

## CHAPTER I

### COLLECTION OF FORMULAS

**1. Formulas for reference.** For the convenience of the student we give the following list of elementary formulas from Algebra, Geometry, Trigonometry, and Analytic Geometry.

**1.** Binomial Theorem ($n$ being a positive integer):

$$(a+b)^n = a^n + na^{n-1}b + \frac{n(n-1)}{\lfloor\!\underline{\,2\,}}a^{n-2}b^2 + \frac{n(n-1)(n-2)}{\lfloor\!\underline{\,3\,}}a^{n-3}b^3 + \cdots$$

$$+ \frac{n(n-1)(n-2)\cdots(n-r+2)}{\lfloor\!\underline{\,r-1\,}}a^{n-r+1}b^{r-1} + \cdots.$$

**2.** $n\,! = \lfloor\!\underline{\,n\,} = 1 \cdot 2 \cdot 3 \cdot 4 \cdots (n-1)\,n.$

**3.** In the quadratic equation $ax^2 + bx + c = 0$,

when $b^2 - 4\,ac > 0$, the roots are real and unequal;
when $b^2 - 4\,ac = 0$, the roots are real and equal;
when $b^2 - 4\,ac < 0$, the roots are imaginary.

**4.** When a quadratic equation is reduced to the form $x^2 + px + q = 0$,

$p$ = sum of roots with sign changed, and $q$ = product of roots.

**5.** In an arithmetical series,

$$l = a + (n-1)\,d\,;\quad s = \frac{n}{2}(a+l) = \frac{n}{2}[2\,a + (n-1)\,d].$$

**6.** In a geometrical series,

$$l = ar^{n-1}\,;\quad s = \frac{rl-a}{r-1} = \frac{a\,(r^n-1)}{r-1}.$$

**7.** $\log ab = \log a + \log b$. **10.** $\log \sqrt[n]{a} = \frac{1}{n}\log a$. **13.** $\log \frac{1}{a} = -\log a$.

**8.** $\log \frac{a}{b} = \log a - \log b$. **11.** $\log 1 = 0$. **14.** Circumference of circle $= 2\pi r$.

**9.** $\log a^n = n \log a$. **12.** $\log_a a = 1$. **15.** Area of circle $= \pi r^2$.

* In formulas 14–25, $r$ denotes radius, $a$ altitude, $B$ area of base, and $s$ slant height.

**16.** Volume of prism $= Ba$.

**17.** Volume of pyramid $= \frac{1}{3} Ba$.

**18.** Volume of right circular cylinder $= \pi r^2 a$.

**19.** Lateral surface of right circular cylinder $= 2\pi r a$.

**20.** Total surface of right circular cylinder $= 2\pi r(r + a)$.

**21.** Volume of right circular cone $= \frac{1}{3}\pi r^2 a$.

**22.** Lateral surface of right circular cone $= \pi r s$.

**23.** Total surface of right circular cone $= \pi r(r + s)$.

**24.** Volume of sphere $= \frac{4}{3}\pi r^3$.

**25.** Surface of sphere $= 4\pi r^2$.

**26.** $\sin x = \dfrac{1}{\csc x}$; $\cos x = \dfrac{1}{\sec x}$; $\tan x = \dfrac{1}{\cot x}$.

**27.** $\tan x = \dfrac{\sin x}{\cos x}$; $\cot x = \dfrac{\cos x}{\sin x}$.

**28.** $\sin^2 x + \cos^2 x = 1$; $1 + \tan^2 x = \sec^2 x$; $1 + \cot^2 x = \csc^2 x$.

**29.** $\sin x = \cos\left(\dfrac{\pi}{2} - x\right)$;

$\cos x = \sin\left(\dfrac{\pi}{2} - x\right)$;

$\tan x = \cot\left(\dfrac{\pi}{2} - x\right)$.

**30.** $\sin(\pi - x) = \sin x$;

$\cos(\pi - x) = -\cos x$;

$\tan(\pi - x) = -\tan x$.

**31.** $\sin(x + y) = \sin x \cos y + \cos x \sin y$.

**32.** $\sin(x - y) = \sin x \cos y - \cos x \sin y$.

**33.** $\cos(x \pm y) = \cos x \cos y \mp \sin x \sin y$.

**34.** $\tan(x + y) = \dfrac{\tan x + \tan y}{1 - \tan x \tan y}$.

**35.** $\tan(x - y) = \dfrac{\tan x - \tan y}{1 + \tan x \tan y}$.

**36.** $\sin 2x = 2 \sin x \cos x$; $\cos 2x = \cos^2 x - \sin^2 x$; $\tan 2x = \dfrac{2 \tan x}{1 - \tan^2 x}$.

**37.** $\sin x = 2 \sin \dfrac{x}{2} \cos \dfrac{x}{2}$; $\cos x = \cos^2 \dfrac{x}{2} - \sin^2 \dfrac{x}{2}$; $\tan x = \dfrac{2 \tan \frac{1}{2} x}{1 - \tan^2 \frac{1}{2} x}$.

**38.** $\cos^2 x = \frac{1}{2} + \frac{1}{2} \cos 2x$; $\sin^2 x = \frac{1}{2} - \frac{1}{2} \cos 2x$.

**39.** $1 + \cos x = 2 \cos^2 \dfrac{x}{2}$; $1 - \cos x = 2 \sin^2 \dfrac{x}{2}$.

**40.** $\sin \dfrac{x}{2} = \pm \sqrt{\dfrac{1 - \cos x}{2}}$; $\cos \dfrac{x}{2} = \pm \sqrt{\dfrac{1 + \cos x}{2}}$; $\tan \dfrac{x}{2} = \pm \sqrt{\dfrac{1 - \cos x}{1 + \cos x}}$.

**41.** $\sin x + \sin y = 2 \sin \frac{1}{2}(x + y) \cos \frac{1}{2}(x - y)$.

**42.** $\sin x - \sin y = 2 \cos \frac{1}{2}(x + y) \sin \frac{1}{2}(x - y)$.

**43.** $\cos x + \cos y = 2 \cos \frac{1}{2}(x + y) \cos \frac{1}{2}(x - y)$.

**44.** $\cos x - \cos y = -2 \sin \frac{1}{2}(x + y) \sin \frac{1}{2}(x - y)$.

**45.** $\dfrac{a}{\sin A} = \dfrac{b}{\sin B} = \dfrac{c}{\sin C}$; Law of Sines.

**46.** $a^2 = b^2 + c^2 - 2bc \cos A$; Law of Cosines.

**47.** $d = \sqrt{(x_1 - x_2)^2 + (y_1 - y_2)^2}$; distance between points $(x_1, y_1)$ and $(x_2, y_2)$.

**48.** $d = \dfrac{Ax_1 + By_1 + C}{\pm\sqrt{A^2 + B^2}}$; distance from line $Ax + By + C = 0$ to $(x_1, y_1)$.

49. $x = \frac{x_1 + x_2}{2}$, $y = \frac{y_1 + y_2}{2}$; coördinates of middle point.

50. $x = x_0 + x'$, $y = y_0 + y'$; transforming to new origin $(x_0, y_0)$.

51. $x = x' \cos\theta - y' \sin\theta$, $y = x' \sin\theta + y' \cos\theta$; transforming to new axes making the angle $\theta$ with old.

52. $x = \rho \cos\theta$, $y = \rho \sin\theta$; transforming from rectangular to polar coördinates.

53. $\rho = \sqrt{x^2 + y^2}$, $\theta = \text{arc} \tan \frac{y}{x}$; transforming from polar to rectangular coördinates.

54. Different forms of equation of a straight line:

(a) $\frac{y - y_1}{x - x_1} = \frac{y_2 - y_1}{x_2 - x_1}$, two-point form;

(b) $\frac{x}{a} + \frac{y}{b} = 1$, intercept form;

(c) $y - y_1 = m(x - x_1)$, slope-point form;
(d) $y = mx + b$, slope-intercept form;
(e) $x \cos\alpha + y \sin\alpha = p$, normal form;
(f) $Ax + By + C = 0$, general form.

55. $\tan\theta = \frac{m_1 - m_2}{1 + m_1 m_2}$, angle between two lines whose slopes are $m_1$ and $m_2$.

$m_1 = m_2$ when lines are parallel,

and $m_1 = -\frac{1}{m_2}$ when lines are perpendicular.

56. $(x - \alpha)^2 + (y - \beta)^2 = r^2$, equation of circle with center $(\alpha, \beta)$ and radius $r$.

## 2. Greek alphabet.

| Letters | Names | Letters | Names | Letters | Names |
|---|---|---|---|---|---|
| Α α | Alpha | Ι ι | Iota | Ρ ρ | Rho |
| Β β | Beta | Κ κ | Kappa | Σ σ ς | Sigma |
| Γ γ | Gamma | Λ λ | Lambda | Τ τ | Tau |
| Δ δ | Delta | Μ μ | Mu | Υ υ | Upsilon |
| Ε ϵ | Epsilon | Ν ν | Nu | Φ ϕ | Phi |
| Ζ ζ | Zeta | Ξ ξ | Xi | Χ χ | Chi |
| Η η | Eta | Ο ο | Omicron | Ψ ψ | Psi |
| Θ θ | Theta | Π π | Pi | Ω ω | Omega |

## 3. Rules for signs of the trigonometric functions.

| Quadrant | Sin | Cos | Tan | Cot | Sec | Csc |
|---|---|---|---|---|---|---|
| First . . . . | + | + | + | + | + | + |
| Second . . . . | + | − | − | − | − | + |
| Third . . . . | − | − | + | + | − | − |
| Fourth . . . . | − | + | − | − | + | − |

**4. Natural values of the trigonometric functions.**

| Angle in Radians | Angle in Degrees | Sin | Cos | Tan | Cot | Sec | Csc |
|---|---|---|---|---|---|---|---|
| 0 | 0° | 0 | 1 | 0 | $\infty$ | 1 | $\infty$ |
| $\frac{\pi}{2}$ | 90° | 1 | 0 | $\infty$ | 0 | $\infty$ | 1 |
| $\pi$ | 180° | 0 | $-1$ | 0 | $\infty$ | $-1$ | $\infty$ |
| $\frac{3\pi}{2}$ | 270° | $-1$ | 0 | $\infty$ | 0 | $\infty$ | $-1$ |
| $2\pi$ | 360° | 0 | 1 | 0 | $\infty$ | 1 | $\infty$ |

| Angle in Radians | Angle in Degrees | Sin | Cos | Tan | Cot | Sec | Csc |
|---|---|---|---|---|---|---|---|
| 0 | 0° | 0 | 1 | 0 | $\infty$ | 1 | $\infty$ |
| $\frac{\pi}{6}$ | 30° | $\frac{1}{2}$ | $\frac{\sqrt{3}}{2}$ | $\frac{\sqrt{3}}{3}$ | $\sqrt{3}$ | $\frac{2\sqrt{3}}{3}$ | 2 |
| $\frac{\pi}{4}$ | 45° | $\frac{\sqrt{2}}{2}$ | $\frac{\sqrt{2}}{2}$ | 1 | 1 | $\sqrt{2}$ | $\sqrt{2}$ |
| $\frac{\pi}{3}$ | 60° | $\frac{\sqrt{3}}{2}$ | $\frac{1}{2}$ | $\sqrt{3}$ | $\frac{\sqrt{3}}{3}$ | 2 | $\frac{2\sqrt{3}}{3}$ |
| $\frac{\pi}{2}$ | 90° | 1 | 0 | $\infty$ | 0 | $\infty$ | 1 |

| Angle in Radians | Angle in Degrees | Sin | Cos | Tan | Cot | | |
|---|---|---|---|---|---|---|---|
| .0000 | 0° | .0000 | 1.0000 | .0000 | $\infty$ | 90° | 1.5708 |
| .0175 | 1° | .0175 | .9998 | .0175 | 57.290 | 89° | 1.5533 |
| .0349 | 2° | .0349 | .9994 | .0349 | 28.636 | 88° | 1.5359 |
| .0524 | 3° | .0523 | .9986 | .0524 | 19.081 | 87° | 1.5184 |
| .0698 | 4° | .0698 | .9976 | .0699 | 14.300 | 86° | 1.5010 |
| .0873 | 5° | .0872 | .9962 | .0875 | 11.430 | 85° | 1.4835 |
| .1745 | 10° | .1736 | .9848 | .1763 | 5.671 | 80° | 1.3963 |
| .2618 | 15° | .2588 | .9659 | .2679 | 3.732 | 75° | 1.3090 |
| .3491 | 20° | .3420 | .9397 | .3640 | 2.747 | 70° | 1.2217 |
| .4363 | 25° | .4226 | .9063 | .4663 | 2.145 | 65° | 1.1345 |
| .5236 | 30° | .5000 | .8660 | .5774 | 1.732 | 60° | 1.0472 |
| .6109 | 35° | .5736 | .8192 | .7002 | 1.428 | 55° | .9599 |
| .6981 | 40° | .6428 | .7660 | .8391 | 1.192 | 50° | .8727 |
| .7854 | 45° | .7071 | .7071 | 1.0000 | 1.000 | 45° | .7854 |
| | | Cos | Sin | Cot | Tan | Angle in Degrees | Angle in Radians |

## 5. Logarithms of numbers and trigonometric functions.

Table of Mantissas of the Common Logarithms of Numbers

| No. | 0 | 1 | 2 | 3 | 4 | 5 | 6 | 7 | 8 | 9 |
|---|---|---|---|---|---|---|---|---|---|---|
| 1 | 0000 | 0414 | 0792 | 1139 | 1461 | 1761 | 2041 | 2304 | 2553 | 2788 |
| 2 | 3010 | 3222 | 3424 | 3617 | 3802 | 3979 | 4150 | 4314 | 4472 | 4624 |
| 3 | 4771 | 4914 | 5051 | 5185 | 5315 | 5441 | 5563 | 5682 | 5798 | 5911 |
| 4 | 6021 | 6128 | 6232 | 6335 | 6435 | 6532 | 6028 | 6721 | 6812 | 6902 |
| 5 | 6990 | 7076 | 7160 | 7243 | 7324 | 7404 | 7482 | 7559 | 7634 | 7709 |
| 6 | 7782 | 7853 | 7924 | 7993 | 8062 | 8129 | 8195 | 8261 | 8325 | 8388 |
| 7 | 8451 | 8513 | 8573 | 8633 | 8692 | 8751 | 8808 | 8865 | 8921 | 8976 |
| 8 | 9031 | 9085 | 9138 | 9191 | 9243 | 9294 | 9345 | 9395 | 9445 | 9494 |
| 9 | 9542 | 9590 | 9638 | 9685 | 9731 | 9777 | 9823 | 9868 | 9912 | 9956 |
| 10 | 0000 | 0043 | 0086 | 0128 | 0170 | 0212 | 0253 | 0294 | 0334 | 0374 |
| 11 | 0414 | 0453 | 0492 | 0531 | 0569 | 0607 | 0645 | 0682 | 0719 | 0755 |
| 12 | 0792 | 0828 | 0864 | 0899 | 0934 | 0969 | 1004 | 1038 | 1072 | 1106 |
| 13 | 1139 | 1173 | 1206 | 1239 | 1271 | 1303 | 1335 | 1367 | 1399 | 1480 |
| 14 | 1461 | 1492 | 1523 | 1553 | 1584 | 1614 | 1644 | 1673 | 1703 | 1732 |
| 15 | 1761 | 1790 | 1818 | 1847 | 1875 | 1903 | 1931 | 1959 | 1987 | 2014 |
| 16 | 2041 | 2068 | 2095 | 2122 | 2148 | 2175 | 2201 | 2227 | 2253 | 2279 |
| 17 | 2304 | 2330 | 2355 | 2380 | 2405 | 2430 | 2455 | 2480 | 2504 | 2529 |
| 18 | 2553 | 2577 | 2601 | 2625 | 2648 | 2672 | 2695 | 2718 | 2742 | 2765 |
| 19 | 2788 | 2810 | 2833 | 2856 | 2878 | 2900 | 2923 | 2945 | 2967 | 2989 |

Table of Logarithms of the Trigonometric Functions

| Angle in Radians | Angle in Degrees | log sin | log cos | log tan | log cot | | |
|---|---|---|---|---|---|---|---|
| .0000 | 0° | . . . . | 0.000 | . . . . | . . . . | 90° | 1.5708 |
| .0175 | 1° | 8.2419 | 9.9999 | 8.2419 | 1.7581 | 89° | 1.5538 |
| .0349 | 2° | 8.5428 | 9.9997 | 8.5431 | 1.4569 | 88° | 1.5359 |
| .0524 | 3° | 8.7188 | 9.9994 | 8.7194 | 1.2806 | 87° | 1.5184 |
| .0698 | 4° | 8.8436 | 9.9989 | 8.8446 | 1.1554 | 86° | 1.5010 |
| .0873 | 5° | 8.9403 | 9.9983 | 8.9420 | 1.0580 | 85° | 1.4835 |
| .1745 | 10° | 9.2397 | 9.9934 | 9.2463 | 0.7537 | 80° | 1.3963 |
| .2618 | 15° | 9.4130 | 9.9849 | 9.4281 | 0.5719 | 75° | 1.3090 |
| .3491 | 20° | 9.5341 | 9.9730 | 9.5611 | 0.4389 | 70° | 1.2217 |
| .4363 | 25° | 9.6259 | 9.9573 | 9.6687 | 0.3313 | 65° | 1.1345 |
| .5236 | 30° | 9.6990 | 9.9375 | 9.7614 | 0.2386 | 60° | 1.0472 |
| .6109 | 35° | 9.7586 | 9.9134 | 9.8452 | 0.1548 | 55° | 0.9599 |
| .6981 | 40° | 9.8081 | 9.8843 | 9.9238 | 0.0762 | 50° | 0.8727 |
| .7854 | 45° | 9.8495 | 9.8495 | 0.0000 | 0.0000 | 45° | 0.7854 |
| | | log cos | log sin | log cot | log tan | Angle in Degrees | Angle in Radians |

# CHAPTER II

## VARIABLES AND FUNCTIONS

**6. Variables and constants.** A *variable* is a quantity to which an unlimited number of values can be assigned. Variables are denoted by the later letters of the alphabet. Thus, in the equation of a straight line,

$$\frac{x}{a}+\frac{y}{b}=1,$$

$x$ and $y$ may be considered as the variable coördinates of a point moving along the line.

A quantity whose value remains unchanged is called a *constant.*

*Numerical* or *absolute constants* retain the same values in all problems, as 2, 5, $\sqrt{7}$, $\pi$, etc.

*Arbitrary constants*, or *parameters*, are constants to which any one of an unlimited set of numerical values may be assigned, and they are supposed to have these assigned values throughout the investigation. They are usually denoted by the earlier letters of the alphabet. Thus, for every pair of values arbitrarily assigned to $a$ and $b$, the equation

$$\frac{x}{a}+\frac{y}{b}=1$$

represents some particular straight line.

**7. Interval of a variable.** Very often we confine ourselves to a portion only of the number system. For example, we may restrict our variable so that it shall take on only such values as lie between $a$ and $b$, where $a$ and $b$ may be included, or either or both excluded. We shall employ the symbol $[a, b]$, $a$ being less than $b$, to represent the numbers $a$, $b$, and all the numbers between them, unless otherwise stated. This symbol $[a, b]$ is read *the interval from a to b.*

**8. Continuous variation.** A variable $x$ is said to vary continuously through an interval $[a, b]$, when $x$ starts with the value $a$ and increases until it takes on the value $b$ in such a manner as to assume the value of every number between $a$ and $b$ in the order of their magnitudes. This may be illustrated geometrically as follows:

The origin being at $O$, lay off on the straight line the points $A$ and $B$ corresponding to the numbers $a$ and $b$. Also let the point $P$ correspond to a particular value of the variable $x$. Evidently the interval $[a, b]$ is represented by the segment $AB$. Now as $x$ varies continuously from $a$ to $b$ inclusive, i.e. through the interval $[a, b]$, the point $P$ generates the segment $AB$.

**9. Functions.** *When two variables are so related that the value of the first variable depends on the value of the second variable, then the first variable is said to be a function of the second variable.*

Nearly all scientific problems deal with quantities and relations of this sort, and in the experiences of everyday life we are continually meeting conditions illustrating the dependence of one quantity on another. For instance, the *weight* a man is able to lift depends on his *strength*, other things being equal. Similarly, the *distance* a boy can run may be considered as depending on the *time*. Or, we may say that the *area* of a square is a function of the *length* of a side, and the *volume* of a sphere is a function of its *diameter*.

**10. Independent and dependent variables.** The second variable, to which values may be assigned at pleasure within limits depending on the particular problem, is called the *independent variable*, or *argument;* and the first variable, whose value is determined as soon as the value of the independent variable is fixed, is called the *dependent variable*, or *function*.

Frequently, when we are considering two related variables, it is in our power to fix upon whichever we please as the *independent variable;* but having once made the choice, no change of independent variable is allowed without certain precautions and transformations.

One quantity (the dependent variable) may be a function of two or more other quantities (the independent variables, or arguments). For example, the *cost* of cloth is a function of both the *quality* and *quantity;* the *area* of a triangle is a function of the *base* and *altitude;* the volume of a rectangular parallelepiped is a function of its *three dimensions*.

**11. Notation of functions.** The symbol $f(x)$ is used to denote a function of $x$, and is read $f$ *of* $x$. In order to distinguish between different functions, the prefixed letter is changed, as $F(x)$, $\phi(x)$, $f'(x)$, etc.

During any investigation the same functional symbol always indicates the same law of dependence of the function upon the variable. In the simpler cases this law takes the form of a series of analytical operations upon that variable. Hence, in such a case, the same functional symbol will indicate the same operations or series of operations, even though applied to different quantities. Thus, if

$$f(x) = x^2 - 9x + 14,$$

then $$f(y) = y^2 - 9y + 14.$$

Also $$f(a) = a^2 - 9a + 14,$$

$$f(b+1) = (b+1)^2 - 9(b+1) + 14 = b^2 - 7b + 6,$$

$$f(0) = 0^2 - 9 \cdot 0 + 14 = 14,$$

$$f(-1) = (-1)^2 - 9(-1) + 14 = 24,$$

$$f(3) = 3^2 - 9 \cdot 3 + 14 = -4,$$

$$f(7) = 7^2 - 9 \cdot 7 + 14 = 0, \text{ etc.}$$

Similarly, $\phi(x, y)$ denotes a function of $x$ and $y$, and is read $\phi$ *of* $x$ *and* $y$.

If $$\phi(x, y) = \sin(x + y),$$

then $$\phi(a, b) = \sin(a + b),$$

and $$\phi\left(\frac{\pi}{2}, 0\right) = \sin\frac{\pi}{2} = 1.$$

Again, if $$F(x, y, z) = 2x + 3y - 12z,$$

then $$F(m, -m, m) = 2m - 3m - 12m = -13m,$$

and $$F(3, 2, 1) = 2 \cdot 3 + 3 \cdot 2 - 12 \cdot 1 = 0.$$

Evidently this system of notation may be extended indefinitely.

**12. Values of the independent variable for which a function is defined.** Consider the functions

$$x^2 - 2x + 5, \quad \sin x, \quad \text{arc tan } x$$

of the independent variable $x$. Denoting the dependent variable in each case by $y$, we may write

$$y = x^2 - 2x + 5, \quad y = \sin x, \quad y = \text{arc tan } x.$$

In each case $y$ (the value of the function) is known, or, as we say, *defined*, for all values of $x$. This is not by any means true of all functions, as the following examples illustrating the more common exceptions will show.

(1) $y = \frac{a}{x-b}$.

Here the value of $y$ (i.e. the function) is *defined* for all values of $x$ except $x = b$. When $x = b$ the divisor becomes zero and the value of $y$ cannot be computed from (1).* Any value might be assigned to the function for this value of the argument.

(2) $y = \sqrt{x}$.

In this case the function is *defined* only for positive values of $x$. Negative values of $x$ give imaginary values for $y$, and these must be excluded here, where we are confining ourselves to real numbers only.

(3) $y = \log_a x$. $\qquad a > 0$

Here $y$ is *defined* only for positive values of $x$. For negative values of $x$ this function does not exist (see § 19).

(4) $y = \text{arc} \sin x$, $y = \text{arc} \cos x$.

Since sines and cosines cannot become greater than $+1$ nor less than $-1$, it follows that the above functions are *defined* for all values of $x$ ranging from $-1$ to $+1$ inclusive, but for no other values.

## EXAMPLES

1. Given $f(x) = x^3 - 10x^2 + 31x - 30$; show that

$$f(0) = -30, \qquad f(y) = y^3 - 10y^2 + 31y - 30,$$
$$f(2) = 0, \qquad f(a) = a^3 - 10a^2 + 31a - 30,$$
$$f(3) = f(5), \qquad f(yz) = y^3z^3 - 10y^2z^2 + 31yz - 30,$$
$$f(1) > f(-3), \qquad f(x-2) = x^3 - 16x^2 + 83x - 140,$$
$$f(-1) = -6f(6).$$

2. If $f(x) = x^3 - 3x + 2$, find $f(0)$, $f(1)$, $f(-1)$, $f(-\frac{1}{2})$, $f(1\frac{1}{3})$.

3. If $f(x) = x^3 - 10x^2 + 31x - 30$, and $\phi(x) = x^4 - 55x^2 - 210x - 216$, show that $f(2) = \phi(-2)$, $f(3) = \phi(-3)$, $f(5) = \phi(-4)$, $f(0) + \phi(0) + 246 = 0$.

4. If $F(x) = 2^x$, find $F(0)$, $F(-3)$, $F(\frac{1}{3})$, $F(-1)$.

5. Given $F(x) = x(x-1)(x+6)(x-\frac{1}{2})(x+\frac{4}{5})$; show that

$$F(0) = F(1) = F(-6) = F(\tfrac{1}{2}) = F(-\tfrac{4}{5}) = 0.$$

* See § 14, p. 12.

6. If $f(m_1) = \frac{m_1 - 1}{m_1 + 1}$, show that

$$\frac{f(m_1) - f(m_2)}{1 + f(m_1)f(m_2)} = \frac{m_1 - m_2}{1 + m_1 m_2}.$$

7. If $\phi(x) = a^x$, show that $\phi(y) \cdot \phi(z) = \phi(y + z)$.

8. Given $\phi(x) = \log \frac{1 - x}{1 + x}$; show that

$$\phi(x) + \phi(y) = \phi\left(\frac{x + y}{1 + xy}\right).$$

9. If $f(\phi) = \cos \phi$, show that

$$f(\phi) = f(-\phi) = -f(\pi - \phi) = -f(\pi + \phi).$$

10. If $F(\theta) = \tan \theta$, show that

$$F(2\theta) = \frac{2F(\theta)}{1 - [F(\theta)]^2}.$$

11. Given $\psi(x) = x^{2n} + x^{2m} + 1$; show that

$$\psi(1) = 3, \quad \psi(0) = 1, \quad \psi(a) = \psi(-a).$$

12. If $f(x) = \frac{2x - 3}{x + 7}$, find $f(\sqrt{2})$. *Ans.* $-.0204$.

# CHAPTER III

## THEORY OF LIMITS

**13. Limit of a variable.** If a variable $v$ takes on successively a series of values that approach nearer and nearer to a constant value $l$ in such a manner that $|v-l|$* becomes and remains less than any assigned arbitrarily small positive quantity, then $v$ is said to *approach the limit l*, or to *converge to the limit l*. Symbolically this is written

$$\text{limit } v = l, \quad \text{or,} \quad v \doteq l.$$

The following familiar examples illustrate what is meant:

(1) As the number of sides of a regular inscribed polygon is indefinitely increased, the limit of the area of the polygon is the area of the circle. In this case *the variable is always less than its limit.*

(2) Similarly, the limit of the area of the circumscribed polygon is also the area of the circle, but now *the variable is always greater than its limit.*

(3) Consider the series

$$(A) \qquad 1-\frac{1}{2}+\frac{1}{4}-\frac{1}{8}+\cdots.$$

The sum of any even number $(2n)$ of the first terms of this series is

$$S_{2n}=1-\frac{1}{2}+\frac{1}{4}-\frac{1}{8}+\cdots+\frac{1}{2^{2n-2}}-\frac{1}{2^{2n-1}},$$

$$(B) \qquad S_{2n}=\frac{\frac{1}{2^{2n}}-1}{-\frac{1}{2}-1}=\frac{2}{3}-\frac{1}{3\cdot 2^{2n-1}}. \qquad \text{By 6, p. 1}$$

Similarly, the sum of any odd number $(2n+1)$ of the first terms of the series is

$$S_{2n+1}=1-\frac{1}{2}+\frac{1}{4}-\frac{1}{8}+\cdots-\frac{1}{2^{2n-1}}+\frac{1}{2^{2n}},$$

$$(C) \qquad S_{2n+1}=\frac{-\frac{1}{2^{2n+1}}-1}{-\frac{1}{2}-1}=\frac{2}{3}+\frac{1}{3\cdot 2^{2n}}. \qquad \text{By 6, p. 1}$$

* To be read *the numerical value of the difference between v and l.*

Writing (*B*) and (*C*) in the forms

$$\frac{2}{3}-S_{2n}=\frac{1}{3\cdot 2^{2n-1}},\quad S_{2n+1}-\frac{2}{3}=\frac{1}{3\cdot 2^{2n}},$$

we have

$$\lim_{n=\infty}\left(\frac{2}{3}-S_{2n}\right)=\lim_{n=\infty}\frac{1}{3\cdot 2^{2n-1}}=0,$$

and

$$\lim_{n=\infty}\left(S_{2n+1}-\frac{2}{3}\right)=\lim_{n=\infty}\frac{1}{3\cdot 2^{2n}}=0.$$

Hence, by definition of the limit of a variable, it is seen that both $S_{2n}$ and $S_{2n+1}$ are variables approaching $\frac{2}{3}$ as a limit as the number of terms increases without limit.

Summing up the first two, three, four, etc., terms of (*A*), the sums are found by (*B*) and (*C*) to be alternately less and greater than $\frac{2}{3}$, illustrating the case when *the variable*, in this case the sum of the terms of (*A*), *is alternately less and greater than its limit.*

In the examples shown *the variable never reaches its limit.* This is not by any means always the case, for from the definition of the *limit of a variable* it is clear that the essence of the definition is simply that the numerical value of the difference between the variable and its limit shall ultimately become and remain less than any positive number we may choose, however small.

(4) As an example illustrating the fact that the variable may reach its limit, consider the following. Let a series of regular polygons be inscribed in a circle, the number of sides increasing indefinitely. Choosing any one of these, construct the circumscribed polygon whose sides touch the circle at the vertices of the inscribed polygon. Let $p_n$ and $P_n$ be the perimeters of the inscribed and circumscribed polygons of $n$ sides, and $C$ the circumference of the circle, and suppose the values of a variable $x$ to be as follows:

$$P_n,\quad p_{n+1},\quad C,\quad P_{n+1},\quad p_{n+2},\quad C,\quad P_{n+2},\quad \text{etc.}$$

Then, evidently,

$$\lim_{n=\infty} x = C,$$

and *the limit is reached by the variable, every third value of the variable being C.*

**14. Division by zero excluded.** $\frac{0}{0}$ is indeterminate. For the quotient of two numbers is that number which multiplied by the divisor will give the dividend. But any number whatever multiplied by zero gives zero, and the quotient is indeterminate; that is, any number whatever may be considered as the quotient, a result which is of no value.

$\frac{a}{0}$ has no meaning, $a$ being different from zero, for there exists no number such that if it be multiplied by zero, the product will equal $a$.

Therefore *division by zero is not an admissible operation.*

Care should be taken not to divide by zero inadvertently. The following fallacy is an illustration.

| | |
|---|---|
| Assume that | $a = b.$ |
| Then evidently | $ab = a^2.$ |
| Subtracting $b^2$, | $ab - b^2 = a^2 - b^2.$ |
| Factoring, | $b(a-b) = (a+b)(a-b).$ |
| Dividing by $a-b$, | $b = a + b.$ |
| But | $a = b,$ |
| therefore | $b = 2b,$ |
| or, | $1 = 2.$ |

The result is absurd, and is caused by the fact that we divided by $a - b = 0$.

**15. Infinitesimals.** A variable $v$ whose limit is zero is called an *infinitesimal.** This is written

$$\text{limit } v = 0, \text{ or, } v \doteq 0,$$

and means that the successive numerical values of $v$ ultimately become and remain less than any positive number however small. Such a variable is said to *become indefinitely small* or to *ultimately vanish.*

If $$\text{limit } v = l, \text{ then limit}(v - l) = 0;$$

that is, *the difference between a variable and its limit is an infinitesimal.*

Conversely, *if the difference between a variable and a constant is an infinitesimal,* then the variable approaches the constant as a limit.

**16. The concept of infinity ($\infty$).** If a variable $v$ ultimately becomes and remains greater than any assigned positive number however large, we say *$v$ increases without limit,* and write

$$\text{limit } v = +\infty, \text{ or, } v \doteq +\infty.$$

If a variable $v$ ultimately becomes and remains algebraically less than any assigned negative number, we say *$v$ decreases without limit,* and write

$$\text{limit } v = -\infty, \text{ or, } v \doteq -\infty.$$

* Hence a constant, no matter how small it may be, is not an infinitesimal.

If a variable $v$ ultimately becomes and remains in numerical value greater than any assigned positive number however large, we say $v$, *in numerical value, increases without limit*, or $v$ *becomes infinitely great*,* and write

$$\text{limit } v = \infty, \text{ or, } v \doteq \infty.$$

Infinity ($\infty$) is not a number; it simply serves to characterize a particular mode of variation of a variable by virtue of which it increases or decreases without limit.

**17. Limiting value of a function.** Given a function $f(x)$.

If the independent variable $x$ takes on any series of values such that

$$\text{limit } x = a,$$

and at the same time the dependent variable $f(x)$ takes on a series of corresponding values such that

$$\text{limit } f(x) = A,$$

then as a single statement this is written

$$\underset{x=a}{\text{limit}} f(x) = A,$$

and is read *the limit of $f(x)$, as $x$ approaches the limit $a$ in any manner, is $A$.*

**18. Continuous and discontinuous functions.** A function $f(x)$ is said to be *continuous for $x = a$* if the limiting value of the function when $x$ approaches the limit $a$ in any manner is the value assigned to the function for $x = a$. In symbols, if

$$\underset{x=a}{\text{limit}} f(x) = f(a),$$

then $f(x)$ is *continuous* for $x = a$.

The function is said to be *discontinuous* for $x = a$ if this condition is not satisfied. For example, if

$$\underset{x=a}{\text{limit}} f(x) = \infty,$$

the function is discontinuous for $x = a$.

The attention of the student is now called to the following cases which occur frequently.

* On account of the notation used and for the sake of uniformity, the expression $v \doteq +\infty$ is sometimes read *$v$ approaches the limit plus infinity*. Similarly, $v \doteq -\infty$ is read *$v$ approaches the limit minus infinity*, and $v \doteq \infty$ is read *$v$, in numerical value, approaches the limit infinity*.

While the above notation is convenient to use in this connection, the student must not forget that infinity is not a limit in the sense in which we defined a limit on p. 11, for infinity is not a number at all.

CASE I. As an example illustrating a simple case of a function continuous for a particular value of the variable, consider the function

$$f(x)=\frac{x^2-4}{x-2}.$$

For $x=1, f(x)=f(1)=3$. Moreover, if $x$ approaches the limit 1 in any manner, the function $f(x)$ approaches 3 as a limit. Hence the function is continuous for $x=1$.

CASE II. The definition of a continuous function assumes that the function is already defined for $x=a$. If this is not the case, however, it is sometimes possible to assign such a value to the function for $x=a$ that the condition of continuity shall be satisfied. The following theorem covers these cases.

**Theorem.** *If $f(x)$ is not defined for $x=a$, and if*

$$\underset{x=a}{\text{limit}}\, f(x)=B,$$

*then $f(x)$ will be continuous for $x=a$, if $B$ is assumed as the value of $f(x)$ for $x=a$.* Thus the function

$$\frac{x^2-4}{x-2}$$

is not defined for $x=2$ (since then there would be division by zero). But for every other value of $x$,

$$\frac{x^2-4}{x-2}=x+2;$$

and $$\underset{x=2}{\text{limit}}\,(x+2)=4;$$

therefore $$\underset{x=2}{\text{limit}}\,\frac{x^2-4}{x-2}=4.$$

Although the function is not defined for $x=2$, if we arbitrarily assign it the value 4 for $x=2$, it then becomes continuous for this value.

*A function $f(x)$ is said to be continuous in an interval when it is continuous for all values of $x$ in this interval.**

* In this book we shall deal only with functions which are in general continuous, that is, continuous for all values of $x$, with the possible exception of certain isolated values, our results in general being understood as valid only for such values of $x$ for which the function in question is actually continuous. Unless special attention is called thereto, we shall as a rule pay no attention to the possibilities of such exceptional values of $x$ for which the function is discontinuous. The definition of a continuous function $f(x)$ is sometimes roughly (but imperfectly) summed up in the statement that *a small change in $x$ shall produce a small change in $f(x)$*. We shall not consider functions having an infinite number of oscillations in a limited region.

**19. Continuity and discontinuity of functions illustrated by their graphs.**

(1) Consider the function $x^2$, and let

$$(A) \qquad y = x^2.$$

If we assume values for $x$ and calculate the corresponding values of $y$, we can plot a series of points. Drawing a smooth line free-hand through these points, a good representation of the general behavior of the function may be obtained. This picture or image of the function is called its *graph*. It is evidently the locus of all points satisfying equation $(A)$.

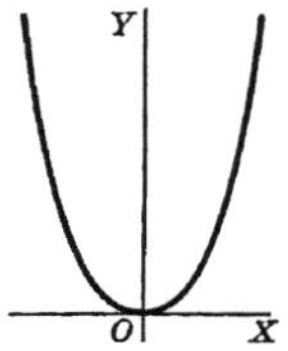


Such a series or assemblage of points is also called a curve. Evidently we may assume values of $x$ so near together as to bring the values of $y$ (and therefore the points of the curve) as near together as we please. In other words, there are no breaks in the curve, and the function $x^2$ is continuous for all values of $x$.

(2) The graph of the continuous function $\sin x$ is plotted by drawing the locus of

$$y = \sin x.$$

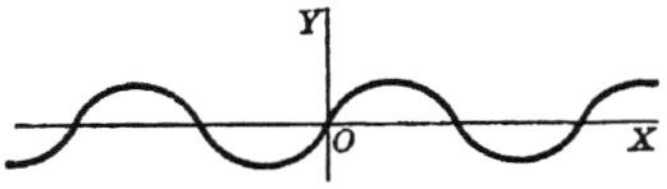


It is seen that no break in the curve occurs anywhere.

(3) The continuous function $e^x$ is of very frequent occurrence in the Calculus. If we plot its graph from

$$y = e^x, \qquad (e = 2.718 \cdots)$$

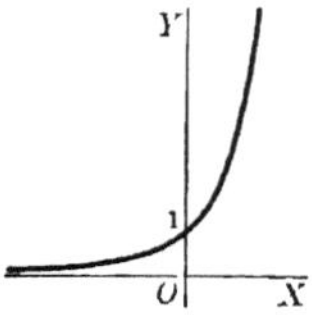


we get a smooth curve as shown. From this it is clearly seen that,

(a) when $x = 0$, $\underset{x=0}{\text{limit}}\, y (= e^x) = 1$;

(b) when $x > 0$, $y (= e^x)$ is positive and increases as we pass towards the right from the origin;

(c) when $x < 0$, $y (= e^x)$ is still positive and decreases as we pass towards the left from the origin.

(4) The function $\log_e x$ is closely related to the last one discussed. In fact, if we plot its graph from

$$y = \log_e x,$$

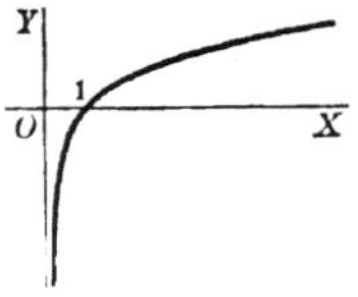


it will be seen that its graph has the same relation to $OX$ and $OY$ as the graph of $e^x$ has to $OY$ and $OX$.

Here we see the following facts pictured:

(a) For $x = 1$, $\log_e x = \log_e 1 = 0$.

(b) For $x > 1$, $\log_e x$ is positive and increases as $x$ increases.

(c) For $1 > x > 0$, $\log_e x$ is negative and increases in *numerical value* as $x$ diminishes, that is, $\underset{x=0}{\text{limit}} \log x = -\infty$.

(d) For $x \leqq 0$, $\log_e x$ is not defined; hence the entire graph lies to the right of $OY$.

(5) Consider the function $\frac{1}{x}$, and set

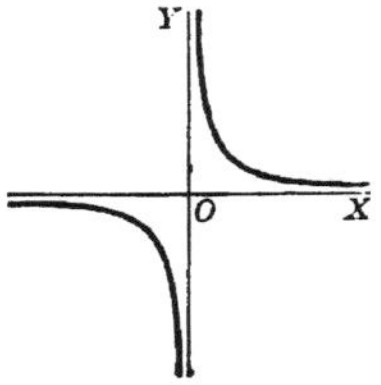


$$y = \frac{1}{x}.$$

If the graph of this function be plotted, it will be seen that as $x$ approaches the value zero from the left (negatively), the points of the curve ultimately drop down an infinitely great distance, and as $x$ approaches the value zero from the right, the curve extends upward infinitely far.

The curve then does not form a continuous branch from one side to the other of the axis of $Y$, showing graphically that the function is discontinuous for $x = 0$, but continuous for all other values of $x$.

(6) From the graph of

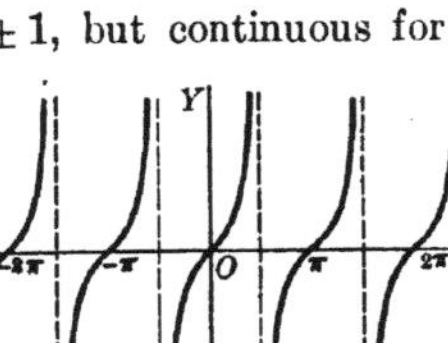


$$y = \frac{2x}{1 - x^2}$$

it is seen that the function

$$\frac{2x}{1 - x^2}$$

is discontinuous for the two values $x = \pm 1$, but continuous for all other values of $x$.

(7) The graph of

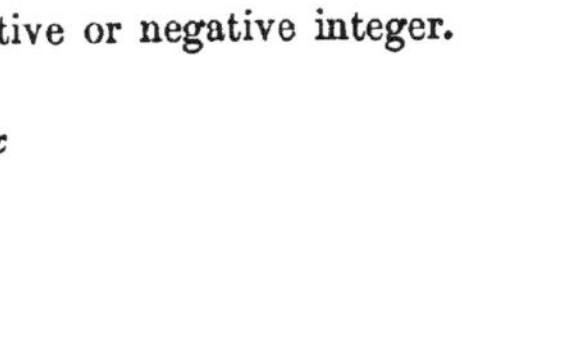


$$y = \tan x$$

shows that the function $\tan x$ is discontinuous for infinitely many values of the independent variable $x$, namely, $x = \frac{n\pi}{2}$, where $n$ denotes any odd positive or negative integer.

(8) The function

$$\text{arc} \tan x$$

has infinitely many values for a given value of $x$, the graph of equation

$$y = \text{arc} \tan x$$

consisting of infinitely many branches. If, however, we confine ourselves to any single branch, the function is continuous. For instance,

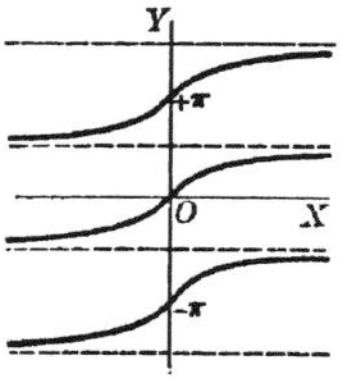


if we say that $y$ shall be the arc of smallest numerical value whose tangent is $x$, that is, $y$ shall take on only values between $-\frac{\pi}{2}$ and $\frac{\pi}{2}$, then we are limited to the branch passing through the origin, and the condition for continuity is satisfied.

(9) Similarly,

$$\text{arc} \tan \frac{1}{x},$$

is found to be a many-valued function. Confining ourselves to one branch of the graph of

$$y = \text{arc} \tan \frac{1}{x},$$

we see that as $x$ approaches zero from the left, $y$ approaches the limit $-\frac{\pi}{2}$, and as $x$ approaches zero from the right, $y$ approaches the limit $+\frac{\pi}{2}$. Hence the function is discontinuous when $x = 0$. Its value for $x = 0$ can be assigned at pleasure.

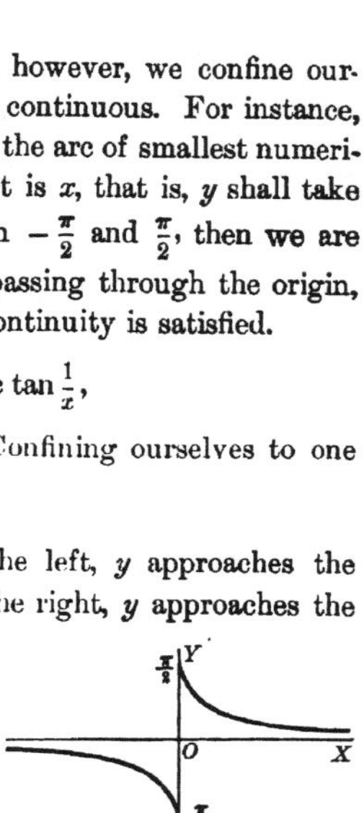


Functions exist which are discontinuous for every value of the independent variable within a certain range. In the ordinary applications of the Calculus, however, we deal with functions which are discontinuous (if at all) only for certain isolated values of the independent variable; such functions are therefore in general continuous, and are the only ones considered in this book.

**20. Fundamental theorems on limits.** In problems involving limits the use of one or more of the following theorems is usually implied. It is assumed that the limit of each variable exists and is finite.

**Theorem I.** *The limit of the algebraic sum of a finite number of variables is equal to the like algebraic sum of the limits of the several variables.*

**Theorem II.** *The limit of the product of a finite number of variables is equal to the product of the limits of the several variables.*

**Theorem III.** *The limit of the quotient of two variables is equal to the quotient of the limits of the separate variables, provided the limit of the denominator is not zero.*

Before proving these theorems it is necessary to establish the following properties of infinitesimals.

(1) *The sum of a finite number of infinitesimals is an infinitesimal.* To prove this we must show that the numerical value of this sum can be made less than any small positive quantity (as $\epsilon$) that may be assigned (§ 15). That this is possible is evident, for, the limit of each infinitesimal being zero, each one can be made numerically less than $\frac{\epsilon}{n}$ ($n$ being the number of infinitesimals), and therefore their sum can be made numerically less than $\epsilon$.

(2) *The product of a constant c and an infinitesimal is an infinitesimal.* For the numerical value of the product can always be made less than any small positive quantity (as $\epsilon$) by making the numerical value of the infinitesimal less than $\frac{\epsilon}{c}$.

(3) *The product of any finite number of infinitesimals is an infinitesimal.* For the numerical value of the product may be made less than any small positive quantity that can be assigned. If the given product contains $n$ factors, then since each infinitesimal may be assumed less than the $n$th root of $\epsilon$, the product can be made less than $\epsilon$ itself.

(4) *If v is a variable which approaches a limit l different from zero, then the quotient of an infinitesimal by v is also an infinitesimal.* For if limit $v = l$, and $k$ is any number numerically less than $l$, then, by definition of a limit, $v$ will ultimately become and remain numerically greater than $k$. Hence the quotient $\frac{\epsilon}{v}$, where $\epsilon$ is an infinitesimal, will ultimately become and remain numerically less than $\frac{\epsilon}{k}$, and is therefore by (2) an infinitesimal.

*Proof of Theorem I.* Let $v_1, v_2, v_3, \cdots$ be the variables, and $l_1, l_2, l_3, \cdots$ their respective limits. We may then write

$$v_1 - l_1 = \epsilon_1,$$
$$v_2 - l_2 = \epsilon_2,$$
$$v_3 - l_3 = \epsilon_3,$$
$$\cdot \quad \cdot \quad \cdot \quad \cdot \quad \cdot$$

where $\epsilon_1, \epsilon_2, \epsilon_3, \cdots$ are infinitesimals (i.e. variables having zero for a limit). Adding

$$(A) \quad (v_1 + v_2 + v_3 + \cdots) - (l_1 + l_2 + l_3 + \cdots) = (\epsilon_1 + \epsilon_2 + \epsilon_3 + \cdots).$$

Since the right-hand member is an infinitesimal by (1), p. 19, we have, from the converse theorem on p. 13,

$$\text{limit}\,(v_1 + v_2 + v_3 + \cdots) = l_1 + l_2 + l_3 + \cdots,$$

or, $$\text{limit}\,(v_1 + v_2 + v_3 + \cdots) = \text{limit}\, v_1 + \text{limit}\, v_2 + \text{limit}\, v_3 + \cdots,$$

which was to be proved.

*Proof of Theorem II.* Let $v_1$ and $v_2$ be the variables, $l_1$ and $l_2$ their respective limits, and $\epsilon_1$ and $\epsilon_2$ infinitesimals; then

$$v_1 = l_1 + \epsilon_1$$

and $$v_2 = l_2 + \epsilon_2.$$

Multiplying, $$v_1 v_2 = (l_1 + \epsilon_1)(l_2 + \epsilon_2)$$
$$= l_1 l_2 + l_1 \epsilon_2 + l_2 \epsilon_1 + \epsilon_1 \epsilon_2,$$

or,

(B) $$v_1 v_2 - l_1 l_2 = l_1 \epsilon_2 + l_2 \epsilon_1 + \epsilon_1 \epsilon_2.$$

Since the right-hand member is an infinitesimal by (1) and (2), p. 19, we have, as before,

$$\text{limit}\,(v_1 v_2) = l_1 l_2 = \text{limit}\, v_1 \cdot \text{limit}\, v_2,$$

which was to be proved.

*Proof of Theorem III.* Using the same notation as before,

$$\frac{v_1}{v_2} = \frac{l_1 + \epsilon_1}{l_2 + \epsilon_2} = \frac{l_1}{l_2} + \left(\frac{l_1 + \epsilon_1}{l_2 + \epsilon_2} - \frac{l_1}{l_2}\right),$$

or,

(C) $$\frac{v_1}{v_2} - \frac{l_1}{l_2} = \frac{l_2 \epsilon_1 - l_1 \epsilon_2}{l_2 (l_2 + \epsilon_2)}.$$

Here again the right-hand member is an infinitesimal by (4), p. 19, if $l_2 \neq 0$; hence

$$\text{limit}\left(\frac{v_1}{v_2}\right) = \frac{l_1}{l_2} = \frac{\text{limit}\, v_1}{\text{limit}\, v_2},$$

which was to be proved.

It is evident that if any of the variables be replaced by constants, our reasoning still holds, and the above theorems are true.

**21. Special limiting values.** The following examples are of special importance in the study of the Calculus. In the following examples $a > 0$ and $c \neq 0$.

| *Written in the form of limits.* | *Abbreviated form often used.* |
|---|---|
| (1) $\underset{x=0}{\text{limit}} \dfrac{c}{x} = \infty;$ | $\dfrac{c}{0} = \infty.$ |
| (2) $\underset{x=\infty}{\text{limit}}\, cx = \infty;$ | $c \cdot \infty = \infty.$ |
| (3) $\underset{x=\infty}{\text{limit}} \dfrac{x}{c} = \infty;$ | $\dfrac{\infty}{c} = \infty.$ |
| (4) $\underset{x=\infty}{\text{limit}} \dfrac{c}{x} = 0;$ | $\dfrac{c}{\infty} = 0.$ |
| (5) $\underset{x=-\infty}{\text{limit}}\, a^x = +\infty,$ when $a<1;$ | $a^{-\infty} = +\infty.$ |
| (6) $\underset{x=+\infty}{\text{limit}}\, a^x = 0,$ when $a<1;$ | $a^{+\infty} = 0.$ |
| (7) $\underset{x=-\infty}{\text{limit}}\, a^x = 0,$ when $a>1;$ | $a^{-\infty} = 0.$ |
| (8) $\underset{x=+\infty}{\text{limit}}\, a^x = +\infty,$ when $a>1;$ | $a^{+\infty} = +\infty.$ |
| (9) $\underset{x=0}{\text{limit}}\, \log_a x = +\infty,$ when $a<1;$ | $\log_a 0 = +\infty.$ |
| (10) $\underset{x=+\infty}{\text{limit}}\, \log_a x = -\infty,$ when $a<1;$ | $\log_a(+\infty) = -\infty.$ |
| (11) $\underset{x=0}{\text{limit}}\, \log_a x = -\infty,$ when $a>1;$ | $\log_a 0 = -\infty.$ |
| (12) $\underset{x=+\infty}{\text{limit}}\, \log_a x = +\infty,$ when $a>1;$ | $\log_a(+\infty) = +\infty.$ |

The expressions in the second column are not to be considered as expressing numerical equalities ($\infty$ not being a number); they are merely ***symbolical equations*** implying the relations indicated in the first column, and should be so understood.

**22. Show that** $\underset{x=0}{\text{limit}} \dfrac{\sin x}{x} = 1.$*

Let $O$ be the center of a circle whose radius is unity.

Let arc $AM =$ arc $AM' = x$, and let $MT$ and $M'T$ be tangents drawn to the circle at $M$ and $M'$. From Geometry,

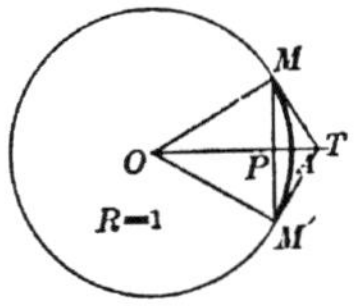


$$MPM' < MAM' < MTM';$$

or
$$2 \sin x < 2x < 2 \tan x.$$

Dividing through by $2 \sin x$, we get

$$1 < \frac{x}{\sin x} < \frac{1}{\cos x}.$$

* If we refer to the table on p. 4, it will be seen that for all angles less than 10° the angle in radians and the sine of the angle are equal to three decimal places. If larger tables are consulted, five-place, say, it will be seen that for all angles less than 2.2° the sine of the angle and the angle itself are equal to four decimal places. From this we may well suspect that

$$\underset{x=0}{\text{limit}} \frac{\sin x}{x} = 1.$$

If now $x$ approaches the limit zero,

$$\underset{x=0}{\text{limit}} \frac{x}{\sin x}$$

must lie between the constant 1 and $\underset{x=0}{\text{limit}} \frac{1}{\cos x}$, which is also 1.

Therefore $\underset{x=0}{\text{limit}} \frac{x}{\sin x}=1$, or, $\underset{x=0}{\text{limit}} \frac{\sin x}{x}=1$. Th. III, p. 18

It is interesting to note the behavior of this function from its graph, the locus of equation

$$y=\frac{\sin x}{x}.$$

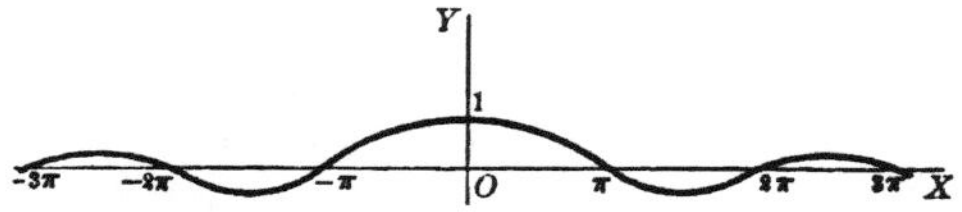


Although the function is not defined for $x=0$, yet it is not discontinuous when $x=0$ if we define

$$\frac{\sin 0}{0}=1.$$ Case II, p. 15

**23. The number *e*.** One of the most important limits in the Calculus is

$$\underset{x=0}{\text{limit}} (1+x)^{\frac{1}{x}}=2.71828\cdots=e.$$

To prove rigorously that such a limit $e$ exists, is beyond the scope of this book. For the present we shall content ourselves by plotting the locus of the equation

$$y=(1+x)^{\frac{1}{x}}$$

and show graphically that, as $x \doteq 0$, the function $(1+x)^{\frac{1}{x}}(=y)$

| $x$ | $y$ | $x$ | $y$ |
|---|---|---|---|
| 10 | 1.2710 | | |
| 5 | 1.4310 | | |
| 2 | 1.7320 | | |
| 1 | 2.0000 | | |
| .5 | 2.2500 | −.5 | 4.0000 |
| .1 | 2.5937 | −.1 | 2.8680 |
| .01 | 2.7048 | −.01 | 2.7320 |
| .001 | 2.7169 | −.001 | 2.7195 |

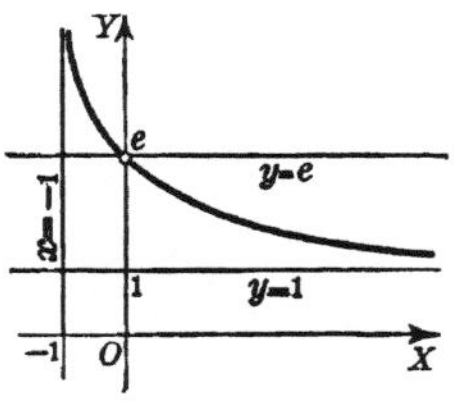


takes on values in the near neighborhood of $2.718\cdots$, and therefore $e=2.718\cdots$ approximately.

As $x \doteq 0$ from the left, $y$ decreases and approaches $e$ as a limit. As $x \doteq 0$ from the right, $y$ increases and also approaches $e$ as a limit.

As $x \doteq \infty$, $y$ approaches the limit 1; and as $x \doteq -1$ from the right, $y$ increases without limit.

In Chap. XVIII, Ex. 15, p. 233, we will show how to calculate the value of $e$ to any number of decimal places.

*Natural logarithms* are those which have the number $e$ for base. These logarithms play a very important rôle in mathematics. When the base is not indicated explicitly, the base $e$ is always understood in what follows in this book. Thus $\log_e v$ is written simply $\log v$.

Natural logarithms possess the following characteristic property: If $x \doteq 0$ in any way whatever,

$$\text{limit}\frac{\log(1+x)}{x} = \text{limit}\log(1+x)^{\frac{1}{x}} = \log e = 1.$$

**24. Expressions assuming the form $\frac{\infty}{\infty}$.** As $\infty$ is not a number, the expression $\infty \div \infty$ is indeterminate. To evaluate a fraction assuming this form, the numerator and denominator being algebraic functions, we shall find useful the following

RULE. *Divide both numerator and denominator by the highest power of the variable occurring in either. Then substitute the value of the variable.*

ILLUSTRATIVE EXAMPLE 1. Evaluate $\underset{x=\infty}{\text{limit}}\frac{2x^3-3x^2+4}{5x-x^2-7x^3}$.

**Solution.** Substituting directly, we get $\underset{x=\infty}{\text{limit}}\frac{2x^3-3x^2+4}{5x-x^2-7x^3} = \frac{\infty}{\infty}$, which is indeterminate. Hence, following the above rule, we divide both numerator and denominator by $x^3$. Then

$$\underset{x=\infty}{\text{limit}}\frac{2x^3-3x^2+4}{5x-x^2-7x^3} = \underset{x=\infty}{\text{limit}}\frac{2-\frac{3}{x}+\frac{4}{x^3}}{\frac{5}{x^2}-\frac{1}{x}-7} = -\frac{2}{7}. \quad \textit{Ans.}$$

## EXAMPLES

Prove the following:

1. $\underset{x=\infty}{\text{limit}}\left(\frac{x+1}{x}\right) = 1.$

**Proof.** $\underset{x=\infty}{\text{limit}}\left(\frac{x+1}{x}\right) = \underset{x=\infty}{\text{limit}}\left(1+\frac{1}{x}\right)$

$= \underset{x=\infty}{\text{limit}}(1) + \underset{x=\infty}{\text{limit}}\left(\frac{1}{x}\right)$ Th. I, p. 18

$= 1 + 0 = 1.$

2. $\lim_{x=\infty}\left(\frac{x^2+2x}{5-3x^2}\right)=-\frac{1}{3}.$

**Proof.** $\lim_{x=\infty}\left(\frac{x^2+2x}{5-3x^2}\right)=\lim_{x=\infty}\left(\frac{1+\frac{2}{x}}{\frac{5}{x^2}-3}\right)$

[Dividing both numerator and denominator by $x^2$.]

$$=\frac{\lim_{x=\infty}\left(1+\frac{2}{x}\right)}{\lim_{x=\infty}\left(\frac{5}{x^2}-3\right)} \qquad \text{Th. III, p. 18}$$

$$=\frac{\lim_{x=\infty}(1)+\lim_{x=\infty}\left(\frac{2}{x}\right)}{\lim_{x=\infty}\left(\frac{5}{x^2}\right)-\lim_{x=\infty}(3)} \qquad \text{Th. I, p. 18}$$

$$=\frac{1+0}{0-3}=-\frac{1}{3}.$$

3. $\lim_{x=1}\frac{x^2-2x+5}{x^2+7}=\frac{1}{2}.$

4. $\lim_{x=0}\frac{3x^3+6x^2}{2x^4-15x^2}=-\frac{2}{5}.$

5. $\lim_{x=-2}\frac{x^2+1}{x+3}=5.$

6. $\lim_{h=0}(3ax^2-2hx+5h^2)=3ax^2.$

7. $\lim_{x=\infty}(ax^2+bx+c)=\infty.$

8. $\lim_{k=0}\frac{(x-k)^2-2kx^3}{x(x+k)}=1.$

9. $\lim_{x=\infty}\frac{x^2+1}{3x^2+2x-1}=\frac{1}{3}.$

10. $\lim_{x=\infty}\frac{3+2x}{x^2-5x}=0.$

11. $\lim_{\alpha=\frac{\pi}{2}}\frac{\cos(\alpha-a)}{\cos(2\alpha-a)}=-\tan a.$

12. $\lim_{x=\infty}\frac{ax^2+bx+c}{dx^2+ex+f}=\frac{a}{d}.$

13. $\lim_{z=0}\frac{a}{2}\left(e^{\frac{x}{a}}+e^{-\frac{x}{a}}\right)=a.$

14. $\lim_{x=0}\frac{2x^3+3x^2}{x^3}=\infty.$

15. $\lim_{x=\infty}\frac{5x^2-2x}{x}=\infty.$

16. $\lim_{y=\infty}\frac{y}{y+1}=1.$

17. $\lim_{n=\infty}\frac{n(n+1)}{(n+2)(n+3)}=1.$

18. $\lim_{s=1}\frac{s^3-1}{s-1}=3.$

19. $\lim_{h=0}\frac{(x+h)^n-x^n}{h}=nx^{n-1}.$

20. $\lim_{h=0}\left[\cos(\theta+h)\frac{\sin h}{h}\right]=\cos\theta.$

21. $\lim_{x=\infty}\frac{4x^2-x}{4-3x^2}=-\frac{4}{3}.$

22. $\lim_{\theta=0}\frac{1-\cos\theta}{\theta^2}=\frac{1}{2}.$

23. $\lim_{x=a}\frac{1}{x-a}=-\infty$, if $x$ is increasing as it approaches the value $a$.

24. $\lim_{x=a}\frac{1}{x-a}=+\infty$, if $x$ is decreasing as it approaches the value $a$.

# CHAPTER IV

## DIFFERENTIATION

**25. Introduction.** We shall now proceed to investigate the manner in which a function changes in value as the independent variable changes. The fundamental problem of the Differential Calculus is to establish a measure of this change in the function with mathematical precision. It was while investigating problems of this sort, dealing with continuously varying quantities, that Newton * was led to the discovery of the fundamental principles of the Calculus, the most scientific and powerful tool of the modern mathematician.

**26. Increments.** The *increment* of a variable in changing from one numerical value to another is the *difference* found by subtracting the first value from the second. An increment of $x$ is denoted by the symbol $\Delta x$, read *delta x*.

The student is warned against reading this symbol *delta times x*, it having no such meaning. Evidently this increment may be either positive or negative † according as the variable in changing is increasing or decreasing in value. Similarly,

$\Delta y$ denotes an increment of $y$,

$\Delta \phi$ denotes an increment of $\phi$,

$\Delta f(x)$ denotes an increment of $f(x)$, etc.

If in $y = f(x)$ the independent variable $x$ takes on an increment $\Delta x$, then $\Delta y$ is always understood to denote the corresponding increment of the function $f(x)$ (or dependent variable $y$).

The increment $\Delta y$ is always assumed to be reckoned from a definite initial value of $y$ corresponding to the arbitrarily fixed initial value of $x$ from which the increment $\Delta x$ is reckoned. For instance, consider the function

$$y = x^2.$$

* Sir Isaac Newton (1642–1727), an Englishman, was a man of the most extraordinary genius. He developed the science of the Calculus under the name of Fluxions. Although Newton had discovered and made use of the new science as early as 1670, his first published work in which it occurs is dated 1687, having the title *Philosophiae Naturalis Principia Mathematica*. This was Newton's principal work. Laplace said of it, "It will always remain preëminent above all other productions of the human mind." See frontispiece.

† Some writers call a *negative increment* a *decrement*.

Assuming $x = 10$ for the initial value of $x$ fixes $y = 100$ as the initial value of $y$.

Suppose $x$ increases to $x = 12$, that is, $\Delta x = 2$;
then $y$ increases to $y = 144$, and $\Delta y = 44$.
Suppose $x$ decreases to $x = 9$, that is, $\Delta x = -1$;
then $y$ decreases to $y = 81$, and $\Delta y = -19$.

It may happen that as $x$ increases, $y$ decreases, or the reverse; in either case $\Delta x$ and $\Delta y$ will have opposite signs.

It is also clear (as illustrated in the above example) that if $y = f(x)$ is a continuous function and $\Delta x$ is decreasing in numerical value, then $\Delta y$ also decreases in numerical value.

**27. Comparison of increments.** Consider the function

$$(A) \qquad y = x^2.$$

Assuming a fixed initial value for $x$, let $x$ take on an increment $\Delta x$. Then $y$ will take on a corresponding increment $\Delta y$, and we have

$$y + \Delta y = (x + \Delta x)^2,$$

or,

$$y + \Delta y = x^2 + 2x \cdot \Delta x + (\Delta x)^2.$$

Subtracting $(A)$,

$$y \qquad = x^2$$

$$(B) \qquad \Delta y = \qquad 2x \cdot \Delta x + (\Delta x)^2$$

we get the increment $\Delta y$ in terms of $x$ and $\Delta x$.

To find the ratio of the increments, divide $(B)$ by $\Delta x$, giving

$$\frac{\Delta y}{\Delta x} = 2x + \Delta x.$$

If the initial value of $x$ is 4, it is evident that

$$\underset{\Delta x = 0}{\text{limit}} \frac{\Delta y}{\Delta x} = 8.$$

Let us carefully note the behavior of the ratio of the increments of $x$ and $y$ as the increment of $x$ diminishes.

| Initial value of $x$ | New value of $x$ | Increment $\Delta x$ | Initial value of $y$ | New value of $y$ | Increment $\Delta y$ | $\frac{\Delta y}{\Delta x}$ |
|---|---|---|---|---|---|---|
| 4 | 5.0 | 1.0 | 16 | 25. | 9. | 9. |
| 4 | 4.8 | 0.8 | 16 | 23.04 | 7.04 | 8.8 |
| 4 | 4.6 | 0.6 | 16 | 21.16 | 5.16 | 8.6 |
| 4 | 4.4 | 0.4 | 16 | 19.36 | 3.36 | 8.4 |
| 4 | 4.2 | 0.2 | 16 | 17.64 | 1.64 | 8.2 |
| 4 | 4.1 | 0.1 | 16 | 16.81 | 0.81 | 8.1 |
| 4 | 4.01 | 0.01 | 16 | 16.0801 | 0.0801 | 8.01 |

It is apparent that as $\Delta x$ decreases, $\Delta y$ also diminishes, but their ratio takes on the successive values 9, 8.8, 8.6, 8.4, 8.2, 8.1, 8.01; illustrating the fact that $\frac{\Delta y}{\Delta x}$ can be brought as near to 8 in value as we please by making $\Delta x$ small enough. Therefore

$$\underset{\Delta x = 0}{\text{limit}} \frac{\Delta y}{\Delta x} = 8.*$$

**28. Derivative of a function of one variable.** The fundamental definition of the Differential Calculus is:

*The derivative* † *of a function is the limit of the ratio of the increment of the function to the increment of the independent variable, when the latter increment varies and approaches the limit zero.*

When the limit of this ratio exists, the function is said to be *differentiable*, or to *possess a derivative.*

The above definition may be given in a more compact form symbolically as follows: Given the function

$$(A) \qquad y = f(x),$$

and consider $x$ to have a fixed value.

Let $x$ take on an increment $\Delta x$; then the function $y$ takes on an increment $\Delta y$, the new value of the function being

$$(B) \qquad y + \Delta y = f(x + \Delta x).$$

To find the increment of the function, subtract $(A)$ from $(B)$, giving

$$(C) \qquad \Delta y = f(x + \Delta x) - f(x).$$

Dividing by the increment of the variable, $\Delta x$, we get

$$(D) \qquad \frac{\Delta y}{\Delta x} = \frac{f(x + \Delta x) - f(x)}{\Delta x}.$$

The limit of this ratio when $\Delta x$ approaches the limit zero is, from our definition, the *derivative* and is *denoted by the symbol* $\frac{dy}{dx}$. Therefore

$$(E) \qquad \frac{dy}{dx} = \underset{\Delta x = 0}{\text{limit}} \frac{f(x + \Delta x) - f(x)}{\Delta x}$$

defines the *derivative of* $y$ [*or* $f(x)$] *with respect to* $x$.

* The student should guard against the common error of concluding that because the numerator and denominator of a fraction are each approaching zero as a limit, the limit of the value of the fraction (or ratio) is zero. The limit of the ratio may take on any numerical value. In the above example the limit is 8.

† Also called the *differential coefficient* or the *derived function.*

From ($D$) we also get

$$\frac{dy}{dx} = \underset{\Delta x = 0}{\text{limit}} \frac{\Delta y}{\Delta x}.$$

The process of finding the derivative of a function is called *differentiation.*

It should be carefully noted that the derivative is the *limit of the ratio,* not the ratio of the limits. The latter ratio would assume the form $\frac{0}{0}$, which is indeterminate (§ 14, p. 12).

**29. Symbols for derivatives.** Since $\Delta y$ and $\Delta x$ are always finite and have definite values, the expression

$$\frac{\Delta y}{\Delta x}$$

is really a fraction. The symbol

$$\frac{dy}{dx},$$

however, is to be regarded *not as a fraction but as the limiting value of a fraction.* In many cases it will be seen that this symbol does possess fractional properties, and later on we shall show how meanings may be attached to $dy$ and $dx$, but for the present the symbol $\frac{dy}{dx}$ is to be considered as a whole.

Since the derivative of a function of $x$ is in general also a function of $x$, the symbol $f'(x)$ is also used to denote the derivative of $f(x)$. Hence, if

$$y = f(x),$$

we may write

$$\frac{dy}{dx} = f'(x),$$

which is read *the derivative of $y$ with respect to $x$ equals $f$ prime of $x$* The symbol

$$\frac{d}{dx}$$

when considered by itself is called the *differentiating operator,* and indicates that any function written after it is to be differentiated with respect to $x$. Thus

$\frac{dy}{dx}$ or $\frac{d}{dx}y$ indicates the derivative of $y$ with respect to $x$;

$\frac{d}{dx}f(x)$ indicates the derivative of $f(x)$ with respect to $x$;

$\frac{d}{dx}(2x^2 + 5)$ indicates the derivative of $2x^2 + 5$ with respect to $x$.

$y'$ is an abbreviated form of $\frac{dy}{dx}$.

The symbol $D_x$ is used by some writers instead of $\frac{d}{dx}$. If then

$$y = f(x),$$

we may write the identities

$$y' = \frac{dy}{dx} = \frac{d}{dx}y = \frac{d}{dx}f(x) = D_x f(x) = f'(x).$$

**30. Differentiable functions.** From the Theory of Limits it is clear that if the derivative of a function exists for a certain value of the independent variable, the function itself must be continuous for that value of the variable.

The converse, however, is not always true, functions having been discovered that are continuous and yet possess no derivative. But such functions do not occur often in applied mathematics, and *in this book only differentiable functions are considered*, that is, functions that possess a derivative for all values of the independent variable save at most for isolated values.

**31. General rule for differentiation.** From the definition of a derivative it is seen that the process of differentiating a function $y = f(x)$ consists in taking the following distinct steps:

### General Rule for Differentiation *

First Step. *In the function replace $x$ by $x + \Delta x$, giving a new value of the function, $y + \Delta y$.*

Second Step. *Subtract the given value of the function from the new value in order to find $\Delta y$ (the increment of the function).*

Third Step. *Divide the remainder $\Delta y$ (the increment of the function) by $\Delta x$ (the increment of the independent variable).*

Fourth Step. *Find the limit of this quotient, when $\Delta x$ (the increment of the independent variable) varies and approaches the limit zero. This is the derivative required.*

The student should become thoroughly familiar with this rule by applying the process to a large number of examples. Three such examples will now be worked out in detail.

Illustrative Example 1. Differentiate $3x^2 + 5$.

**Solution.** Applying the successive steps in the *General Rule*, we get, after placing

$$y = 3x^2 + 5,$$

*First step.*

$$y + \Delta y = 3(x + \Delta x)^2 + 5$$
$$= 3x^2 + 6x \cdot \Delta x + 3(\Delta x)^2 + 5.$$

* Also called the *Four-step Rule.*

*Second step.*

$$\begin{aligned} y + \Delta y &= 3x^2 + 6x \cdot \Delta x + 3(\Delta x)^2 + 5 \\ y &= 3x^2 \qquad\qquad\qquad\qquad + 5 \\ \hline \Delta y &= \qquad 6x \cdot \Delta x + 3(\Delta x)^2. \end{aligned}$$

*Third step.* $\dfrac{\Delta y}{\Delta x} = 6x + 3 \cdot \Delta x.$

*Fourth step.* $\dfrac{dy}{dx} = 6x.$ *Ans.*

We may also write this

$$\frac{d}{dx}(3x^2 + 5) = 6x.$$

ILLUSTRATIVE EXAMPLE 2. Differentiate $x^3 - 2x + 7$.

**Solution.** Place $y = x^3 - 2x + 7.$

*First step.*

$$\begin{aligned} y + \Delta y &= (x + \Delta x)^3 - 2(x + \Delta x) + 7 \\ &= x^3 + 3x^2 \cdot \Delta x + 3x \cdot (\Delta x)^2 + (\Delta x)^3 - 2x - 2 \cdot \Delta x + 7. \end{aligned}$$

*Second step.*

$$\begin{aligned} y + \Delta y &= x^3 + 3x^2 \cdot \Delta x + 3x \cdot (\Delta x)^2 + (\Delta x)^3 - 2x - 2 \cdot \Delta x + 7 \\ y &= x^3 \qquad\qquad\qquad\qquad\qquad\qquad - 2x \qquad\quad + 7 \\ \hline \Delta y &= \qquad 3x^2 \cdot \Delta x + 3x \cdot (\Delta x)^2 + (\Delta x)^3 \qquad - 2 \cdot \Delta x. \end{aligned}$$

*Third step.* $\dfrac{\Delta y}{\Delta x} = 3x^2 + 3x \cdot \Delta x + (\Delta x)^2 - 2.$

*Fourth step.* $\dfrac{dy}{dx} = 3x^2 - 2.$ *Ans.*

Or, $\dfrac{d}{dx}(x^3 - 2x + 7) = 3x^2 - 2.$

ILLUSTRATIVE EXAMPLE 3. Differentiate $\dfrac{c}{x^2}$.

**Solution.** Place $y = \dfrac{c}{x^2}.$

*First step.* $y + \Delta y = \dfrac{c}{(x + \Delta x)^2}.$

*Second step.*

$$\begin{aligned} y + \Delta y &= \frac{c}{(x + \Delta x)^2} \\ y &= \frac{c}{x^2} \\ \hline \Delta y &= \frac{c}{(x + \Delta x)^2} - \frac{c}{x^2} = \frac{-c \cdot \Delta x(2x + \Delta x)}{x^2(x + \Delta x)^2}. \end{aligned}$$

*Third step.* $\dfrac{\Delta y}{\Delta x} = -c \cdot \dfrac{2x + \Delta x}{x^2(x + \Delta x)^2}.$

*Fourth step.*

$$\begin{aligned} \frac{dy}{dx} &= -c \cdot \frac{2x}{x^2(x)^2} \\ &= -\frac{2c}{x^3}. \quad \textit{Ans.} \end{aligned}$$

Or, $\dfrac{d}{dx}\left(\dfrac{c}{x^2}\right) = \dfrac{-2c}{x^3}.$

**EXAMPLES**

Use the *General Rule*, p. 29, in differentia'ing the following functions:

1. $y = 3x^2$. *Ans.* $\frac{dy}{dx} = 6x$.
2. $y = x^2 + 2$. $\frac{dy}{dx} = 2x$.
3. $y = 5 - 4x$. $\frac{dy}{dx} = -4$.
4. $s = 2t^2 - 4$. $\frac{ds}{dt} = 4t$.
5. $y = \frac{1}{x}$. $\frac{dy}{dx} = -\frac{1}{x^2}$.
6. $y = \frac{x+2}{x}$. $\frac{dy}{dx} = -\frac{2}{x^2}$.
7. $y = x^3$. *Ans.* $\frac{dy}{dx} = 3x^2$.
8. $y = 2x^2 - 3$. $\frac{dy}{dx} = 4x$.
9. $y = 1 - 2x^3$. $\frac{dy}{dx} = -6x^2$.
10. $\rho = a\theta^2$. $\frac{d\rho}{d\theta} = 2a\theta$.
11. $y = \frac{2}{x^2}$. $\frac{dy}{dx} = -\frac{4}{x^3}$.
12. $y = \frac{3}{x^2 - 1}$. $\frac{dy}{dx} = -\frac{6x}{(x^2 - 1)^2}$.
13. $y = 7x^2 + x$.
14. $s = at^2 - 2bt$.
15. $r = 8t + 3t^2$.
16. $y = \frac{3}{x^2}$.
17. $s = -\frac{a}{2t + 3}$.
18. $y = bx^3 - cx$.
19. $\rho = 3\theta^3 - 2\theta^2$.
20. $y = \frac{3}{4}x^2 - \frac{1}{2}x$.
21. $y = \frac{x^2 - 5}{x}$.
22. $\rho = \frac{\theta^2}{1 + \theta}$.
23. $y = \frac{1}{2}x^2 + 2x$.
24. $z = 4x - 3x^2$.
25. $\rho = 3\theta + \theta^2$.
26. $y = \frac{ax + b}{x^2}$.
27. $z = \frac{x^3 + 2}{x}$.
28. $y = x^2 - 3x + 6$. *Ans.* $y' = 2x - 3$.
29. $s = 2t^2 + 5t - 8$. $s' = 4t + 5$.
30. $\rho = 5\theta^3 - 2\theta + 6$. $\rho' = 15\theta^2 - 2$.
31. $y = ax^2 + bx + c$. $y' = 2ax + b$.

**32. Applications of the derivative to Geometry.** We shall now consider a theorem which is fundamental in all applications of the Differential Calculus to Geometry. Let

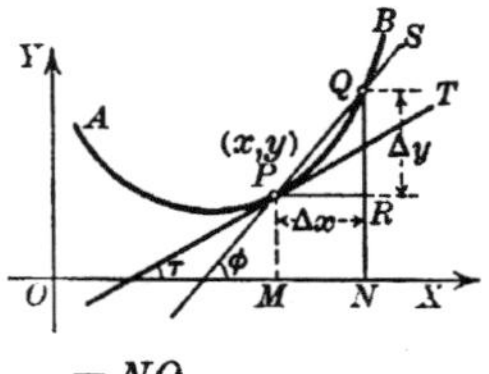


(*A*) $$y = f(x)$$

be the equation of a curve *AB*.

Now differentiate (*A*) by the *General Rule* and interpret each step geometrically.

FIRST STEP. $$y + \Delta y = f(x + \Delta x) \qquad = NQ$$

SECOND STEP. $$y + \Delta y = f(x + \Delta x) \qquad = NQ$$
$$\underline{y \qquad = f(x) \qquad\qquad = MP = NR}$$
$$\Delta y = f(x + \Delta x) - f(x) = RQ.$$

THIRD STEP. $$\frac{\Delta y}{\Delta x} = \frac{f(x + \Delta x) - f(x)}{\Delta x} = \frac{RQ}{MN} = \frac{RQ}{PR}$$
$$= \tan RPQ = \tan \phi$$
$$= \text{slope of secant line } PQ.$$

FOURTH STEP: $$\lim_{\Delta x = 0} \frac{\Delta y}{\Delta x} = \lim_{\Delta x = 0} \frac{f(x+\Delta x) - f(x)}{\Delta x}$$

(B) $$= \frac{dy}{dx} = \text{value of the derivative at } P.$$

But when we let $\Delta x \doteq 0$, the point $Q$ will move along the curve and approach nearer and nearer to $P$, the secant will turn about $P$ and approach the tangent as a limiting position, and we have also

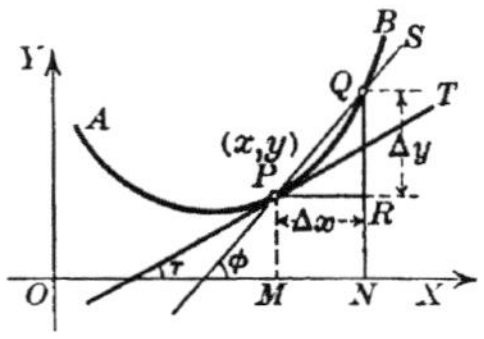


$$\lim_{\Delta x = 0} \frac{\Delta y}{\Delta x} = \lim_{\Delta x = 0} \tan \phi = \tan \tau$$

(C) $$= \text{slope of the tangent at } P.$$

Hence from (B) and (C),

$\frac{dy}{dx}$ = **slope of the tangent line $PT$.** Therefore

**Theorem.** *The value of the derivative at any point of a curve is equal to the slope of the line drawn tangent to the curve at that point.*

It was this tangent problem that led Leibnitz* to the discovery of the Differential Calculus.

ILLUSTRATIVE EXAMPLE 1. Find the slopes of the tangents to the parabola $y = x^2$ at the vertex, and at the point where $x = \frac{1}{2}$.

**Solution.** Differentiating by *General Rule*, p. 29, we get

(A) $$\frac{dy}{dx} = 2x = \text{slope of tangent line at } any \text{ point on curve.}$$

To find slope of tangent at vertex, substitute $x = 0$ in (A),

giving $$\frac{dy}{dx} = 0.$$

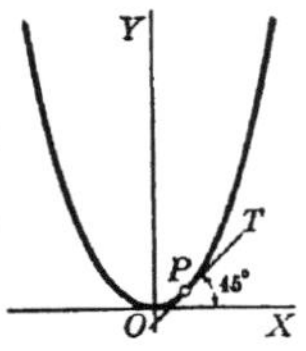


Therefore the tangent at vertex has the slope zero; that is, it is parallel to the axis of $x$ and in this case coincides with it.

To find slope of tangent at the point $P$, where $x = \frac{1}{2}$, substitute in (A), giving $$\frac{dy}{dx} = 1;$$

that is, the tangent at the point $P$ makes an angle of 45° with the axis of $x$.

* Gottfried Wilhelm Leibnitz (1646–1716) was a native of Leipzig. His remarkable abilities were shown by original investigations in several branches of learning. He was first to publish his discoveries in Calculus in a short essay appearing in the periodical *Acta Eruditorum* at Leipzig in 1684. It is known, however, that manuscripts on Fluxions written by Newton were already in existence, and from these some claim Leibnitz got the new ideas. The decision of modern times seems to be that both Newton and Leibnitz invented the Calculus independently of each other. The notation used to-day was introduced by Leibnitz. **See frontispiece.**

### EXAMPLES

Find by differentiation the slopes of the tangents to the following curves at the points indicated. Verify each result by drawing the curve and its tangent.

1. $y = x^2 - 4$, where $x = 2$. *Ans.* 4.

2. $y = 6 - 3x^2$ where $x = 1$. $-6$.

3. $y = x^3$, where $x = -1$. 3.

4. $y = \frac{2}{x}$, where $x = -2$. $-\frac{1}{2}$.

5. $y = x - x^2$, where $x = 0$. 1.

6. $y = \frac{1}{x-1}$, where $x = 3$. $-\frac{1}{4}$.

7. $y = \frac{1}{2}x^2$, where $x = 4$. 4.

8. $y = x^2 - 2x + 3$, where $x = 1$. 0.

9. $y = 9 - x^2$, where $x = -3$. 6.

**10.** Find the slope of the tangent to the curve $y = 2x^3 - 6x + 5$, (a) at the point where $x = 1$; (b) at the point where $x = 0$. *Ans.* (a) 0; (b) $-6$.

**11.** (a) Find the slopes of the tangents to the two curves $y = 3x^2 - 1$ and $y = 2x^2 + 3$ at their points of intersection. (b) At what angle do they intersect?
*Ans.* (a) $\pm 12$, $\pm 8$; (b) arc tan $\frac{4}{97}$.

**12.** The curves on a railway track are often made parabolic in form. Suppose that a track has the form of the parabola $y = x^2$ (last figure, p. 32), the directions $OX$ and $OY$ being east and north respectively, and the unit of measurement 1 mile. If the train is going east when passing through $O$, in what direction will it be going

(a) when $\frac{1}{2}$ mi. east of $OY$? *Ans.* Northeast.
(b) when $\frac{1}{2}$ mi. west of $OY$? Southeast.
(c) when $\frac{\sqrt{3}}{2}$ mi. east of $OY$? N. 30° E.
(d) when $\frac{1}{12}$ mi. north of $OX$? E. 30° S., or E. 30° N.

**13.** A street-car track has the form of the cubical parabola $y = x^3$. Assume the same directions and unit as in the last example. If a car is going west when passing through $O$, in what direction will it be going

(a) when $\frac{1}{\sqrt{3}}$ mi. east of $OY$? *Ans.* Southwest.
(b) when $\frac{1}{\sqrt{3}}$ mi. west of $OY$? Southwest.
(c) when $\frac{1}{2}$ mi. north of $OX$? S. 27° 53′ W.
(d) when 2 mi. south of $OX$?
(e) when equidistant from $OX$ and $OY$?

# CHAPTER V

## RULES FOR DIFFERENTIATING STANDARD ELEMENTARY FORMS

**33. Importance of General Rule.** The *General Rule* for differentiation, given in the last chapter, p. 29, is fundamental, being found directly from the definition of a derivative, and it is very important that the student should be thoroughly familiar with it. However, the process of applying the rule to examples in general has been found too tedious or difficult; consequently special rules have been derived from the *General Rule* for differentiating certain standard forms of frequent occurrence in order to facilitate the work.

It has been found convenient to express these special rules by means of formulas, a list of which follows. The student should not only memorize each formula when deduced, but should be able to state the corresponding rule in words.

In these formulas *u*, *v*, and *w* denote *variable* quantities which are functions of *x*, and are differentiable.

### FORMULAS FOR DIFFERENTIATION

I $$\frac{dc}{dx} = 0.$$

II $$\frac{dx}{dx} = 1.$$

III $$\frac{d}{dx}(u + v - w) = \frac{du}{dx} + \frac{dv}{dx} - \frac{dw}{dx}.$$

IV $$\frac{d}{dx}(cv) = c\frac{dv}{dx}.$$

V $$\frac{d}{dx}(uv) = u\frac{dv}{dx} + v\frac{du}{dx}.$$

VI $$\frac{d}{dx}(v^n) = nv^{n-1}\frac{dv}{dx}.$$

VI *a* $$\frac{d}{dx}(x^n) = nx^{n-1}.$$

VII $$\frac{d}{dx}\left(\frac{u}{v}\right) = \frac{v\frac{du}{dx} - u\frac{dv}{dx}}{v^2}.$$

VII *a* $$\frac{d}{dx}\left(\frac{u}{c}\right) = \frac{\frac{du}{dx}}{c}.$$

VIII $$\frac{d}{dx}(\log_a v) = \log_a e \cdot \frac{\frac{dv}{dx}}{v}.$$

VIII *a* $$\frac{d}{dx}(\log v) = \frac{\frac{dv}{dx}}{v}.$$

IX $$\frac{d}{dx}(a^v) = a^v \log a \frac{dv}{dx}.$$

IX *a* $$\frac{d}{dx}(e^v) = e^v \frac{dv}{dx}.$$

X $$\frac{d}{dx}(u^v) = vu^{v-1}\frac{du}{dx} + \log u \cdot u^v \frac{dv}{dx}.$$

XI $$\frac{d}{dx}(\sin v) = \cos v \frac{dv}{dx}.$$

XII $$\frac{d}{dx}(\cos v) = -\sin v \frac{dv}{dx}.$$

XIII $$\frac{d}{dx}(\tan v) = \sec^2 v \frac{dv}{dx}.$$

XIV $$\frac{d}{dx}(\cot v) = -\csc^2 v \frac{dv}{dx}.$$

XV $$\frac{d}{dx}(\sec v) = \sec v \tan v \frac{dv}{dx}.$$

XVI $$\frac{d}{dx}(\csc v) = -\csc v \cot v \frac{dv}{dx}.$$

XVII $$\frac{d}{dx}(\text{vers } v) = \sin v \frac{dv}{dx}.$$

XVIII $$\frac{d}{dx}(\text{arc sin } v) = \frac{\frac{dv}{dx}}{\sqrt{1-v^2}}.$$

XIX $$\frac{d}{dx}(\text{arc cos } v) = -\frac{\frac{dv}{dx}}{\sqrt{1-v^2}}.$$

XX $$\frac{d}{dx}(\text{arc tan } v) = \frac{\frac{dv}{dx}}{1+v^2}.$$

XXI $$\frac{d}{dx}(\text{arc cot } v) = -\frac{\frac{dv}{dx}}{1+v^2}.$$

XXII $$\frac{d}{dx}(\text{arc sec } v) = \frac{\frac{dv}{dx}}{v\sqrt{v^2-1}}.$$

XXIII $$\frac{d}{dx}(\text{arc csc } v) = -\frac{\frac{dv}{dx}}{v\sqrt{v^2-1}}.$$

XXIV $$\frac{d}{dx}(\text{arc vers } v) = \frac{\frac{dv}{dx}}{\sqrt{2v-v^2}}.$$

XXV $$\frac{dy}{dx} = \frac{dy}{dv} \cdot \frac{dv}{dx}, \text{ } y \text{ being a function of } v.$$

XXVI $$\frac{dy}{dx} = \frac{1}{\frac{dx}{dy}}, \text{ } y \text{ being a function of } x.$$

**34. Differentiation of a constant.** A function that is known to have the same value for every value of the independent variable is constant, and we may denote it by

$$y = c.$$

As $x$ takes on an increment $\Delta x$, the function does not change in value, that is, $\Delta y = 0$, and

$$\frac{\Delta y}{\Delta x} = 0.$$

But $$\underset{\Delta x = 0}{\text{limit}}\left(\frac{\Delta y}{\Delta x}\right) = \frac{dy}{dx} = 0.$$

I $$\therefore \frac{dc}{dx} = 0.$$

*The derivative of a constant is zero.*

**35. Differentiation of a variable with respect to itself.**

Let $$y = x.$$

Following the *General Rule*, p. 29, we have

FIRST STEP. $$y + \Delta y = x + \Delta x.$$

SECOND STEP. $$\Delta y = \Delta x.$$

THIRD STEP. $$\frac{\Delta y}{\Delta x} = 1.$$

FOURTH STEP. $$\frac{dy}{dx} = 1.$$

II $$\therefore \frac{dx}{dx} = 1.$$

*The derivative of a variable with respect to itself is unity.*

**36. Differentiation of a sum.**

Let $$y = u + v - w.$$

By the *General Rule*,

FIRST STEP. $$y + \Delta y = u + \Delta u + v + \Delta v - w - \Delta w.$$

SECOND STEP. $$\Delta y = \Delta u + \Delta v - \Delta w.$$

THIRD STEP. $$\frac{\Delta y}{\Delta x} = \frac{\Delta u}{\Delta x} + \frac{\Delta v}{\Delta x} - \frac{\Delta w}{\Delta x}.$$

FOURTH STEP. $$\frac{dy}{dx} = \frac{du}{dx} + \frac{dv}{dx} - \frac{dw}{dx}.$$

[Applying Th. I, p. 18.]

III $$\therefore \frac{d}{dx}(u + v - w) = \frac{du}{dx} + \frac{dv}{dx} - \frac{dw}{dx}.$$

Similarly, for the algebraic sum of any finite number of functions.

*The derivative of the algebraic sum of a finite number of functions is equal to the same algebraic sum of their derivatives.*

**37. Differentiation of the product of a constant and a function.**

Let $$y = cv.$$

By the *General Rule*,

FIRST STEP. $$y + \Delta y = c(v + \Delta v) = cv + c\Delta v.$$

SECOND STEP. $$\Delta y = c \cdot \Delta v.$$

THIRD STEP. $$\frac{\Delta y}{\Delta x} = c\frac{\Delta v}{\Delta x}.$$

FOURTH STEP. $$\frac{dy}{dx} = c\frac{dv}{dx}.$$

[Applying Th. II, p. 18.]

**IV** $$\therefore \frac{d}{dx}(cv) = c\frac{dv}{dx}.$$

*The derivative of the product of a constant and a function is equal to the product of the constant and the derivative of the function.*

**38. Differentiation of the product of two functions.**

Let $$y = uv.$$

By the *General Rule*,

FIRST STEP. $$y + \Delta y = (u + \Delta u)(v + \Delta v)$$

Multiplying out this becomes

$$y + \Delta y = uv + u \cdot \Delta v + v \cdot \Delta u + \Delta u \cdot \Delta v.$$

SECOND STEP. $$\Delta y = u \cdot \Delta v + v \cdot \Delta u + \Delta u \cdot \Delta v.$$

THIRD STEP. $$\frac{\Delta y}{\Delta x} = u\frac{\Delta v}{\Delta x} + v\frac{\Delta u}{\Delta x} + \Delta u\frac{\Delta v}{\Delta x}.$$

FOURTH STEP. $$\frac{dy}{dx} = u\frac{dv}{dx} + v\frac{du}{dx}.$$

$\left[\text{Applying Th. II, p. 18, since when } \Delta x \doteq 0, \Delta u \doteq 0, \text{ and } \left(\Delta u \frac{\Delta v}{\Delta x}\right) \doteq 0.\right]$

**V** $$\therefore \frac{d}{dx}(uv) = u\frac{dv}{dx} + v\frac{du}{dx}.$$

*The derivative of the product of two functions is equal to the first function times the derivative of the second, plus the second function times the derivative of the first.*

**39. Differentiation of the product of any finite number of functions.**

Now in dividing both sides of **V** by $uv$, this formula assumes the form

$$\frac{\frac{d}{dx}(uv)}{uv} = \frac{\frac{du}{dx}}{u} + \frac{\frac{dv}{dx}}{v}.$$

If then we have the product of $n$ functions

$$y = v_1 v_2 \cdots v_n,$$

we may write

$$\frac{\frac{d}{dx}(v_1 v_2 \cdots v_n)}{v_1 v_2 \cdots v_n} = \frac{\frac{dv_1}{dx}}{v_1} + \frac{\frac{d}{dx}(v_2 v_3 \cdots v_n)}{v_2 v_3 \cdots v_n}$$

$$= \frac{\frac{dv_1}{dx}}{v_1} + \frac{\frac{dv_2}{dx}}{v_2} + \frac{\frac{d}{dx}(v_3 v_4 \cdots v_n)}{v_3 v_4 \cdots v_n}$$

$$= \frac{\frac{dv_1}{dx}}{v_1} + \frac{\frac{dv_2}{dx}}{v_2} + \frac{\frac{dv_3}{dx}}{v_3} + \cdots + \frac{\frac{dv_n}{dx}}{v_n}.$$

Multiplying both sides by $v_1 v_2 \cdots v_n$, we get

$$\frac{d}{dx}(v_1 v_2 \cdots v_n) = (v_2 v_3 \cdots v_n)\frac{dv_1}{dx} + (v_1 v_3 \cdots v_n)\frac{dv_2}{dx} + \cdots + (v_1 v_2 \cdots v_{n-1})\frac{dv_n}{dx}.$$

*The derivative of the product of a finite number of functions is equal to the sum of all the products that can be formed by multiplying the derivative of each function by all the other functions.*

**40. Differentiation of a function with a constant exponent.** If the $n$ factors in the above result are each equal to $v$, we get

$$\frac{\frac{d}{dx}(v^n)}{v^n} = n\frac{\frac{dv}{dx}}{v}.$$

VI $$\therefore \frac{d}{dx}(v^n) = nv^{n-1}\frac{dv}{dx}.$$

When $v = x$ this becomes

VI *a* $$\frac{d}{dx}(x^n) = nx^{n-1}.$$

We have so far proven **VI** only for the case when $n$ is a positive integer. In § 46, however, it will be shown that this formula holds true for any value of $n$, and we shall make use of this general result now.

*The derivative of a function with a constant exponent is equal to the product of the exponent, the function with the exponent diminished by unity, and the derivative of the function.*

**41. Differentiation of a quotient.**

Let $$y=\frac{u}{v}, \qquad v \neq 0.$$

By the *General Rule*,

FIRST STEP. $$y+\Delta y=\frac{u+\Delta u}{v+\Delta v}.$$

SECOND STEP. $$\Delta y=\frac{u+\Delta u}{v+\Delta v}-\frac{u}{v}=\frac{v\cdot\Delta u-u\cdot\Delta v}{v(v+\Delta v)}.$$

THIRD STEP. $$\frac{\Delta y}{\Delta x}=\frac{v\dfrac{\Delta u}{\Delta x}-u\dfrac{\Delta v}{\Delta x}}{v(v+\Delta v)}.$$

FOURTH STEP. $$\frac{dy}{dx}=\frac{v\dfrac{du}{dx}-u\dfrac{dv}{dx}}{v^2}.$$

[Applying Theorems II and III, p. 18.]

VII $$\therefore\ \frac{d}{dx}\left(\frac{u}{v}\right)=\frac{v\dfrac{du}{dx}-u\dfrac{dv}{dx}}{v^2}.$$

*The derivative of a fraction is equal to the denominator times the derivative of the numerator, minus the numerator times the derivative of the denominator, all divided by the square of the denominator.*

When the denominator is constant, set $v=c$ in **VII**, giving

VII *a* $$\frac{d}{dx}\left(\frac{u}{c}\right)=\frac{\dfrac{du}{dx}}{c}.$$

$\left[\text{Since } \frac{dv}{dx}=\frac{dc}{dx}=0.\right]$

We may also get **VII *a*** from **IV** as follows:

$$\frac{d}{dx}\left(\frac{u}{c}\right)=\frac{1}{c}\frac{du}{dx}=\frac{\dfrac{du}{dx}}{c}.$$

*The derivative of the quotient of a function by a constant is equal to the derivative of the function divided by the constant.*

All explicit algebraic functions of one independent variable may be differentiated by following the rules we have deduced so far.

**EXAMPLES***

Differentiate the following:

**1.** $y = x^3.$

**Solution.** $\frac{dy}{dx} = \frac{d}{dx}(x^3) = 3x^2.$ *Ans.* By VI *a*

$[n = 3.]$

**2.** $y = ax^4 - bx^2.$

**Solution.** $\frac{dy}{dx} = \frac{d}{dx}(ax^4 - bx^2) = \frac{d}{dx}(ax^4) - \frac{d}{dx}(bx^2)$ by III

$= a\frac{d}{dx}(x^4) - b\frac{d}{dx}(x^2)$ by IV

$= 4ax^3 - 2bx.$ *Ans.* By VI *a*

**3.** $y = x^{\frac{4}{3}} + 5.$

**Solution.** $\frac{dy}{dx} = \frac{d}{dx}(x^{\frac{4}{3}}) + \frac{d}{dx}(5)$ by III

$= \frac{4}{3}x^{\frac{1}{3}}.$ *Ans.* By VI *a* and I

**4.** $y = \frac{3x^3}{\sqrt[5]{x^2}} - \frac{7x}{\sqrt[3]{x^4}} + 8\sqrt[7]{x^3}.$

**Solution.** $\frac{dy}{dx} = \frac{d}{dx}(3x^{\frac{13}{5}}) - \frac{d}{dx}(7x^{-\frac{1}{3}}) + \frac{d}{dx}(8x^{\frac{3}{7}})$ by III

$= \frac{39}{5}x^{\frac{8}{5}} + \frac{7}{3}x^{-\frac{4}{3}} + \frac{24}{7}x^{-\frac{4}{7}}.$ *Ans.* By IV and VI *a*

**5.** $y = (x^2 - 3)^5.$

**Solution.** $\frac{dy}{dx} = 5(x^2 - 3)^4\frac{d}{dx}(x^2 - 3)$ by VI

$[v = x^2 - 3 \text{ and } n = 5.]$

$= 5(x^2 - 3)^4 \cdot 2x = 10x(x^2 - 3)^4.$ *Ans.*

We might have expanded this function by the Binomial Theorem and then applied III, etc., but the above process is to be preferred.

**6.** $y = \sqrt{a^2 - x^2}.$

**Solution.** $\frac{dy}{dx} = \frac{d}{dx}(a^2 - x^2)^{\frac{1}{2}} = \frac{1}{2}(a^2 - x^2)^{-\frac{1}{2}}\frac{d}{dx}(a^2 - x^2)$ by VI

$[v = a^2 - x^2, \text{ and } n = \frac{1}{2}.]$

$= \frac{1}{2}(a^2 - x^2)^{-\frac{1}{2}}(-2x) = -\frac{x}{\sqrt{a^2 - x^2}}.$ *Ans.*

**7.** $y = (3x^2 + 2)\sqrt{1 + 5x^2}.$

**Solution.** $\frac{dy}{dx} = (3x^2 + 2)\frac{d}{dx}(1 + 5x^2)^{\frac{1}{2}} + (1 + 5x^2)^{\frac{1}{2}}\frac{d}{dx}(3x^2 + 2)$ by V

$[u = 3x^2 + 2, \text{ and } v = (1 + 5x^2)^{\frac{1}{2}}.]$

$= (3x^2 + 2)\frac{1}{2}(1 + 5x^2)^{-\frac{1}{2}}\frac{d}{dx}(1 + 5x^2) + (1 + 5x^2)^{\frac{1}{2}}6x$ by VI, etc.

$= (3x^2 + 2)(1 + 5x^2)^{-\frac{1}{2}}5x + 6x(1 + 5x^2)^{\frac{1}{2}}$

$= \frac{5x(3x^2 + 2)}{\sqrt{1 + 5x^2}} + 6x\sqrt{1 + 5x^2} = \frac{45x^3 + 16x}{\sqrt{1 + 5x^2}}.$ *Ans.*

* When learning to differentiate, the student should have oral drill in differentiating simple functions.

8. $y = \dfrac{a^2 + x^2}{\sqrt{a^2 - x^2}}.$

**Solution.** $\dfrac{dy}{dx} = \dfrac{(a^2 - x^2)^{\frac{1}{2}} \dfrac{d}{dx}(a^2 + x^2) - (a^2 + x^2) \dfrac{d}{dx}(a^2 - x^2)^{\frac{1}{2}}}{a^2 - x^2}$ by VII

$= \dfrac{2x(a^2 - x^2) + x(a^2 + x^2)}{(a^2 - x^2)^{\frac{3}{2}}}$

[Multiplying both numerator and denominator by $(a^2 - x^2)^{\frac{1}{2}}$.]

$= \dfrac{3a^2x - x^3}{(a^2 - x^2)^{\frac{3}{2}}}.$ *Ans.*

9. $y = 5x^4 + 3x^2 - 6.$ $\qquad \dfrac{dy}{dx} = 20x^3 + 6x.$

10. $y = 3cx^2 - 8dx + 5e.$ $\qquad \dfrac{dy}{dx} = 6cx - 8d.$

11. $y = x^{a+b}.$ $\qquad \dfrac{dy}{dx} = (a + b)x^{a+b-1}.$

12. $y = x^n + nx + n.$ $\qquad \dfrac{dy}{dx} = nx^{n-1} + n.$

13. $f(x) = \frac{2}{3}x^3 - \frac{3}{2}x^2 + 5.$ $\qquad f'(x) = 2x^2 - 3x.$

14. $f(x) = (a + b)x^2 + cx + d.$ $\qquad f'(x) = 2(a + b)x + c.$

15. $\dfrac{d}{dx}(a + bx + cx^2) = b + 2cx.$

16. $\dfrac{d}{dy}(5y^m - 3y + 6) = 5my^{m-1} - 3.$

17. $\dfrac{d}{dx}(2x^{-2} + 3x^{-3}) = -4x^{-3} - 9x^{-4}.$

18. $\dfrac{d}{ds}(3s^{-4} - s) = -12s^{-5} - 1.$

19. $\dfrac{d}{dx}(4x^{\frac{1}{2}} + x^2) = 2x^{-\frac{1}{2}} + 2x.$

20. $\dfrac{d}{dy}(y^{-2} - 4y^{-\frac{1}{2}}) = -2y^{-3} + 2y^{-\frac{3}{2}}.$

21. $\dfrac{d}{dx}(2x^3 + 5) = 6x^2.$

22. $\dfrac{d}{dt}(3t^5 - 2t^2) = 15t^4 - 4t.$

23. $\dfrac{d}{d\theta}(a\theta^4 + b\theta) = 4a\theta^3 + b.$

24. $\dfrac{d}{d\alpha}(5 - 2\alpha^{\frac{3}{2}}) = -3\alpha^{\frac{1}{2}}.$

25. $\dfrac{d}{dt}(9t^{\frac{5}{3}} + t^{-1}) = 15t^{\frac{2}{3}} - t^{-2}.$

26. $\dfrac{d}{dx}(2x^{12} - x^9) = 24x^{11} - 9x^8$

27. $r = c\theta^3 + d\theta^2 + c\theta.$ $\qquad r' = 3c\theta^2 + 2d\theta + e.$

28. $y = 6x^{\frac{7}{2}} + 4x^{\frac{5}{2}} + 2x^{\frac{3}{2}}.$ $\qquad y' = 21x^{\frac{5}{2}} + 10x^{\frac{3}{2}} + 3x^{\frac{1}{2}}.$

29. $y = \sqrt{3x} + \sqrt[3]{x} + \dfrac{1}{x}.$ $\qquad y' = \dfrac{3}{2\sqrt{3x}} + \dfrac{1}{3\sqrt[3]{x^2}} - \dfrac{1}{x^2}$

30. $y = \dfrac{a + bx + cx^2}{x}.$ $\qquad y' = c - \dfrac{a}{x^2}.$

31. $y = \dfrac{(x - 1)^3}{x^{\frac{1}{3}}}.$ $\qquad y' = \frac{8}{3}x^{\frac{5}{3}} - 5x^{\frac{2}{3}} + 2x^{-\frac{1}{3}} + \frac{1}{3}x^{-\frac{4}{3}}.$

32. $y = \dfrac{x^{\frac{5}{2}} - x - x^{\frac{1}{2}} + a}{x^{\frac{3}{2}}}.$ $\qquad y' = \dfrac{2x^{\frac{5}{2}} + x + 2x^{\frac{1}{2}} - 3a}{2x^{\frac{5}{2}}}.$

33. $y = (2x^3 + x^2 - 5)^3.$ $\qquad y' = 6x(3x + 1)(2x^3 + x^2 - 5)^2.$

**34.** $f(x) = (a + bx^2)^{\frac{5}{4}}$. $\qquad f'(x) = \dfrac{5\,bx}{2}(a + bx^2)^{\frac{1}{4}}$.

**35.** $f(x) = (1 + 4\,x^3)(1 + 2\,x^2)$. $\qquad f'(x) = 4\,x(1 + 3\,x + 10\,x^3)$.

**36.** $f(x) = (a + x)\sqrt{a - x}$. $\qquad f'(x) = \dfrac{a - 3\,x}{2\sqrt{a - x}}$.

**37.** $f(x) = (a + x)^m(b + x)^n$. $\qquad f'(x) = (a + x)^m(b + x)^n\left[\dfrac{m}{a + x} + \dfrac{n}{b + x}\right]$.

**38.** $y = \dfrac{1}{x^n}$. $\qquad \dfrac{dy}{dx} = -\dfrac{n}{x^{n+1}}$.

**39.** $y = x(a^2 + x^2)\sqrt{a^2 - x^2}$. $\qquad \dfrac{dy}{dx} = \dfrac{a^4 + a^2x^2 - 4\,x^4}{\sqrt{a^2 - x^2}}$.

**40.** Differentiate the following functions:

(a) $\dfrac{d}{dx}(2\,x^3 - 4\,x + 6)$.

(b) $\dfrac{d}{dt}(at^7 + bt^5 - 9)$.

(c) $\dfrac{d}{d\theta}(3\,\theta^{\frac{3}{2}} - 2\,\theta^{\frac{1}{2}} + 6\,\theta)$.

(d) $\dfrac{d}{dx}(2\,x^3 + x)^{\frac{5}{3}}$.

(e) $\dfrac{d}{dt}(b + at^2)^{\frac{1}{2}}$.

(f) $\dfrac{d}{dx}(x^2 - a^2)^{\frac{3}{2}}$.

(g) $\dfrac{d}{d\phi}(4 - \phi^{\frac{2}{3}})$.

(h) $\dfrac{d}{dt}\sqrt{1 + 9\,t^2}$.

(i) $\dfrac{d}{dx}(x^{\frac{2}{3}} - a^{\frac{2}{3}})$.

(j) $\dfrac{d}{dt}(5 + 2\,t)^{\frac{9}{2}}$.

(k) $\dfrac{d}{ds}\sqrt{a + b\sqrt{s}}$.

(l) $\dfrac{d}{dx}(2\,x^{\frac{1}{3}} + 2\,x^{\frac{5}{3}})$.

**41.** $y = \dfrac{2\,x^4}{b^2 - x^2}$. $\qquad \dfrac{dy}{dx} = \dfrac{8\,b^2x^3 - 4\,x^5}{(b^2 - x^2)^2}$.

**42.** $y = \dfrac{a - x}{a + x}$. $\qquad \dfrac{dy}{dx} = -\dfrac{2\,a}{(a + x)^2}$.

**43.** $s = \dfrac{t^3}{(1 + t)^2}$. $\qquad \dfrac{ds}{dt} = \dfrac{3\,t^2 + t^3}{(1 + t)^3}$.

**44.** $f(s) = \dfrac{(s + 4)^2}{s + 3}$. $\qquad f'(s) = \dfrac{(s + 2)(s + 4)}{(s + 3)^2}$.

**45.** $f(\theta) = \dfrac{\theta}{\sqrt{a - b\theta^2}}$. $\qquad f'(\theta) = \dfrac{a}{(a - b\theta^2)^{\frac{3}{2}}}$.

**46.** $F(r) = \sqrt{\dfrac{1 + r}{1 - r}}$. $\qquad F'(r) = \dfrac{1}{(1 - r)\sqrt{1 - r^2}}$.

**47.** $\psi(y) = \left(\dfrac{y}{1 - y}\right)^m$. $\qquad \psi'(y) = \dfrac{my^{m-1}}{(1 - y)^{m+1}}$.

**48.** $\phi(x) = \dfrac{2\,x^2 - 1}{x\sqrt{1 + x^2}}$. $\qquad \phi'(x) = \dfrac{1 + 4\,x^2}{x^2(1 + x^2)^{\frac{3}{2}}}$.

**49.** $y = \sqrt{2\,px}$. $\qquad y' = \dfrac{p}{y}$.

**50.** $y = \dfrac{b}{a}\sqrt{a^2 - x^2}$. $\qquad y' = -\dfrac{b^2x}{a^2y}$.

**51.** $y = (a^{\frac{2}{3}} - x^{\frac{2}{3}})^{\frac{3}{2}}$. $\qquad y' = -\sqrt[3]{\dfrac{y}{x}}$.

52. $r = \sqrt{a\phi} + c\sqrt{\phi^3}$. $\qquad r' = \dfrac{\sqrt{a} + 3c\phi}{2\sqrt{\phi}}$.

53. $u = \dfrac{v^c + v^d}{cd}$. $\qquad u' = \dfrac{v^{c-1}}{d} + \dfrac{v^{d-1}}{c}$.

54. $p = \dfrac{(q+1)^{\frac{3}{2}}}{\sqrt{q-1}}$. $\qquad p' = \dfrac{(q-2)\sqrt{q+1}}{(q-1)^{\frac{3}{2}}}$.

55. Differentiate the following functions:

(a) $\dfrac{d}{dx}\left(\dfrac{a^2 - x^2}{a^2 + x^2}\right)$.

(b) $\dfrac{d}{dx}\left(\dfrac{x^3}{1 + x^4}\right)$.

(c) $\dfrac{d}{dx}\left(\dfrac{1 + x}{\sqrt{1 - x}}\right)$.

(d) $\dfrac{d}{dy}\left(\dfrac{ay^2}{b + y^3}\right)$.

(e) $\dfrac{d}{ds}\left(\dfrac{a^2 - s^2}{\sqrt{a^2 + s^2}}\right)$.

(f) $\dfrac{d}{dx}\dfrac{\sqrt{4 - 2x^3}}{x}$.

(g) $\dfrac{d}{dx}\dfrac{x^2}{\sqrt{1 - x^2}}$.

(h) $\dfrac{d}{dx}\dfrac{1 + x^2}{(1 - x^2)^{\frac{3}{2}}}$.

(i) $\dfrac{d}{dt}\sqrt{\dfrac{1 + t^2}{1 - t^2}}$.

**42. Differentiation of a function of a function.** It sometimes happens that $y$, instead of being defined directly as a function of $x$, is given as a function of another variable $v$, which is defined as a function of $x$. In that case $y$ is a function of $x$ through $v$ and is called a *function of a function.*

For example, if
$$y = \frac{2v}{1 - v^2},$$
and
$$v = 1 - x^2,$$
then $y$ is a function of a function. By eliminating $v$ we may express $y$ directly as a function of $x$, but in general this is not the best plan when we wish to find $\dfrac{dy}{dx}$.

If $y = f(v)$ and $v = \phi(x)$, then $y$ is a function of $x$ through $v$. Hence, when we let $x$ take on an increment $\Delta x$, $v$ will take on an increment $\Delta v$ and $y$ will also take on a corresponding increment $\Delta y$. Keeping this in mind, let us apply the *General Rule* simultaneously to the two functions
$$y = f(v) \text{ and } v = \phi(x).$$

FIRST STEP. $\quad y + \Delta y = f(v + \Delta v) \qquad v + \Delta v = \phi(x + \Delta x)$

SECOND STEP.
$$\begin{array}{ll} y + \Delta y = f(v + \Delta v) & v + \Delta v = \phi(x + \Delta x) \\ y \qquad\; = f(v) & v \qquad\; = \phi(x) \\ \hline \Delta y = f(v + \Delta v) - f(v), & \Delta v = \phi(x + \Delta x) - \phi(x) \end{array}$$

THIRD STEP. $\quad \dfrac{\Delta y}{\Delta v} = \dfrac{f(v + \Delta v) - f(v)}{\Delta v}, \qquad \dfrac{\Delta v}{\Delta x} = \dfrac{\phi(x + \Delta x) - \phi(x)}{\Delta x}.$

The left-hand members show one form of the ratio of the increment of each function to the increment of the corresponding variable, and the right-hand members exhibit the same ratios in another form. Before passing to the limit let us form a product of these two ratios, choosing the left-hand forms for this purpose.

This gives $\frac{\Delta y}{\Delta v}\cdot\frac{\Delta v}{\Delta x}$, which equals $\frac{\Delta y}{\Delta x}$.

Write this $$\frac{\Delta y}{\Delta x}=\frac{\Delta y}{\Delta v}\cdot\frac{\Delta v}{\Delta x}.$$

FOURTH STEP. Passing to the limit,

(A) $$\frac{dy}{dx}=\frac{dy}{dv}\cdot\frac{dv}{dx}.$$ Th. II, p. 18

This may also be written

(B) $$\frac{dy}{dx}=f'(v)\cdot\phi'(x).$$

*If $y=f(v)$ and $v=\phi(x)$, the derivative of $y$ with respect to $x$ equals the product of the derivative of $y$ with respect to $v$ and the derivative of $v$ with respect to $x$.*

**43. Differentiation of inverse functions.** Let $y$ be given as a function of $x$ by means of the relation $y=f(x)$.

It is usually possible in the case of functions considered in this book to solve this equation for $x$, giving

$$x=\phi(y);$$

that is, to consider $y$ as the independent and $x$ as the dependent variable. In that case $f(x)$ and $\phi(y)$ are said to be *inverse functions*. When we wish to distinguish between the two it is customary to call the first one given the *direct function* and the second one the *inverse function*. Thus, in the examples which follow, if the second members in the first column are taken as the direct functions, then the corresponding members in the second column will be respectively their *inverse functions*.

$$y=x^2+1, \qquad x=\pm\sqrt{y-1}.$$
$$y=a^x, \qquad x=\log_a y.$$
$$y=\sin x, \qquad x=\text{arc}\sin y.$$

Let us now differentiate the inverse functions

$$y=f(x) \quad\text{and}\quad x=\phi(y)$$

simultaneously by the *General Rule.*

FIRST STEP. $y+\Delta y=f(x+\Delta x)$ $\qquad x+\Delta x=\phi(y+\Delta y)$

SECOND STEP.
$$\begin{array}{l} y+\Delta y=f(x+\Delta x) \\ y \qquad\ \ =f(x) \\ \hline \Delta y=f(x+\Delta x)-f(x) \end{array}, \qquad \begin{array}{l} x+\Delta x=\phi(y+\Delta y) \\ x \qquad\ \ =\phi(y) \\ \hline \Delta x=\phi(y+\Delta y)-\phi(y) \end{array}$$

THIRD STEP.
$$\frac{\Delta y}{\Delta x}=\frac{f(x+\Delta x)-f(x)}{\Delta x}, \qquad \frac{\Delta x}{\Delta y}=\frac{\phi(y+\Delta y)-\phi(y)}{\Delta y}$$

Taking the product of the left-hand forms of these ratios, we get

$$\frac{\Delta y}{\Delta x}\cdot\frac{\Delta x}{\Delta y}=1,$$

or,

$$\frac{\Delta y}{\Delta x}=\frac{1}{\dfrac{\Delta x}{\Delta y}}.$$

FOURTH STEP. Passing to the limit,

$$(C) \qquad \frac{dy}{dx}=\frac{1}{\dfrac{dx}{dy}},$$

or,

$$(D) \qquad f'(x)=\frac{1}{\phi'(y)}.$$

*The derivative of the inverse function is equal to the reciprocal of the derivative of the direct function.*

**44. Differentiation of a logarithm.**

Let

$$y=\log_a v.^*$$

Differentiating by the *General Rule*, p. 29, considering $v$ as the independent variable, we have

FIRST STEP. $$y+\Delta y=\log_a(v+\Delta v).$$

SECOND STEP. $$\Delta y=\log_a(v+\Delta v)-\log_a v\ \dagger$$
$$=\log_a\left(\frac{v+\Delta v}{v}\right)=\log_a\left(1+\frac{\Delta v}{v}\right).$$

[By 8, p. 1.]

* The student must not forget that this function is defined only for positive values of the base $a$ and the variable $v$.

† If we take the third and fourth steps without transforming the right-hand member, there results:

*Third step.* $$\frac{\Delta y}{\Delta v}=\frac{\log_a(v+\Delta v)-\log_a v}{\Delta v}.$$

*Fourth step.* $\frac{dy}{dv}=\frac{0}{0}$, which is indeterminate. Hence the limiting value of the right-hand member in the third step cannot be found by direct substitution, and the above transformation is necessary.

THIRD STEP. $$\frac{\Delta y}{\Delta v}=\frac{1}{\Delta v}\log_a\left(1+\frac{\Delta v}{v}\right)=\log_a\left(1+\frac{\Delta v}{v}\right)^{\frac{1}{\Delta v}}$$
$$=\frac{1}{v}\log_a\left(1+\frac{\Delta v}{v}\right)^{\frac{v}{\Delta v}}.$$

[Dividing the logarithm by $v$ and at the same time multiplying the exponent of the parenthesis by $v$ changes the form of the expression but not its value (see 9, p. 1).]

FOURTH STEP. $$\frac{dy}{dv}=\frac{1}{v}\log_a e.$$

[When $\Delta v \doteq 0$, $\frac{\Delta v}{v}\doteq 0$. Therefore $\underset{\Delta v \doteq 0}{\text{limit}}\left(1+\frac{\Delta v}{v}\right)^{\frac{v}{\Delta v}}=e$, from p. 22, placing $x=\frac{\Delta v}{v}$.]

Hence

(*A*) $$\frac{dy}{dv}=\frac{d}{dv}\left(\log_a v\right)=\log_a e\cdot\frac{1}{v}.$$

Since $v$ is a function of $x$ and it is required to differentiate $\log_a v$ with respect to $x$, we must use formula (*A*), § 42, for differentiating a *function of a function*, namely,

$$\frac{dy}{dx}=\frac{dy}{dv}\cdot\frac{dv}{dx}.$$

Substituting value of $\frac{dy}{dv}$ from (*A*), we get

$$\frac{dy}{dx}=\log_a e\cdot\frac{1}{v}\cdot\frac{dv}{dx}.$$

**VIII** $$\therefore\ \frac{d}{dx}(\log_a v)=\log_a e\cdot\frac{\frac{dv}{dx}}{v}.$$

When $a=e$, $\log_a e=\log_e e=1$, and **VIII** becomes

**VIII *a*** $$\frac{d}{dx}(\log v)=\frac{\frac{dv}{dx}}{v}.$$

***The derivative of the logarithm of a function is equal to the product of the modulus* of the system of logarithms and the derivative of the function, divided by the function.***

* The logarithm of $e$ to any base $a$ ($=\log_a e$) is called the *modulus* of the system whose base is $a$. In Algebra it is shown that we may find the logarithm of a number $N$ to any base $a$ by means of the formula

$$\log_a N=\log_a e\cdot\log_e N=\frac{\log_e N}{\log_e a}.$$

The modulus of the common or Briggs system with base 10 is

$$\log_{10} e=.434294\cdots.$$

**45. Differentiation of the simple exponential function.**

Let $$y = a^v. \qquad a > 0$$

Taking the logarithm of both sides to the base $e$, we get

$$\log y = v \log a,$$

or,

$$v = \frac{\log y}{\log a}$$

$$= \frac{1}{\log a} \cdot \log y.$$

Differentiate with respect to $y$ by formula **VIII a**,

$$\frac{dv}{dy} = \frac{1}{\log a} \cdot \frac{1}{y};$$

and from (*C*), § 43, relating to *inverse functions*, we get

$$\frac{dy}{dv} = \log a \cdot y,$$

or,

(*A*) $$\frac{dy}{dv} = \log a \cdot a^v.$$

Since $v$ is a function of $x$ and it is required to differentiate $a^v$ with respect to $x$, we must use formula (*A*), § 42, for differentiating a *function of a function*, namely,

$$\frac{dy}{dx} = \frac{dy}{dv} \cdot \frac{dv}{dx}.$$

Substituting the value of $\frac{dy}{dv}$ from (*A*), we get

$$\frac{dy}{dx} = \log a \cdot a^v \cdot \frac{dv}{dx}.$$

**IX** $$\therefore \quad \frac{d}{dx}(a^v) = \log a \cdot a^v \cdot \frac{dv}{dx}.$$

When $a = e$, $\log a = \log e = 1$, and **IX** becomes

**IX a** $$\frac{d}{dx}(e^v) = e^v \frac{dv}{dx}.$$

*The derivative of a constant with a variable exponent is equal to the product of the natural logarithm of the constant, the constant with the variable exponent, and the derivative of the exponent.*

**46. Differentiation of the general exponential function.**

Let $$y = u^v.^*$$

Taking the logarithm of both sides to the base $e$,

$$\log_e y = v \log_e u,$$

or, $$y = e^{v \log u}.$$

Differentiating by formula **IX a**,

$$\frac{dy}{dx} = e^{v \log u}\frac{d}{dx}(v \log u)$$

$$= e^{v \log u}\left(\frac{v}{u}\frac{du}{dx} + \log u \frac{dv}{dx}\right) \qquad \text{by V}$$

$$= u^v\left(\frac{v}{u}\frac{du}{dx} + \log u \frac{dv}{dx}\right).$$

**X** $$\therefore \frac{d}{dx}(u^v) = vu^{v-1}\frac{du}{dx} + \log u \cdot u^v \frac{dv}{dx}.$$

*The derivative of a function with a variable exponent is equal to the sum of the two results obtained by first differentiating by* **VI**, *regarding the exponent as constant; and again differentiating by* **IX**, *regarding the function as constant.*

Let $v = n$, any constant; then **X** reduces to

$$\frac{d}{dx}(u^n) = nu^{n-1}\frac{du}{dx}.$$

But this is the form differentiated in § 40; therefore **VI** holds true for any value of $n$.

ILLUSTRATIVE EXAMPLE 1. Differentiate $y = \log(x^2 + a)$.

**Solution.** $$\frac{dy}{dx} = \frac{\frac{d}{dx}(x^2 + a)}{x^2 + a} \qquad \text{by VIII } a$$

$$[v = x^2 + a.]$$

$$= \frac{2x}{x^2 + a}. \quad \textit{Ans.}$$

ILLUSTRATIVE EXAMPLE 2. Differentiate $y = \log\sqrt{1 - x^2}$.

**Solution.** $$\frac{dy}{dx} = \frac{\frac{d}{dx}(1 - x^2)^{\frac{1}{2}}}{(1 - x^2)^{\frac{1}{2}}} \qquad \text{by VIII } a$$

$$= \frac{\frac{1}{2}(1 - x^2)^{-\frac{1}{2}}(-2x)}{(1 - x^2)^{\frac{1}{2}}} \qquad \text{by VI}$$

$$= \frac{x}{x^2 - 1}. \quad \textit{Ans.}$$

* $u$ can here assume only positive values.

ILLUSTRATIVE EXAMPLE 3. Differentiate $y = a^{3x^2}$.

**Solution.**

$$\frac{dy}{dx} = \log a \cdot a^{3x^2} \frac{d}{dx}(3x^2) \qquad \text{by IX}$$

$$= 6x \log a \cdot a^{3x^2}. \quad Ans.$$

ILLUSTRATIVE EXAMPLE 4. Differentiate $y = be^{c^2+x^2}$.

**Solution.**

$$\frac{dy}{dx} = b\frac{d}{dx}(e^{c^2+x^2}) \qquad \text{by IV}$$

$$= be^{c^2+x^2}\frac{d}{dx}(c^2+x^2) \qquad \text{by IX } a$$

$$= 2bxe^{c^2+x^2}. \quad Ans.$$

ILLUSTRATIVE EXAMPLE 5. Differentiate $y = x^{e^x}$.

**Solution.**

$$\frac{dy}{dx} = e^x x^{e^x-1}\frac{d}{dx}(x) + x^{e^x}\log x\frac{d}{dx}(e^x) \qquad \text{by X}$$

$$= e^x x^{e^x-1} + x^{e^x}\log x \cdot e^x$$

$$= e^x x^{e^x}\left(\frac{1}{x} + \log x\right). \quad Ans.$$

**47. Logarithmic differentiation.** Instead of applying **VIII** and **VIII *a*** at once in differentiating logarithmic functions, we may sometimes simplify the work by first making use of one of the formulas 7–10 on p. 1. Thus above Illustrative Example 2 may be solved as follows:

ILLUSTRATIVE EXAMPLE 1. Differentiate $y = \log\sqrt{1-x^2}$.

**Solution.** By using 10, p. 1, we may write this in a form free from radicals as follows:

$$y = \tfrac{1}{2}\log(1-x^2).$$

Then

$$\frac{dy}{dx} = \frac{1}{2}\,\frac{\frac{d}{dx}(1-x^2)}{1-x^2} \qquad \text{by VIII } a$$

$$= \frac{1}{2}\cdot\frac{-2x}{1-x^2} = \frac{x}{x^2-1}. \quad Ans.$$

ILLUSTRATIVE EXAMPLE 2. Differentiate $y = \log\sqrt{\dfrac{1+x^2}{1-x^2}}$.

**Solution.** Simplifying by means of 10 and 8, p. 1,

$$y = \tfrac{1}{2}\left[\log(1+x^2) - \log(1-x^2)\right].$$

$$\frac{dy}{dx} = \frac{1}{2}\left[\frac{\frac{d}{dx}(1+x^2)}{1+x^2} - \frac{\frac{d}{dx}(1-x^2)}{1-x^2}\right] \qquad \text{by VIII } a\text{, etc.}$$

$$= \frac{x}{1+x^2} + \frac{x}{1-x^2} = \frac{2x}{1-x^4}. \quad Ans.$$

In differentiating an exponential function, especially a variable with a variable exponent, the best plan is first to take the logarithm of the function and then differentiate. Thus Illustrative Example 5, p. 50, is solved more elegantly as follows:

ILLUSTRATIVE EXAMPLE 3. Differentiate $y = x^{e^x}$.

**Solution.** Taking the logarithm of both sides,

$$\log y = e^x \log x. \qquad \text{By 9, p. 1}$$

Now differentiate both sides with respect to $x$.

$$\frac{\frac{dy}{dx}}{y} = e^x \frac{d}{dx}(\log x) + \log x \frac{d}{dx}(e^x) \qquad \text{by VIII and V}$$

$$= e^x \cdot \frac{1}{x} + \log x \cdot e^x,$$

or,

$$\frac{dy}{dx} = e^x \cdot y\left(\frac{1}{x} + \log x\right)$$

$$= e^x x^{e^x}\left(\frac{1}{x} + \log x\right). \quad Ans.$$

ILLUSTRATIVE EXAMPLE 4. Differentiate $y = (4x^2 - 7)^{2+\sqrt{x^2-5}}$.

**Solution.** Taking the logarithm of both sides,

$$\log y = (2 + \sqrt{x^2 - 5})\log(4x^2 - 7).$$

Differentiating both sides with respect to $x$,

$$\frac{1}{y}\frac{dy}{dx} = (2 + \sqrt{x^2 - 5})\frac{8x}{4x^2 - 7} + \log(4x^2 - 7)\cdot\frac{x}{\sqrt{x^2 - 5}}.$$

$$\frac{dy}{dx} = x(4x^2 - 7)^{2+\sqrt{x^2-5}}\left[\frac{8(2 + \sqrt{x^2 - 5})}{4x^2 - 7} + \frac{\log(4x^2 - 7)}{\sqrt{x^2 - 5}}\right]. \quad Ans.$$

In the case of a function consisting of a number of factors it is sometimes convenient to take the logarithm before differentiating. Thus,

ILLUSTRATIVE EXAMPLE 5. Differentiate $y = \sqrt{\frac{(x-1)(x-2)}{(x-3)(x-4)}}$.

**Solution.** Taking the logarithm of both sides,

$$\log y = \tfrac{1}{2}[\log(x-1) + \log(x-2) - \log(x-3) - \log(x-4)].$$

Differentiating both sides with respect to $x$,

$$\frac{1}{y}\frac{dy}{dx} = \frac{1}{2}\left[\frac{1}{x-1} + \frac{1}{x-2} - \frac{1}{x-3} - \frac{1}{x-4}\right]$$

$$= -\frac{2x^2 - 10x + 11}{(x-1)(x-2)(x-3)(x-4)},$$

or,

$$\frac{dy}{dx} = -\frac{2x^2 - 10x - 11}{(x-1)^{\frac{1}{2}}(x-2)^{\frac{1}{2}}(x-3)^{\frac{3}{2}}(x-4)^{\frac{3}{2}}}. \quad Ans.$$

**EXAMPLES**

**Differentiate the following:**

1. $y = \log(x + a)$. $\qquad \frac{dy}{dx} = \frac{1}{x + a}$.

2. $y = \log(ax + b)$. $\qquad \frac{dy}{dx} = \frac{a}{ax + b}$.

3. $y = \log\frac{1 + x^2}{1 - x^2}$. $\qquad \frac{dy}{dx} = \frac{4x}{1 - x^4}$.

4. $y = \log(x^2 + x)$. $\qquad y' = \frac{2x + 1}{x^2 + x}$.

5. $y = \log(x^3 - 2x + 5)$. $\qquad y' = \frac{3x^2 - 2}{x^3 - 2x + 5}$.

6. $y = \log_a(2x + x^3)$. $\qquad y' = \log_a e \cdot \frac{2 + 3x^2}{2x + x^3}$.

7. $y = x \log x$. $\qquad y' = \log x + 1$.

8. $f(x) = \log x^3$. $\qquad f'(x) = \frac{3}{x}$.

9. $f(x) = \log^3 x$. $\qquad f'(x) = \frac{3\log^2 x}{x}$.

HINT. $\log^3 x = (\log x)^3$. Use first VI, $v = \log x$, $n = 3$; and then VIII *a*.

10. $f(x) = \log\frac{a + x}{a - x}$. $\qquad f'(x) = \frac{2a}{a^2 - x^2}$.

11. $f(x) = \log(x + \sqrt{1 + x^2})$. $\qquad f'(x) = \frac{1}{\sqrt{1 + x^2}}$.

12. $\frac{d}{dx} e^{ax} = ae^{ax}$.

13. $\frac{d}{dx} e^{4x+5} = 4e^{4x+5}$.

14. $\frac{d}{dx} a^{3x} = 3a^{3x}\log a$.

15. $\frac{d}{dt}\log(3 - 2t^2) = \frac{4t}{2t^2 - 3}$.

16. $\frac{d}{dy}\log\frac{1 + y}{1 - y} = \frac{2}{1 - y^2}$.

17. $\frac{d}{dx} e^{b^2 + x^2} = 2xe^{b^2 + x^2}$.

18. $\frac{d}{d\theta} a^{\log\theta} = \frac{1}{\theta} a^{\log\theta}\log a$.

19. $\frac{d}{ds} b^{s^2} = 2s\log b \cdot b^{s^2}$.

20. $\frac{d}{dv} ae^{\sqrt{v}} = \frac{ae^{\sqrt{v}}}{2\sqrt{v}}$.

21. $\frac{d}{dx} a^{e^x} = \log a \cdot a^{e^x} \cdot e^x$.

22. $y = 7^{x^2 + 2x}$. $\qquad y' = 2\log 7 \cdot (x + 1)\, 7^{x^2 + 2x}$.

23. $y = c^{a^2 - x^2}$. $\qquad y' = -2x\log c \cdot c^{a^2 - x^2}$.

24. $y = \log\frac{e^x}{1 + e^x}$. $\qquad \frac{dy}{dx} = \frac{1}{1 + e^x}$.

25. $\frac{d}{dx}[e^x(1 - x^2)] = e^x(1 - 2x - x^2)$.

26. $\frac{d}{dx}\left(\frac{e^x - 1}{e^x + 1}\right) = \frac{2e^x}{(e^x + 1)^2}$.

27. $\frac{d}{dx}(x^2 e^{ax}) = xe^{ax}(ax + 2)$.

28. $y = \frac{a}{2}(e^{\frac{x}{a}} - e^{-\frac{x}{a}})$. $\frac{dy}{dx} = \frac{1}{2}(e^{\frac{x}{a}} + e^{-\frac{x}{a}})$.

29. $y = \frac{e^x - e^{-x}}{e^x + e^{-x}}$. $\frac{dy}{dx} = \frac{4}{(e^x + e^{-x})^2}$.

30. $y = x^n a^x$. $y' = a^x x^{n-1}(n + x \log a)$.

31. $y = x^x$. $y' = x^x(\log x + 1)$.

32. $y = x^{\frac{1}{x}}$. $y' = \frac{x^{\frac{1}{x}}(1 - \log x)}{x^2}$.

33. $y = x^{\log x}$. $y' = \log x^2 \cdot x^{\log x - 1}$.

34. $f(y) = \log y \cdot e^y$. $f'(y) = e^y\left(\log y + \frac{1}{y}\right)$.

35. $f(s) = \frac{\log s}{e^s}$. $f'(s) = \frac{1 - s\log s}{se^s}$.

36. $f(x) = \log(\log x)$. $f'(x) = \frac{1}{x \log x}$.

37. $F(x) = \log^4(\log x)$. $F'(x) = \frac{4\log^3(\log x)}{x\log x}$.

38. $\phi(x) = \log(\log^4 x)$. $\phi'(x) = \frac{4}{x \log x}$.

39. $\psi(y) = \log\sqrt{\frac{1+y}{1-y}}$. $\psi'(y) = \frac{1}{1-y^2}$.

40. $f(x) = \log\frac{\sqrt{x^2+1} - x}{\sqrt{x^2+1} + x}$. $f'(x) = -\frac{2}{\sqrt{1+x^2}}$.

HINT. First rationalize the denominator.

41. $y = x^{\frac{1}{\log x}}$. $\frac{dy}{dx} = 0$.

42. $y = e^{x^x}$. $\frac{dy}{dx} = e^{x^x}(1 + \log x)x^x$.

43. $y = \frac{c^x}{x^x}$. $\frac{dy}{dx} = \left(\frac{c}{x}\right)^x\left(\log\frac{c}{x} - 1\right)$.

44. $y = \left(\frac{x}{n}\right)^{nx}$. $\frac{dy}{dx} = n\left(\frac{x}{n}\right)^{nx}\left(1 + \log\frac{x}{n}\right)$.

45. $w = v^{e^v}$. $\frac{dw}{dv} = v^{e^v}e^v\left(\frac{1 + v\log v}{v}\right)$.

46. $z = \left(\frac{a}{t}\right)^t$. $\frac{dz}{dt} = \left(\frac{a}{t}\right)^t(\log a - \log t - 1)$.

47. $y = x^{x^n}$. $\frac{dy}{dx} = x^{x^n + n - 1}(n\log x + 1)$.

48. $y = x^{x^x}$. $\frac{dy}{dx} = x^{x^x}x^x\left(\log x + \log^2 x + \frac{1}{x}\right)$

49. $y = a^{\frac{1}{\sqrt{a^2 - x^2}}}$. $\frac{dy}{dx} = \frac{xy \log a}{(a^2 - x^2)^{\frac{3}{2}}}$.

**50.** Differentiate the following functions:

(a) $\frac{d}{dx}x^2\log x$. (f) $\frac{d}{dx}e^x\log x$. (k) $\frac{d}{dx}\log(a^x+b^x)$.

(b) $\frac{d}{dx}(e^{2x}-1)^4$. (g) $\frac{d}{dx}x^3 3^x$. (l) $\frac{d}{dx}\log_{10}(x^2+5x)$.

(c) $\frac{d}{dx}\log\frac{3x+1}{x+3}$. (h) $\frac{d}{dx}\frac{1}{x\log x}$. (m) $\frac{d}{dx}\frac{2+x^2}{e^{3x}}$.

(d) $\frac{d}{dx}\log\frac{1-x^2}{\sqrt{1+x}}$. (i) $\frac{d}{dx}\log x^3\sqrt{1+x^2}$. (n) $\frac{d}{dx}(x^2+a^2)e^{x^2+a^2}$.

(e) $\frac{d}{dx}x^{\sqrt{x}}$. (j) $\frac{d}{dx}\left(\frac{1}{x}\right)^x$. (o) $\frac{d}{dx}(x^2+4)^x$.

**51.** $y=\frac{(x+1)^2}{(x+2)^3(x+3)^4}$. $\frac{dy}{dx}=-\frac{(x+1)(5x^2+14x+5)}{(x+2)^4(x+3)^5}$.

HINT. Take logarithm of both sides before differentiating in this and the following examples.

**52.** $y=\frac{(x-1)^{\frac{5}{2}}}{(x-2)^{\frac{3}{4}}(x-3)^{\frac{7}{3}}}$. $\frac{dy}{dx}=-\frac{(x-1)^{\frac{3}{2}}(7x^2+30x-97)}{12(x-2)^{\frac{7}{4}}(x-3)^{\frac{10}{3}}}$.

**53.** $y=x\sqrt{1-x}(1+x)$. $\frac{dy}{dx}=\frac{2+x-5x^2}{2\sqrt{1-x}}$.

**54.** $y=\frac{x(1+x^2)}{\sqrt{1-x^2}}$. $\frac{dy}{dx}=\frac{1+3x^2-2x^4}{(1-x^2)^{\frac{3}{2}}}$.

**55.** $y=x^5(a+3x)^3(a-2x)^2$. $\frac{dy}{dx}=5x^4(a+3x)^2(a-2x)(a^2+2ax-12x^2)$.

## 48. Differentiation of sin v.

Let $y=\sin v$.

By *General Rule*, p. 29, considering $v$ as the independent variable, we have

FIRST STEP. $y+\Delta y=\sin(v+\Delta v)$.

SECOND STEP.
$$\Delta y=\sin(v+\Delta v)-\sin v^*$$
$$=2\cos\left(v+\frac{\Delta v}{2}\right)\cdot\sin\frac{\Delta v}{2}.\dagger$$

* If we take the third and fourth steps without transforming the right-hand member, there results:

*Third step.* $\frac{\Delta y}{\Delta v}=\frac{\sin(v+\Delta v)-\sin v}{\Delta v}$.

*Fourth step.* $\frac{dy}{dv}=\frac{0}{0}$, which is indeterminate (see footnote, p. 46).

† Let $A=v+\Delta v$
and $B=v$

Adding, $A+B=2v+\Delta v$ Subtracting, $A-B=\Delta v$

Therefore $\frac{1}{2}(A+B)=v+\frac{\Delta v}{2}$. $\frac{1}{2}(A-B)=\frac{\Delta v}{2}$.

Substituting these values of $A$, $B$, $\frac{1}{2}(A+B)$, $\frac{1}{2}(A-B)$ in terms of $v$ and $\Delta v$ in the formula from Trigonometry (42, p. 2),

$$\sin A-\sin B=2\cos\tfrac{1}{2}(A+B)\sin\tfrac{1}{2}(A-B),$$

we get $\sin(v+\Delta v)-\sin v=2\cos\left(v+\frac{\Delta v}{2}\right)\sin\frac{\Delta v}{2}$.

THIRD STEP. $$\frac{\Delta y}{\Delta v} = \cos\left(v + \frac{\Delta v}{2}\right)\left(\frac{\sin\frac{\Delta v}{2}}{\frac{\Delta v}{2}}\right).$$

FOURTH STEP. $$\frac{dy}{dv} = \cos v.$$

$$\left[\text{Since} \quad \underset{\Delta v=0}{\text{limit}}\left(\frac{\sin\frac{\Delta v}{2}}{\frac{\Delta v}{2}}\right) = 1, \text{ by § 22, p. 21, and } \underset{\Delta v=0}{\text{limit}} \cos\left(v + \frac{\Delta v}{2}\right) = \cos v.\right]$$

Since $v$ is a function of $x$ and it is required to differentiate $\sin v$ with respect to $x$, we must use formula ($A$), § 42, for differentiating a *function of a function*, namely,

$$\frac{dy}{dx} = \frac{dy}{dv} \cdot \frac{dv}{dx}.$$

Substituting value $\frac{dy}{dv}$ from Fourth Step, we get

$$\frac{dy}{dx} = \cos v \frac{dv}{dx}.$$

**XI** $$\therefore \frac{d}{dx}(\sin v) = \cos v \frac{dv}{dx}.$$

The statement of the corresponding rules will now be left to the student.

**49. Differentiation of cos $v$.**

Let $$y = \cos v.$$

By 29, p. 2, this may be written

$$y = \sin\left(\frac{\pi}{2} - v\right).$$

Differentiating by formula **XI**,

$$\frac{dy}{dx} = \cos\left(\frac{\pi}{2} - v\right)\frac{d}{dx}\left(\frac{\pi}{2} - v\right)$$

$$= \cos\left(\frac{\pi}{2} - v\right)\left(-\frac{dv}{dx}\right)$$

$$= -\sin v \frac{dv}{dx}.$$

$$\left[\text{Since } \cos\left(\frac{\pi}{2} - v\right) = \sin v, \text{ by 29, p. 2.}\right]$$

**XII** $$\therefore \frac{d}{dx}(\cos v) = -\sin v \frac{dv}{dx}.$$

**50. Differentiation of tan $v$.**

Let $$y = \tan v.$$

By 27, p. 2, this may be written

$$y = \frac{\sin v}{\cos v}.$$

Differentiating by formula **VII**,

$$\frac{dy}{dx} = \frac{\cos v \frac{d}{dx}(\sin v) - \sin v \frac{d}{dx}(\cos v)}{\cos^2 v}$$

$$= \frac{\cos^2 v \frac{dv}{dx} + \sin^2 v \frac{dv}{dx}}{\cos^2 v}$$

$$= \frac{\frac{dv}{dx}}{\cos^2 v} = \sec^2 v \frac{dv}{dx}.$$

XIII $$\therefore \frac{d}{dx}(\tan v) = \sec^2 v \frac{dv}{dx}.$$

**51. Differentiation of cot $v$.**

Let $$y = \cot v.$$

By 26, p. 2, this may be written

$$y = \frac{1}{\tan v}.$$

Differentiating by formula **VII**,

$$\frac{dy}{dx} = -\frac{\frac{d}{dx}(\tan v)}{\tan^2 v}$$

$$= -\frac{\sec^2 v \frac{dv}{dx}}{\tan^2 v} = -\csc^2 v \frac{dv}{dx}.$$

XIV $$\therefore \frac{d}{dx}(\cot v) = -\csc^2 v \frac{dv}{dx}.$$

**52. Differentiation of sec $v$.**

Let $$y = \sec v.$$

By 26, p. 2, this may be written

$$y = \frac{1}{\cos v}.$$

Differentiating by formula **VII**,

$$\frac{dy}{dx} = -\frac{\frac{d}{dx}(\cos v)}{\cos^2 v}$$

$$= \frac{\sin v \frac{dv}{dx}}{\cos^2 v}$$

$$= \frac{1}{\cos v} \frac{\sin v}{\cos v} \frac{dv}{dx}$$

$$= \sec v \tan v \frac{dv}{dx}.$$

**XV** $$\therefore \frac{d}{dx}(\sec v) = \sec v \tan v \frac{dv}{dx}.$$

**53. Differentiation of csc *v*.**

Let $$y = \csc v.$$

By 26, p. 2, this may be written

$$y = \frac{1}{\sin v}.$$

Differentiating by formula **VII**,

$$\frac{dy}{dx} = -\frac{\frac{d}{dx}(\sin v)}{\sin^2 v}$$

$$= -\frac{\cos v \frac{dv}{dx}}{\sin^2 v}$$

$$= -\csc v \cot v \frac{dv}{dx}.$$

**XVI** $$\therefore \frac{d}{dx}(\csc v) = -\csc v \cot v \frac{dv}{dx}.$$

**54. Differentiation of vers *v*.**

Let $$y = \text{vers } v.$$

By Trigonometry this may be written

$$y = 1 - \cos v.$$

Differentiating,

$$\frac{dy}{dx} = \sin v \frac{dv}{dx}.$$

**XVII** $$\therefore \frac{d}{dx}(\text{vers}\, v) = \sin v \frac{dv}{dx}.$$

In the derivation of our formulas so far it has been necessary to apply the *General Rule*, p. 29 (i.e. the four steps), only for the following:

III $$\frac{d}{dx}(u + v - w) = \frac{du}{dx} + \frac{dv}{dx} - \frac{dw}{dx}.$$ Algebraic sum.

V $$\frac{d}{dx}(uv) = u\frac{dv}{dx} + v\frac{du}{dx}.$$ Product.

VII $$\frac{d}{dx}\left(\frac{u}{v}\right) = \frac{v\frac{du}{dx} - u\frac{dv}{dx}}{v^2}.$$ Quotient.

VIII $$\frac{d}{dx}(\log_a v) = \log_a e \frac{\frac{dv}{dx}}{v}.$$ Logarithm.

XI $$\frac{d}{dx}(\sin v) = \cos v \frac{dv}{dx}.$$ Sine.

XXV $$\frac{dy}{dx} = \frac{dy}{dv} \cdot \frac{dv}{dx}.$$ Function of a function.

XXVI $$\frac{dy}{dx} = \frac{1}{\frac{dx}{dy}}.$$ Inverse functions.

Not only do all the other formulas we have deduced depend on these, but all we shall deduce hereafter depend on them as well. Hence it follows that the derivation of the fundamental formulas for differentiation involves the calculation of only two limits of any difficulty, viz.,

$$\underset{v=0}{\text{limit}} \frac{\sin v}{v} = 1$$ by § 22, p. 21

and $$\underset{v=0}{\text{limit}} (1 + v)^{\frac{1}{v}} = e.$$ By § 23, p. 22

**EXAMPLES**

Differentiate the following:

**1.** $y = \sin ax^2$.

$$\frac{dy}{dx} = \cos ax^2 \frac{d}{dx}(ax^2) \qquad \text{by XI}$$

$[v = ax^2.]$

$$= 2\,ax \cos ax^2.$$

**2.** $y = \tan \sqrt{1-x}$.

$$\frac{dy}{dx} = \sec^2 \sqrt{1-x}\frac{d}{dx}(1-x)^{\frac{1}{2}} \qquad \text{by XIII}$$

$[v = \sqrt{1-x}.]$

$$= \sec^2 \sqrt{1-x} \cdot \tfrac{1}{2}(1-x)^{-\frac{1}{2}}(-1)$$

$$= -\frac{\sec^2 \sqrt{1-x}}{2\sqrt{1-x}}.$$

**3.** $y = \cos^3 x$.

This may also be written

$$y = (\cos x)^3.$$

$$\frac{dy}{dx} = 3(\cos x)^2 \frac{d}{dx}(\cos x) \qquad \text{by VI}$$

$[v = \cos x \text{ and } n = 3.]$

$$= 3\cos^2 x(-\sin x) \qquad \text{by XII}$$

$$= -3\sin x \cos^2 x.$$

**4.** $y = \sin nx \sin^n x$.

$$\frac{dy}{dx} = \sin nx \frac{d}{dx}(\sin x)^n + \sin^n x \frac{d}{dx}(\sin nx) \qquad \text{by V}$$

$[u = \sin nx \text{ and } v = \sin^n x.]$

$$= \sin nx \cdot n(\sin x)^{n-1}\frac{d}{dx}(\sin x) + \sin^n x \cos nx \frac{d}{dx}(nx) \qquad \text{by VI and XI}$$

$$= n \sin nx \cdot \sin^{n-1} x \cos x + n \sin^n x \cos nx$$

$$= n \sin^{n-1} x(\sin nx \cos x + \cos nx \sin x)$$

$$= n \sin^{n-1} x \sin(n+1)x.$$

**5.** $y = \sec ax$. *Ans.* $\frac{dy}{dx} = a \sec ax \tan ax.$

**6.** $y = \tan(ax+b)$. $\frac{dy}{dx} = a\sec^2(ax+b).$

**7.** $s = \cos 3\,ax$. $\frac{ds}{dx} = -3\,a \sin 3\,ax.$

**8.** $s = \cot(2\,t^2+3)$. $\frac{ds}{dt} = -4\,t\csc^2(2\,t^2+3).$

**9.** $f(y) = \sin 2\,y \cos y$. $f'(y) = 2\cos 2\,y \cos y - \sin 2\,y \sin y$

**10.** $F(x) = \cot^2 5\,x$. $F'(x) = -10\cot 5\,x \csc^2 5\,x.$

**11.** $F(\theta) = \tan\theta - \theta$. $F'(\theta) = \tan^2\theta.$

**12.** $f(\phi) = \phi \sin\phi + \cos\phi$. $f'(\phi) = \phi\cos\phi.$

**13.** $f(t) = \sin^3 t \cos t$. $f'(t) = \sin^2 t(3\cos^2 t - \sin^2 t).$

**14.** $r = a\cos 2\theta$. $\frac{dr}{d\theta} = -2\,a \sin 2\theta.$

15. $\frac{d}{dx}\sin^2 x = \sin 2x.$

16. $\frac{d}{dx}\cos^3 x^2 = -6x\cos^2 x^2 \sin x^2.$

17. $\frac{d}{dt}\csc\frac{t^2}{2} = -t\csc\frac{t^2}{2}\cot\frac{t^2}{2}.$

18. $\frac{d}{ds}a\sqrt{\cos 2s} = -\frac{a\sin 2s}{\sqrt{\cos 2s}}.$

19. $\frac{d}{d\theta}a(1-\cos\theta) = a\sin\theta.$

20. $\frac{d}{dx}(\log\cos x) = -\tan x.$

21. $\frac{d}{dx}(\log\tan x) = \frac{2}{\sin 2x}.$

22. $\frac{d}{dx}(\log\sin^2 x) = 2\cot x.$

23. $\frac{d}{dt}\cos\frac{a}{t} = \frac{a}{t^2}\sin\frac{a}{t}.$

24. $\frac{d}{d\theta}\sin\frac{1}{\theta^2} = -\frac{2}{\theta^3}\cos\frac{1}{\theta^2}.$

25. $\frac{d}{dx}e^{\sin x} = e^{\sin x}\cos x.$

26. $\frac{d}{dx}\sin(\log x) = \frac{\cos(\log x)}{x}.$

27. $\frac{d}{dx}\tan(\log x) = \frac{\sec^2(\log x)}{x}.$

28. $\frac{d}{d\theta}a\sin^3\frac{\theta}{3} = a\sin^2\frac{\theta}{3}\cos\frac{\theta}{3}$

29. $\frac{d}{d\alpha}\sin(\cos\alpha) = -\sin\alpha\cos(\cos\alpha)$

30. $\frac{d}{dx}\frac{\tan x - 1}{\sec x} = \sin x + \cos x.$

31. $y = \log\sqrt{\frac{1+\sin x}{1-\sin x}}.$ $\qquad \frac{dy}{dx} = \frac{1}{\cos x}.$

32. $y = \log\tan\left(\frac{\pi}{4}+\frac{x}{2}\right).$ $\qquad \frac{dy}{dx} = \frac{1}{\cos x}.$

33. $f(x) = \sin(x+a)\cos(x-a).$ $\qquad f'(x) = \cos 2x.$

34. $y = a^{\tan nx}.$ $\qquad y' = na^{\tan nx}\sec^2 nx \log a.$

35. $y = e^{\cos x}\sin x.$ $\qquad y' = e^{\cos x}(\cos x - \sin^2 x).$

36. $y = e^x \log\sin x.$ $\qquad y' = e^x(\cot x + \log\sin x).$

37. Differentiate the following functions:

(a) $\frac{d}{dx}\sin 5x^2.$

(b) $\frac{d}{dx}\cos(a - bx).$

(c) $\frac{d}{dx}\tan\frac{ax}{b}.$

(d) $\frac{d}{dx}\cot\sqrt{ax}.$

(e) $\frac{d}{dx}\sec e^{3x}.$

(f) $\frac{d}{dx}\csc(\log x).$

(g) $\frac{d}{dx}\sin^3 2x.$

(h) $\frac{d}{dx}\cos^2(\log x).$

(i) $\frac{d}{dx}\tan^2\sqrt{1-x^2}.$

(j) $\frac{d}{dx}\log(\sin^2 ax).$

(k) $\frac{d}{dt}e^{a-b\cos t}.$

(l) $\frac{d}{dt}\sin\frac{t}{3}\cos^2\frac{t}{3}.$

(m) $\frac{d}{d\theta}\cot\frac{b}{\theta^2}.$

(n) $\frac{d}{d\phi}\sqrt{1+\cos^2\phi}.$

(o) $\frac{d}{ds}\log\sqrt{1-2\sin^2 s}.$

38. $\frac{d}{dx}(x^n e^{\sin x}) = x^{n-1}e^{\sin x}(n + x\cos x).$

39. $\frac{d}{dx}(e^{ax}\cos mx) = e^{ax}(a\cos mx - m\sin mx).$

40. $f(\theta) = \frac{1+\cos\theta}{1-\cos\theta}.$ $\qquad f'(\theta) = -\frac{2\sin\theta}{(1-\cos\theta)^2}.$

41. $f(\phi) = \frac{e^{a\phi}(a\sin\phi - \cos\phi)}{a^2+1}.$ $\qquad f'(\phi) = e^{a\phi}\sin\phi.$

42. $f(s) = (s\cot s)^2.$ $\qquad f'(s) = 2s\cot s(\cot s - s\csc^2 s).$

**43.** $r = \frac{1}{3}\tan^3\theta - \tan\theta + \theta.$ $\qquad \frac{dr}{d\theta} = \tan^4\theta.$

**44.** $y = x^{\sin x}.$ $\qquad \frac{dy}{dx} = x^{\sin x}\left(\frac{\sin x}{x} + \log x \cos x\right).$

**45.** $y = (\sin x)^x.$ $\qquad y' = (\sin x)^x\,[\log \sin x + x \cot x].$

**46.** $y = (\sin x)^{\tan x}.$ $\qquad y' = (\sin x)^{\tan x}(1 + \sec^2 x \log \sin x)$

**47.** Prove $\frac{d}{dx}\cos v = -\sin v \frac{dv}{dx}$, using the *General Rule.*

**48.** Prove $\frac{d}{dx}\cot v = -\csc^2 v \frac{dv}{dx}$ by replacing $\cot v$ by $\frac{\cos v}{\sin v}$.

### 55. Differentiation of arc sin $v$.

Let $\qquad y = \text{arc} \sin v;$*

then $\qquad v = \sin y.$

Differentiating with respect to $y$ by **XI**,

$$\frac{dv}{dy} = \cos y;$$

therefore
$$\frac{dy}{dv} = \frac{1}{\cos y}. \qquad \text{By } (C), \text{ p. 46}$$

But since $v$ is a function of $x$, this may be substituted in

$$\frac{dy}{dx} = \frac{dy}{dv}\cdot\frac{dv}{dx}, \qquad (A), \text{ p. 45}$$

giving
$$\frac{dy}{dx} = \frac{1}{\cos y}\cdot\frac{dv}{dx}$$

$$= \frac{1}{\sqrt{1-v^2}}\frac{dv}{dx}.$$

[$\cos y = \sqrt{1-\sin^2 y} = \sqrt{1-v^2}$, the positive sign of the radical being taken, since $\cos y$ is positive for all values of $y$ between $-\frac{\pi}{2}$ and $\frac{\pi}{2}$ inclusive.]

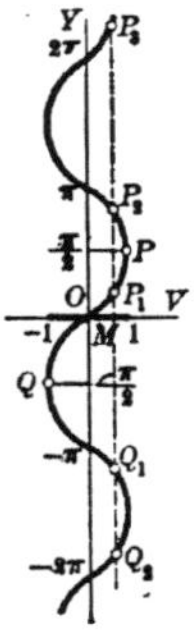


**XVIII** $\qquad \therefore\ \frac{d}{dx}(\text{arc}\sin v) = \frac{\frac{dv}{dx}}{\sqrt{1-v^2}}.$

* It should be remembered that this function is defined only for values of $v$ between $-1$ and $+1$ inclusive and that $y$ (the function) is many-valued, there being infinitely many arcs whose sines all equal $v$. Thus, in the figure (the locus of $y = \text{arc}\sin v$), when $v = OM$, $y = MP_1, MP_2, MP_3, \cdots, MQ_1, MQ_2, \cdots$. In the above discussion, in order to make the function single-valued, only values of $y$ between $-\frac{\pi}{2}$ and $\frac{\pi}{2}$ inclusive (points on arc $QOP$) are considered; that is, the arc of smallest numerical value whose sine is $v$.

**56. Differentiation of arc cos $v$.**

Let $$y = \text{arc} \cos v;\text{*}$$
then $$v = \cos y.$$

Differentiating with respect to $y$ by **XII**,

$$\frac{dv}{dy} = -\sin y;$$

therefore $$\frac{dy}{dv} = -\frac{1}{\sin y}.$$ By (*C*), p. 46

But since $v$ is a function of $x$, this may be substituted in the formula

$$\frac{dy}{dx} = \frac{dy}{dv} \cdot \frac{dv}{dx},$$ (*A*), p. 45

giving $$\frac{dy}{dx} = -\frac{1}{\sin y} \cdot \frac{dv}{dx}$$

$$= -\frac{1}{\sqrt{1-v^2}} \frac{dv}{dx}.$$

[$\sin y = \sqrt{1-\cos^2 y} = \sqrt{1-v^2}$, the plus sign of the radical being taken, since $\sin y$ is positive for all values of $y$ between 0 and $\pi$ inclusive.]

**XIX** $$\therefore \frac{d}{dx}(\text{arc} \cos v) = -\frac{\frac{dv}{dx}}{\sqrt{1-v^2}}.$$

**57. Differentiation of arc tan $v$.**

Let $$y = \text{arc} \tan v;\dagger$$
then $$v = \tan y.$$

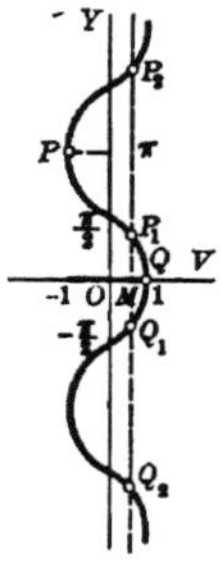


Differentiating with respect to $y$ by **XIV**,

$$\frac{dv}{dy} = \sec^2 y;$$

therefore $$\frac{dy}{dv} = \frac{1}{\sec^2 y}.$$ By (*C*), p. 46

* This function is defined only for values of $v$ between $-1$ and $+1$ inclusive, and is many-valued. In the figure (the locus of $y = \text{arc} \cos v$), when $v = OM$, $y = MP_1, MP_2, \cdots, MQ_1, MQ_2, \cdots$.

In order to make the function single-valued, only values of $y$ between 0 and $\pi$ inclusive are considered; that is, the smallest positive arc whose cosine is $v$. Hence we confine ourselves to arc $QP$ of the graph.

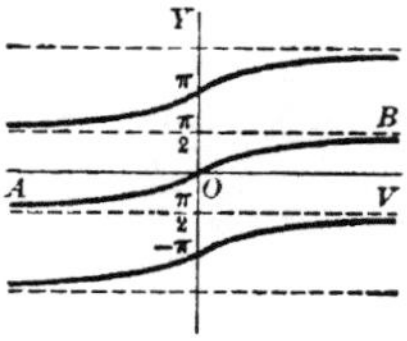


† This function is defined for all values of $v$, and is many-valued, as is clearly shown by its graph. In order to make it single-valued, only values of $y$ between $-\frac{\pi}{2}$ and $\frac{\pi}{2}$ are considered; that is, the arc of smallest numerical value whose tangent is $v$ (branch $AOB$).

But since $v$ is a function of $x$, this may be substituted in the formula

$$\frac{dy}{dx}=\frac{dy}{dv}\cdot\frac{dv}{dx}, \qquad (A), \text{ p. } 45$$

giving

$$\frac{dy}{dx}=\frac{1}{\sec^2 y}\cdot\frac{dv}{dx}$$

$$=\frac{1}{1+v^2}\frac{dv}{dx}.$$

$[\sec^2 y = 1+\tan^2 y = 1+v^2.]$

**XX** $$\therefore\ \frac{d}{dx}(\text{arc}\tan v)=\frac{\frac{dv}{dx}}{1+v^2}.$$

**58. Differentiation of arc cot $v$.***

Following the method of the last section, we get

**XXI** $$\frac{d}{dx}(\text{arc}\cot v)=-\frac{\frac{dv}{dx}}{1+v^2}.$$

**59. Differentiation of arc sec $v$.**

Let $$y=\text{arc}\sec v;\dagger$$

then $$v=\sec y.$$

* This function is defined for all values of $v$, and is many-valued, as is seen from its graph (Fig. $a$). In order to make it single-valued, only values of $y$ between 0 and $\pi$ are considered; that is, the smallest positive arc whose cotangent is $v$. Hence we confine ourselves to branch $AB$.

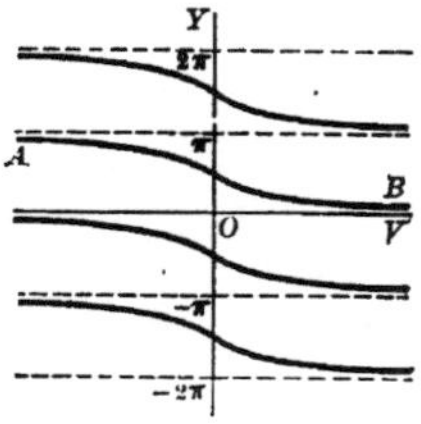


FIG. $a$

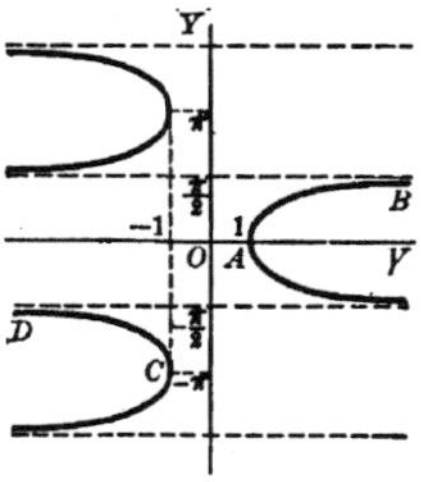


FIG. $b$

† This function is defined for all values of $v$ except those lying between −1 and +1, and is seen to be many-valued. To make the function single-valued, $y$ is taken as the arc of smallest numerical value whose secant is $v$. This means that if $v$ is positive, we confine ourselves to points on arc $AB$ (Fig. $b$), $y$ taking on values between 0 and $\frac{\pi}{2}$ (0 may be included); and if $v$ is negative, we confine ourselves to points on arc $DC$, $y$ taking on values between $-\pi$ and $-\frac{\pi}{2}$ ($-\pi$ may be included).

Differentiating with respect to $y$ by **XV**,

$$\frac{dv}{dy} = \sec y \tan y;$$

therefore $$\frac{dy}{dv} = \frac{1}{\sec y \tan y}.$$ By (*C*), p. 46

But since $v$ is a function of $x$, this may be substituted in the formula

$$\frac{dy}{dx} = \frac{dy}{dv} \cdot \frac{dv}{dx},$$ (*A*), p. 45

giving $$\frac{dy}{dx} = \frac{1}{\sec y \tan y}\frac{dv}{dx}$$

$$= \frac{1}{v\sqrt{v^2-1}}\frac{dv}{dx}.$$

[$\sec y = v$, and $\tan y = \sqrt{\sec^2 y - 1} = \sqrt{v^2 - 1}$, the plus sign of the radical being taken, since $\tan y$ is positive for all values of $y$ between 0 and $\frac{\pi}{2}$ and between $-\pi$ and $-\frac{\pi}{2}$, including 0 and $-\pi$.]

**XXII** $$\therefore \frac{d}{dx}(\text{arc} \sec v) = \frac{\frac{dv}{dx}}{v\sqrt{v^2 - 1}}.$$

**60. Differentiation of arc csc $v$.***

Let $$y = \text{arc} \csc v;$$

then $$v = \csc y.$$

Differentiating with respect to $y$ by **XVI** and following the method of the last section, we get

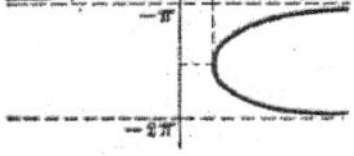

Fig. *a*

**XXIII** $$\frac{d}{dx}(\text{arc} \csc v) = -\frac{\frac{dv}{dx}}{v\sqrt{v^2 - 1}}.$$

* This function is defined for all values of $v$ except those lying between $-1$ and $+1$, and is seen to be many-valued. To make the function single-valued, $y$ is taken as the arc of smallest numerical value whose cosecant is $v$. This means that if $v$ is positive, we confine ourselves to points on the arc $AB$ (Fig. *a*), $y$ taking on values between 0 and $\frac{\pi}{2}$ $\left(\frac{\pi}{2}\text{ may be included}\right)$; and if $v$ is negative, we confine ourselves to points on the arc $CD$, $y$ taking on values between $-\pi$ and $-\frac{\pi}{2}$ $\left(-\frac{\pi}{2}\text{ may be included}\right)$.

**61. Differentiation of arc vers $v$.**

Let $$y = \text{arc vers } v;\text{*}$$
then $$v = \text{vers } y.$$

Differentiating with respect to $y$ by **XVII**,

$$\frac{dv}{dy} = \sin y;$$

therefore $$\frac{dy}{dv} = \frac{1}{\sin y}. \qquad \text{By } (C), \text{ p. } 46$$

But since $v$ is a function of $x$, this may be substituted in the formula

$$\frac{dy}{dx} = \frac{dy}{dv} \cdot \frac{dv}{dx}, \qquad (A), \text{ p. } 45$$

giving $$\frac{dy}{dx} = \frac{1}{\sin y} \cdot \frac{dv}{dx}$$

$$= \frac{1}{\sqrt{2v - v^2}} \frac{dv}{dx}.$$

[$\sin y = \sqrt{1 - \cos^2 y} = \sqrt{1 - (1 - \text{vers } y)^2} = \sqrt{2v - v^2}$, the plus sign of the radical being taken, since $\sin y$ is positive for all values of $y$ between 0 and $\pi$ inclusive.]

**XXIV** $$\therefore \frac{d}{dx}(\text{arc vers } v) = \frac{\frac{dv}{dx}}{\sqrt{2v - v^2}}.$$

**EXAMPLES**

Differentiate the following:

1. $y = \text{arc tan } ax^2$.

Solution. $$\frac{dy}{dx} = \frac{\frac{d}{dx}(ax^2)}{1 + (ax^2)^2} \qquad \text{by XX}$$

$[v = ax^2.]$

$$= \frac{2ax}{1 + a^2x^4}.$$

2. $y = \text{arc sin } (3x - 4x^3)$.

Solution. $$\frac{dy}{dx} = \frac{\frac{d}{dx}(3x - 4x^3)}{\sqrt{1 - (3x - 4x^3)^2}} \qquad \text{by XVIII}$$

$[v = 3x - 4x^3.]$

$$= \frac{3 - 12x^2}{\sqrt{1 - 9x^2 + 24x^4 - 16x^6}} = \frac{3}{\sqrt{1 - x^2}}.$$

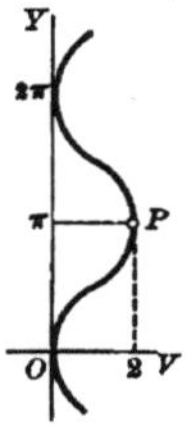

FIG. $a$

* Defined only for values of $v$ between 0 and 2 inclusive, and is many-valued. To make the function continuous, $y$ is taken as the smallest positive arc whose versed sine is $v$; that is, $y$ lies between 0 and $\pi$ inclusive. Hence we confine ourselves to arc $OP$ of the graph (Fig. $a$).

3. $y = \text{arc sec}\,\dfrac{x^2+1}{x^2-1}$.

**Solution.**
$$\frac{dy}{dx} = \frac{\dfrac{d}{dx}\left(\dfrac{x^2+1}{x^2-1}\right)}{\dfrac{x^2+1}{x^2-1}\sqrt{\left(\dfrac{x^2+1}{x^2-1}\right)^2-1}} \quad \text{by XXII}$$
$$\left[v = \frac{x^2+1}{x^2-1}.\right]$$
$$= \frac{\dfrac{(x^2-1)\,2x-(x^2+1)\,2x}{(x^2-1)^2}}{\dfrac{x^2+1}{x^2-1}\cdot\dfrac{2x}{x^2-1}} = -\frac{2}{x^2+1}.$$

4. $\dfrac{d}{dx}\text{arc sin}\,\dfrac{x}{a} = \dfrac{1}{\sqrt{a^2-x^2}}$.

5. $\dfrac{d}{dx}\text{arc cot}\,(x^2-5) = \dfrac{-2x}{1+(x^2-5)^2}$.

6. $\dfrac{d}{dx}\text{arc tan}\,\dfrac{2x}{1-x^2} = \dfrac{2}{1+x^2}$.

7. $\dfrac{d}{dx}\text{arc cosec}\,\dfrac{1}{2x^2-1} = \dfrac{2}{\sqrt{1-x^2}}$.

8. $\dfrac{d}{dx}\text{arc vers}\,2x^2 = \dfrac{2}{\sqrt{1-x^2}}$.

9. $\dfrac{d}{dx}\text{arc tan}\,\sqrt{1-x} = -\dfrac{1}{2\sqrt{1-x}(2-x)}$.

10. $\dfrac{d}{dx}\text{arc cosec}\,\dfrac{3}{2x} = \dfrac{2}{\sqrt{9-4x^2}}$.

11. $\dfrac{d}{dx}\text{arc vers}\,\dfrac{2x^2}{1+x^2} = \dfrac{2}{1+x^2}$.

12. $\dfrac{d}{dx}\text{arc tan}\,\dfrac{x}{a} = \dfrac{a}{a^2+x^2}$.

13. $\dfrac{d}{dx}\text{arc sin}\,\dfrac{x+1}{\sqrt{2}} = \dfrac{1}{\sqrt{1-2x-x^2}}$.

14. $f(x) = x\sqrt{a^2-x^2} + a^2\,\text{arc sin}\,\dfrac{x}{a}$. $\qquad f'(x) = 2\sqrt{a^2-x^2}$.

15. $f(x) = \sqrt{a^2-x^2} + a\,\text{arc sin}\,\dfrac{x}{a}$. $\qquad f'(x) = \left(\dfrac{a-x}{a+x}\right)^{\frac{1}{2}}$.

16. $x = r\,\text{arc vers}\,\dfrac{y}{r} - \sqrt{2ry-y^2}$. $\qquad \dfrac{dx}{dy} = \dfrac{y}{\sqrt{2ry-y^2}}$.

17. $\theta = \text{arc sin}\,(3r-1)$. $\qquad \dfrac{d\theta}{dr} = \dfrac{3}{\sqrt{6r-9r^2}}$.

18. $\phi = \text{arc tan}\,\dfrac{r+a}{1-ar}$. $\qquad \dfrac{d\phi}{dr} = \dfrac{1}{1+r^2}$.

19. $s = \text{arc sec}\,\dfrac{1}{\sqrt{1-t^2}}$. $\qquad \dfrac{ds}{dt} = \dfrac{1}{\sqrt{1-t^2}}$.

20. $\dfrac{d}{dx}(x\,\text{arc sin}\,x) = \text{arc sin}\,x + \dfrac{x}{\sqrt{1-x^2}}$.

21. $\dfrac{d}{d\theta}(\tan\theta\,\text{arc tan}\,\theta) = \sec^2\theta\,\text{arc tan}\,\theta + \dfrac{\tan\theta}{1+\theta^2}$.

22. $\dfrac{d}{dt}[\log(\text{arc cos}\,t)] = -\dfrac{1}{\text{arc cos}\,t\sqrt{1-t^2}}$.

23. $f(y) = \text{arc cos}\,(\log y)$. $\qquad f'(y) = -\dfrac{1}{y\sqrt{1-(\log y)^2}}$.

24. $f(\theta) = \text{arc sin}\,\sqrt{\sin\theta}$. $\qquad f'(\theta) = \frac{1}{2}\sqrt{1+\csc\theta}$.

**25.** $f(\phi) = \text{arc tan}\sqrt{\dfrac{1-\cos\phi}{1+\cos\phi}}.$ $\qquad f'(\phi) = \dfrac{1}{2}.$

**26.** $p = e^{\text{arc tan}\, q}.$ $\qquad \dfrac{dp}{dq} = \dfrac{e^{\text{arc tan}\, q}}{1+q^2}.$

**27.** $u = \text{arc tan}\dfrac{e^v - e^{-v}}{2}.$ $\qquad \dfrac{du}{dv} = \dfrac{2}{e^v + e^{-v}}.$

**28.** $s = \text{arc cos}\dfrac{e^t - e^{-t}}{e^t + e^{-t}}.$ $\qquad \dfrac{ds}{dt} = -\dfrac{2}{e^t + e^{-t}}.$

• **29.** $y = x^{\text{arc sin}\, x}.$ $\qquad y' = x^{\text{arc sin}\, x}\left(\dfrac{\text{arc sin}\, x}{x} + \dfrac{\log x}{\sqrt{1-x^2}}\right).$

**30.** $y = e^{x^x}\text{arc tan}\, x.$ $\qquad y' = e^{x^x}\left[\dfrac{1}{1+x^2} + x^x \text{arc tan}\, x\,(1+\log x)\right].$

**31.** $y = \text{arc sin}(\sin x).$ $\qquad y' = 1.$

**32.** $y = \text{arc tan}\dfrac{4\sin x}{3+5\cos x}.$ $\qquad y' = \dfrac{4}{5+3\cos x}.$

**33.** $y = \text{arc cot}\dfrac{a}{x} + \log\sqrt{\dfrac{x-a}{x+a}}.$ $\qquad y' = \dfrac{2ax^2}{x^4-a^4}.$

**34.** $y = \log\left(\dfrac{1+x}{1-x}\right)^{\frac{1}{4}} - \dfrac{1}{2}\text{arc tan}\, x.$ $\qquad y' = \dfrac{x^2}{1-x^4}.$

**35.** $y = \sqrt{1-x^2}\,\text{arc sin}\, x - x.$ $\qquad y' = -\dfrac{x\,\text{arc sin}\, x}{\sqrt{1-x^2}}.$

**36.** **Differentiate the following functions:**

(a) $\dfrac{d}{dx}\text{arc sin}\, 2x^2.$ (f) $\dfrac{d}{dt}t^2\,\text{arc sin}\dfrac{t}{3}.$ (k) $\dfrac{d}{dy}\text{arc sin}\sqrt{1-y^2}.$

(b) $\dfrac{d}{dx}\text{arc tan}\, a^2x.$ (g) $\dfrac{d}{dt}e^{\text{arc tan}\, at}.$ (l) $\dfrac{d}{dx}\text{arc tan}(\log 3\,ax).$

(c) $\dfrac{d}{dx}\text{arc sec}\dfrac{x}{a}.$ (h) $\dfrac{d}{d\phi}\tan\phi^2\cdot\text{arc tan}\,\phi^{\frac{1}{2}}.$ (m) $\dfrac{d}{ds}(a^2+s^2)\,\text{arc sec}\dfrac{s}{2}.$

(d) $\dfrac{d}{dx}x\,\text{arc cos}\, x.$ (i) $\dfrac{d}{d\theta}\text{arc sin}\, a^\theta.$ (n) $\dfrac{d}{d\alpha}\text{arc cot}\dfrac{2\alpha}{3}.$

(e) $\dfrac{d}{dx}x^2\,\text{arc cot}\, ax.$ (j) $\dfrac{d}{d\theta}\text{arc tan}\sqrt{1+\theta^2}.$ (o) $\dfrac{d}{dt}\sqrt{1-t^2}\,\text{arc sin}\, t.$

Formulas (*A*), p. 45, for differentiating a *function of a function*, and (*C*), p. 46, for differentiating *inverse functions*, have been added to the list of formulas at the beginning of this chapter as XXV and XXVI respectively.

In the next eight examples, first find $\dfrac{dy}{dv}$ and $\dfrac{dv}{dx}$ by differentiation and then substitute the results in

$$\frac{dy}{dx} = \frac{dy}{dv}\cdot\frac{dv}{dx} \qquad \text{by XXV}$$

to find $\dfrac{dy}{dx}$.*

* As was pointed out on p. 44, it might be possible to eliminate $v$ between the two given expressions so as to find $y$ directly as a function of $x$, but in most cases the above method is to be preferred.

In general our results should be expressed explicitly in terms of the independent variable; that is, $\frac{dy}{dx}$ in terms of $x$, $\frac{dx}{dy}$ in terms of $y$, $\frac{d\phi}{d\theta}$ in terms of $\theta$, etc.

**37.** $y = 2v^2 - 4$, $v = 3x^2 + 1$.

$$\frac{dy}{dv} = 4v;\ \frac{dv}{dx} = 6x;\ \text{substituting in XXV,}$$

$$\frac{dy}{dx} = 4v \cdot 6x = 24x(3x^2 + 1).$$

**38.** $y = \tan 2v$, $v = \text{arc} \tan (2x - 1)$.

$$\frac{dy}{dv} = 2\sec^2 2v;\ \frac{dv}{dx} = \frac{1}{2x^2 - 2x + 1};\ \text{substituting in XXV,}$$

$$\frac{dy}{dx} = \frac{2\sec^2 2v}{2x^2 - 2x + 1} = 2\frac{\tan^2 2v + 1}{2x^2 - 2x + 1} = \frac{2x^2 - 2x + 1}{2(x - x^2)^2}.$$

$\left[\text{Since } v = \text{arc} \tan (2x - 1),\ \tan v = 2x - 1,\ \tan 2v = \frac{2x - 1}{2x - 2x^2}.\right]$

**39.** $y = 3v^2 - 4v + 5$, $v = 2x^3 - 5$. $\qquad \frac{dy}{dx} = 72x^5 - 204x^2$.

**40.** $y = \frac{2v}{3v - 2}$, $v = \frac{x}{2x - 1}$. $\qquad \frac{dy}{dx} = \frac{4}{(x - 2)^2}$.

**41.** $y = \log (a^2 - v^2)$, $v = a \sin x$. $\qquad \frac{dy}{dx} = -2\tan x$.

**42.** $y = \text{arc} \tan (a + v)$, $v = e^x$. $\qquad \frac{dy}{dx} = \frac{e^x}{1 + (a + e^x)^2}$.

**43.** $r = e^{2s} + e^s$, $s = \log (t - t^2)$. $\qquad \frac{dr}{dt} = 4t^3 - 6t^2 + 1$.

In the following examples first find $\frac{dx}{dy}$ by differentiation and then substitute in

$$\frac{dy}{dx} = \frac{1}{\frac{dx}{dy}} \qquad \text{by XXVI}$$

to find $\frac{dy}{dx}$.

**44.** $x = y\sqrt{1 + y}$. $\qquad \frac{dy}{dx} = \frac{2\sqrt{1 + y}}{2 + 3y} = \frac{2x}{2y + 3y^2}$.

**45.** $x = \sqrt{1 + \cos y}$. $\qquad \frac{dy}{dx} = -\frac{2\sqrt{1 + \cos y}}{\sin y} = -\frac{2}{\sqrt{2 - x^2}}$.

**46.** $x = \frac{y}{1 + \log y}$. $\qquad \frac{dy}{dx} = \frac{(1 + \log y)^2}{\log y}$.

**47.** $x = a \log \frac{a + \sqrt{a^2 - y^2}}{y}$. $\qquad \frac{dy}{dx} = -\frac{y\sqrt{a^2 - y^2}}{a^2}$.

**48.** $x = r \,\text{arc vers} \frac{y}{r} - \sqrt{2ry - y^2}$. $\qquad \frac{dy}{dx} = \sqrt{\frac{2r - y}{y}}$.

**49.** Show that the geometrical significance of XXVI is that the tangent makes complementary angles with the two coördinate axes.

**62. Implicit functions.** When a relation between $x$ and $y$ is given by means of an equation *not solved for y*, then $y$ is called an *implicit function* of $x$. For example, the equation

$$x^2 - 4y = 0$$

defines $y$ as an implicit function of $x$. Evidently $x$ is also defined by means of this equation as an implicit function of $y$. Similarly,

$$x^2 + y^2 + z^2 - a^2 = 0$$

defines any one of the three variables as an implicit function of the other two.

It is sometimes possible to solve the equation defining an implicit function for one of the variables and thus change it into an explicit function. For instance, the above two implicit functions may be solved for $y$, giving

$$y = \frac{x^2}{4}$$

and

$$y = \pm\sqrt{a^2 - x^2 - z^2};$$

the first showing $y$ as an explicit function of $x$, and the second as an explicit function of $x$ and $z$. In a given case, however, such a solution may be either impossible or too complicated for convenient use.

The two implicit functions used in this article for illustration may be respectively denoted by

$$f(x, y) = 0$$

and

$$F(x, y, z) = 0.$$

**63. Differentiation of implicit functions.** When $y$ is defined as an implicit function of $x$ by means of an equation in the form

$$(A) \qquad f(x, y) = 0,$$

it was explained in the last section how it might be inconvenient to solve for $y$ in terms of $x$; that is, to find $y$ as an explicit function of $x$ so that the formulas we have deduced in this chapter may be applied directly. Such, for instance, would be the case for the equation

$$(B) \qquad ax^6 + 2x^3y - y^7x - 10 = 0.$$

We then follow the rule:

*Differentiate, regarding y as a function of x, and put the result equal to zero.** That is,

$$(C) \qquad \frac{d}{dx}f(x, y) = 0.$$

* This process will be justified in §127. Only corresponding values of $x$ and $y$ which satisfy the given equation may be substituted in the derivative.

Let us apply this rule in finding $\frac{dy}{dx}$ from $(B)$.

$$\frac{d}{dx}(ax^6 + 2x^3y - y^7x - 10) = 0; \qquad \text{by } (C)$$

$$\frac{d}{dx}(ax^6) + \frac{d}{dx}(2x^3y) - \frac{d}{dx}(y^7x) - \frac{d}{dx}(10) = 0;$$

$$6ax^5 + 2x^3\frac{dy}{dx} + 6x^2y - y^7 - 7xy^6\frac{dy}{dx} = 0;$$

$$(2x^3 - 7xy^6)\frac{dy}{dx} = y^7 - 6ax^5 - 6x^2y;$$

$$\frac{dy}{dx} = \frac{y^7 - 6ax^5 - 6x^2y}{2x^3 - 7xy^6}. \quad \textit{Ans.}$$

The student should observe that in general the result will contain both $x$ and $y$.

**EXAMPLES**

**Differentiate the following by the above rule:**

1. $y^2 = 4px$. $\qquad \frac{dy}{dx} = \frac{2p}{y}$.
2. $x^2 + y^2 = r^2$. $\qquad \frac{dy}{dx} = -\frac{x}{y}$.
3. $b^2x^2 + a^2y^2 = a^2b^2$. $\qquad \frac{dy}{dx} = -\frac{b^2x}{a^2y}$.
4. $y^3 - 3y + 2ax = 0$. $\qquad \frac{dy}{dx} = \frac{2a}{3(1 - y^2)}$.
5. $x^{\frac{1}{2}} + y^{\frac{1}{2}} = a^{\frac{1}{2}}$. $\qquad \frac{dy}{dx} = -\sqrt{\frac{y}{x}}$.
6. $x^{\frac{2}{3}} + y^{\frac{2}{3}} = a^{\frac{2}{3}}$. $\qquad \frac{dy}{dx} = -\sqrt[3]{\frac{y}{x}}$.
7. $\left(\frac{x}{a}\right)^2 + \left(\frac{y}{b}\right)^{\frac{2}{3}} = 1$. $\qquad \frac{dy}{dx} = -\frac{3b^{\frac{2}{3}}xy^{\frac{1}{3}}}{a^2}$.
8. $y^2 - 2xy + b^2 = 0$. $\qquad \frac{dy}{dx} = \frac{y}{y - x}$.
9. $x^3 + y^3 - 3axy = 0$. $\qquad \frac{dy}{dx} = \frac{ay - x^2}{y^2 - ax}$.
10. $x^y = y^x$. $\qquad \frac{dy}{dx} = \frac{y^2 - xy\log y}{x^2 - xy\log x}$.
11. $\rho^2 = a^2\cos 2\theta$. $\qquad \frac{d\rho}{d\theta} = -\frac{a^2\sin 2\theta}{\rho}$.
12. $\rho^2\cos\theta = a^2\sin 3\theta$. $\qquad \frac{d\rho}{d\theta} = \frac{3a^2\cos 3\theta + \rho^2\sin\theta}{2\rho\cos\theta}$.
13. $\cos(uv) = cv$. $\qquad \frac{du}{dv} = \frac{c + u\sin(uv)}{-v\sin(uv)}$.
14. $\theta = \cos(\theta + \phi)$. $\qquad \frac{d\theta}{d\phi} = -\frac{\sin(\theta + \phi)}{1 + \sin(\theta + \phi)}$.

**15.** Find $\frac{dy}{dx}$ from the following equations:

(a) $x^2 = ay$.
(b) $x^2 + 4y^2 = 16$.
(c) $b^2x^2 - a^2y^2 = a^2b^2$.
(d) $y^2 = x^3 + a$.
(e) $x^2 - y^2 = 16$.
(f) $xy + y^2 + 4x = 0$.
(g) $yx^2 - y^3 = 5$.
(h) $x^3 - 2x^2 = y^2$.
(i) $x^2y^2 + 4y = 0$.
(j) $y^2 = \sin 2x$.
(k) $\tan x + y^2 = 0$.
(l) $\cos y + 3x^2 = 0$.
(m) $x \cot y + y = 0$.
(n) $y^2 = \log x$.
(o) $e^{x^2} + 2y^2 = 0$.

**16.** A race track has the form of the circle $x^2 + y^2 = 2500$. The directions $OX$ and $OY$ are east and north respectively, and the unit is 1 rod. If a runner starts east at the extreme north point, in what direction will he be going

(a) when $25\sqrt{2}$ rods east of $OY$? *Ans.* Southeast or southwest.
(b) when $25\sqrt{2}$ rods north of $OX$? Southeast or northeast.
(c) when 30 rods west of $OY$? E. 36° 52′ 12″ N. or W. 36° 52′ 12″ N.
(d) when 40 rods south of $OX$?
(e) when 10 rods east of $OY$?

**17.** An automobile course is elliptic in form, the major axis being 6 miles long and running east and west, while the minor axis is 2 miles long. If a car starts north at the extreme east point of the course, in what direction will the car be going

(a) when 2 miles west of the starting point?
(b) when $\frac{1}{2}$ mile north of the starting point?

## MISCELLANEOUS EXAMPLES

Differentiate the following functions:

1. $\arcsin\sqrt{1-4x^2}$. *Ans.* $\frac{-2}{\sqrt{1-4x^2}}$.
2. $xe^{x^2}$. $e^{x^2}(2x^2+1)$.
3. $\log\sin\frac{v}{2}$. $\frac{1}{2}\cot\frac{v}{2}$.
4. $\arccos\frac{a}{y}$. $\frac{a}{y\sqrt{y^2-a^2}}$.
5. $\frac{x}{\sqrt{a^2-x^2}}$. $\frac{a^2}{(a^2-x^2)^{\frac{3}{2}}}$.
6. $\frac{x}{1+\log x}$. $\frac{\log x}{(1+\log x)^2}$.
7. $\log\sec(1-2x)$. $-2\tan(1-2x)$.
8. $x^2e^{2-3x}$. $xe^{2-3x}(2-3x)$.
9. $\log\sqrt{\frac{1-\cos t}{1+\cos t}}$. $\csc t$.
10. $\arcsin\sqrt{\frac{1}{2}(1-\cos x)}$. $\frac{1}{2}$.
11. $\arctan\frac{2s}{\sqrt{s^2-1}}$. $\frac{2}{(1-5s^2)\sqrt{s^2-1}}$.
12. $(2x-1)\sqrt[3]{\frac{2}{1+x}}$. $\frac{7+4x}{3(1+x)}\sqrt[3]{\frac{2}{1+x}}$.
13. $\frac{x^3\arcsin x}{3} + \frac{(x^2+2)\sqrt{1-x^2}}{9}$. $x^2\arcsin x$.

14. $\tan^2\dfrac{\theta}{3} + \log\sec^2\dfrac{\theta}{3}$.

15. $\text{arc}\tan\frac{1}{2}(e^{2x} + e^{-2x})$.

16. $\left(\dfrac{3}{x}\right)^{2x}$.

17. $x^{\tan x}$.

18. $\dfrac{(x+2)^{\frac{1}{5}}(x^2-1)^{\frac{2}{5}}}{x^{\frac{3}{2}}}$.

19. $e^{\sec(1-3x)}$.

20. $\text{arc}\tan\sqrt{1-x^2}$.

21. $\dfrac{z^2}{\cos z}$.

22. $e^{\tan x^2}$.

23. $\log\sin^2\frac{1}{2}\theta$.

24. $e^{ax}\log\sin ax$.

25. $\sin^3\phi\cos\phi$.

26. $\dfrac{a}{2\sqrt{(b-cx^n)^m}}$.

27. $\dfrac{m+x}{1+m^2}\cdot\dfrac{e^{m\,\text{arc}\tan x}}{\sqrt{1+x^2}}$.

28. $\tan^2 x - \log\sec^2 x$.

29. $\dfrac{3\log(2\cos x + 3\sin x) + 2x}{13}$.

30. $\text{arc}\cot\dfrac{a}{x} + \log\sqrt{\dfrac{x-a}{x+a}}$.

31. $(\log\tan 3 - x^2)^3$.

32. $\dfrac{2 - 3t^{\frac{1}{2}} + 4t^{\frac{1}{3}} + t^2}{t}$.

33. $\dfrac{(1+x)(1-2x)(2+x)}{(3+x)(2-3x)}$.

34. $\text{arc}\tan(\log 3x)$.

35. $\sqrt[3]{(b-ax^m)^n}$.

36. $\log\sqrt{(a^2-bx^2)^m}$.

37. $\log\sqrt{\dfrac{y^2+1}{y^2-1}}$.

38. $e^{\text{arc}\sec 2\theta}$.

39. $\sqrt{\dfrac{(2-3x)^3}{1+4x}}$.

40. $\dfrac{\sqrt[3]{a^2-x^2}}{\cos x}$.

41. $e^x\log\sin x$.

42. $\text{arc}\sin\dfrac{x}{\sqrt{1+x^2}}$.

43. $\text{arc}\tan a^x$.

44. $a^{\sin^2 mx}$.

45. $\cot^3(\log ax)$.

46. $(1-3x^2)e^{\frac{1}{x}}$.

47. $\log\dfrac{\sqrt{1-x^2}}{\sqrt[4]{1+x^3}}$.

# CHAPTER VI

## SIMPLE APPLICATIONS OF THE DERIVATIVE

**64. Direction of a curve.** It was shown in § 32, p. 31, that if

$$y = f(x)$$

is the equation of a curve (see figure), then

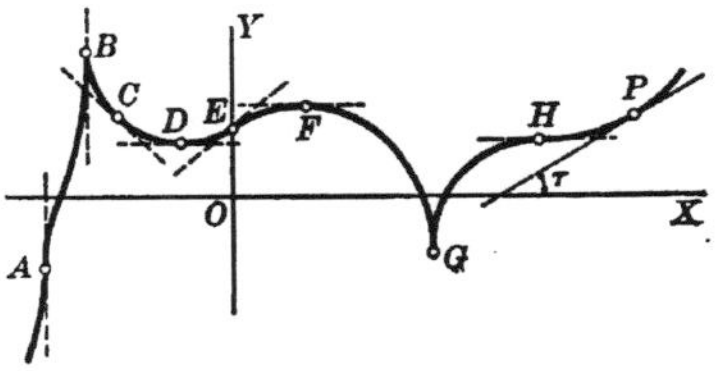


$$\frac{dy}{dx} = \tan\tau = \text{slope of line tangent to the curve at any point } P.$$

The *direction of a curve* at any point is defined to be the same as the direction of the line tangent to the curve at that point. From this it follows at once that

$$\frac{dy}{dx} = \tan\tau = \text{slope of the curve at any point } P.$$

At a particular point whose coördinates are known we write

$$\left[\frac{dy}{dx}\right]_{\substack{x=x_1\\y=y_1}} = \text{slope of the curve (or tangent) at point } (x_1, y_1).$$

At points such as $D$, $F$, $H$, where the curve (or tangent) is *parallel to the axis of X*,

$$\tau = 0^\circ;\ \text{therefore } \frac{dy}{dx} = 0.$$

At points such as $A$, $B$, $G$, where the curve (or tangent) is *perpendicular to the axis of X*,

$$\tau = 90^\circ;\ \text{therefore } \frac{dy}{dx} = \infty.$$

At points such as $E$, where the *curve is rising*,*

$$\tau = \textbf{an acute angle; therefore } \frac{dy}{dx} = \textbf{a positive number.}$$

The curve (or tangent) has a positive slope to the left of $B$, between $D$ and $F$, and to the right of $G$.

At points such as $C$, where the *curve is falling*,*

$$\tau = \textbf{an obtuse angle; therefore } \frac{dy}{dx} = \textbf{a negative number.}$$

The curve (or tangent) has a negative slope between $B$ and $D$, and between $F$ and $G$.

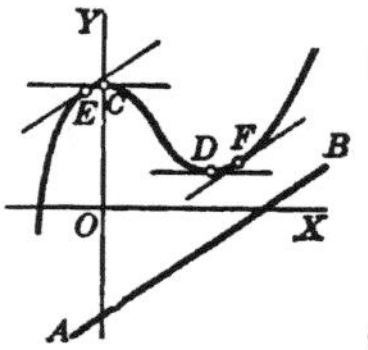


ILLUSTRATIVE EXAMPLE 1. Given the curve $y = \frac{x^3}{3} - x^2 + 2$ (see figure).

(a) Find $\tau$ when $x = 1$.
(b) Find $\tau$ when $x = 3$.
(c) Find the points where the curve is parallel to $OX$.
(d) Find the points where $\tau = 45°$.
(e) Find the points where the curve is parallel to the line $2x - 3y = 6$ (line $AB$).

**Solution.** Differentiating, $\frac{dy}{dx} = x^2 - 2x =$ slope at any point.

(a) $\tan \tau = \left[\frac{dy}{dx}\right]_{x=1} = 1 - 2 = -1$; therefore $\tau = 135°$. *Ans.*

(b) $\tan \tau = \left[\frac{dy}{dx}\right]_{x=3} = 9 - 6 = 3$; therefore $\tau = \text{arc tan } 3$. *Ans.*

(c) $\tau = 0°$, $\tan \tau = \frac{dy}{dx} = 0$; therefore $x^2 - 2x = 0$. Solving this equation, we find that $x = 0$ or $2$, giving points $C$ and $D$ where the curve (or tangent) is parallel to $OX$.

(d) $\tau = 45°$, $\tan \tau = \frac{dy}{dx} = 1$; therefore $x^2 - 2x = 1$. Solving, we get $x = 1 \pm \sqrt{2}$, giving two points where the slope of the curve (or tangent) is unity.

(e) Slope of line $= \frac{2}{3}$; therefore $x^2 - 2x = \frac{2}{3}$. Solving, we get $x = 1 \pm \sqrt{\frac{5}{3}}$, giving points $E$ and $F$ where curve (or tangent) is parallel to line $AB$.

Since a curve at any point has the same direction as its tangent at that point, the angle between two curves at a common point will be the angle between their tangents at that point.

ILLUSTRATIVE EXAMPLE 2. Find the angle of intersection of the circles

$$(A) \qquad x^2 + y^2 - 4x = 1,$$

$$(B) \qquad x^2 + y^2 - 2y = 9.$$

* When moving from left to right on curve.

**Solution.** Solving simultaneously, we find the points of intersection to be (3, 2) and (1, − 2).

$$\frac{dy}{dx} = \frac{2-x}{y} \text{ from } (A). \qquad \text{By § 63, p. 69}$$

$$\frac{dy}{dx} = \frac{x}{1-y} \text{ from } (B). \qquad \text{By § 63, p. 69}$$

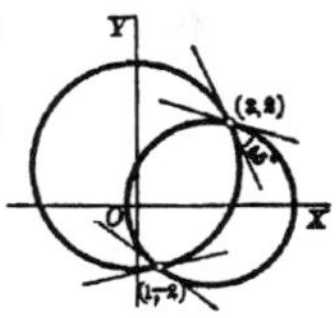


$$\left[\frac{2-x}{y}\right]_{\substack{x=3\\y=2}} = -\tfrac{1}{2} = \text{slope of tangent to } (A) \text{ at } (3, 2).$$

$$\left[\frac{x}{1-y}\right]_{\substack{x=3\\y=2}} = -3 = \text{slope of tangent to } (B) \text{ at } (3, 2).$$

The formula for finding the angle between two lines whose slopes are $m_1$ and $m_2$ is

$$\tan\theta = \frac{m_1 - m_2}{1 + m_1 m_2}. \qquad 55, \text{p. 3}$$

Substituting, $\tan\theta = \dfrac{-\frac{1}{2}+3}{1+\frac{3}{2}} = 1$; therefore $\theta = 45°$. *Ans.*

This is also the angle of intersection at the point (1, − 2).

## EXAMPLES

The corresponding figure should be drawn in each of the following examples:

1. Find the slope of $y = \dfrac{x}{1+x^2}$ at the origin. *Ans.* $1 = \tan\tau$.

2. What angle does the tangent to the curve $x^2y^2 = a^3(x+y)$ at the origin make with the axis of $X$? *Ans.* $\tau = 135°$.

3. What is the direction in which the point generating the graph of $y = 3x^2 - x$ tends to move at the instant when $x = 1$? *Ans.* Parallel to a line whose slope is 5.

4. Show that $\dfrac{dy}{dx}$ (or slope) is constant for a straight line.

5. Find the points where the curve $y = x^3 - 3x^2 - 9x + 5$ is parallel to the axis of $X$. *Ans.* $x = 3$, $x = -1$.

6. At what point on $y^2 = 2x^3$ is the slope equal to 3? *Ans.* (2, 4).

7. At what points on the circle $x^2 + y^2 = r^2$ is the slope of the tangent line equal to $-\frac{3}{4}$? *Ans.* $\left(\pm\dfrac{3r}{5}, \pm\dfrac{4r}{5}\right)$.

8. Where will a point moving on the parabola $y = x^2 - 7x + 3$ be moving parallel to the line $y = 5x + 2$? *Ans.* (6, − 3).

9. Find the points where a particle moving on the circle $x^2 + y^2 = 169$ moves perpendicular to the line $5x + 12y = 60$. *Ans.* $(\pm 12, \mp 5)$.

10. Show that all the curves of the system $y = \log kx$ have the same slope; i.e. the slope is independent of $k$.

11. The path of the projectile from a mortar cannon lies on the parabola $y = 2x - x^2$; the unit is 1 mile, $OX$ being horizontal and $OY$ vertical, and the origin being the point of projection. Find the direction of motion of the projectile
    (a) at instant of projection;
    (b) when it strikes a vertical cliff $1\frac{1}{2}$ miles distant.
    (c) Where will the path make an inclination of 45° with the horizontal?
    (d) Where will the projectile travel horizontally?

*Ans.* (a) arc tan 2; (b) 135°; (c) $(\frac{1}{2}, \frac{3}{4})$; (d) (1, 1).

**12.** If the cannon in the preceding example was situated on a hillside of inclination 45°, at what angle would a shot fired up strike the hillside? *Ans.* 45°.

**13.** At what angles does a road following the line $3y - 2x - 8 = 0$ intersect a railway track following the parabola $y^2 = 8x$. *Ans.* arc tan $\frac{1}{5}$, and arc tan $\frac{1}{2}$.

**14.** Find the angle of intersection between the parabola $y^2 = 6x$ and the circle $x^2 + y^2 = 16$. *Ans.* arc tan $\frac{5}{3}\sqrt{3}$.

**15.** Show that the hyperbola $x^2 - y^2 = 5$ and the ellipse $\frac{x^2}{18} + \frac{y^2}{8} = 1$ intersect at right angles.

**16.** Show that the circle $x^2 + y^2 = 8ax$ and the cissoid $y^2 = \frac{x^3}{2a - x}$

(a) are perpendicular at the origin;

(b) intersect at an angle of 45° at two other points.

**17.** Find the angle of intersection of the parabola $x^2 = 4ay$ and the witch $y = \frac{8a^3}{x^2 + 4a^2}$. *Ans.* arc tan 3 = 71° 33′.9.

**18.** Show that the tangents to the folium of Descartes $x^3 + y^3 = 3axy$ at the points where it meets the parabola $y^2 = ax$ are parallel to the axis of $Y$.

**19.** At how many points will a particle moving on the curve $y = x^3 - 2x^2 + x - 4$ be moving parallel to the axis of $X$? What are the points?

*Ans.* Two; at $(1, -4)$ and $(\frac{1}{3}, -\frac{104}{27})$.

**20.** Find the angle at which the parabolas $y = 3x^2 - 1$ and $y = 2x^2 + 3$ intersect.

*Ans.* arc tan $\frac{4}{97}$.

**21.** Find the relation between the coefficients of the conics $a_1x^2 + b_1y^2 = 1$ and $a_2x^2 + b_2y^2 = 1$ when they intersect at right angles. *Ans.* $\frac{1}{a_1} - \frac{1}{b_1} = \frac{1}{a_2} - \frac{1}{b_2}$.

**65. Equations of tangent and normal, lengths of subtangent and subnormal. Rectangular coördinates.** The equation of a straight line passing through the point $(x_1, y_1)$ and having the slope $m$ is

$$y - y_1 = m(x - x_1). \quad 54, (c), p. 3$$

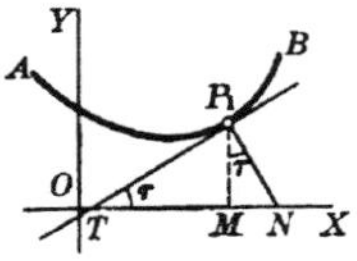


If this line is tangent to the curve $AB$ at the point $P(x_1, y_1)$, then from § 64, p. 73,

$$m = \tan\tau = \left[\frac{dy}{dx}\right]_{\substack{x = x_1 \\ y = y_1}} = \frac{dy_1}{dx_1}.^*$$

Hence at point of contact $P_1(x_1, y_1)$ the *equation of the tangent line* $TP_1$ is

$$(1) \qquad y - y_1 = \frac{dy_1}{dx_1}(x - x_1).$$

* By this notation is meant that we should first find $\frac{dy}{dx}$, then in the result substitute $x_1$ for $x$ and $y_1$ for $y$. The student is warned against interpreting the symbol $\frac{dy_1}{dx_1}$ to mean *the derivative of* $y_1$ *with respect to* $x_1$, for that has no meaning whatever, since $x_1$ and $y_1$ are both constants.

The normal being perpendicular to tangent, its slope is

$$-\frac{1}{m}=-\frac{dx_1}{dy_1}. \qquad \text{By 55, p. 3}$$

And since it also passes through the point of contact $P_1(x_1, y_1)$, we have for the *equation of the normal* $P_1N$

$$(2) \qquad y-y_1=-\frac{dx_1}{dy_1}(x-x_1).$$

That portion of the tangent which is intercepted between the point of contact and $OX$ is called the *length of the tangent* $(=TP_1)$, and its projection on the axis of $X$ is called the *length of the subtangent* $(=TM)$. Similarly, we have the *length of the normal* $(=P_1N)$ and the *length of the subnormal* $(=MN)$.

In the triangle $TP_1M$, $\tan\tau=\frac{MP_1}{TM}$; therefore

$$(3) \qquad TM^{*}=\frac{MP_1}{\tan\tau}=y_1\frac{dx_1}{dy_1}=\text{length of subtangent.}$$

In the triangle $MP_1N$, $\tan\tau=\frac{MN}{MP_1}$; therefore

$$(4) \qquad MN^{\dagger}=MP_1\tan\tau=y_1\frac{dy_1}{dx_1}=\text{length of subnormal.}$$

The length of tangent $(=TP_1)$ and the length of normal $(=P_1N)$ may then be found directly from the figure, each being the hypotenuse of a right triangle having the two legs known. Thus

$$TP_1=\sqrt{\overline{TM}^2+\overline{MP_1}^2}=\sqrt{\left(y_1\frac{dx_1}{dy_1}\right)^2+(y_1)^2}$$

$$(5) \qquad =y_1\sqrt{\left(\frac{dx_1}{dy_1}\right)^2+1}=\text{length of tangent.}$$

$$P_1N=\sqrt{\overline{MP_1}^2+\overline{MN}^2}=\sqrt{(y_1)^2+\left(y_1\frac{dy_1}{dx_1}\right)^2}$$

$$(6) \qquad =y_1\sqrt{1+\left(\frac{dy_1}{dx_1}\right)^2}=\text{length of normal.}$$

The student is advised to get the lengths of the tangent and of the normal directly from the figure rather than by using (5) and (6).

When the length of subtangent or subnormal at a point on a curve is determined, the tangent and normal may be easily constructed.

* If subtangent extends to the right of $T$, we consider it positive; if to the left, negative.
† If subnormal extends to the right of $M$, we consider it positive; if to the left, negative.

## EXAMPLES

1. Find the equations of tangent and normal, lengths of subtangent, subnormal tangent, and normal at the point $(a, a)$ on the cissoid $y^2 = \frac{x^3}{2a - x}$.

**Solution.**

$$\frac{dy}{dx} = \frac{3ax^2 - x^3}{y(2a - x)^2}.$$

Hence

$$\frac{dy_1}{dx_1} = \left[\frac{dy}{dx}\right]_{\substack{x=a\\y=a}} = \frac{3a^3 - a^3}{a(2a - a)^2} = 2 = \text{slope of tangent.}$$

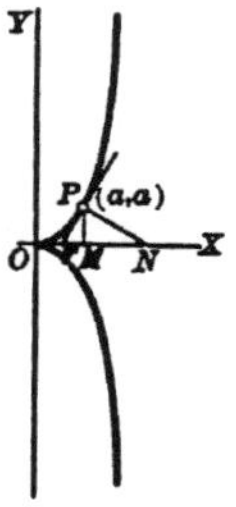


Substituting in (1) gives

$$y = 2x - a, \text{ equation of tangent.}$$

Substituting in (2) gives

$$2y + x = 3a, \text{ equation of normal.}$$

Substituting in (3) gives

$$TM = \frac{a}{2} = \text{length of subtangent.}$$

Substituting in (4) gives

$$MN = 2a = \text{length of subnormal.}$$

Also $PT = \sqrt{(TM)^2 + (MP)^2} = \sqrt{\frac{a^2}{4} + a^2} = \frac{a}{2}\sqrt{5}$ = length of tangent,

and $PN = \sqrt{(MN)^2 + (MP)^2} = \sqrt{4a^2 + a^2} = a\sqrt{5}$ = length of normal.

2. Find equations of tangent and normal to the ellipse $x^2 + 2y^2 - 2xy - x = 0$ at the points where $x = 1$.
*Ans.* At $(1, 0)$, $2y = x - 1$, $y + 2x = 2$.
At $(1, 1)$, $2y = x + 1$, $y + 2x = 3$.

3. Find equations of tangent and normal, lengths of subtangent and subnormal at the point $(x_1, y_1)$ on the circle $x^2 + y^2 = r^2$.*

*Ans.* $x_1x + y_1y = r^2$, $x_1y - y_1x = 0$, $-\frac{y_1^2}{x_1}$, $-x_1$.

4. Show that the subtangent to the parabola $y^2 = 4px$ is bisected at the vertex, and that the subnormal is constant and equal to $2p$.

5. Find the equation of the tangent at $(x_1, y_1)$ to the ellipse $\frac{x^2}{a^2} + \frac{y^2}{b^2} = 1$.

*Ans.* $\frac{x_1x}{a^2} + \frac{y_1y}{b^2} = 1$.

6. Find equations of tangent and normal to the witch $y = \frac{8a^3}{4a^2 + x^2}$ at the point where $x = 2a$.

*Ans.* $x + 2y = 4a$, $y = 2x - 3a$.

7. Prove that at any point on the catenary $y = \frac{a}{2}\left(e^{\frac{x}{a}} + e^{-\frac{x}{a}}\right)$ the lengths of subnormal and normal are $\frac{a}{4}\left(e^{\frac{2x}{a}} - e^{-\frac{2x}{a}}\right)$ and $\frac{y^2}{a}$ respectively.

8. Find equations of tangent and normal, lengths of subtangent and subnormal, to each of the following curves at the points indicated:

(a) $y = x^3$ at $(\frac{1}{2}, \frac{1}{8})$.
(b) $y^2 = 4x$ at $(9, -6)$.
(c) $x^2 + 5y^2 = 14$ where $y = 1$.
(d) $x^2 + y^2 = 25$ at $(-3, -4)$.
(e) $y = 9 - x^2$ at $(-3, 0)$.
(f) $x^2 = 6y$ where $x = -6$.
(g) $x^2 - xy + 2x - 9 = 0$, $(3, 2)$.
(h) $2x^2 - y^2 = 14$ at $(3, -2)$.

* In Exs. 3 and 5 the student should notice that if we drop the subscripts in equations of tangents, they reduce to the equations of the curves themselves.

**9.** Prove that the length of subtangent to $y = a^x$ is constant and equal to $\frac{1}{\log a}$.

**10.** Get the equation of tangent to the parabola $y^2 = 20x$ which makes an angle of 45° with the axis of $X$. *Ans.* $y = x + 5$.

HINT. First find point of contact by method of Illustrative Example 1, (d), p. 74.

**11.** Find equations of tangents to the circle $x^2 + y^2 = 52$ which are parallel to the line $2x + 3y = 6$. *Ans.* $2x + 3y \pm 26 = 0$.

**12.** Find equations of tangents to the hyperbola $4x^2 - 9y^2 + 36 = 0$ which are perpendicular to the line $2y + 5x = 10$. *Ans.* $2x - 5y \pm 8 = 0$.

**13.** Show that in the equilateral hyperbola $2xy = a^2$ the area of the triangle formed by a tangent and the coördinate axes is constant and equal to $a^2$.

**14.** Find equations of tangents and normals to the curve $y^2 = 2x^2 - x^3$ at the points where $x = 1$. *Ans.* At $(1, 1)$, $2y = x + 1$, $y + 2x = 3$.
At $(1, -1)$, $2y = -x - 1$, $y - 2x = -3$.

**15.** Show that the sum of the intercepts of the tangent to the parabola

$$x^{\frac{1}{2}} + y^{\frac{1}{2}} = a^{\frac{1}{2}}$$

on the coördinate axes is constant and equal to $a$.

**16.** Find the equation of tangent to the curve $x^2(x + y) = a^2(x - y)$ at the origin. *Ans.* $y = x$.

**17.** Show that for the hypocycloid $x^{\frac{2}{3}} + y^{\frac{2}{3}} = a^{\frac{2}{3}}$ that portion of the tangent included between the coördinate axes is constant and equal to $a$.

**18.** Show that the curve $y = ae^{\frac{x}{c}}$ has a constant subtangent.

**66. Parametric equations of a curve.** Let the equation of a curve be

$$(A) \qquad F(x, y) = 0.$$

If $x$ is given as a function of a third variable, $t$ say, called a *parameter*, then by virtue of $(A)$ $y$ is also a function of $t$, and the same functional relation $(A)$ between $x$ and $y$ may generally be expressed by means of equations in the form

$$(B) \qquad \begin{cases} x = f(t), \\ y = \phi(t); \end{cases}$$

each value of $t$ giving a value of $x$ and a value of $y$. Equations $(B)$ are called *parametric equations* of the curve. If we eliminate $t$ between equations $(B)$, it is evident that the relation $(A)$ must result. For example, take equation of circle

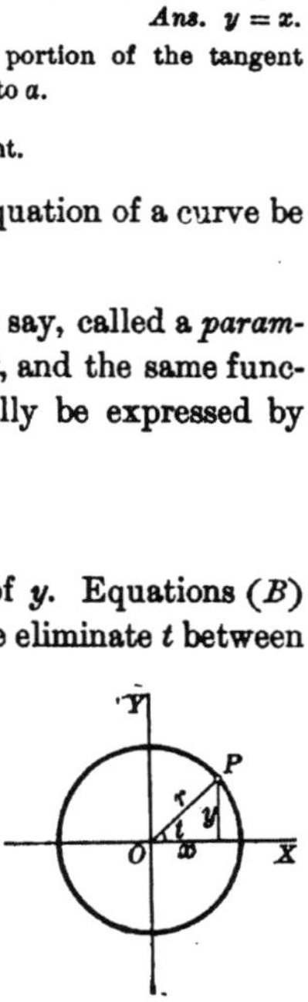


$$x^2 + y^2 = r^2, \text{ or } y = \sqrt{r^2 - x^2}.$$

Let $x = r\cos t$; then

$y = r\sin t$, and we have

$$(C) \qquad \begin{cases} x = r\cos t, \\ y = r\sin t, \end{cases}$$

as parametric equations of the circle in the figure, $t$ being the parameter.

If we eliminate $t$ between equations $(C)$ by squaring and adding the results, we have

$$x^2 + y^2 = r^2(\cos^2 t + \sin^2 t) = r^2,$$

the rectangular equation of the circle. It is evident that if $t$ varies from 0 to $2\pi$, the point $P(x, y)$ will describe a complete circumference.

In § 71 we shall discuss the motion of a point $P$, which motion is defined by equations such as

$$\begin{cases} x = f(t), \\ y = \phi(t). \end{cases}$$

We call these the parametric equations of the path, the time $t$ being the parameter. Thus in Ex. 2, p. 93, we see that

$$\begin{cases} x = v_0 \cos\alpha \cdot t, \\ y = -\frac{1}{2}gt^2 + v_0 \sin\alpha \cdot t \end{cases}$$

are really the parametric equations of the trajectory of a projectile, the time $t$ being the parameter. The elimination of $t$ gives the rectangular equation of the trajectory

$$y = x\tan\alpha - \frac{gx^2}{2v_0^2\cos^2\alpha}.$$

Since from $(B)$ $y$ is given as a function of $t$, and $t$ as a function of $x$, we have

$$\frac{dy}{dx} = \frac{dy}{dt}\cdot\frac{dt}{dx} \qquad \text{by XXV}$$

$$= \frac{dy}{dt}\cdot\frac{1}{\frac{dx}{dt}}; \qquad \text{by XXVI}$$

that is,

$$(D) \qquad \frac{dy}{dx} = \frac{\frac{dy}{dt}}{\frac{dx}{dt}} = \frac{\phi'(t)}{f'(t)}.$$

Hence, if the parametric equations of a curve are given, we can find equations of tangent and normal, lengths of subtangent and subnormal at a given point on the curve, by first finding the value of $\frac{dy}{dx}$ at that point from $(D)$ and then substituting in formulas (1), (2), (3), (4) of the last section.

ILLUSTRATIVE EXAMPLE 1. **Find equations of tangent and normal, lengths of subtangent and subnormal to the ellipse**

$$(E)\qquad \begin{cases} x = a\cos\phi, \\ y = b\sin\phi,^* \end{cases}$$

**at the point where** $\phi = \frac{\pi}{4}$.

**Solution.** The parameter being $\phi$, $\frac{dx}{d\phi} = -a\sin\phi$,

$$\frac{dy}{d\phi} = b\cos\phi.$$

Substituting in (*D*), $\frac{dy}{dx} = -\frac{b\cos\phi}{a\sin\phi}$ = slope at any point.

Substituting $\phi = \frac{\pi}{4}$ in the given equations (*E*), we get $\left(\frac{a}{\sqrt{2}}, \frac{b}{\sqrt{2}}\right)$ as the point of contact. And

$$\frac{dy}{dx} \text{ becomes } \frac{dy_1}{dx_1} = -\frac{b}{a}, \text{ when } \phi = \frac{\pi}{4}.$$

Substituting in (1), p. 76, $y - \frac{b}{\sqrt{2}} = -\frac{b}{a}\left(x - \frac{a}{\sqrt{2}}\right)$,

or, $bx + ay = \sqrt{2}\,ab$, equation of tangent.

Substituting in (2), p. 77, $y - \frac{b}{\sqrt{2}} = \frac{a}{b}\left(x - \frac{a}{\sqrt{2}}\right)$,

or, $\sqrt{2}\,(ax - by) = a^2 - b^2$, equation of normal.

Substituting in (3) and (4), p. 77,

$$\frac{b}{\sqrt{2}}\left(-\frac{b}{a}\right) = -\frac{b^2}{a\sqrt{2}} = \text{length of subnormal.}$$

$$\frac{b}{\sqrt{2}}\left(-\frac{a}{b}\right) = -\frac{a}{\sqrt{2}} = \text{length of subtangent.}$$

* As in the figure draw the major and minor auxiliary circles of the ellipse. Through two points *B* and *C* on the same radius draw lines parallel to the axes of coördinates. These lines will intersect in a point *P* $(x, y)$ on the ellipse, because

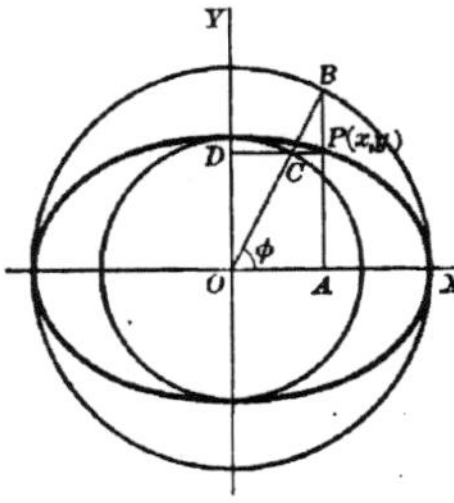


$$x = OA = OB\cos\phi = a\cos\phi$$

and $y = AP = OD = OC\sin\phi = b\sin\phi$,

or, $\frac{x}{a} = \cos\phi$ and $\frac{y}{b} = \sin\phi$.

Now squaring and adding, we get

$$\frac{x^2}{a^2} + \frac{y^2}{b^2} = \cos^2\phi + \sin^2\phi = 1,$$

the rectangular equation of the ellipse. $\phi$ is sometimes called the eccentric angle of the ellipse at the point *P*.

ILLUSTRATIVE EXAMPLE 2. Given equation of the cycloid* in parametric form

$$\begin{cases} x = a(\theta - \sin\theta), \\ y = a(1 - \cos\theta), \end{cases}$$

$\theta$ being the variable parameter; find lengths of subtangent, subnormal, tangent, and normal at the point where $\theta = \frac{\pi}{2}$.

**Solution.** $$\frac{dx}{d\theta} = a(1 - \cos\theta),\ \frac{dy}{d\theta} = a\sin\theta.$$

Substituting in (*D*), p. 80, $$\frac{dy}{dx} = \frac{\sin\theta}{1 - \cos\theta} = \text{slope at any point.}$$

Since $\theta = \frac{\pi}{2}$, the point of contact is $\left(\frac{\pi a}{2} - a,\ a\right)$, and $\frac{dy_1}{dx_1} = 1$.

Substituting in (3), (4), (5), (6) of the last section, we get

length of subtangent $= a$, length of subnormal $= a$,

length of tangent $= a\sqrt{2}$, length of normal $= a\sqrt{2}$. *Ans.*

**EXAMPLES**

Find equations of tangent and normal, lengths of subtangent and subnormal to each of the following curves at the point indicated:

| | *Tangent* | *Normal* | *Subt.* | *Subn.* |
|---|---|---|---|---|
| 1. $x = t^2$, $2y = t$; $t = 1$. | $x - 4y + 1 = 0$, | $8x + 2y - 9 = 0$, | 2, | $\frac{1}{8}$. |
| 2. $x = t$, $y = t^3$; $t = 2$. | $12x - y - 16 = 0$, | $x + 12y - 98 = 0$, | $\frac{2}{3}$, | 96. |
| 3. $x = t^2$, $y = t^3$; $t = 1$. | $3x - 2y - 1 = 0$, | $2x + 3y - 5 = 0$, | $\frac{2}{3}$, | $\frac{3}{2}$. |
| 4. $x = 2e^t$, $y = e^{-t}$; $t = 0$. | $x + 2y - 4 = 0$, | $2x - y - 3 = 0$, | $-2$, | $-\frac{1}{2}$. |
| 5. $x = \sin t$, $y = \cos 2t$; $t = \frac{\pi}{6}$. | $2y + 4x - 3 = 0$, | $4y - 2x - 1 = 0$, | $-\frac{1}{4}$, | $-1$. |

* The path described by a point on the circumference of a circle which rolls without sliding on a fixed straight line is called the cycloid. Let the radius of the rolling circle be $a$, $P$ the generating point, and $M$ the point of contact with the fixed line $OX$, which is called the

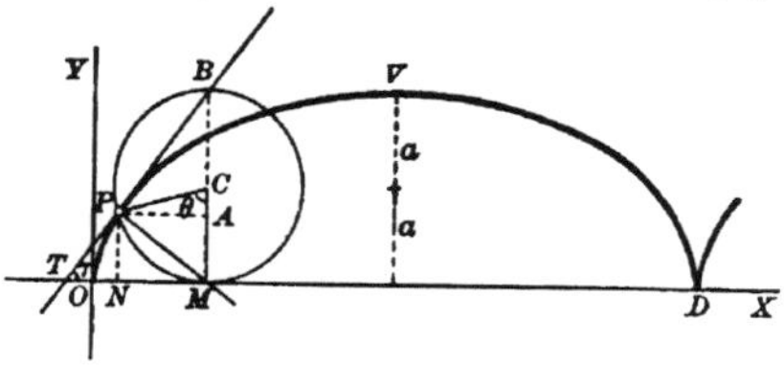


base. If arc $PM$ equals $OM$ in length, then $P$ will touch at $O$ if the circle is rolled to the left. We have, denoting angle $PCM$ by $\theta$,

$$x = OM - NM = a\theta - a\sin\theta = a(\theta - \sin\theta),$$
$$y = PN = MC - AC = a - a\cos\theta = a(1 - \cos\theta),$$

the parametric equations of the cycloid, the angle $\theta$ through which the rolling circle turns being the parameter. $OD = 2\pi a$ is called the base of one arch of the cycloid, and the point $V$ is called the vertex. Eliminating $\theta$, we get the rectangular equation

$$x = a \operatorname{arc\,cos}\left(\frac{a - y}{a}\right) - \sqrt{2ay - y^2}.$$

6. $x = 1 - t,\ y = t^2;\ t = 3.$

7. $x = 3t,\ y = 6t - t^2;\ t = 0.$

8. $x = t^3,\ y = t;\ t = 2.$

9. $x = t^3,\ y = t^2;\ t = -1.$

10. $x = 2 - t,\ y = 3t^2;\ t = 1.$

11. $x = \cos t,\ y = \sin 2t;\ t = \frac{\pi}{3}.$

12. $x = 3e^{-t},\ y = 2e^t;\ t = 0.$

13. $x = \sin t,\ y = 2\cos t;\ t = \frac{\pi}{4}.$

14. $x = 4\cos t,\ y = 3\sin t;\ t = \frac{\pi}{2}.$

15. $x = \log(t + 2),\ y = t;\ t = 2.$

In the following curves find lengths of (a) subtangent, (b) subnormal, (c) tangent, (d) normal, at any point:

16. The curve $\begin{cases} x = a(\cos t + t\sin t), \\ y = a(\sin t - t\cos t). \end{cases}$

*Ans.* (a) $y\cot t$, (b) $y\tan t$, (c) $\frac{y}{\sin t}$, (d) $\frac{y}{\cos t}$.

17. The hypocycloid (astroid) $\begin{cases} x = 4a\cos^3 t, \\ y = 4a\sin^3 t. \end{cases}$

*Ans.* (a) $-y\cot t$, (b) $-y\tan t$, (c) $\frac{y}{\sin t}$, (d) $\frac{y}{\cos t}$.

18. The circle $\begin{cases} x = r\cos t, \\ y = r\sin t. \end{cases}$

19. The cardioid $\begin{cases} x = a(2\cos t - \cos 2t), \\ y = a(2\sin t - \sin 2t). \end{cases}$

20. The folium $\begin{cases} x = \dfrac{3t}{1 + t^3}, \\ y = \dfrac{3t^2}{1 + t^3}. \end{cases}$

21. The hyperbolic spiral $\begin{cases} x = \dfrac{a}{t}\cos t, \\ y = \dfrac{a}{t}\sin t. \end{cases}$

**67. Angle between the radius vector drawn to a point on a curve and the tangent to the curve at that point.** Let the equation of the curve in polar coördinates be $\rho = f(\theta)$.

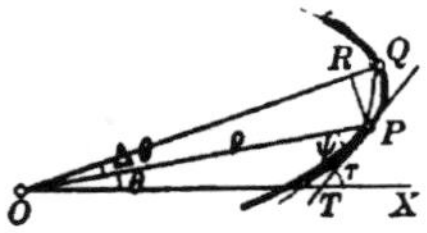


Let $P$ be any fixed point $(\rho, \theta)$ on the curve. If $\theta$, which we assume as the independent variable, takes on an increment $\Delta\theta$, then $\rho$ will take on a corresponding increment $\Delta\rho$. Denote by $Q$ the point $(\rho + \Delta\rho,\ \theta + \Delta\theta)$. Draw $PR$ perpendicular to $OQ$. Then $OQ = \rho + \Delta\rho$, $PR = \rho\sin\Delta\theta$, and $OR = \rho\cos\Delta\theta$. Also,

$$\tan PQR = \frac{PR}{RQ} = \frac{PR}{OQ - OR} = \frac{\rho\sin\Delta\theta}{\rho + \Delta\rho - \rho\cos\Delta\theta}.$$

Denote by $\psi$ the angle between the radius vector $OP$ and the tangent $PT$. If we now let $\Delta\theta$ approach the limit zero, then

(a) *the point $Q$ will approach indefinitely near $P$;*

(b) *the secant $PQ$ will approach the tangent $PT$ as a limiting position; and*

(c) *the angle $PQR$ will approach $\psi$ as a limit.*

Hence

$$\tan\psi = \underset{\Delta\theta=0}{\text{limit}} \frac{\rho\sin\Delta\theta}{\rho+\Delta\rho-\rho\cos\Delta\theta}$$

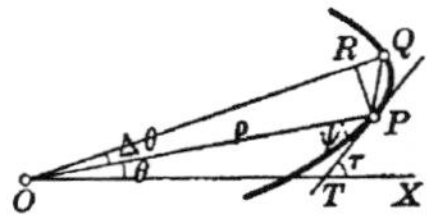


$$= \underset{\Delta\theta=0}{\text{limit}} \frac{\rho\sin\Delta\theta}{2\rho\sin^2\frac{\Delta\theta}{2}+\Delta\rho}$$

[Since from 39, p. 2, $\rho-\rho\cos\Delta\theta=\rho(1-\cos\Delta\theta)=2\rho\sin^2\frac{\Delta\theta}{2}$.]

$$= \underset{\Delta\theta=0}{\text{limit}} \frac{\dfrac{\rho\sin\Delta\theta}{\Delta\theta}}{\dfrac{2\rho\sin^2\frac{\Delta\theta}{2}}{\Delta\theta}+\dfrac{\Delta\rho}{\Delta\theta}}$$

[Dividing both numerator and denominator by $\Delta\theta$.]

$$= \underset{\Delta\theta=0}{\text{limit}} \frac{\rho\cdot\dfrac{\sin\Delta\theta}{\Delta\theta}}{\rho\sin\dfrac{\Delta\theta}{2}\cdot\dfrac{\sin\frac{\Delta\theta}{2}}{\frac{\Delta\theta}{2}}+\dfrac{\Delta\rho}{\Delta\theta}}.$$

Since $\underset{\Delta\theta=0}{\text{limit}}\left(\frac{\Delta\rho}{\Delta\theta}\right)=\frac{d\rho}{d\theta}$ and $\underset{\Delta\theta=0}{\text{limit}}\left(\sin\frac{\Delta\theta}{2}\right)=0$, also $\underset{\Delta\theta=0}{\text{limit}}\left(\frac{\sin\Delta\theta}{\Delta\theta}\right)=1$

and $\underset{\Delta\theta=0}{\text{limit}}\dfrac{\sin\frac{\Delta\theta}{2}}{\frac{\Delta\theta}{2}}=1$ by § 22, p. 21, we have

$$(A)\qquad \tan\psi=\frac{\rho}{\dfrac{d\rho}{d\theta}}.$$

From the triangle $OPT$ we get

$$(B)\qquad \tau=\theta+\psi.$$

Having found $\tau$, we may then find $\tan\tau$, the slope of the tangent to the curve at $P$. Or since, from **(B)**,

$$\tan\tau = \tan(\theta+\psi) = \frac{\tan\theta + \tan\psi}{1-\tan\theta\tan\psi},$$

we may calculate $\tan\psi$ from **(A)** and substitute in the formula

$$(C) \qquad \textbf{slope of tangent} = \tan\tau = \frac{\tan\theta + \tan\psi}{1-\tan\theta\tan\psi}.$$

ILLUSTRATIVE EXAMPLE 1. Find $\psi$ and $\tau$ in the cardioid $\rho = a(1-\cos\theta)$. Also find the slope at $\theta = \frac{\pi}{6}$.

**Solution.** $\frac{d\rho}{d\theta} = a\sin\theta$. Substituting in $(A)$ gives

$$\tan\psi = \frac{\rho}{\frac{d\rho}{d\theta}} = \frac{a(1-\cos\theta)}{a\sin\theta} = \frac{2a\sin^2\frac{\theta}{2}}{2a\sin\frac{\theta}{2}\cos\frac{\theta}{2}} = \tan\frac{\theta}{2}. \quad \text{By 39, p. 2, and 37, p. 2}$$

Since $\tan\psi = \tan\frac{\theta}{2}$, $\psi = \frac{\theta}{2}$. *Ans.* Substituting in $(B)$, $\tau = \theta + \frac{\theta}{2} = \frac{3\theta}{2}$. *Ans.*

$\tan\tau = \tan\frac{\pi}{4} = 1$. *Ans.*

To find the angle of intersection $\phi$ of two curves $C$ and $C'$ whose equations are given in polar coördinates, we may proceed as follows:

angle $TPT'$ = angle $OPT'$ − angle $OPT$,

or, $\qquad \phi = \psi' - \psi$. Hence

$$(D) \quad \tan\phi = \frac{\tan\psi' - \tan\psi}{1+\tan\psi'\tan\psi},$$

where $\tan\psi'$ and $\tan\psi$ are calculated by **(A)** from the two curves and evaluated for the point of intersection.

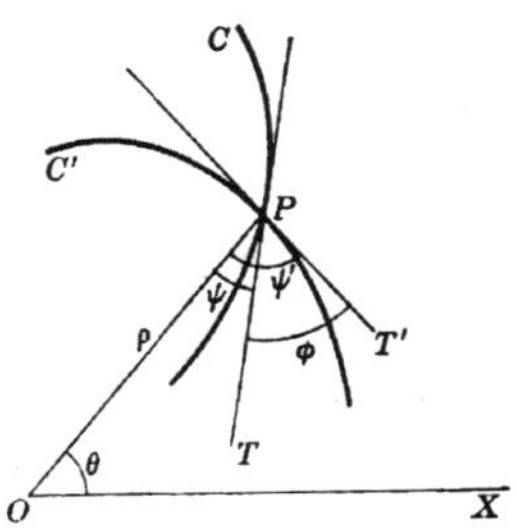


ILLUSTRATIVE EXAMPLE 2. Find the angle of intersection of the curves $\rho = a\sin 2\theta$, $\rho = a\cos 2\theta$.

**Solution.** Solving the two equations simultaneously, we get at the point of intersection

$$\tan 2\theta = 1,\ 2\theta = 45^\circ,\ \theta = 22\tfrac{1}{2}^\circ.$$

From the first curve, using $(A)$,

$$\tan\psi' = \tfrac{1}{2}\tan 2\theta = \tfrac{1}{2}, \text{ for } \theta = 22\tfrac{1}{2}^\circ.$$

From the second curve,

$$\tan\psi = -\tfrac{1}{2}\cot 2\theta = -\tfrac{1}{2}, \text{ for } \theta = 22\tfrac{1}{2}^\circ.$$

Substituting in $(D)$,

$$\tan\phi = \frac{\frac{1}{2}+\frac{1}{2}}{1-\frac{1}{4}} = \tfrac{4}{3}. \quad \therefore \phi = \operatorname{arc}\tan\tfrac{4}{3}. \ \textit{Ans.}$$

**68. Lengths of polar subtangent and polar subnormal.** Draw a line $NT$ through the origin perpendicular to the radius vector of the point $P$ on the curve. If $PT$ is the tangent and $PN$ the normal to the curve at $P$, then

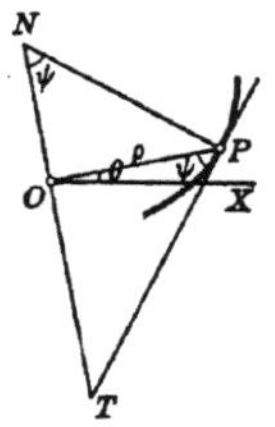

$$OT = \textit{length of polar subtangent},$$

and $$ON = \textit{length of polar subnormal}$$

of the curve at $P$.

In the triangle $OPT$, $\tan\psi = \frac{OT}{\rho}$. Therefore

(7) $$OT = \rho\tan\psi = \rho^2\frac{d\theta}{d\rho} = \textbf{length of polar subtangent.*}$$

In the triangle $OPN$, $\tan\psi = \frac{\rho}{ON}$. Therefore

(8) $$ON = \frac{\rho}{\tan\psi} = \frac{d\rho}{d\theta} = \textbf{length of polar subnormal.}$$

The *length of the polar tangent* ($=PT$) and the *length of the polar normal* ($=PN$) may be found from the figure, each being the hypotenuse of a right triangle.

ILLUSTRATIVE EXAMPLE 3. Find lengths of polar subtangent and subnormal to the lemniscate $\rho^2 = a^2\cos 2\theta$.

**Solution.** Differentiating the equation of the curve as an implicit function with respect to $\theta$,

$$2\rho\frac{d\rho}{d\theta} = -2a^2\sin 2\theta,$$

or, $$\frac{d\rho}{d\theta} = -\frac{a^2\sin 2\theta}{\rho}.$$

Substituting in (7) and (8), we get

$$\text{length of polar subtangent} = -\frac{\rho^3}{a^2\sin 2\theta},$$

$$\text{length of polar subnormal} = -\frac{a^2\sin 2\theta}{\rho}.$$

If we wish to express the results in terms of $\theta$, find $\rho$ in terms of $\theta$ from the given equation and substitute. Thus, in the above, $\rho = \pm a\sqrt{\cos 2\theta}$; therefore length of polar subtangent $= \pm a\cot 2\theta\sqrt{\cos 2\theta}$.

* When $\theta$ increases with $\rho$, $\frac{d\theta}{d\rho}$ is positive and $\psi$ is an acute angle, as in the above figure. Then the subtangent $OT$ is positive and is measured to the right of an observer placed at $O$ and looking along $OP$. When $\frac{d\theta}{d\rho}$ is negative, the subtangent is negative and is measured to the left of the observer.

## EXAMPLES

1. In the circle $\rho = r \sin\theta$, find $\psi$ and $\tau$ in terms of $\theta$. *Ans.* $\psi = \theta$, $\tau = 2\theta$.

2. In the parabola $\rho = a\sec^2\frac{\theta}{2}$, show that $\tau + \psi = \pi$.

3. In the curve $\rho^2 = a^2 \cos 2\theta$, show that $2\psi = \pi + 4\theta$.

4. Show that $\psi$ is constant in the logarithmic spiral $\rho = e^{a\theta}$. Since the tangent makes a constant angle with the radius vector, this curve is also called the equiangular spiral.

5. Given the curve $\rho = a\sin^3\frac{\theta}{3}$, prove that $\tau = 4\psi$.

6. Show that $\tan\psi = \theta$ in the spiral of Archimedes $\rho = a\theta$. Find values of $\psi$ when $\theta = 2\pi$ and $4\pi$. *Ans.* $\psi = 80° 57'$ and $85° 27'$.

7. Find the angle between the straight line $\rho\cos\theta = 2a$ and the circle $\rho = 5a\sin\theta$. *Ans.* arc tan $\frac{3}{4}$.

8. Show that the parabolas $\rho = a\sec^2\frac{\theta}{2}$ and $\rho = b\csc^2\frac{\theta}{2}$ intersect at right angles.

9. Find the angle of intersection of $\rho = a\sin\theta$ and $\rho = a\sin 2\theta$.
*Ans.* At origin 0°; at two other points arc tan $3\sqrt{3}$.

10. Find the slopes of the following curves at the points designated:

| | | | |
|---|---|---|---|
| (a) | $\rho = a(1 - \cos\theta)$. | $\theta = \frac{\pi}{2}$. | *Ans.* $-1$. |
| (b) | $\rho = a\sec^2\theta$. | $\rho = 2a$. | 3. |
| (c) | $\rho = a\sin 4\theta$. | origin. | $0, 1, \infty, -1$. |
| (d) | $\rho^2 = a^2\sin 4\theta$. | origin. | $0, 1, \infty, -1$. |
| (e) | $\rho = a\sin 3\theta$. | origin. | $0, \sqrt{3}, -\sqrt{3}$. |
| (f) | $\rho = a\cos 3\theta$. | origin. | |
| (g) | $\rho = a\cos 2\theta$. | origin. | |
| (h) | $\rho = a\sin 2\theta$. | $\theta = \frac{\pi}{4}$. | |
| (i) | $\rho = a\sin 3\theta$. | $\theta = \frac{\pi}{6}$. | |
| (j) | $\rho = a\theta$. | $\theta = \frac{\pi}{2}$. | |
| (k) | $\rho\theta = a$. | $\theta = \frac{\pi}{2}$. | |
| (l) | $\rho = e^{\theta}$. | $\theta = 0$. | |

11. Prove that the spiral of Archimedes $\rho = a\theta$, and the reciprocal spiral $\rho = \frac{a}{\theta}$, intersect at right angles.

12. Find the angle between the parabola $\rho = a\sec^2\frac{\theta}{2}$ and the straight line $\rho\sin\theta = 2a$. *Ans.* 45°.

13. Show that the two cardioids $\rho = a(1 + \cos\theta)$ and $\rho = a(1 - \cos\theta)$ cut each other perpendicularly.

14. Find lengths of subtangent, subnormal, tangent, and normal of the spiral of Archimedes $\rho = a\theta$.
*Ans.* subt. $= \frac{\rho^2}{a}$, tan. $= \frac{\rho}{a}\sqrt{a^2 + \rho^2}$,
subn. $= a$, nor. $= \sqrt{a^2 + \rho^2}$.

The student should note the fact that the subnormal is constant.

**15.** Get lengths of subtangent, subnormal, tangent, and normal in the logarithmic spiral $\rho = a^{\theta}$.

*Ans.* subt. $= \dfrac{\rho}{\log a}$, tan. $= \rho\sqrt{1 + \dfrac{1}{\log^2 a}}$,

subn. $= \rho \log a$, nor. $= \rho\sqrt{1 + \log^2 a}$.

When $a = e$, we notice that subt. = subn., and tan. = nor.

**16.** Find the angles between the curves $\rho = a(1 + \cos\theta)$, $\rho = b(1 - \cos\theta)$.

*Ans.* 0 and $\frac{\pi}{2}$.

**17.** Show that the reciprocal spiral $\rho = \frac{a}{\theta}$ has a constant subtangent.

**18.** Show that the equilateral hyperbolas $\rho^2 \sin 2\theta = a^2$, $\rho^2 \cos 2\theta = b^2$ intersect at right angles.

**69. Solution of equations having multiple roots.** Any root which occurs more than once in an equation is called a *multiple root.* Thus 3, 3, 3, $-2$ are the roots of

$$(A) \qquad x^4 - 7x^3 + 9x^2 + 27x - 54 = 0;$$

hence 3 is a multiple root occurring three times.

Evidently $(A)$ may also be written in the form

$$(x - 3)^3(x + 2) = 0.$$

Let $f(x)$ denote an integral rational function of $x$ having a multiple root $a$, and suppose it occurs $m$ times. Then we may write

$$(B) \qquad f(x) = (x - a)^m \phi(x),$$

where $\phi(x)$ is the product of the factors corresponding to all the roots of $f(x)$ differing from $a$. Differentiating $(B)$,

$$f'(x) = (x - a)^m \phi'(x) + \phi(x)\, m(x - a)^{m-1},$$

or,

$$(C) \qquad f'(x) = (x - a)^{m-1}[(x - a)\phi'(x) + \phi(x)\, m].$$

Therefore $f'(x)$ contains the factor $(x - a)$ repeated $m - 1$ times and no more; that is, the highest common factor (H.C.F.) of $f(x)$ and $f'(x)$ has $m - 1$ roots equal to $a$.

In case $f(x)$ has a second multiple root $\beta$ occurring $r$ times, it is evident that the H.C.F. would also contain the factor $(x - \beta)^{r-1}$, and so on for any number of different multiple roots, each occurring once more in $f(x)$ than in the H.C.F.

We may then state a **rule for finding the multiple roots** of an equation $f(x) = 0$ as follows:

FIRST STEP. *Find $f'(x)$.*

SECOND STEP. *Find the H.C.F. of $f(x)$ and $f'(x)$.*

THIRD STEP. *Find the roots of the H.C.F. Each different root of the H.C.F. will occur once more in $f(x)$ than it does in the H.C.F.*

If it turns out that the H.C.F. does not involve $x$, then $f(x)$ has no multiple roots and the above process is of no assistance in the solution of the equation, but it may be of interest to know that the equation has no *equal*, i.e. *multiple*, roots.

ILLUSTRATIVE EXAMPLE 1. Solve the equation $x^3 - 8x^2 + 13x - 6 = 0$.

**Solution.** Place $f(x) = x^3 - 8x^2 + 13x - 6.$

*First step.* $f'(x) = 3x^2 - 16x + 13.$

*Second step.* $\text{H.C.F.} = x - 1.$

*Third step.* $x - 1 = 0. \quad \therefore x = 1.$

Since 1 occurs once as a root in the H.C.F., it will occur twice in the given equation; that is, $(x-1)^2$ will occur there as a factor. Dividing $x^3 - 8x^2 + 13x - 6$ by $(x-1)^2$ gives the only remaining factor $(x - 6)$, yielding the root 6. The roots of our equation are then 1, 1, 6. Drawing the graph of the function, we see that at the double root $x = 1$ the graph touches $OX$ but does not cross it.*

## EXAMPLES

Solve the first ten equations by the method of this section:

1. $x^3 - 7x^2 + 16x - 12 = 0.$ *Ans.* 2, 2, 3.
2. $x^4 - 6x^2 - 8x - 3 = 0.$ $-1, -1, -1, 3.$
3. $x^4 - 7x^3 + 9x^2 + 27x - 54 = 0.$ $3, 3, 3, -2.$
4. $x^4 - 5x^3 - 9x^2 + 81x - 108 = 0.$ $3, 3, 3, -4.$
5. $x^4 + 6x^3 + x^2 - 24x + 16 = 0.$ $1, 1, -4, -4.$
6. $x^4 - 9x^3 + 23x^2 - 3x - 36 = 0.$ $3, 3, -1, 4.$
7. $x^4 - 6x^3 + 10x^2 - 8 = 0.$ $2, 2, 1 \pm \sqrt{3}.$
8. $x^5 - x^4 - 5x^3 + x^2 + 8x + 4 = 0.$ $-1, -1, -1, 2, 2.$
9. $x^5 - 15x^3 + 10x^2 + 60x - 72 = 0.$ $2, 2, 2, -3, -3.$
10. $x^5 - 3x^4 - 5x^3 + 13x^2 + 24x + 10 = 0.$ $-1, -1, -1, 3 \pm \sqrt{-1}.$

Show that the following four equations have no multiple (equal) roots:

11. $x^3 + 9x^2 + 2x - 48 = 0.$
12. $x^4 - 15x^2 - 10x + 24 = 0.$
13. $x^4 - 3x^3 - 6x^2 + 14x + 12 = 0.$
14. $x^n - a^n = 0.$
15. Show that the condition that the equation

$$x^3 + 3qx + r = 0$$

shall have a double root is $4q^3 + r^2 = 0$.

16. Show that the condition that the equation

$$x^3 + 3px^2 + r = 0$$

shall have a double root is $r(4p^3 + r) = 0$.

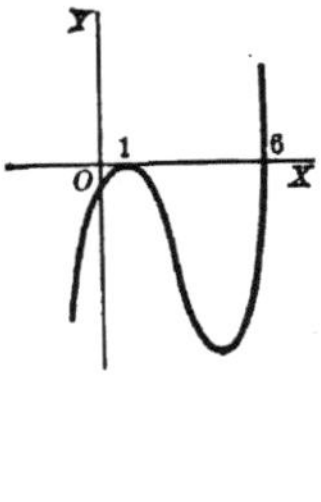


* Since the first derivative vanishes for every multiple root, it follows that the axis of $X$ is tangent to the graph at all points corresponding to multiple roots. If a multiple root occurs an even number of times, the graph will not cross the axis of $X$ at such a point (see figure); if it occurs an odd number of times, the graph will cross.

**70. Applications of the derivative in mechanics. Velocity. Rectilinear motion.** Consider the motion of a point $P$ on the straight line $AB$. Let $s$ be the distance measured from some fixed point as $A$ to any position of $P$, and let $t$ be the corresponding elapsed time. To each value of $t$ corresponds a position of $P$ and therefore a distance (or space) $s$. Hence $s$ will be a function of $t$, and we may write

$$s = f(t).$$

Now let $t$ take on an increment $\Delta t$; then $s$ takes on an increment $\Delta s$,* and

$$(A) \qquad \frac{\Delta s}{\Delta t} = \textit{the average velocity}$$

of $P$ during the time interval $\Delta t$. If $P$ moves with uniform motion, the above ratio will have the same value for every interval of time and is the *velocity* at any instant.

For the general case of any kind of motion, uniform or not, we define the *velocity* (time rate of change of $s$) at any instant as the limit of the ratio $\frac{\Delta s}{\Delta t}$ as $\Delta t$ approaches the limit zero; that is,

$$v = \underset{\Delta t = 0}{\text{limit}} \frac{\Delta s}{\Delta t},$$

or,

$$(9) \qquad v = \frac{ds}{dt}.$$

*The velocity is the derivative of the distance (= space) with respect to the time.*

To show that this agrees with the conception we already have of velocity, let us find the velocity of a falling body at the end of two seconds.

By experiment it has been found that a body falling freely from rest in a vacuum near the earth's surface follows approximately the law

$$(B) \qquad s = 16.1\, t^2,$$

where $s$ = space fallen in feet, $t$ = time in seconds. Apply the *General Rule*, p. 29, to $(B)$.

* $\Delta s$ being the space or distance passed over in the time $\Delta t$.

FIRST STEP. $s+\Delta s=16.1(t+\Delta t)^2=16.1t^2+32.2\,t\cdot\Delta t+16.1(\Delta t)^2.$

SECOND STEP. $\Delta s=32.2\,t\cdot\Delta t+16.1\,(\Delta t)^2.$

THIRD STEP. $\frac{\Delta s}{\Delta t}=32.2\,t+16.1\,\Delta t=$ *average velocity throughout the time interval* $\Delta t$.

Placing $t=2$,

(C) $\frac{\Delta s}{\Delta t}=64.4+16.1\,\Delta t=$ *average velocity throughout the time interval* $\Delta t$ *after two seconds of falling.*

Our notion of velocity tells us at once that (C) does not give us the actual velocity *at the end of two seconds;* for even if we take $\Delta t$ very small, say $\frac{1}{100}$ or $\frac{1}{1000}$ of a second, (C) still gives only the *average velocity* during the corresponding small interval of time. But what we do mean by the velocity at the end of two seconds is *the limit of the average velocity when* $\Delta t$ *diminishes towards zero;* that is, the velocity at the end of two seconds is from (C), 64.4 ft. per second. Thus even the everyday notion of velocity which we get from experience involves the idea of a limit, or in our notation

$$v=\underset{\Delta t=0}{\text{limit}}\left(\frac{\Delta s}{\Delta t}\right)=64.4 \text{ ft. per second.}$$

The above example illustrates well the notion of a limiting value. The student should be impressed with the idea that a *limiting value* is a *definite, fixed* value, not something that is only approximated. Observe that it does not make any difference how small $16.1\,\Delta t$ may be taken; it is only the *limiting value* of

$$64.4+16.1\,\Delta t,$$

when $\Delta t$ diminishes towards zero, that is of importance, and that value is *exactly* 64.4.

**71. Component velocities. Curvilinear motion.** The coördinates $x$ and $y$ of a point $P$ moving on a locus in the $XY$-plane are also functions of the time, and the motion may be defined by means of two equations,

$$x=f(t), \qquad y=\phi(t).*$$

These are the parametric equations of the path (see § 66, p. 79).

* The equation of the path in rectangular coördinates may be found by eliminating $t$ between these equations.

The horizontal component $v_x$ of $v$ * is the velocity along $OX$ of the projection $M$ of $P$, and is therefore the time rate of change of $x$. Hence, from (9), p. 90, when $s$ is replaced by $x$, we get

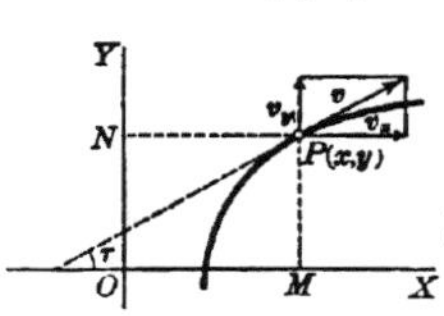


(10) $$v_x = \frac{dx}{dt}.$$

In the same way we get the vertical component, or time rate of change of $y$,

(11) $$v_y = \frac{dy}{dt}.$$

Representing the velocity and its components by vectors, we have at once from the figure

$$v^2 = v_x^2 + v_y^2,$$

or,

(12) $$v = \frac{ds}{dt} = \sqrt{\left(\frac{dx}{dt}\right)^2 + \left(\frac{dy}{dt}\right)^2},$$

giving the magnitude of the velocity at any instant.

If $\tau$ be the angle which the direction of the velocity makes with the axis of $X$, we have from the figure, using (9), (10), (11),

(13) $$\sin\tau = \frac{v_y}{v} = \frac{\frac{dy}{dt}}{\frac{ds}{dt}};\quad \cos\tau = \frac{v_x}{v} = \frac{\frac{dx}{dt}}{\frac{ds}{dt}};\quad \tan\tau = \frac{v_y}{v_x} = \frac{\frac{dy}{dt}}{\frac{dx}{dt}}.$$

**72. Acceleration. Rectilinear motion.** In general, $v$ will be a function of $t$, and we may write

$$v = \psi(t).$$

Now let $t$ take on an increment $\Delta t$, then $v$ takes on an increment $\Delta v$, and

$\frac{\Delta v}{\Delta t}$ = *the average acceleration* of $P$ during the time interval $\Delta t$.

We define *the acceleration* $a$ at any instant as the limit of the ratio $\frac{\Delta v}{\Delta t}$ as $\Delta t$ approaches the limit zero; that is,

$$a = \underset{\Delta t = 0}{\text{limit}} \left(\frac{\Delta v}{\Delta t}\right),$$

or,

(14) $$a = \frac{dv}{dt}.$$

*The acceleration is the derivative of the velocity with respect to the time.*

* The direction of $v$ is along the tangent to the path.

**73. Component accelerations. Curvilinear motion.** In treatises on Mechanics it is shown that in curvilinear motion the acceleration is not, like the velocity, directed along the tangent, but toward the concave side of the path of motion. It may be resolved into a tangential component, $\alpha_t$, and a normal component, $\alpha_n$, where

$$\text{(14 a)} \qquad a_t = \frac{dv}{dt}; \quad a_n = \frac{v^2}{R}.$$

(*R* is the radius of curvature. See § 103.)

The acceleration may also be resolved into components parallel to the axes of the path of motion. Following the same plan used in § 71 for finding component velocities, we define the *component accelerations* parallel to $OX$ and $OY$,

$$\text{(15)} \qquad a_x = \frac{dv_x}{dt}; \quad a_y = \frac{dv_y}{dt}. \quad \text{Also,}$$

$$\text{(16)} \qquad a = \sqrt{\left(\frac{dv_x}{dt}\right)^2 + \left(\frac{dv_y}{dt}\right)^2},$$

which gives the magnitude of the acceleration at any instant.

**EXAMPLES**

1. By experiment it has been found that a body falling freely from rest in a vacuum near the earth's surface follows approximately the law $s = 16.1\,t^2$, where $s$ = space (height) in feet, $t$ = time in seconds. Find the velocity and acceleration (a) at any instant; (b) at end of the first second; (c) at end of the fifth second.

**Solution.** $(A)$ $s = 16.1\,t^2$.

(a) Differentiating, $(B)$ $\frac{ds}{dt} = 32.2\,t$, or, from (9), $v = 32.2\,t$ ft. per sec.

Differentiating again, $(C)$ $\frac{dv}{dt} = 32.2$, or, from (14), $\alpha = 32.2$ ft. per (sec.)$^2$,

which tells us that the acceleration of a falling body is constant; in other words, the velocity increases 32.2 ft. per sec. every second it keeps on falling.

(b) To find $v$ and $\alpha$ at the end of the first second, substitute $t = 1$ in $(B)$ and $(C)$;

$$v = 32.2 \text{ ft. per sec.,} \quad \alpha = 32.2 \text{ ft. per (sec.)}^2.$$

(c) To find $v$ and $\alpha$ at the end of the fifth second, substitute $t = 5$ in $(B)$ and $(C)$;

$$v = 161 \text{ ft. per sec.,} \quad \alpha = 32.2 \text{ ft. per (sec.)}^2.$$

2. Neglecting the resistance of the air, the equations of motion for a projectile are

$$x = v_1 \cos\phi \cdot t, \quad y = v_1 \sin\phi \cdot t - 16.1\,t^2;$$

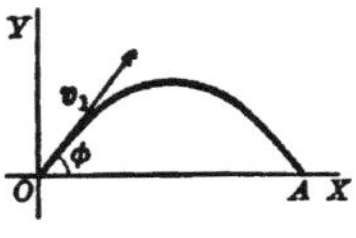


where $v_1$ = initial velocity, $\phi$ = angle of projection with horizon, $t$ = time of flight in seconds, $x$ and $y$ being measured in feet. Find the velocity, acceleration, component velocities, and component accelerations (a) at any instant; (b) at the end of the first second, having given $v_1 = 100$ ft. per sec., $\phi = 30°$; (c) find direction of motion at the end of the first second.

**Solution.** From (10) and (11),

(a) $v_x = v_1 \cos\phi;\ v_y = v_1 \sin\phi - 32.2\,t.$

Also, from (12), $v = \sqrt{v_1^2 - 64.4\,tv_1 \sin\phi + 1036.8\,t^2}.$

From (15) and (16), $\alpha_x = 0;\ \alpha_y = -32.2;\ \alpha = -32.2.$

(b) Substituting $t = 1,\ v_1 = 100,\ \phi = 30°$ in these results, we get

$v_x = 86.6$ ft. per sec. $\qquad \alpha_x = 0.$

$v_y = 17.8$ ft. per sec. $\qquad \alpha_y = -32.2$ ft. per (sec.)$^2$.

$v = 88.4$ ft. per sec. $\qquad \alpha = -32.2$ ft. per (sec.)$^2$.

(c) $\tau = \text{arc tan}\,\frac{v_y}{v_x} = \text{arc tan}\,\frac{17.8}{86.6} = 11°\ 36'.6 =$ angle of direction of motion with the horizontal.

**3.** Given the following equations of rectilinear motion. Find the distance, velocity, and acceleration at the instant indicated:

(a) $s = t^3 + 2t^2;\ t = 2.$ *Ans.* $s = 16,\ v = 20,\ \alpha = 16.$

(b) $s = t^2 + 2t;\ t = 3.$ $s = 15,\ v = 8,\ \alpha = 2.$

(c) $s = 3 - 4t;\ t = 4.$ $s = -13,\ v = -4,\ \alpha = 0.$

(d) $x = 2t - t^2;\ t = 1.$ $x = 1,\ v = 0,\ \alpha = -2.$

(e) $y = 2t - t^3;\ t = 0.$ $y = 0,\ v = 2,\ \alpha = 0.$

(f) $h = 20t + 16t^2;\ t = 10.$ $h = 1800,\ v = 340,\ \alpha = 32.$

(g) $s = 2\sin t;\ t = \frac{\pi}{4}.$ $s = \sqrt{2},\ v = \sqrt{2},\ \alpha = -\sqrt{2}.$

(h) $y = a\cos\frac{\pi t}{3};\ t = 1.$ $y = \frac{a}{2},\ v = -\frac{\pi a\sqrt{3}}{6},\ \alpha = -\frac{\pi^2 a}{18}.$

(i) $s = 2e^{3t};\ t = 0.$ $s = 2,\ v = 6,\ \alpha = 18.$

(j) $s = 2t^3 - 3t;\ t = 2.$

(k) $x = 4 + t^3;\ t = 3.$

(l) $y = 5\cos 2t;\ t = \frac{\pi}{6}.$

(m) $s = b\sin\frac{\pi t}{4};\ t = 2.$

(n) $x = ae^{-2t};\ t = 1.$

(o) $s = \frac{a}{t} + bt^2;\ t = t_0.$

(p) $s = 10\log\frac{4}{4+t};\ t = 1.$

**4.** If a projectile be given an initial velocity of 200 ft. per sec. in a direction inclined 45° with the horizontal, find

(a) the velocity and direction of motion at the end of the third and sixth seconds;

(b) the component velocities at the same instants.

Conditions are the same as for Ex. 2.

*Ans.* (a) When $t = 3$, $v = 148.3$ ft. per sec., $\tau = 17°\ 35'$,
when $t = 6$, $v = 150.5$ ft. per sec., $\tau = 159°\ 53'$;
(b) when $t = 3$, $v_x = 141.4$ ft. per sec., $v_y = 44.8$ ft. per sec.
when $t = 6$, $v_x = 141.4$ ft. per sec., $v_y = -51.8$ ft. per sec.

**5.** The height ($= s$) in feet reached in $t$ seconds by a body projected vertically upwards with a velocity of $v_1$ ft. per sec. is given by the formula

$$s = v_1 t - 16.1\,t^2.$$

Find (a) velocity and acceleration at any instant; and, if $v_1 = 300$ ft. per sec., find velocity and acceleration (b) at end of 2 seconds; (c) at end of 15 seconds. Resistance of air is neglected. *Ans.* (a) $v = v_1 - 32.2t$, $\alpha = -32.2$;
(b) $v = 235.6$ ft. per sec. upwards, $\alpha = 32.2$ ft. per (sec.)$^2$ downwards;
(c) $v = 183$ ft. per sec. downwards, $\alpha = 32.2$ ft. per (sec.)$^2$ downwards.

**6.** A cannon ball is fired vertically upwards with a muzzle velocity of 644 ft. per sec. Find (a) its velocity at the end of 10 seconds; (b) for how long it will continue to rise. Conditions same as for Ex. 5. *Ans.* (a) 322 ft. per sec. upwards;
(b) 20 seconds.

**7.** A train left a station and in $t$ hours was at a distance (space) of

$$s = t^3 + 2t^2 + 3t$$

miles from the starting point. Find its acceleration (a) at the end of $t$ hours; (b) at the end of 2 hours. *Ans.* (a) $\alpha = 6t + 4$;
(b) $\alpha = 16$ miles per (hour)$^2$.

**8.** In $t$ hours a train had reached a point at the distance of $\frac{1}{4}t^4 - 4t^3 + 16t^2$ miles from the starting point. (a) Find its velocity and acceleration. (b) When will the train stop to change the direction of its motion? (c) Describe the motion during the first 10 hours. *Ans.* (a) $v = t^3 - 12t^2 + 32t$, $\alpha = 3t^2 - 24t + 32$;
(b) at end of fourth and eighth hours;
(c) forward first 4 hours, backward the next 4 hours, forward again after 8 hours.

**9.** The space in feet described in $t$ seconds by a point is expressed by the formula

$$s = 48t - 16t^2.$$

Find the velocity and acceleration at the end of $1\frac{1}{2}$ seconds.
*Ans.* $v = 0$, $\alpha = -32$ ft. per (sec.)$^2$.

**10.** Find the acceleration, having given

(a) $v = t^2 + 2t$; $t = 3$. *Ans.* $\alpha = 8$.
(b) $v = 3t - t^3$; $t = 2$. $\alpha = -9$.
(c) $v = 4\sin\frac{t}{2}$; $t = \frac{\pi}{3}$. $\alpha = \sqrt{3}$.
(d) $v = a\cos 3t$; $t = \frac{\pi}{6}$. $\alpha = -3a$.
(e) $v = 5e^{2t}$; $t = 1$. $\alpha = 10e^2$.

**11.** At the end of $t$ seconds a body has a velocity of $3t^2 + 2t$ ft. per sec.; find its acceleration (a) in general; (b) at the end of 4 seconds.
*Ans.* (a) $\alpha = 6t + 2$ ft. per (sec.)$^2$; (b) $\alpha = 26$ ft. per (sec.)$^2$.

**12.** The vertical component of velocity of a point at the end of $t$ seconds is

$$v_y = 3t^2 - 2t + 6 \text{ ft. per sec.}$$

Find the vertical component of acceleration (a) at any instant; (b) at the end of 2 seconds. *Ans.* (a) $\alpha_y = 6t - 2$; (b) 10 ft. per (sec.)$^2$.

**13.** If a point moves in a fixed path so that

$$s = \sqrt{t},$$

show that the acceleration is negative and proportional to the cube of the velocity.

**14.** If the space described is given by

$$s = ae^t + be^{-t},$$

show that the acceleration is always equal in magnitude to the space passed over.

**15.** If a point referred to rectangular coördinates moves so that

$$x = a\cos t + b, \text{ and } y = a\sin t + c,$$

show that its velocity has a constant magnitude.

**16.** If the path of a moving point is the sine curve

$$\begin{cases} x = at, \\ y = b\sin at, \end{cases}$$

show (a) that the $x$-component of the velocity is constant; (b) that the acceleration of the point at any instant is proportional to its distance from the axis of $X$.

**17.** Given the following equations of curvilinear motion, find at the given instant $v_x$, $v_y$, $v$; $\alpha_x$, $\alpha_y$, $\alpha$; position of point (coördinates); direction of motion. Also find the equation of the path in rectangular coördinates.

(a) $x = t^2$, $y = t$; $t = 2$.

(b) $x = t$, $y = t^3$; $t = 1$.

(c) $x = t^2$, $y = t^3$; $t = 3$.

(d) $x = 2t$, $y = t^2 + 3$; $t = 0$.

(e) $x = 1 - t^2$, $y = 2t$; $t = 2$.

(f) $x = a\sin t$, $y = a\cos t$; $t = \frac{3\pi}{4}$.

(g) $x = 2\sin t$, $y = 3\cos t$; $t = \pi$.

(h) $x = \sin t$, $y = \cos 2t$; $t = \frac{\pi}{4}$.

(i) $x = 2t$, $y = 3e^t$; $t = 0$.

(j) $x = 3t$, $y = \log t$; $t = 1$.

(k) $x = t$, $y = 12t^{-1}$; $t = 3$.

# CHAPTER VII

## SUCCESSIVE DIFFERENTIATION

**74. Definition of successive derivatives.** We have seen that the derivative of a function of $x$ is in general also a function of $x$. This new function may also be differentiable, in which case the derivative of the *first derivative* is called the *second derivative* of the original function. Similarly, the derivative of the second derivative is called the *third derivative*; and so on to the $n$th *derivative*. Thus, if

$$y = 3x^4,$$

$$\frac{dy}{dx} = 12x^3,$$

$$\frac{d}{dx}\left(\frac{dy}{dx}\right) = 36x^2,$$

$$\frac{d}{dx}\left[\frac{d}{dx}\left(\frac{dy}{dx}\right)\right] = 72x, \text{ etc.}$$

**75. Notation.** The symbols for the successive derivatives are usually abbreviated as follows:

$$\frac{d}{dx}\left(\frac{dy}{dx}\right) = \frac{d^2y}{dx^2},$$

$$\frac{d}{dx}\left[\frac{d}{dx}\left(\frac{dy}{dx}\right)\right] = \frac{d}{dx}\left(\frac{d^2y}{dx^2}\right) = \frac{d^3y}{dx^3},$$

$$\cdot \quad \cdot \quad \cdot \quad \cdot \quad \cdot \quad \cdot$$

$$\frac{d}{dx}\left(\frac{d^{n-1}y}{dx^{n-1}}\right) = \frac{d^ny}{dx^n}.$$

If $y = f(x)$, the successive derivatives are also denoted by

$$f'(x),\ f''(x),\ f'''(x),\ f^{\text{iv}}(x),\ \cdots,\ f^{(n)}(x);$$

$$y',\ y'',\ y''',\ y^{\text{iv}},\ \cdots,\ y^{(n)};$$

or, $$\frac{d}{dx}f(x),\ \frac{d^2}{dx^2}f(x),\ \frac{d^3}{dx^3}f(x),\ \frac{d^4}{dx^4}f(x),\ \cdots,\ \frac{d^n}{dx^n}f(x).$$

**76. The $n$th derivative.** For certain functions a general expression involving $n$ may be found for the $n$th *derivative*. The usual plan is to find a number of the first successive derivatives, as many as may be necessary to discover their law of formation, and then by induction write down the $n$th derivative.

ILLUSTRATIVE EXAMPLE 1. Given $y = e^{ax}$, find $\frac{d^ny}{dx^n}$.

Solution.

$$\frac{dy}{dx} = ae^{ax},$$

$$\frac{d^2y}{dx^2} = a^2e^{ax},$$

$$\cdot \quad \cdot \quad \cdot \quad \cdot$$

$$\therefore \frac{d^ny}{dx^n} = a^ne^{ax}. \quad \textit{Ans.}$$

ILLUSTRATIVE EXAMPLE 2. Given $y = \log x$, find $\frac{d^ny}{dx^n}$.

Solution.

$$\frac{dy}{dx} = \frac{1}{x},$$

$$\frac{d^2y}{dx^2} = -\frac{1}{x^2},$$

$$\frac{d^3y}{dx^3} = \frac{1 \cdot 2}{x^3},$$

$$\frac{d^4y}{dx^4} = -\frac{1 \cdot 2 \cdot 3}{x^4},$$

$$\cdot \quad \cdot \quad \cdot \quad \cdot \quad \cdot$$

$$\therefore \frac{d^ny}{dx^n} = (-1)^{n-1}\frac{\lfloor n-1}{x^n}. \quad \textit{Ans.}$$

ILLUSTRATIVE EXAMPLE 3. Given $y = \sin x$, find $\frac{d^ny}{dx^n}$.

Solution.

$$\frac{dy}{dx} = \cos x = \sin\left(x + \frac{\pi}{2}\right),$$

$$\frac{d^2y}{dx^2} = \frac{d}{dx}\sin\left(x + \frac{\pi}{2}\right) = \cos\left(x + \frac{\pi}{2}\right) = \sin\left(x + \frac{2\pi}{2}\right),$$

$$\frac{d^3y}{dx^3} = \frac{d}{dx}\sin\left(x + \frac{2\pi}{2}\right) = \cos\left(x + \frac{2\pi}{2}\right) = \sin\left(x + \frac{3\pi}{2}\right),$$

$$\cdot \quad \cdot \quad \cdot \quad \cdot \quad \cdot \quad \cdot \quad \cdot \quad \cdot \quad \cdot \quad \cdot \quad \cdot \quad \cdot$$

$$\therefore \frac{d^ny}{dx^n} = \sin\left(x + \frac{n\pi}{2}\right). \quad \textit{Ans.}$$

**77. Leibnitz's Formula for the $n$th derivative of a product.** This formula expresses the $n$th derivative of the product of two variables in terms of the variables themselves and their successive derivatives.

If $u$ and $v$ are functions of $x$, we have, from **V**,

$$\frac{d}{dx}(uv) = \frac{du}{dx}v + u\frac{dv}{dx}.$$

Differentiating again with respect to $x$,

$$\frac{d^2}{dx^2}(uv) = \frac{d^2u}{dx^2}v + \frac{du}{dx}\frac{dv}{dx} + \frac{du}{dx}\frac{dv}{dx} + u\frac{d^2v}{dx^2} = \frac{d^2u}{dx^2}v + 2\frac{du}{dx}\frac{dv}{dx} + u\frac{d^2v}{dx^2}.$$

Similarly,

$$\frac{d^3}{dx^3}(uv) = \frac{d^3u}{dx^3}v + \frac{d^2u}{dx^2}\frac{dv}{dx} + 2\frac{d^2u}{dx^2}\frac{dv}{dx} + 2\frac{du}{dx}\frac{d^2v}{dx^2} + \frac{du}{dx}\frac{d^2v}{dx^2} + u\frac{d^3v}{dx^3}$$

$$= \frac{d^3u}{dx^3}v + 3\frac{d^2u}{dx^2}\frac{dv}{dx} + 3\frac{du}{dx}\frac{d^2v}{dx^2} + u\frac{d^3v}{dx^3}.$$

However far this process may be continued, it will be seen that the numerical coefficients follow the same law as those of the Binomial Theorem, and the indices of the derivatives correspond to the exponents of the Binomial Theorem.* Reasoning then by mathematical induction from the $m$th to the $(m+1)$th derivative of the product, we can prove *Leibnitz's Formula*

$$(17)\qquad \frac{d^n}{dx^n}(uv) = \frac{d^nu}{dx^n}v + n\frac{d^{n-1}u}{dx^{n-1}}\frac{dv}{dx} + \frac{n(n-1)}{\underline{|2}}\frac{d^{n-2}u}{dx^{n-2}}\frac{d^2v}{dx^2} + \cdots + n\frac{du}{dx}\frac{d^{n-1}v}{dx^{n-1}} + u\frac{d^nv}{dx^n}.$$

ILLUSTRATIVE EXAMPLE 1. Given $y = e^x \log x$, find $\frac{d^3y}{dx^3}$ by Leibnitz's Formula.

**Solution.** Let $u = e^x$, and $v = \log x$;

then

$$\frac{du}{dx} = e^x, \qquad \frac{dv}{dx} = \frac{1}{x},$$

$$\frac{d^2u}{dx^2} = e^x, \qquad \frac{d^2v}{dx^2} = -\frac{1}{x^2},$$

$$\frac{d^3u}{dx^3} = e^x, \qquad \frac{d^3v}{dx^3} = \frac{2}{x^3}.$$

Substituting in (17), we get

$$\frac{d^3y}{dx^3} = e^x \log x + \frac{3e^x}{x} - \frac{3e^x}{x^2} + \frac{2e^x}{x^3} = e^x\left(\log x + \frac{3}{x} - \frac{3}{x^2} + \frac{2}{x^3}\right).$$

* To make this correspondence complete, $u$ and $v$ are considered as $\frac{d^0u}{dx^0}$ and $\frac{d^0v}{dx^0}$.

ILLUSTRATIVE EXAMPLE 2. Given $y = x^2e^{ax}$, find $\frac{d^ny}{dx^n}$ by Leibnitz's Formula.

**Solution.** Let $u = x^2$, and $v = e^{ax}$;

then
$$\frac{du}{dx} = 2x, \quad \frac{dv}{dx} = ae^{ax},$$
$$\frac{d^2u}{dx^2} = 2, \quad \frac{d^2v}{dx^2} = a^2e^{ax},$$
$$\frac{d^3u}{dx^3} = 0, \quad \frac{d^3v}{dx^3} = a^3e^{ax},$$
$$\cdot \quad \cdot \quad \cdot \quad \cdot \quad \cdot \quad \cdot \quad \cdot$$
$$\frac{d^nu}{dx^n} = 0, \quad \frac{d^nv}{dx^n} = a^ne^{ax}.$$

Substituting in (17), we get
$$\frac{d^ny}{dx^n} = x^2a^ne^{ax} + 2na^{n-1}xe^{ax} + n(n-1)a^{n-2}e^{ax} = a^{n-2}e^{ax}[x^2a^2 + 2nax + n(n-1)].$$

**78. Successive differentiation of implicit functions.** To illustrate the process we shall find $\frac{d^2y}{dx^2}$ from the equation of the hyperbola
$$b^2x^2 - a^2y^2 = a^2b^2.$$

Differentiating with respect to $x$, as in § 63, p. 69,
$$2b^2x - 2a^2y\frac{dy}{dx} = 0,$$
or,
$$(A) \qquad \frac{dy}{dx} = \frac{b^2x}{a^2y}.$$

Differentiating again, remembering that $y$ is a function of $x$,
$$\frac{d^2y}{dx^2} = \frac{a^2yb^2 - b^2xa^2\frac{dy}{dx}}{a^4y^2}.$$

Substituting for $\frac{dy}{dx}$ its value from $(A)$,
$$\frac{d^2y}{dx^2} = \frac{a^2b^2y - a^2b^2x\left(\frac{b^2x}{a^2y}\right)}{a^4y^2} = -\frac{b^2(b^2x^2 - a^2y^2)}{a^4y^3}.$$

But from the given equation, $b^2x^2 - a^2y^2 = a^2b^2$.
$$\therefore \frac{d^2y}{dx^2} = -\frac{b^4}{a^2y^3}.$$

## EXAMPLES

Differentiate the following:

1. $y = 4x^3 - 6x^2 + 4x + 7.$ $\qquad \dfrac{d^2y}{dx^2} = 12(2x-1).$

2. $f(x) = \dfrac{x^3}{1-x}.$ $\qquad f^{iv}(x) = \dfrac{\underline{|4}}{(1-x)^5}.$

3. $f(y) = y^6.$ $\qquad f^{vi}(y) = \underline{|6}.$

4. $y = x^3 \log x.$ $\qquad \dfrac{d^4y}{dx^4} = \dfrac{6}{x}.$

5. $y = \dfrac{c}{x^n}.$ $\qquad y'' = \dfrac{n(n+1)c}{x^{n+2}}.$

6. $y = (x-3)e^{2x} + 4xe^x + x.$ $\qquad y'' = 4e^x[(x-2)e^x + x + 2].$

7. $y = \dfrac{a}{2}(e^{\frac{x}{a}} + e^{-\frac{x}{a}}).$ $\qquad y'' = \dfrac{1}{2a}(e^{\frac{x}{a}} + e^{-\frac{x}{a}}) = \dfrac{y}{a^2}.$

8. $f(x) = ax^2 + bx + c.$ $\qquad f'''(x) = 0.$

9. $f(x) = \log(x+1).$ $\qquad f^{iv}(x) = -\dfrac{6}{(x+1)^4}.$

10. $f(x) = \log(e^x + e^{-x}).$ $\qquad f'''(x) = -\dfrac{8(e^x - e^{-x})}{(e^x + e^{-x})^3}.$

11. $r = \sin a\theta.$ $\qquad \dfrac{d^4r}{d\theta^4} = a^4 \sin a\theta = a^4 r.$

12. $r = \tan \phi.$ $\qquad \dfrac{d^3r}{d\phi^3} = 6\sec^4\phi - 4\sec^2\phi.$

13. $r = \log \sin \phi.$ $\qquad r''' = 2\cot\phi \csc^2\phi.$

14. $f(t) = e^{-t}\cos t.$ $\qquad f^{iv}(t) = -4e^{-t}\cos t = -4f(t).$

15. $f(\theta) = \sqrt{\sec 2\theta}.$ $\qquad f''(\theta) = 3[f(\theta)]^5 - f(\theta).$

16. $p = (q^2 + a^2)\,\text{arc}\tan\dfrac{q}{a}.$ $\qquad \dfrac{d^3p}{dq^3} = \dfrac{4a^3}{(a^2+q^2)^2}.$

17. $y = a^x.$ $\qquad \dfrac{d^ny}{dx^n} = (\log a)^n a^x.$

18. $y = \log(1+x).$ $\qquad \dfrac{d^ny}{dx^n} = (-1)^{n-1}\dfrac{\underline{|n-1}}{(1+x)^n}.$

19. $y = \cos ax.$ $\qquad \dfrac{d^ny}{dx^n} = a^n \cos\left(ax + \dfrac{n\pi}{2}\right).$

20. $y = x^{n-1}\log x.$ $\qquad \dfrac{d^ny}{dx^n} = \dfrac{\underline{|n-1}}{x}.$

[$n$ = a positive integer.]

21. $y = \dfrac{1-x}{1+x}.$ $\qquad \dfrac{d^ny}{dx^n} = 2(-1)^n \dfrac{\underline{|n}}{(1+x)^{n+1}}.$

HINT. Reduce fraction to form $-1 + \dfrac{2}{1+x}$ before differentiating.

22. If $y = e^x \sin x$, prove that $\dfrac{d^2y}{dx^2} - 2\dfrac{dy}{dx} + 2y = 0.$

23. If $y = a\cos(\log x) + b\sin(\log x)$, prove that $x^2\dfrac{d^2y}{dx^2} + x\dfrac{dy}{dx} + y = 0.$

Use Leibnitz's Formula in the next four examples:

**24.** $y = x^2a^x$. $\qquad \dfrac{d^ny}{dx^n} = a^x(\log a)^{n-2}[(x\log a + n)^2 - n]$.

**25.** $y = e^xx$. $\qquad \dfrac{d^ny}{dx^n} = e^x(x + n)$.

**26.** $f(x) = e^x \sin x$. $\qquad f^{(n)}(x) = (\sqrt{2})^n e^x \sin\left(x + \dfrac{n\pi}{4}\right)$.

**27.** $f(\theta) = \cos a\theta \cos b\theta$.

$$f^{(n)}(\theta) = \frac{(a+b)^n}{2}\cos\left[(a+b)\theta + \frac{n\pi}{2}\right] + \frac{(a-b)^n}{2}\cos\left[(a-b)\theta + \frac{n\pi}{2}\right]$$

**28.** Show that the formulas for acceleration, (14), (15), p. 92, may be written

$$\alpha = \frac{d^2s}{dt^2},\quad \alpha_x = \frac{d^2x}{dt^2},\quad \alpha_y = \frac{d^2y}{dt^2}.$$

**29.** $y^2 = 4ax$. $\qquad \dfrac{d^2y}{dx^2} = -\dfrac{4a^2}{y^3}$.

**30.** $b^2x^2 + a^2y^2 = a^2b^2$. $\qquad \dfrac{d^2y}{dx^2} = -\dfrac{b^4}{a^2y^3}$; $\dfrac{d^3y}{dx^3} = -\dfrac{3b^6x}{a^4y^5}$.

**31.** $x^2 + y^2 = r^2$. $\qquad \dfrac{d^2y}{dx^2} = -\dfrac{r^2}{y^3}$.

**32.** $y^2 + y = x^2$. $\qquad \dfrac{d^3y}{dx^3} = -\dfrac{24x}{(1+2y)^5}$.

**33.** $ax^2 + 2hxy + by^2 = 1$. $\qquad \dfrac{d^2y}{dx^2} = \dfrac{h^2 - ab}{(hx + by)^3}$.

**34.** $y^2 - 2xy = a^2$. $\qquad \dfrac{d^2y}{dx^2} = \dfrac{a^2}{(y-x)^3}$; $\dfrac{d^3y}{dx^3} = -\dfrac{3a^2x}{(y-x)^5}$.

**35.** $\sec\phi \cos\theta = c$. $\qquad \dfrac{d^2\theta}{d\phi^2} = \dfrac{\tan^2\theta - \tan^2\phi}{\tan^3\theta}$.

**36.** $\theta = \tan(\phi + \theta)$. $\qquad \dfrac{d^3\theta}{d\phi^3} = -\dfrac{2(5 + 8\theta^2 + 3\theta^4)}{\theta^8}$.

**37.** Find the second derivative in the following:

(a) $\log(u + v) = u - v$.
(b) $e^u + u = e^v + v$.
(c) $s = 1 + te^s$.
(d) $e^s + st - e = 0$.
(e) $y^3 + x^3 - 3axy = 0$.
(f) $y^2 - 2mxy + x^2 - a = 0$.
(g) $y = \sin(x + y)$.
(h) $e^{x+y} = xy$.

# CHAPTER VIII

## MAXIMA AND MINIMA. POINTS OF INFLECTION. CURVE TRACING

**79. Introduction.** A great many practical problems occur where we have to deal with functions of such a nature that they have a greatest (maximum) value or a least (minimum) value,* and it is very important to know what particular value of the variable gives such a value of the function. For instance, suppose that it is required to find the dimensions of the rectangle of greatest area that can be inscribed in a circle of radius 5 inches. Consider the circle in the following figure:

Inscribe any rectangle, as $BD$.

Let $CD = x$; then $DE = \sqrt{100 - x^2}$, and the area of the rectangle is evidently

$$A = x\sqrt{100 - x^2}. \tag{1}$$

That a rectangle of maximum area must exist may be seen as follows: Let the base $CD$ $(= x)$ increase to 10 inches (the diameter); then the altitude $DE = \sqrt{100 - x^2}$ will decrease to zero and the area will become zero. Now let the base decrease to zero; then the altitude will increase to 10 inches and the area will again become zero. It is therefore intuitionally evident that there exists a greatest rectangle. By a careful study of the figure we might suspect that when the rectangle becomes a square its area would be the greatest, but this would at best be mere guesswork. A better way would evidently be to plot the graph of the function (1) and note its behavior. To aid us in drawing the graph of (1), we observe that

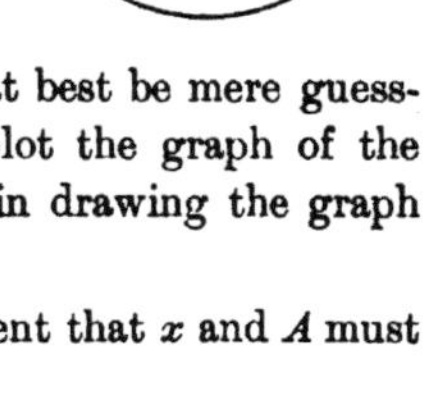


(a) from the nature of the problem it is evident that $x$ and $A$ must both be positive; and

(b) the values of $x$ range from zero to 10 inclusive.

* There may be more than one of each, as illustrated on p. 109.

Now construct a table of values and draw the graph.

*What do we learn from the graph?*

| $x$ | $A$ |
|---|---|
| 0 | 0 |
| 1 | 9.9 |
| 2 | 19.6 |
| 3 | 28.6 |
| 4 | 36.6 |
| 5 | 43.0 |
| 6 | 48.0 |
| 7 | 49.7 |
| 8 | 48.0 |
| 9 | 39.6 |
| 10 | 0.0 |

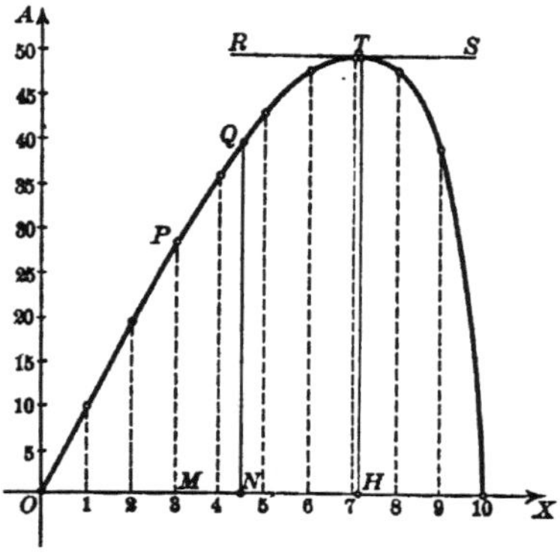


(a) If carefully drawn, we may find quite accurately the area of the rectangle corresponding to any value of $x$ by measuring the length of the corresponding ordinate. Thus,

when $\qquad x = OM = 3$ inches,

then $\qquad A = MP = 28.6$ square inches;

and when $\qquad x = ON = 4\frac{1}{2}$ inches,

then $\qquad A = NQ =$ about 39.8 sq. in. (found by measurement).

(b) There is one horizontal tangent ($RS$). The ordinate $TH$ from its point of contact $T$ is greater than any other ordinate. Hence this discovery: *One of the inscribed rectangles has evidently a greater area than any of the others.* In other words, we may infer from this that the function defined by (1) has a *maximum value.* We cannot find this value ($= HT$) exactly by measurement, but it is very easy to find, using Calculus methods. We observed that at $T$ the tangent was horizontal; hence the slope will be zero at that point (Illustrative Example 1, p. 74). To find the abscissa of $T$ we then find the first derivative of (1), place it equal to zero, and solve for $x$. Thus

$$(1) \qquad A = x\sqrt{100 - x^2},$$

$$\frac{dA}{dx} = \frac{100 - 2x^2}{\sqrt{100 - x^2}}.$$

$$\frac{100 - 2x^2}{\sqrt{100 - x^2}} = 0.$$

Solving, $x = 5\sqrt{2}.$

Substituting back, we get $DE = \sqrt{100 - x^2} = 5\sqrt{2}.$

Hence the rectangle of maximum area inscribed in the circle is a square of area

$A = CD \times DE = 5\sqrt{2} \times 5\sqrt{2} = 50$ square inches. The length of $HT$ is therefore 50.

Take another example. A wooden box is to be built to contain 108 cu. ft. It is to have an open top and a square base. What must be its dimensions in order that the amount of material required shall be a minimum; that is, what dimensions will make the cost the least?

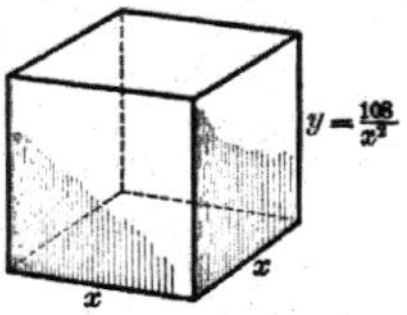


Let $x$ = length of side of square base in feet,
and $y$ = height of box.

Since the volume of the box is given, however, $y$ may be found in terms of $x$. Thus

$$\text{volume} = x^2y = 108; \therefore y = \frac{108}{x^2}.$$

We may now express the number ($= M$) of square feet of lumber required as a function of $x$ as follows:

$$\text{area of base} = x^2 \text{ sq. ft.},$$

$$\text{and area of four sides} = 4\,xy = \frac{432}{x} \text{ sq. ft. Hence}$$

$$(2) \qquad M = x^2 + \frac{432}{x}$$

| $X$ | $M$ |
|---|---|
| 1 | 433 |
| 2 | 220 |
| 3 | 153 |
| 4 | 124 |
| 5 | 111 |
| 6 | 108 |
| 7 | 111 |
| 8 | 118 |
| 9 | 129 |
| 10 | 143 |

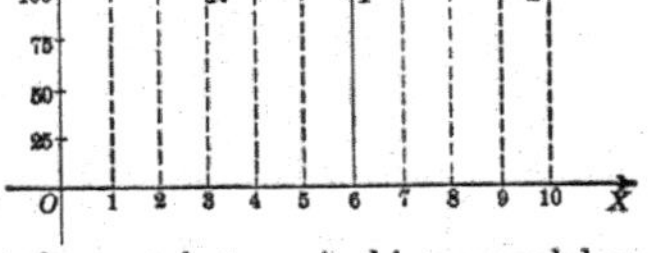


is a formula giving the number of square feet required in any such box having a capacity of 108 cu. ft. Draw a graph of (2).

*What do we learn from the graph?*

(a) If carefully drawn, we may measure the ordinate corresponding to any length $(=x)$ of the side of the square base and so determine the number of square feet of lumber required.

(b) There is one horizontal tangent $(RS)$. The ordinate from its point of contact $T$ is less than any other ordinate. Hence this discovery: *One of the boxes evidently takes less lumber than any of the others.* In other words, we may infer that the function defined by (2) has a *minimum value.* Let us find this point on the graph exactly, using our Calculus. Differentiating (2) to get the slope at any point, we have

$$\frac{dM}{dx}=2x-\frac{432}{x^2}.$$

At the lowest point $T$ the slope will be zero. Hence

$$2x-\frac{432}{x^2}=0;$$

that is, when $x=6$ the least amount of lumber will be needed.

Substituting in (2), we see that this is

$$M=108 \text{ sq. ft.}$$

The fact that a least value of $M$ exists is also shown by the following reasoning. Let the base increase from a very small square to a very large one. In the former case the height must be very great and therefore the amount of lumber required will be large. In the latter case, while the height is small, the base will take a great deal of lumber. Hence $M$ varies from a large value, grows less, then increases again to another large value. It follows, then, that the graph must have a "lowest" point corresponding to the dimensions which require the least amount of lumber, and therefore would involve the least cost.

We will now proceed to the treatment in detail of the subject of maxima and minima.

**80. Increasing and decreasing functions.*** A function is said to be *increasing* when it increases as the variable increases and decreases as the variable decreases. A function is said to be *decreasing* when it decreases as the variable increases and increases as the variable decreases.

* The proofs given here depend chiefly on geometric intuition. The subject of Maxima and Minima will be treated analytically in § 108, p. 167.

The graph of a function indicates plainly whether it is increasing or decreasing. For instance, consider the function $a^x$ whose graph (Fig. *a*) is the locus of the equation

$$y = a^x. \qquad a > 1$$

As we move along the curve from left to right the curve is ***rising***; that is, as $x$ increases the function ($= y$) always increases. Therefore $a^x$ is an increasing function for all values of $x$.

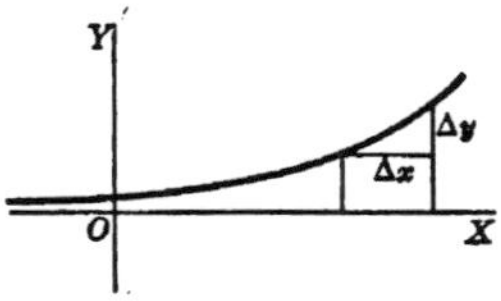


FIG. *a*

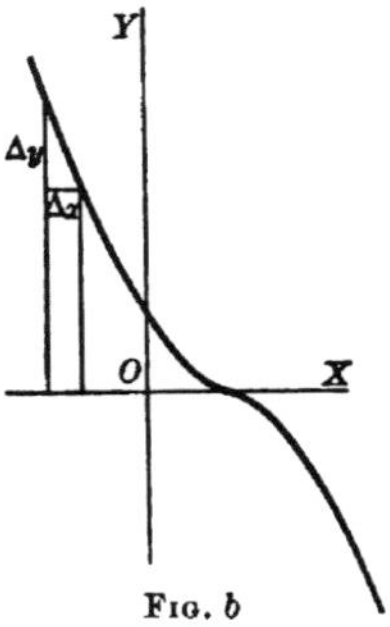


FIG. *b*

On the other hand, consider the function $(a-x)^3$ whose graph (Fig. *b*) is the locus of the equation

$$y = (a-x)^3.$$

Now as we move along the curve from left to right the curve is *falling*; that is, as $x$ increases, the function ($= y$) always decreases. Hence $(a-x)^3$ is a decreasing function for all values of $x$.

That a function may be sometimes increasing and sometimes decreasing is shown by the graph (Fig. *c*) of

$$y = 2x^3 - 9x^2 + 12x - 3.$$

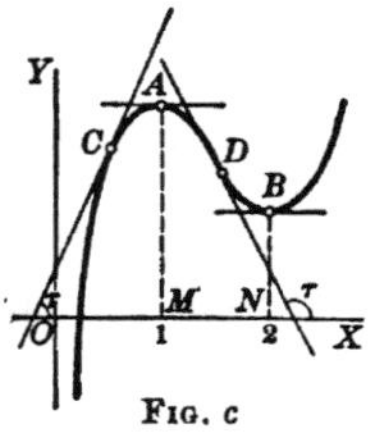


FIG. *c*

As we move along the curve from left to right the curve rises until we reach the point $A$, then it falls from $A$ to $B$, and to the right of $B$ it is always rising. Hence

(a) *from* $x = -\infty$ *to* $x = 1$ *the function is increasing;*
(b) *from* $x = 1$ *to* $x = 2$ *the function is decreasing;*
(c) *from* $x = 2$ *to* $x = +\infty$ *the function is increasing.*

The student should study the curve carefully in order to note the behavior of the function when $x=1$ and $x=2$. Evidently $A$ and $B$ are turning points. At $A$ the function ceases to increase and commences to decrease; at $B$, the reverse is true. At $A$ and $B$ the tangent (or curve) is evidently parallel to the axis of $X$, and therefore the slope is zero.

**81. Tests for determining when a function is increasing and when decreasing.** It is evident from Fig. $c$ that at a point, as $C$, where a function

$$y=f(x)$$

is *increasing*, the tangent in general makes an acute angle with the axis of $X$; hence

$$slope = tan\,\tau = \frac{dy}{dx} = f'(x) = a\ positive\ number.$$

Similarly, at a point, as $D$, where a function is *decreasing*, the tangent in general makes an obtuse angle with the axis of $X$; therefore

$$slope = tan\,\tau = \frac{dy}{dx} = f'(x) = a\ negative\ number.^*$$

In order, then, that the function shall change from an increasing to a decreasing function, or vice versa, it is a necessary and sufficient condition that **the first derivative shall change sign.** But this can only happen for a continuous derivative by passing through the value zero. Thus in Fig. $c$, p. 107, as we pass along the curve the derivative (= slope) changes sign at $A$ and $B$ where it has the value zero. In general, then, we have at **turning points**

$$\frac{dy}{dx} = f'(x) = 0. \tag{18}$$

The derivative is continuous in nearly all our important applications, but it is interesting to note the case when the derivative (= slope) changes sign by passing through $\infty$.† This would evidently

* Conversely, for any given value of $x$,

$$if\ f'(x) = +,\ then\ f(x)\ is\ increasing;$$
$$if\ f'(x) = -,\ then\ f(x)\ is\ decreasing.$$

When $f'(x) = 0$, we cannot decide without further investigation whether $f(x)$ is increasing or decreasing.

† By this is meant that its reciprocal passes through the value zero.

happen at the points $B$, $E$, $G$ in the following figure, where the tangents (and curve) are perpendicular to the axis of $X$. At such exceptional turning points

$$\frac{dy}{dx} = f'(x) = \infty;$$

or, what amounts to the same thing,

$$\frac{1}{f'(x)} = 0.$$

**82. Maximum and minimum values of a function.** A *maximum* value of a function is one that is *greater* than any values immediately preceding or following.

A *minimum* value of a function is one that is *less* than any values immediately preceding or following.

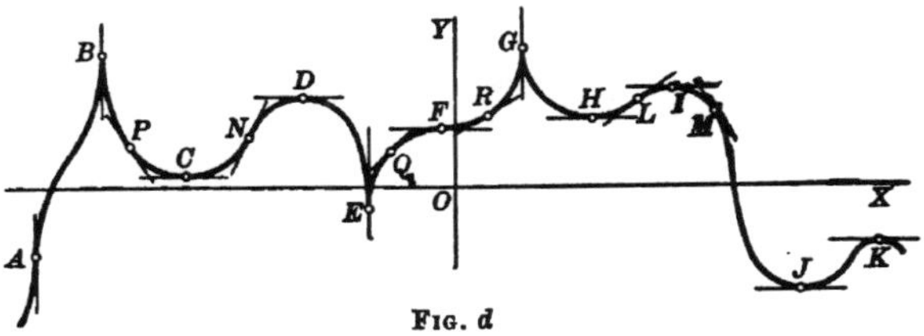


FIG. *d*

For example, in Fig. *c*, p. 107, it is clear that the function has a maximum value $MA$ $(= y = 2)$ when $x = 1$, and a minimum value $NB$ $(= y = 1)$ when $x = 2$.

The student should observe that a maximum value is not necessarily the greatest possible value of a function nor a minimum value the least. For in Fig. *c* it is seen that the function $(= y)$ has values to the right of $B$ that are greater than the maximum $MA$, and values to the left of $A$ that are less than the minimum $NB$.

A function may have several maximum and minimum values. Suppose that the above figure represents the graph of a function $f(x)$.

At $B$, $D$, $G$, $I$, $K$ the function is a maximum, and at $C$, $E$, $H$, $J$ a minimum. That some particular minimum value of a function may be greater than some particular maximum value is shown in the figure, the minimum values at $C$ and $H$ being greater than the maximum value at $K$.

At the ordinary turning points $C$, $D$, $H$, $I$, $J$, $K$ the tangent (or curve) is parallel to $OX$; therefore

$$\textit{slope} = \frac{dy}{dx} = f'(x) = 0.$$

At the exceptional turning points $B$, $E$, $G$ the tangent (or curve) is perpendicular to $OX$, giving

$$\textit{slope} = \frac{dy}{dx} = f'(x) = \infty.$$

One of these two conditions is then necessary in order that the function shall have a maximum or a minimum value. But such a condition is not sufficient; for at $F$ the slope is zero and at $A$ it is infinite, and yet the function has neither a maximum nor a minimum value at either point. It is necessary for us to know, in addition, how the function behaves in the neighborhood of each point. Thus at the points of *maximum value*, $B$, $D$, $G$, $I$, $K$, the function *changes from an increasing to a decreasing function*, and at the points of *minimum value*, $C$, $E$, $H$, $J$, the function *changes from a decreasing to an increasing function*. It therefore follows from § 81 that at *maximum points*

$$\textit{slope} = \frac{dy}{dx} = f'(x) \textit{ must change from } + \textit{ to } -,$$

and at *minimum points*

$$\textit{slope} = \frac{dy}{dx} = f'(x) \textit{ must change from } - \textit{ to } +$$

when we move along the curve from left to right.

At such points as $A$ and $F$ where the slope is zero or infinite, but which are *neither maximum nor minimum points*,

$$\textit{slope} = \frac{dy}{dx} = f'(x) \textit{ does not change sign.}$$

We may then state the conditions in general for maximum and minimum values of $f(x)$ for certain values of the variable as follows:

**(19) $f(x)$ is a maximum if $f'(x) = 0$, and $f'(x)$ changes from $+$ to $-$.**

**(20) $f(x)$ is a minimum if $f'(x) = 0$, and $f'(x)$ changes from $-$ to $+$.**

The values of the variable at the turning points of a function are called *critical values;* thus $x = 1$ and $x = 2$ are the critical values of the variable for the function whose graph is shown in Fig. *a*, p. 107. The critical values at turning points where the tangent is parallel to $OX$ are evidently found by placing the first derivative equal to zero and solving for real values of $x$, just as under § 64, p. 73.*

To determine the sign of the first derivative at points near a particular turning point, substitute in it, first, a value of the variable just a *little less* than the corresponding critical value, and then one a *little greater*.† If the first gives + (as at $L$, Fig. *d*, p. 109) and the second − (as at $M$), then the function $(= y)$ has a maximum value in that interval (as at $I$).

If the first gives − (as at $P$) and the second + (as at $N$), then the function $(= y)$ has a minimum value in that interval (as at $C$).

If the sign is the same in both cases (as at $Q$ and $R$), then the function $(= y)$ has neither a maximum nor a minimum value in that interval (as at $F$).‡

We shall now summarize our results into a compact *working rule*.

**83. First method for examining a function for maximum and minimum values. Working rule.**

First Step. *Find the first derivative of the function.*

Second Step. *Set the first derivative equal to zero*§ *and solve the resulting equation for real roots in order to find the critical values of the variable.*

Third Step. *Write the derivative in factor form; if it is algebraic, write it in linear form.*

Fourth Step. *Considering one critical value at a time, test the first derivative, first for a value a trifle less and then for a value a trifle greater than the critical value. If the sign of the derivative is first + and then −, the function has a maximum value for that particular critical value of the variable; but if the reverse is true, then it has a minimum value. If the sign does not change, the function has neither.*

* Similarly, if we wish to examine a function at exceptional turning points where the tangent is perpendicular to $OX$, we set the reciprocal of the first derivative equal to zero and solve to find critical values.

† In this connection the term "little less," or "trifle less," means any value between the next smaller root (critical value) and the one under consideration; and the term "little greater," or "trifle greater," means any value between the root under consideration and the next larger one.

‡ A similar discussion will evidently hold for the exceptional turning points $B$, $E$, and $A$ respectively.

§ When the first derivative becomes infinite for a certain value of the independent variable, then the function should be examined for such a critical value of the variable, for it may give maximum or minimum values, as at $B$, $E$, or $A$ (Fig. *d*, p. 109). See footnote on p. 108.

In the problem worked out on p. 104 we showed by means of the graph of the function

$$A = x\sqrt{100 - x^2}$$

that the rectangle of maximum area inscribed in a circle of radius 5 inches contained 50 square inches. This may now be proved analytically as follows by applying the above rule.

**Solution.** $$f(x) = x\sqrt{100 - x^2}.$$

*First step.* $$f'(x) = \frac{100 - 2x^2}{\sqrt{100 - x^2}}.$$

*Second step.* $$\frac{100 - 2x^2}{\sqrt{100 - x^2}} = 0,$$

$$x = 5\sqrt{2},$$

which is the critical value. Only the positive sign of the radical is taken, since, from the nature of the problem, the negative sign has no meaning.

*Third step.* $$f'(x) = \frac{2(5\sqrt{2} - x)(5\sqrt{2} + x)}{\sqrt{(10 - x)(10 + x)}}.$$

*Fourth step.* When $x < 5\sqrt{2}$, $$f'(x) = \frac{2(+)(+)}{\sqrt{(+)(+)}} = +.$$

When $x > 5\sqrt{2}$, $$f'(x) = \frac{2(-)(+)}{\sqrt{(+)(+)}} = -.$$

Since the sign of the first derivative changes from + to − at $x = 5\sqrt{2}$, the function has a maximum value

$$f(5\sqrt{2}) = 5\sqrt{2} \cdot 5\sqrt{2} = 50. \quad \textit{Ans.}$$

**84. Second method for examining a function for maximum and minimum values.** From (**19**), p. 110, it is clear that in the vicinity of a maximum value of $f(x)$, in passing along the graph from left to right,

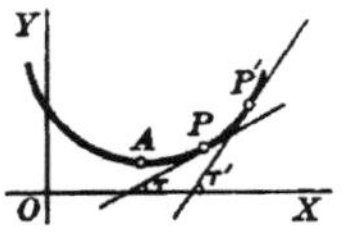


$$f'(x) \textit{ changes from } + \textit{ to } 0 \textit{ to } -.$$

Hence $f'(x)$ is a decreasing function, and by § 81 we know that its derivative, i.e. the second derivative $[=f''(x)]$ of the function itself, is negative or zero.

Similarly, we have, from (**20**), p. 110, that in the vicinity of a minimum value of $f(x)$

$$f'(x) \textit{ changes from } - \textit{ to } 0 \textit{ to } +.$$

Hence $f'(x)$ is an increasing function and by § 81 it follows that $f''(x)$ is positive or zero.

The student should observe that $f''(x)$ is positive not only at minimum points (as at $A$) but also at points such as $P$. For, as a point passes through $P$ in moving from left to right,

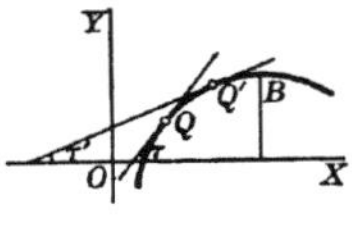


$$\textit{slope} = \tan\tau = \frac{dy}{dx} = f'(x) \text{ is an increasing function.}$$

At such a point the curve is said to be *concave upwards.*

Similarly, $f''(x)$ is negative not only at maximum points (as at $B$) but also at points such as $Q$. For, as a point passes through $Q$,

$$\textit{slope} = \tan\tau = \frac{dy}{dx} = f'(x) \text{ is a decreasing function.}$$

At such a point the curve is said to be *concave downwards.**

We may then state the sufficient conditions for maximum and minimum values of $f(x)$ for certain values of the variable as follows:

**(21) $f(x)$ is a maximum if $f'(x) = 0$ and $f''(x) =$ a negative number.**

**(22) $f(x)$ is a minimum if $f'(x) = 0$ and $f''(x) =$ a positive number.**

Following is the corresponding **working rule.**

FIRST STEP. *Find the first derivative of the function.*

SECOND STEP. *Set the first derivative equal to zero and solve the resulting equation for real roots in order to find the critical values of the variable.*

THIRD STEP. *Find the second derivative.*

FOURTH STEP. *Substitute each critical value for the variable in the second derivative. If the result is negative, then the function is a maximum for that critical value; if the result is positive, the function is a minimum.*

When $f''(x) = 0$, or does not exist, the above process fails, although there may even then be a maximum or a minimum; in that case the first method given in the last section still holds, being fundamental. Usually this second method does apply, and when the process of finding the second derivative is not too long or tedious, it is generally the shortest method.

Let us now apply the above rule to test analytically the function

$$M = x^2 + \frac{432}{x}$$

found in the example worked out on p. 105.

* At a point where the curve is *concave upwards* we sometimes say that the curve has a *positive bending*, and where it is *concave downwards* a *negative bending*.

**Solution.** $$f(x) = x^2 + \frac{432}{x}.$$

*First step.* $$f'(x) = 2x - \frac{432}{x^2}.$$

*Second step.* $$2x - \frac{432}{x^2} = 0,$$
$$x = 6, \text{ critical value.}$$

*Third step.* $$f''(x) = 2 + \frac{864}{x^3}.$$

*Fourth step.* $$f''(6) = +. \text{ Hence}$$
$$f(6) = 108, \text{ minimum value.}$$

The work of finding maximum and minimum values may frequently be simplified by the aid of the following principles, which follow at once from our discussion of the subject.

(a) *The maximum and minimum values of a continuous function must occur alternately.*

(b) *When $c$ is a positive constant, $c \cdot f(x)$ is a maximum or a minimum for such values of $x$, and such only, as make $f(x)$ a maximum or a minimum.*

Hence, in determining the critical values of $x$ and testing for maxima and minima, any constant factor may be omitted.

*When $c$ is negative, $c \cdot f(x)$ is a maximum when $f(x)$ is a minimum, and conversely.*

(c) *If $c$ is a constant,* $$f(x) \text{ and } c + f(x)$$
*have maximum and minimum values for the same values of $x$.*

Hence a constant term may be omitted when finding critical values of $x$ and testing.

In general we must first construct, from the conditions given in the problem, the function whose maximum and minimum values are required, as was done in the two examples worked out on pp. 103–106. This is sometimes a problem of considerable difficulty. No rule applicable in all cases can be given for constructing the function, but in a large number of problems we may be guided by the following

**General directions.**

(a) *Express the function whose maximum or minimum is involved in the problem.*

(b) *If the resulting expression contains more than one variable, the conditions of the problem will furnish enough relations between the variables so that all may be expressed in terms of a single one.*

(c) *To the resulting function of a single variable apply one of our two rules for finding maximum and minimum values.*

(d) *In practical problems it is usually easy to tell which critical value will give a maximum and which a minimum value, so it is not always necessary to apply the fourth step of our rules.*

(e) *Draw the graph of the function (p. 104) in order to check the work.*

## PROBLEMS

**1.** It is desired to make an open-top box of greatest possible volume from a square piece of tin whose side is $a$, by cutting equal squares out of the corners and then folding up the tin to form the sides. What should be the length of a side of the squares cut out?

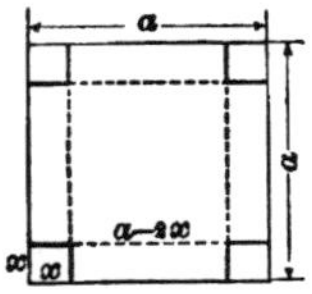


**Solution.** Let $x$ = side of small square = depth of box;

then $a - 2x$ = side of square forming bottom of box,

and volume is $V = (a - 2x)^2 x$;

which is the function to be made a maximum by varying $x$. Applying rule,

*First step.* $\frac{dV}{dx} = (a - 2x)^2 - 4x(a - 2x) = a^2 - 8ax + 12x^2.$

*Second step.* Solving $a^2 - 8ax + 12x^2 = 0$ gives critical values $x = \frac{a}{2}$ and $\frac{a}{6}$.

It is evident from the figure that $x = \frac{a}{2}$ must give a minimum, for then all the tin would be cut away, leaving no material out of which to make a box. By the usual test, $x = \frac{a}{6}$ is found to give a maximum volume $\frac{2a^3}{27}$. Hence the side of the square to be cut out is one sixth of the side of the given square.

The drawing of the graph of the function in this and the following problems is left to the student.

**2.** Assuming that the strength of a beam with rectangular cross section varies directly as the breadth and as the square of the depth, what are the dimensions of the strongest beam that can be sawed out of a round log whose diameter is $d$?

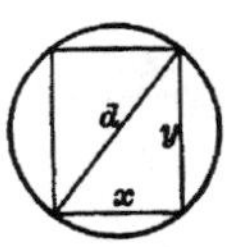


**Solution.** If $x$ = breadth and $y$ = depth, then the beam will have maximum strength when the function $xy^2$ is a maximum. From the figure, $y^2 = d^2 - x^2$; hence we should test the function

$$f(x) = x(d^2 - x^2).$$

*First step.* $f'(x) = -2x^2 + d^2 - x^2 = d^2 - 8x^2.$

*Second step.* $d^2 - 8x^2 = 0. \therefore x = \frac{d}{\sqrt{8}}$ = critical value which gives a maximum.

Therefore, if the beam is cut so that

depth $= \sqrt{\tfrac{2}{3}}$ of diameter of log,

and breadth $= \sqrt{\tfrac{1}{3}}$ of diameter of log,

the beam will have maximum strength.

**3.** What is the width of the rectangle of maximum area that can be inscribed in a given segment $OAA'$ of a parabola?

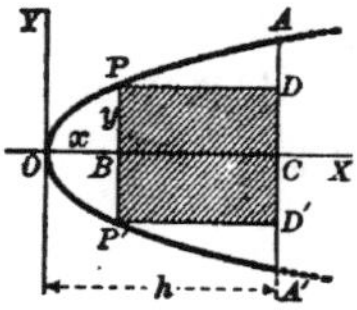


HINT. If $OC = h$, $BC = h - x$ and $PP' = 2y$; therefore the area of rectangle $PDD'P'$ is

$$2(h - x)y.$$

But since $P$ lies on the parabola $y^2 = 2px$, the function to be tested is

$$2(h - x)\sqrt{2px}.$$

*Ans.* Width $= \frac{2}{3}h$.

**4.** Find the altitude of the cone of maximum volume that can be inscribed in a sphere of radius $r$.

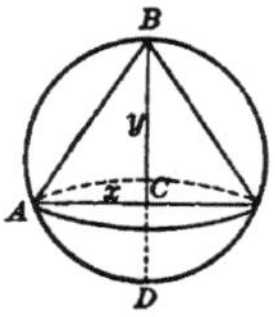


HINT. Volume of cone $= \frac{1}{3}\pi x^2 y$. But $x^2 = BC \times CD = y(2r - y)$; therefore the function to be tested is

$$f(y) = \frac{\pi}{3}y^2(2r - y).$$

*Ans.* Altitude of cone $= \frac{4}{3}r$.

**5.** Find the altitude of the cylinder of maximum volume that can be inscribed in a given right cone.

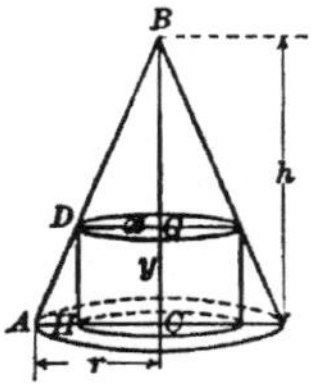


HINT. Let $AC = r$ and $BC = h$. Volume of cylinder $= \pi x^2 y$. But from similar triangles $ABC$ and $DBG$

$$r : x :: h : h - y. \quad \therefore x = \frac{r(h - y)}{h}.$$

Hence the function to be tested is

$$f(y) = \frac{r^2}{h^2}y(h - y)^2.$$

*Ans.* Altitude $= \frac{1}{3}h$.

**6.** Divide $a$ into two parts such that their product is a maximum.

*Ans.* Each part $= \frac{a}{2}$.

**7.** Divide 10 into two such parts that the sum of the double of one and square of the other may be a minimum. *Ans.* 9 and 1.

**8.** Find the number that exceeds its square by the greatest possible quantity.

*Ans.* $\frac{1}{2}$.

**9.** What number added to its reciprocal gives the least possible sum? *Ans.* 1.

**10.** Assuming that the stiffness of a beam of rectangular cross section varies directly as the breadth and the cube of the depth, what must be the breadth of the stiffest beam that can be cut from a log 16 inches in diameter? *Ans.* Breadth = 8 inches.

**11.** A water tank is to be constructed with a square base and open top, and is to hold 64 cubic yards. If the cost of the sides is \$1 a square yard, and of the bottom \$2 a square yard, what are the dimensions when the cost is a minimum? What is the minimum cost? *Ans.* Side of base = 4 yd., height = 4 yd., cost \$96.

**12.** A rectangular tract of land is to be bought for the purpose of laying out a quarter-mile track with straightaway sides and semicircular ends. In addition a strip 35 yards wide along each straightaway is to be bought for grand stands, training quarters, etc. If the land costs \$200 an acre, what will be the maximum cost of the land required? *Ans.* \$856.

**13.** A torpedo boat is anchored 9 miles from the nearest point of a beach, and it is desired to send a messenger in the shortest possible time to a military camp situated 15 miles from that point along the shore. If he can walk 5 miles an hour but row only 4 miles an hour, required the place he must land. *Ans.* 3 miles from the camp.

**14.** A gas holder is a cylindrical vessel closed at the top and open at the bottom, where it sinks into the water. What should be its proportions for a given volume to require the least material (this would also give least weight)?

*Ans.* Diameter = double the height.

**15.** What should be the dimensions and weight of a gas holder of 8,000,000 cubic feet capacity, built in the most economical manner out of sheet iron $\frac{1}{16}$ of an inch thick and weighing $2\frac{1}{2}$ lb. per sq. ft.?

*Ans.* Height = 137 ft., diameter = 273 ft., weight = 220 tons.

**16.** A sheet of paper is to contain 18 sq. in. of printed matter. The margins at the top and bottom are to be 2 inches each and at the sides 1 inch each. Determine the dimensions of the sheet which will require the least amount of paper. *Ans.* 5 in. by 10 in.

**17.** A paper-box manufacturer has in stock a quantity of strawboard 30 inches by 14 inches. Out of this material he wishes to make open-top boxes by cutting equal squares out of each corner and then folding up to form the sides. Find the side of the square that should be cut out in order to give the boxes maximum volume. *Ans.* 3 inches.

**18.** A roofer wishes to make an open gutter of maximum capacity whose bottom and sides are each 4 inches wide and whose sides have the same slope. What should be the width across the top? *Ans.* 8 inches.

**19.** Assuming that the energy expended in driving a steamboat through the water varies as the cube of her velocity, find her most economical rate per hour when steaming against a current running $c$ miles per hour.

HINT. Let $v$ = most economical speed;

then $av^3$ = energy expended each hour, $a$ being a constant depending upon the particular conditions,

and $v - c$ = actual distance advanced per hour.

Hence $\frac{av^3}{v-c}$ is the energy expended per mile of distance advanced, and it is therefore the function whose minimum is wanted.

*Ans.* $v = \frac{3}{2}c$.

**20.** Prove that a conical tent of a given capacity will require the least amount of canvas when the height is $\sqrt{2}$ times the radius of the base. Show that when the canvas is laid out flat it will be a circle with a sector of 152° 9′ cut out. A bell tent 10 ft. high should then have a base of diameter 14 ft. and would require 272 sq. ft. of canvas.

**21.** A cylindrical steam boiler is to be constructed having a capacity of 1000 cu. ft. The material for the side costs \$2 a square foot, and for the ends \$3 a square foot. Find radius when the cost is the least.

*Ans.* $\frac{10}{\sqrt[3]{3\pi}}$ ft.

**22.** In the corner of a field bounded by two perpendicular roads a spring is situated 6 rods from one road and 8 rods from the other. How should a straight road be run by this spring and across the corner so as to cut off as little of the field as possible?

*Ans.* 12 and 16 rods from corner.

What would be the length of the shortest road that could be run across?

*Ans.* $(6^{\frac{2}{3}} + 8^{\frac{2}{3}})^{\frac{3}{2}}$ rods.

**23.** Show that a square is the rectangle of maximum perimeter that can be inscribed in a given circle.

**24.** Two poles of height $a$ and $b$ feet are standing upright and are $c$ feet apart. Find the point on the line joining their bases such that the sum of the squares of the distances from this point to the tops of the poles is a minimum. *Ans.* Midway between the poles.

When will the sum of these distances be a minimum?

**25.** A conical tank with open top is to be built to contain $V$ cubic feet. Determine the shape if the material used is a minimum.

**26.** An isosceles triangle has a base 12 in. long and altitude 10 in. Find the rectangle of maximum area that can be inscribed in it, one side of the rectangle coinciding with the base of the triangle.

**27.** Divide the number 4 into two such parts that the sum of the cube of one part and three times the square of the other shall have a maximum value.

**28.** Divide the number $a$ into two parts such that the product of one part by the fourth power of the other part shall be a maximum.

**29.** A can buoy in the form of a double cone is to be made from two equal circular iron plates of radius $r$. Find the radius of the base of the cone when the buoy has the greatest displacement (maximum volume). *Ans.* $r\sqrt{\frac{2}{3}}$.

**30.** Into a full conical wineglass of depth $a$ and generating angle $\alpha$ there is carefully dropped a sphere of such size as to cause the greatest overflow. Show that the radius of the sphere is

$$\frac{a \sin \alpha}{\sin \alpha + \cos 2\alpha}.$$

**31.** A wall 27 ft. high is 8 ft. from a house. Find the length of the shortest ladder that will reach the house if one end rests on the ground outside of the wall. *Ans.* $13\sqrt{13}$.

**32.** A vessel is anchored 3 miles offshore, and opposite a point 5 miles further along the shore another vessel is anchored 9 miles from the shore. A boat from the first vessel is to land a passenger on the shore and then proceed to the other vessel. What is the shortest course of the boat? *Ans.* 13 miles.

**33.** A steel girder 25 ft. long is moved on rollers along a passageway 12.8 ft. wide and into a corridor at right angles to the passageway. Neglecting the width of the girder, how wide must the corridor be? *Ans.* 5.4 ft.

**34.** A miner wishes to dig a tunnel from a point $A$ to a point $B$ 300 feet below and 500 feet to the east of $A$. Below the level of $A$ it is bed rock and above $A$ is soft earth. If the cost of tunneling through earth is \$1 and through rock \$3 per linear foot, find the minimum cost of a tunnel. *Ans.* \$1348.53.

**35.** A carpenter has 108 sq. ft. of lumber with which to build a box with a square base and open top. Find the dimensions of the largest possible box he can make. *Ans.* $6 \times 6 \times 3$.

**36.** Find the right triangle of maximum area that can be constructed on a line of length $h$ as hypotenuse. *Ans.* $\frac{h}{\sqrt{2}}$ = length of both legs.

**37.** What is the isosceles triangle of maximum area that can be inscribed in a given circle? *Ans.* An equilateral triangle.

**38.** Find the altitude of the maximum rectangle that can be inscribed in a right triangle with base $b$ and altitude $h$. *Ans.* Altitude $= \frac{h}{2}$.

**39.** Find the dimensions of the rectangle of maximum area that can be inscribed in the ellipse $b^2x^2 + a^2y^2 = a^2b^2$. *Ans.* $a\sqrt{2}$ and $b\sqrt{2}$; area $= 2ab$.

**40.** Find the altitude of the right cylinder of maximum volume that can be inscribed in a sphere of radius $r$.

*Ans.* Altitude of cylinder $= \dfrac{2r}{\sqrt{3}}$.

**41.** Find the altitude of the right cylinder of maximum convex (curved) surface that can be inscribed in a given sphere. *Ans.* Altitude of cylinder $= r\sqrt{2}$.

**42.** What are the dimensions of the right hexagonal prism of minimum surface whose volume is 36 cubic feet? *Ans.* Altitude $= 2\sqrt{3}$; side of hexagon $= 2$.

**43.** Find the altitude of the right cone of minimum volume circumscribed about a given sphere. *Ans.* Altitude $= 4r$, and volume $= 2 \times$ vol. of sphere.

**44.** A right cone of maximum volume is inscribed in a given right cone, the vertex of the inside cone being at the center of the base of the given cone. Show that the altitude of the inside cone is one third the altitude of the given cone.

**45.** Given a point on the axis of the parabola $y^2 = 2px$ at a distance $a$ from the vertex; find the abscissa of the point of the curve nearest to it. *Ans.* $x = a - p$.

**46.** What is the length of the shortest line that can be drawn tangent to the ellipse $b^2x^2 + a^2y^2 = a^2b^2$ and meeting the coördinate axes? *Ans.* $a + b$.

**47.** A Norman window consists of a rectangle surmounted by a semicircle. Given the perimeter, required the height and breadth of the window when the quantity of light admitted is a maximum. *Ans.* Radius of circle = height of rectangle.

**48.** A tapestry 7 feet in height is hung on a wall so that its lower edge is 9 feet above an observer's eye. At what distance from the wall should he stand in order to obtain the most favorable view? *Ans.* 12 feet.

HINT. The vertical angle subtended by the tapestry in the eye of the observer must be at a maximum.

**49.** What are the most economical proportions of a tin can which shall have a given capacity, making allowance for waste?

*Ans.* Height $= \dfrac{2\sqrt{3}}{\pi} \times$ diameter of base.

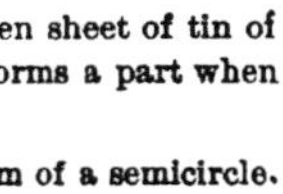

HINT. There is no waste in cutting out tin for the side of the can, but for top and bottom a hexagon of tin circumscribing the circular pieces required is used up.

NOTE 1. If no allowance is made for waste, then height = diameter.

NOTE 2. We know that the shape of a bee cell is hexagonal, giving a certain capacity for honey with the greatest possible economy of wax.

**50.** An open cylindrical trough is constructed by bending a given sheet of tin of breadth $2a$. Find the radius of the cylinder of which the trough forms a part when the capacity of the trough is a maximum.

*Ans.* Rad. $= \dfrac{2a}{\pi}$; i.e. it must be bent in the form of a semicircle.

**51.** A weight $W$ is to be raised by means of a lever with the force $F$ at one end and the point of support at the other. If the weight is suspended from a point at a distance $a$ from the point of support, and the weight of the beam is $w$ pounds per linear foot, what should be the length of the lever in order that the force required to lift it shall be a minimum?

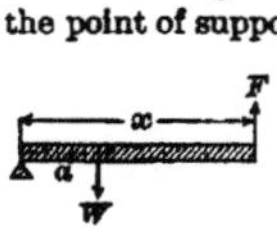


*Ans.* $x = \sqrt{\dfrac{2aW}{w}}$ feet.

52. An electric arc light is to be placed directly over the center of a circular plot of grass 100 feet in diameter. Assuming that the intensity of light varies directly as the sine of the angle under which it strikes an illuminated surface, and inversely as the square of its distance from the surface, how high should the light be hung in order that the best possible light shall fall on a walk along the circumference of the plot?

*Ans.* $\frac{50}{\sqrt{2}}$ feet.

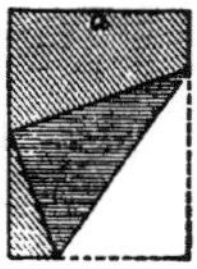

53. The lower corner of a leaf, whose width is $a$, is folded over so as just to reach the inner edge of the page. (a) Find the width of the part folded over when the length of the crease is a minimum. (b) Find the width when the area folded over is a minimum. *Ans.* (a) $\frac{3}{4}a$; (b) $\frac{2}{3}a$.

54. A rectangular stockade is to be built which must have a certain area. If a stone wall already constructed is available for one of the sides, find the dimensions which would make the cost of construction the least.

*Ans.* Side parallel to wall = twice the length of each end.

55. A cow is tethered by a perfectly smooth rope, a slip noose in the rope being thrown over a large square post. If the cow pulls the rope taut in the direction shown in the figure, at what angle will the rope leave the post? *Ans.* 30°.

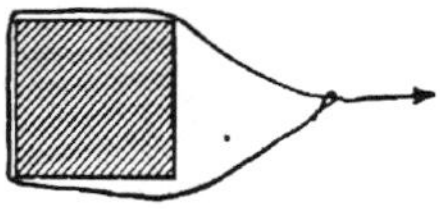

56. When the resistance of air is taken into account, the inclination of a pendulum to the vertical may be given by the formula

$$\theta = ae^{-kt}\cos(nt + \epsilon).$$

Show that the greatest elongations occur at equal intervals $\frac{\pi}{n}$ of time.

57. It is required to measure a certain unknown magnitude $x$ with precision. Suppose that $n$ equally careful observations of the magnitude are made, giving the results

$$a_1, \quad a_2, \quad a_3, \quad \cdots, \quad a_n.$$

The errors of these observations are evidently

$$x - a_1, \quad x - a_2, \quad x - a_3, \quad \cdots, \quad x - a_n,$$

some of which are positive and some negative.

It has been agreed that the most probable value of $x$ is such that it renders the sum of the squares of the errors, namely

$$(x - a_1)^2 + (x - a_2)^2 + (x - a_3)^2 + \cdots + (x - a_n)^2,$$

a minimum. Show that this gives the arithmetical mean of the observations as the most probable value of $x$.

58. The bending moment at $B$ of a beam of length $l$, uniformly loaded, is given by the formula

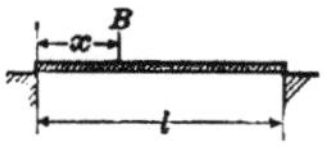


$$M = \frac{1}{2}wlx - \frac{1}{2}wx^2,$$

where $w$ = load per unit length. Show that the maximum bending moment is at the center of the beam.

59. If the total waste per mile in an electric conductor is

$$W = c^2r + \frac{t^2}{r}, \qquad [c = \text{constant}]$$

where $c$ = current in amperes, $r$ = resistance in ohms per mile, and $t$ = a constant depending on the interest on the investment and the depreciation of the plant, what is the relation between $c$, $r$, and $t$ when the waste is a minimum? *Ans.* $cr = t$.

**60.** A submarine telegraph cable consists of a core of copper wires with a covering made of nonconducting material. If $x$ denote the ratio of the radius of the core to the thickness of the covering, it is known that the speed of signaling varies as

$$x^2 \log \frac{1}{x}.$$

Show that the greatest speed is attained when $x = \frac{1}{\sqrt{e}}$.

**61.** Assuming that the power given out by a voltaic cell is given by the formula

$$P = \frac{E^2 R}{(r+R)^2},$$

where $E$ = constant electromotive force, $r$ = constant internal resistance, $R$ = external resistance, prove that $P$ is a maximum when $r = R$.

**62.** The force exerted by a circular electric current of radius $a$ on a small magnet whose axis coincides with the axis of the circle varies as

$$\frac{x}{(a^2 + x^2)^{\frac{5}{2}}},$$

where $x$ = distance of magnet from plane of circle. Prove that the force is a maximum when $x = \frac{a}{2}$.

**63.** We have two sources of heat at $A$ and $B$ with intensities $a$ and $b$ respectively. The total intensity of heat at a distance of $x$ from $A$ is given by the formula

$$I = \frac{a}{x^2} + \frac{b}{(d-x)^2}.$$

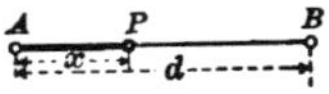


Show that the temperature at $P$ will be the lowest when

$$\frac{d-x}{x} = \frac{\sqrt[3]{b}}{\sqrt[3]{a}};$$

that is, the distances $BP$ and $AP$ have the same ratio as the cube roots of the corresponding heat intensities. The distance of $P$ from $A$ is

$$x = \frac{a^{\frac{1}{3}} d}{a^{\frac{1}{3}} + b^{\frac{1}{3}}}.$$

**64.** The range $OX$ of a projectile in a vacuum is given by the formula

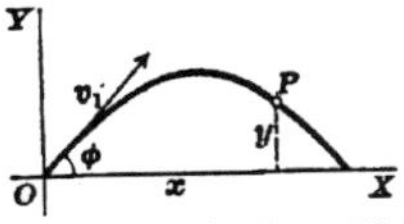


$$R = \frac{v_1^2 \sin 2\phi}{g};$$

where $v_1$ = initial velocity, $g$ = acceleration due to gravity, $\phi$ = angle of projection with the horizontal. Find the angle of projection which gives the greatest range for a given initial velocity.

*Ans.* $\phi = 45°$.

**65.** The total time of flight of the projectile in the last problem is given by the formula

$$T = \frac{2 v_1 \sin \phi}{g}.$$

At what angle should it be projected in order to make the time of flight a maximum?

*Ans.* $\phi = 90°$.

**66.** The time it takes a ball to roll down an inclined plane $AB$ is given by the formula

$$T = 2\sqrt{\frac{a}{g \sin 2\phi}}.$$

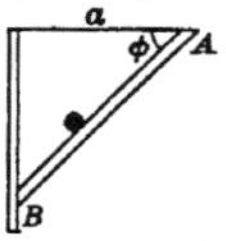


Neglecting friction, etc., what must be the value of $\phi$ to make the quickest descent? *Ans.* $\phi = 45°$.

**67.** Examine the function $(x-1)^2(x+1)^3$ for maximum and minimum values. Use the first method, p. 111.

**Solution.** $f(x) = (x-1)^2(x+1)^3$.

*First step.* $f'(x) = 2(x-1)(x+1)^3 + 3(x-1)^2(x+1)^2 = (x-1)(x+1)^2(5x-1)$.

*Second step.* $(x-1)(x+1)^2(5x-1) = 0$,

$x = 1, -1, \frac{1}{5}$, which are critical values.

*Third step.* $f'(x) = 5(x-1)(x+1)^2(x-\frac{1}{5})$.

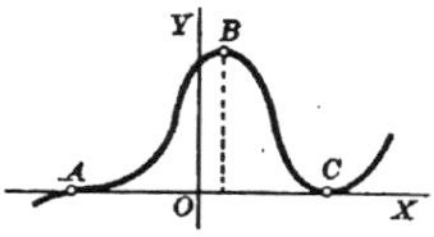


*Fourth step.* Examine first for critical value $x = 1$ ($C$ in figure).

When $x < 1, f'(x) = 5(-)(+)^2(+) = -$.

When $x > 1, f'(x) = 5(+)(+)^2(+) = +$.

Therefore, when $x = 1$ the function has a minimum value $f(1) = 0$ (= ordinate of $C$).

Examine now for the critical value $x = \frac{1}{5}$ ($B$ in figure).

When $x < \frac{1}{5}, f'(x) = 5(-)(+)^2(-) = +$.

When $x > \frac{1}{5}, f'(x) = 5(-)(+)^2(+) = -$.

Therefore, when $x = \frac{1}{5}$ the function has a maximum value $f(\frac{1}{5}) = 1.11$ (= ordinate of $B$).

Examine lastly for the critical value $x = -1$ ($A$ in figure).

When $x < -1, f'(x) = 5(-)(-)^2(-) = +$.

When $x > -1, f'(x) = 5(-)(+)^2(-) = +$.

Therefore, when $x = -1$ the function has neither a maximum nor a minimum value.

**68.** Examine the function $a - b(x-c)^{\frac{2}{3}}$ for maxima and minima.

**Solution.** $f(x) = a - b(x-c)^{\frac{2}{3}}$.

$$f'(x) = -\frac{2b}{3(x-c)^{\frac{1}{3}}}.$$

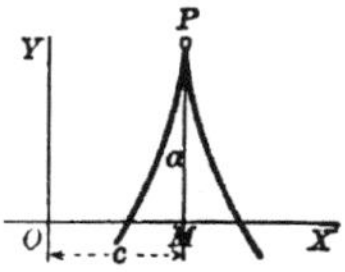


Since $x = c$ is a critical value for which $f'(x) = \infty$, but for which $f(x)$ is not infinite, let us test the function for maximum and minimum values when $x = c$.

When $x < c, f'(x) = +$.

When $x > c, f'(x) = -$.

Hence, when $x = c = OM$ the function has a maximum value $f(c) = a = MP$.

Examine the following functions for maximum and minimum values:

**69.** $(x-3)^2(x-2)$. *Ans.* $x = \frac{7}{3}$, gives max. $= \frac{4}{27}$;
$x = 3$, gives min. $= 0$.

**70.** $(x-1)^3(x-2)^2$. $x = \frac{7}{5}$, gives max. $= .03456$;
$x = 2$, gives min. $= 0$;
$x = 1$, gives neither.

**71.** $(x-4)^5(x+2)^4$. *Ans.* $x=-2$, gives max.;
$x=\frac{2}{3}$, gives min.;
$x=4$, gives neither.

**72.** $(x-2)^5(2x+1)^4$. $x=-\frac{1}{2}$, gives max.;
$x=\frac{11}{18}$, gives min.;
$x=2$, gives neither.

**73.** $(x+1)^{\frac{2}{3}}(x-5)^2$. $x=\frac{1}{2}$, gives max.;
$x=-1$ and $5$, give min.

**74.** $(2x-a)^{\frac{1}{3}}(x-a)^{\frac{2}{3}}$. $x=\frac{2a}{3}$, gives max.;
$x=a$, gives min.;
$x=\frac{a}{2}$, gives neither.

**75.** $x(x-1)^2(x+1)^3$. $x=\frac{1}{2}$, gives max.;
$x=1$ and $-\frac{1}{3}$, give min.;
$x=-1$, gives neither.

**76.** $x(a+x)^2(a-x)^3$. $x=-a$ and $\frac{a}{3}$, give max.;
$x=-\frac{a}{2}$, gives min.;
$x=a$, gives neither.

**77.** $b+c(x-a)^{\frac{2}{3}}$. $x=a$, gives min. $=b$.

**78.** $a-b(x-c)^{\frac{1}{3}}$. No max. or min.

**79.** $\frac{x^2-7x+6}{x-10}$. $x=4$, gives max.;
$x=16$, gives min.

**80.** $\frac{(a-x)^3}{a-2x}$. $x=\frac{a}{4}$, gives min.

**81.** $\frac{1-x+x^2}{1+x-x^2}$. $x=\frac{1}{2}$, gives min.

**82.** $\frac{x^2-3x+2}{x^2+3x+2}$. $x=\sqrt{2}$, gives min. $=12\sqrt{2}-17$;
$x=-\sqrt{2}$, gives max. $=-12\sqrt{2}-17$;
$x=-1,\ -2$, give neither.

**83.** $\frac{(x-a)(b-x)}{x^2}$. $x=\frac{2ab}{a+b}$, gives max. $=\frac{(a-b)^2}{4ab}$.

**84.** $\frac{a^2}{x}+\frac{b^2}{a-x}$. $x=\frac{a^2}{a-b}$, gives min.;
$x=\frac{a^2}{a+b}$, gives max.

**85.** Examine $x^3 - 3x^2 - 9x + 5$ for maxima and minima. Use the second method, p. 113.

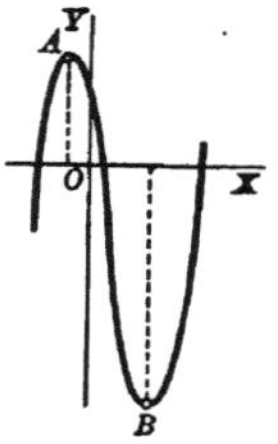


**Solution.** $f(x) = x^3 - 3x^2 - 9x + 5.$

*First step.* $f'(x) = 3x^2 - 6x - 9.$

*Second step.* $3x^2 - 6x - 9 = 0;$

hence the critical values are $x = -1$ and $3.$

*Third step.* $f''(x) = 6x - 6.$

*Fourth step.* $f''(-1) = -12.$

$\therefore f(-1) = 10 =$ (ordinate of $A$) $=$ maximum value.

$f''(3) = +12. \therefore f(3) = -22$ (ordinate of $B$) $=$ minimum value.

**86.** Examine $\sin^2 x \cos x$ for maximum and minimum values.

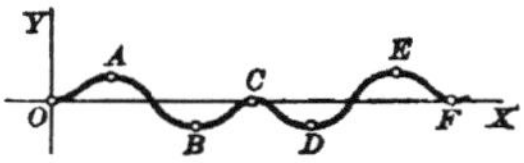


**Solution.** $f(x) = \sin^2 x \cos x.$

*First step.* $f'(x) = 2 \sin x \cos^2 x - \sin^3 x.$

*Second step.* $2 \sin x \cos^2 x - \sin^3 x = 0;$

hence the critical values are $x = n\pi$

and $x = n\pi \pm \text{arc tan} \sqrt{2} = n\pi \pm \alpha.$

*Third step.* $f''(x) = \cos x (2 \cos^2 x - 7 \sin^2 x).$

*Fourth step.* $f''(0) = +. \quad \therefore f(0) = 0 =$ minimum value at $O$.

$f''(\pi) = -. \quad \therefore f(\pi) = 0 =$ maximum value at $C$.

$f''(\alpha) = -. \quad \therefore f(\alpha) =$ maximum value at $A$.

$f''(\pi - \alpha) = +. \quad \therefore f(\pi - \alpha) =$ minimum value at $B$, etc.

Examine the following functions for maximum and minimum values.

**87.** $3x^3 - 9x^2 - 27x + 30.$ *Ans.* $x = -1$, gives max. $= 45$; $x = 3$, gives min. $= -51.$

**88.** $2x^3 - 21x^2 + 36x - 20.$ $x = 1$, gives max. $= -3$; $x = 6$, gives min. $= -128.$

**89.** $\dfrac{x^3}{3} - 2x^2 + 3x + 1.$ $x = 1$, gives max. $= \frac{7}{3}$; $x = 3$, gives min. $= 1.$

**90.** $2x^3 - 15x^2 + 36x + 10.$ $x = 2$, gives max. $= 38$; $x = 3$, gives min. $= 37.$

**91.** $x^3 - 9x^2 + 15x - 3.$ $x = 1$, gives max. $= 4$; $x = 5$, gives min. $= -28.$

**92.** $x^3 - 3x^2 + 6x + 10.$ No max. or min.

**93.** $x^5 - 5x^4 + 5x^3 + 1.$ $x = 1$, gives max. $= 2$; $x = 3$, gives min. $= -26$; $x = 0$, gives neither.

**94.** $3x^5 - 125x^3 + 2160x.$ $x = -4$ and $3$, give max.; $x = -3$ and $4$, give min.

**95.** $2x^3 - 3x^2 - 12x + 4.$

**96.** $x(x - 3)^3.$

**97.** $x^4 - 2x^2 + 10.$

**98.** $x^4 - 4.$

**99.** $x^3 - 8.$

**100.** $4 - x^6.$

101. $\sin x(1+\cos x)$. *Ans.* $x = 2n\pi + \frac{\pi}{3}$, give max. $= \frac{3}{4}\sqrt{3}$;
$x = 2n\pi - \frac{\pi}{3}$, give min. $= -\frac{3}{4}\sqrt{3}$;
$x = n\pi$, give neither.

102. $\frac{x}{\log x}$. $x = e$, gives min. $= e$;
$x = 1$, gives neither.

103. $\log \cos x$. $x = n\pi$, gives max.

104. $ae^{kx} + be^{-kx}$. $x = \frac{1}{k}\log\sqrt{\frac{b}{a}}$, gives min. $= 2\sqrt{ab}$.

105. $x^x$. $x = \frac{1}{e}$, gives min.

106. $x^{\frac{1}{x}}$. $x = e$, gives max.

107. $\cos x + \sin x$. $x = \frac{\pi}{4}$, gives max. $= \sqrt{2}$;
$x = \frac{5\pi}{4}$, gives min. $= -\sqrt{2}$.

108. $\sin 2x - x$. $x = \frac{\pi}{6}$, gives max.;
$x = -\frac{\pi}{6}$, gives min.

109. $x + \tan x$. No max. or min.

110. $\sin^3 x \cos x$. $x = n\pi + \frac{\pi}{3}$, gives max. $= \frac{3}{16}\sqrt{3}$;
$x = n\pi - \frac{\pi}{3}$, gives min. $= -\frac{3}{16}\sqrt{3}$;
$x = n\pi$, gives neither.

111. $x\cos x$. $x = \cot x$, gives max.

112. $\sin x + \cos 2x$. $x = \text{arc}\sin\frac{1}{4}$, gives max.;
$x = \frac{\pi}{2}$, gives min.

113. $2\tan x - \tan^2 x$. $x = \frac{\pi}{4}$, gives max.

114. $\frac{\sin x}{1+\tan x}$. $x = \frac{\pi}{4}$, gives max.

115. $\frac{x}{1+x\tan x}$. $x = \cos x$, gives max.;
$x = -\cos x$, gives min.

**85. Points of inflection. Definition.** *Points of inflection* separate arcs concave upwards from arcs concave downwards.* Thus, if a curve $y=f(x)$ changes (as at $B$) from concave upwards (as at $A$) to concave downwards (as at $C$), or the reverse, then such a point as $B$ is called a point of inflection.

* Points of inflection may also be defined as points where

(a) $\frac{d^2y}{dx^2}=0$ and $\frac{d^2y}{dx^2}$ changes sign,

or (b) $\frac{d^2x}{dy^2}=0$ and $\frac{d^2x}{dy^2}$ changes sign.

From the discussion of § 84 it follows at once that at $A$, $f''(x) = +$, and at $C$, $f''(x) = -$. In order to change sign it must pass through the value zero;* hence we have

$$\textbf{(23)} \qquad \textbf{at points of inflection, } f''(x) = 0.$$

Solving the equation resulting from **(23)** gives the abscissas of the points of inflection. To determine the direction of curving or direction of bending in the vicinity of a point of inflection, test $f''(x)$ for values of $x$, first a trifle less and then a trifle greater than the abscissa at that point.

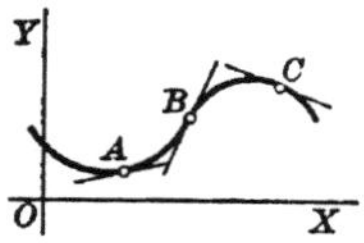


If $f''(x)$ changes sign, we have a point of inflection, and the signs obtained determine if the curve is concave upwards or concave downwards in the neighborhood of each point of inflection.

The student should observe that near a point where the curve is concave upwards (as at $A$) the curve lies above the tangent, and at a point where the curve is concave downwards (as at $C$) the curve lies below the tangent. At a point of inflection (as at $B$) the tangent evidently crosses the curve.

Following is a **rule for finding points of inflection** of the curve whose equation is $y = f(x)$. This rule includes also directions for examining the direction of curvature of the curve in the neighborhood of each point of inflection.

FIRST STEP. *Find* $f''(x)$.

SECOND STEP. *Set* $f''(x) = 0$, *and solve the resulting equation for real roots.*

THIRD STEP. *Write* $f''(x)$ *in factor form.*

FOURTH STEP. *Test* $f''(x)$ *for values of* $x$, *first a trifle less and then a trifle greater than each root found in the second step. If* $f''(x)$ *changes sign, we have a point of inflection.*

*When* $f''(x) = +$, *the curve is concave upwards* $\overset{+}{\smile}$.†

*When* $f''(x) = -$, *the curve is concave downwards* $\underset{-}{\frown}$.

* It is assumed that $f'(x)$ and $f''(x)$ are continuous. The solution of Ex. 2, p. 127, shows how to discuss a case where $f'(x)$ and $f''(x)$ are both infinite. Evidently salient points (see p. 258) are excluded, since at such points $f'(x)$ is discontinuous.

† This may be easily remembered if we say that a vessel shaped like the curve where it is concave upwards will hold (+) water, and where it is concave downwards will spill (−) water.

**EXAMPLES**

Examine the following curves for points of inflection and direction of bending.

1. $y = 3x^4 - 4x^3 + 1$.

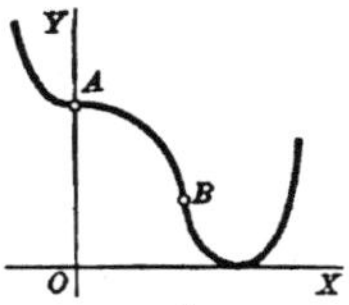


**Solution.** $f(x) = 3x^4 - 4x^3 + 1$.

*First step.* $f''(x) = 36x^2 - 24x$.

*Second step.* $36x^2 - 24x = 0$

$\therefore x = \frac{2}{3}$ and $x = 0$, critical values.

*Third step.* $f''(x) = 36x(x - \frac{2}{3})$.

*Fourth step.* When $x < 0, f''(x) = +$ ; and when $x > 0, f''(x) = -$.

$\therefore$ curve is concave upwards to the left and concave downwards to the right of $x = 0$ ($A$ in figure). When $x < \frac{2}{3}, f''(x) = -$ ; and when $x > \frac{2}{3}, f''(x) = +$.

$\therefore$ curve is concave downwards to the left and concave upwards to the right of $x = \frac{2}{3}$ ($B$ in figure).

The curve is evidently concave upwards everywhere to the left of $A$, concave downwards between $A$ (0, 1) and $B$ $(\frac{2}{3}, \frac{11}{27})$, and concave upwards everywhere to the right of $B$.

2. $(y - 2)^3 = (x - 4)$.

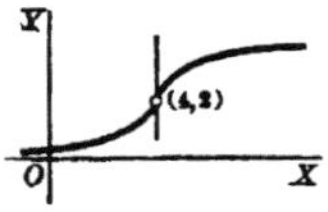


**Solution.** $y = 2 + (x - 4)^{\frac{1}{3}}$.

*First step.* $\frac{dy}{dx} = \frac{1}{3}(x - 4)^{-\frac{2}{3}}$,

$\frac{d^2y}{dx^2} = -\frac{2}{9}(x - 4)^{-\frac{5}{3}}$.

*Second step.* When $x = 4$, both first and second derivatives are infinite.

*Third step.* When $x < 4, \frac{d^2y}{dx^2} = +$ ; but when $x > 4, \frac{d^2y}{dx^2} = -$.

We may therefore conclude that the tangent at (4, 2) is perpendicular to the axis of $X$, that to the left of (4, 2) the curve is concave upwards, and to the right of (4, 2) it is concave downwards. Therefore (4, 2) must be considered a point of inflection.

3. $y = x^2$. *Ans.* Concave upwards everywhere.

4. $y = 5 - 2x - x^2$. Concave downwards everywhere.

5. $y = x^3$. Concave downwards to the left and concave upwards to the right of (0, 0).

6. $y = x^3 - 3x^2 - 9x + 9$. Concave downwards to the left and concave upwards to the right of (1, − 2).

7. $y = a + (x - b)^3$. Concave downwards to the left and concave upwards to the right of ($b$, $a$).

8. $a^2y = \frac{x^3}{3} - ax^2 + 2a^3$. Concave downwards to the left and concave upwards to the right of $\left(a, \frac{4a}{3}\right)$.

9. $y = x^4$. Concave upwards everywhere.

10. $y = x^4 - 12x^3 + 48x^2 - 50$.

Concave upwards to the left of $x = 2$, concave downwards between $x = 2$ and $x = 4$, concave upwards to the right of $x = 4$.

11. $y = \sin x$. Points of inflection are $x = n\pi$, $n$ being any integer.

**12.** $y = \tan x$. *Ans.* Points of inflection are $x = n\pi$, $n$ being any integer.

**13.** Show that no conic section can have a point of inflection.

**14.** Show that the graphs of $e^x$ and $\log x$ have no points of inflection.

**86. Curve tracing.** The elementary method of tracing (or plotting) a curve whose equation is given in rectangular coördinates, and one with which the student is already familiar, is to solve its equation for $y$ (or $x$), assume arbitrary values of $x$ (or $y$), calculate the corresponding values of $y$ (or $x$), plot the respective points, and draw a smooth curve through them, the result being an approximation to the required curve. This process is laborious at best, and in case the equation of the curve is of a degree higher than the second, the solved form of such an equation may be unsuitable for the purpose of computation, or else it may fail altogether, since it is not always possible to solve the equation for $y$ or $x$.

The general form of a curve is usually all that is desired, and the Calculus furnishes us with powerful methods for determining the shape of a curve with very little computation.

The first derivative gives us the slope of the curve at any point; the second derivative determines the intervals within which the curve is concave upward or concave downward, and the points of inflection separate these intervals; the maximum points are the high points and the minimum points are the low points on the curve. As a guide in his work the student may follow the

**Rule for tracing curves. Rectangular coördinates.**

FIRST STEP. *Find the first derivative; place it equal to zero; solving gives the abscissas of maximum and minimum points.*

SECOND STEP. *Find the second derivative; place it equal to zero; solving gives the abscissas of the points of inflection.*

THIRD STEP. *Calculate the corresponding ordinates of the points whose abscissas were found in the first two steps. Calculate as many more points as may be necessary to give a good idea of the shape of the curve. Fill out a table such as is shown in the example worked out.*

FOURTH STEP. *Plot the points determined and sketch in the curve to correspond with the results shown in the table.*

If the calculated values of the ordinates are large, it is best to reduce the scale on the $Y$-axis so that the general behavior of the curve will be shown within the limits of the paper used. Coördinate plotting paper should be employed.

## EXAMPLES

Trace the following curves, making use of the above rule. Also find the equations of the tangent and normal at each point of inflection.

1. $y = x^3 - 9x^2 + 24x - 7.$

**Solution.** Use the above rule.

*First step.*

$$y' = 3x^2 - 18x + 24,$$
$$3x^2 - 18x + 24 = 0,$$
$$x = 2, 4.$$

*Second step.*

$$y'' = 6x - 18,$$
$$6x - 18 = 0,$$
$$x = 3.$$

*Third step.*

| $x$ | $y$ | $y'$ | $y''$ | Remarks | Direction of Curve |
|---|---|---|---|---|---|
| 0 | −7 | + | − | | concave down |
| 2 | 13 | 0 | − | max. | |
| 3 | 11 | − | 0 | pt. of infl. | concave up |
| 4 | 9 | 0 | + | min. | |
| 6 | 29 | + | + | | |

*Fourth step.* Plotting the points and sketching in the curve, we get the figure shown.

To find the equations of the tangent and normal to the curve at the point of inflection $P_1(3, 11)$, use formulas (1), (2), pp. 76, 77. This gives $3x + y = 20$ for the tangent and $3y - x = 30$ for the normal.

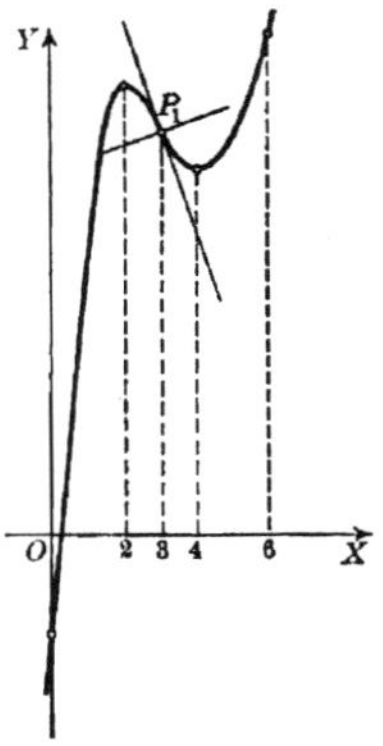


2. $y = x^3 - 6x^2 - 36x + 5.$
*Ans.* Max. $(-2, 45)$; min. $(6, -211)$; pt. of infl. $(2, -83)$; tan. $y + 48x - 13 = 0$; nor. $48y - x + 3986 = 0$.

3. $y = x^4 - 2x^2 + 10.$
*Ans.* Max. $(0, 10)$; min. $(\pm 1, 9)$; pt. of infl. $\left(\pm \frac{1}{\sqrt{3}}, \frac{85}{9}\right)$.

4. $y = \frac{1}{2}x^4 - 3x^2 + 2.$
*Ans.* Max. $(0, 2)$; min. $(\pm\sqrt{3}, -\frac{5}{2})$; pt. of infl. $(\pm 1, -\frac{1}{2})$.

5. $y = \frac{6x}{1+x^2}.$
*Ans.* Max. $(1, 3)$; min. $(-1, -3)$; pt. of infl. $(0, 0)$, $\left(\pm\sqrt{3}, \pm\frac{3\sqrt{3}}{2}\right)$.

6. $y = 12x - x^3.$ *Ans.* Max. $(2, 16)$; min. $(-2, -16)$; pt. of infl. $(0, 0)$.

7. $4y + x^3 - 3x^2 + 4 = 0.$ Max. $(2, 0)$; min. $(0, -1)$.

8. $y = x^3 - 3x^2 - 9x + 9.$

9. $2y + x^3 - 9x + 6 = 0.$

10. $y = x^3 - 6x^2 - 15x + 2.$

11. $y(1 + x^2) = x.$

12. $y = \dfrac{8a^3}{x^2 + 4a^2}.$

13. $y = e^{-x^2}.$

14. $y = \dfrac{4 + x}{x^2}.$

15. $y = (x + 1)^{\frac{2}{3}}(x - 5)^2.$

16. $y = \dfrac{x + 2}{x^3}.$

17. $y = x^3 - 3x^2 - 24x.$

18. $y = 18 + 36x - 3x^2 - 2x^3.$

19. $y = x - 2\cos x.$

20. $y = 3x - x^3.$

21. $y = x^3 - 9x^2 + 15x - 3$

22. $x^2y = 4 + x.$

23. $4y = x^4 - 6x^2 + 5.$

24. $y = \dfrac{x^3}{x^2 + 3a^2}.$

25. $y = \sin x + \dfrac{x}{2}.$

26. $y = \dfrac{x^2 + 4}{x}.$

27. $y = 5x - 2x^2 - \frac{1}{3}x^3.$

28. $y = \dfrac{1 + x^2}{2x}.$

29. $y = x - 2\sin x.$

30. $y = \log\cos x.$

31. $y = \log(1 + x^2).$

# CHAPTER IX

## DIFFERENTIALS

**87. Introduction.** Thus far we have represented the derivative of $y=f(x)$ by the notation

$$\frac{dy}{dx}=f'(x).$$

We have taken special pains to impress on the student that the symbol

$$\frac{dy}{dx}$$

was to be considered not as an ordinary fraction with $dy$ as numerator and $dx$ as denominator, but as a single symbol denoting the limit of the quotient

$$\frac{\Delta y}{\Delta x}$$

as $\Delta x$ approaches the limit zero.

Problems do occur, however, where it is very convenient to be able to give a meaning to $dx$ and $dy$ separately, and it is especially useful in applications of the Integral Calculus. How this may be done is explained in what follows.

**88. Definitions.** If $f'(x)$ is the derivative of $f(x)$ for a particular value of $x$, and $\Delta x$ is an arbitrarily chosen increment of $x$, then the *differential of* $f(x)$, denoted by the symbol $df(x)$, is defined by the equation

$$(A) \qquad df(x)=f'(x)\,\Delta x.$$

If now $f(x)=x$, then $f'(x)=1$, and $(A)$ reduces to

$$dx=\Delta x,$$

showing that when $x$ is the independent variable, the *differential of* $x(=dx)$ is identical with $\Delta x$. Hence, if $y=f(x)$, $(A)$ may in general be written in the form

$$(B) \qquad dy=f'(x)\,dx.^*$$

* On account of the position which the derivative $f'(x)$ here occupies, it is sometimes called the *differential coefficient*.

The student should observe the important fact that, since $dx$ may be given any arbitrary value whatever, *$dx$ is independent of $x$*. Hence, *$dy$ is a function of two independent variables $x$ and $dx$.*

*The differential of a function equals its derivative multiplied by the differential of the independent variable.*

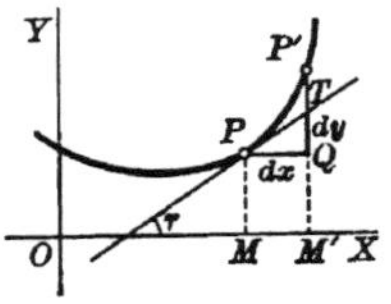


Let us illustrate what this means geometrically.

Let $f'(x)$ be the derivative of $y=f(x)$ at $P$. Take $dx=PQ$, then

$$dy=f'(x)\,dx=\tan\tau\cdot PQ=\frac{QT}{PQ}\cdot PQ=QT.$$

Therefore $dy$, or $df(x)$, is the increment ($=QT$) of the ordinate of the tangent corresponding to $dx$.*

This gives the following interpretation of the derivative as a fraction.

*If an arbitrarily chosen increment of the independent variable $x$ for a point $P(x, y)$ on the curve $y=f(x)$ be denoted by $dx$, then in the derivative*

$$\frac{dy}{dx}=f'(x)=\tan\tau,$$

*$dy$ denotes the corresponding increment of the ordinate drawn to the tangent.*

**89. Infinitesimals.** In the Differential Calculus we are usually concerned with the derivative, that is, with the ratio of the differentials $dy$ and $dx$. In some applications it is also useful to consider $dx$ as an infinitesimal (see § 15, p. 13), that is, as a variable whose values remain numerically small, and which, at some stage of the investigation, approaches the limit zero. Then by (**B**), p. 131, and (2), p. 19, $dy$ is also an infinitesimal.

In problems where several infinitesimals enter we often make use of the following

**Theorem.** *In problems involving the limit of the ratio of two infinitesimals, either infinitesimal may be replaced by an infinitesimal so related to it that the limit of their ratio is unity.*

**Proof.** Let $\alpha$, $\beta$, $\alpha'$, $\beta'$ be infinitesimals so related that

$$(C)\qquad \text{limit}\,\frac{\alpha'}{\alpha}=1 \text{ and limit}\,\frac{\beta'}{\beta}=1.$$

* The student should note especially that the differential ($=dy$) and the increment ($=\Delta y$) of the function corresponding to the same value of $dx$ ($=\Delta x$) are not in general equal. For, in the figure, $dy=QT$, but $\Delta y=QP'$.

We have $$\frac{\alpha}{\beta}=\frac{\alpha'}{\beta'}\cdot\frac{\alpha}{\alpha'}\cdot\frac{\beta'}{\beta}\text{ identically,}$$

and $$\text{limit}\frac{\alpha}{\beta}=\text{limit}\frac{\alpha'}{\beta'}\cdot\text{limit}\frac{\alpha}{\alpha'}\cdot\text{limit}\frac{\beta'}{\beta}\qquad\text{Th. II, p. 18}$$

$$=\text{limit}\frac{\alpha'}{\beta'}\cdot 1\cdot 1.\qquad\text{By }(C)$$

(*D*) $$\therefore\ \text{limit}\frac{\alpha}{\beta}=\text{limit}\frac{\alpha'}{\beta'}.\qquad\text{Q.E.D.}$$

Now let us apply this theorem to the two following important limits.

For the independent variable $x$, we know from the previous section that $\Delta x$ and $dx$ are identical.

Hence their ratio is unity, and also limit $\frac{\Delta x}{dx}=1$. That is, by the above theorem,

(*E*) *In the limit of the ratio of* $\Delta x$ *and a second infinitesimal,* $\Delta x$ *may be replaced by* $dx$.

On the contrary it was shown that, for the dependent variable $y$, $\Delta y$ and $dy$ are in general unequal. But we shall now show, however, that in this case also $$\text{limit}\frac{\Delta y}{dy}=1.$$

Since $\underset{\Delta x=0}{\text{limit}}\frac{\Delta y}{\Delta x}=f'(x)$, we way write
$$\frac{\Delta y}{\Delta x}=f'(x)+\epsilon,$$
where $\epsilon$ is an infinitesimal which approaches zero when $\Delta x \doteq 0$.

Clearing of fractions, remembering that $\Delta x = dx$,
$$\Delta y=f'(x)\,dx+\epsilon\cdot\Delta x,$$
or $$\Delta y=dy+\epsilon\cdot\Delta x.\qquad(B),\text{ p. 131}$$

Dividing both sides by $\Delta y$,
$$1=\frac{dy}{\Delta y}+\epsilon\cdot\frac{\Delta x}{\Delta y},$$
or
$$\frac{dy}{\Delta y}=1-\epsilon\cdot\frac{\Delta x}{\Delta y}.$$
$$\therefore\ \underset{\Delta x=0}{\text{limit}}\frac{dy}{\Delta y}=1,$$
and hence $\underset{\Delta x=0}{\text{limit}}\frac{\Delta y}{dy}=1$. That is, by the above theorem,

(*F*) *In the limit of the ratio of* $\Delta y$ *and a second infinitesimal,* $\Delta y$ *may be replaced by* $dy$.

**90. Derivative of the arc in rectangular coördinates.** Let $s$ be the length* of the arc $AP$ measured from a fixed point $A$ on the curve. Denote the increment of $s$ ($=\text{arc } PQ$) by $\Delta s$. The definition of the length of arc depends on the assumption that, as $Q$ approaches $P$,

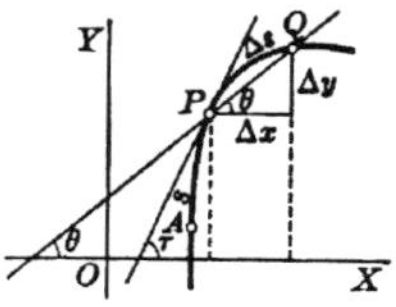


$$\text{limit}\left(\frac{\text{chord } PQ}{\text{arc } PQ}\right)=1.$$

If we now apply the theorem on p. 132 to this, we get

(*G*) *In the limit of the ratio of chord $PQ$ and a second infinitesimal, chord $PQ$ may be replaced by arc $PQ(=\Delta s)$.*

From the above figure

(*H*) $$(\text{chord } PQ)^2=(\Delta x)^2+(\Delta y)^2.$$

Dividing through by $(\Delta x)^2$, we get

(*I*) $$\left(\frac{\text{chord } PQ}{\Delta x}\right)^2=1+\left(\frac{\Delta y}{\Delta x}\right)^2.$$

Now let $Q$ approach $P$ as a limiting position; then $\Delta x \doteq 0$ and we have

$$\left(\frac{ds}{dx}\right)^2=1+\left(\frac{dy}{dx}\right)^2.$$

$$\left[\text{Since } \lim_{\Delta x=0}\left(\frac{\text{chord } PQ}{\Delta x}\right)=\lim_{\Delta x=0}\left(\frac{\Delta s}{\Delta x}\right)=\frac{ds}{dx}, \text{ by } (G).\right]$$

**(24)** $$\therefore \frac{ds}{dx}=\sqrt{1+\left(\frac{dy}{dx}\right)^2}.$$

Similarly, if we divide (*H*) by $(\Delta y)^2$ and pass to the limit, we get

**(25)** $$\frac{ds}{dy}=\sqrt{\left(\frac{dx}{dy}\right)^2+1}.$$

Also, from the above figure,

$$\cos\theta=\frac{\Delta x}{\text{chord } PQ}, \qquad \sin\theta=\frac{\Delta y}{\text{chord } PQ}.$$

Now as $Q$ approaches $P$ as a limiting position $\theta \doteq \tau$, and we get

**(26)** $$\cos\tau=\frac{dx}{ds}, \qquad \sin\tau=\frac{dy}{ds}.$$

$$\left[\text{Since from } (G) \text{ limit } \frac{\Delta x}{\text{chord } PQ}=\text{limit }\frac{\Delta x}{\Delta s}=\frac{dx}{ds}, \text{ and limit } \frac{\Delta y}{\text{chord } PQ}=\text{limit }\frac{\Delta y}{\Delta s}=\frac{dy}{ds}.\right]$$

* Defined in § 209.

Using the notation of differentials, formulas (**24**) and (**25**) may be written

$$\text{(27)}\qquad ds=\left[1+\left(\frac{dy}{dx}\right)^2\right]^{\frac{1}{2}}dx.$$

$$\text{(28)}\qquad ds=\left[\left(\frac{dx}{dy}\right)^2+1\right]^{\frac{1}{2}}dy.$$

Substituting the value of $ds$ from (**27**) in (**26**),

$$\text{(29)}\qquad \cos\tau=\frac{1}{\left[1+\left(\frac{dy}{dx}\right)^2\right]^{\frac{1}{2}}},\qquad \sin\tau=\frac{\frac{dy}{dx}}{\left[1+\left(\frac{dy}{dx}\right)^2\right]^{\frac{1}{2}}}.$$

An easy way to remember the relations (**24**)–(**26**) between the differentials $dx$, $dy$, $ds$ is to note that they are correctly represented by a right triangle whose hypotenuse is $ds$, whose sides are $dx$ and $dy$, and whose angle at the base is $\tau$. Then

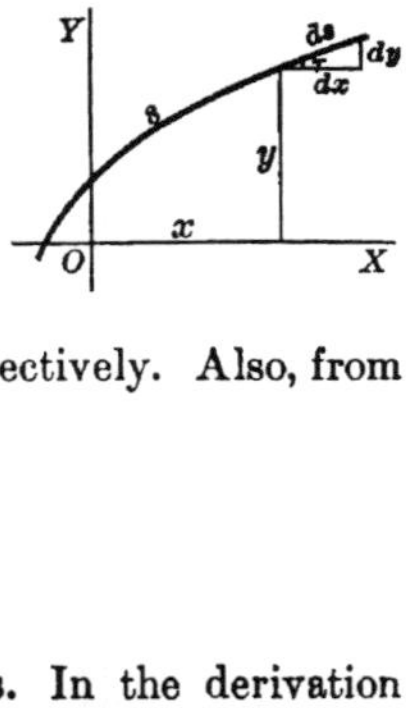


$$ds=\sqrt{(dx)^2+(dy)^2},$$

and, dividing by $dx$ or $dy$, gives (**24**) or (**25**) respectively. Also, from the figure,

$$\cos\tau=\frac{dx}{ds},\qquad \sin\tau=\frac{dy}{ds};$$

the same relations given by (**26**).

**91. Derivative of the arc in polar coördinates.** In the derivation which follows we shall employ the same figure and the same notation used on pp. 83, 84.

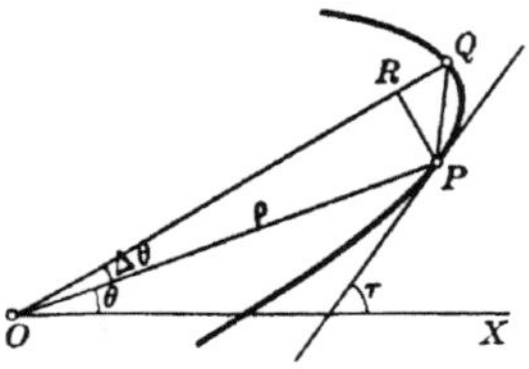


From the right triangle $PRQ$

$$(\text{chord } PQ)^2 = (PR)^2+(RQ)^2 = (\rho\sin\Delta\theta)^2+(\rho+\Delta\rho-\rho\cos\Delta\theta)^2.$$

Dividing throughout by $(\Delta\theta)^2$, we get

$$\left(\frac{\text{chord } PQ}{\Delta\theta}\right)^2=\rho^2\left(\frac{\sin\Delta\theta}{\Delta\theta}\right)^2+\left(\frac{\Delta\rho}{\Delta\theta}+\rho\cdot\frac{1-\cos\Delta\theta}{\Delta\theta}\right)^2.$$

Passing to the limit as $\Delta\theta$ diminishes towards zero, we get *

$$\left(\frac{ds}{d\theta}\right)^2 = \rho^2 + \left(\frac{d\rho}{d\theta}\right)^2.$$

(30) $$\therefore \frac{ds}{d\theta} = \sqrt{\rho^2 + \left(\frac{d\rho}{d\theta}\right)^2}.$$

In the notation of differentials this becomes

(31) $$ds = \left[\rho^2 + \left(\frac{d\rho}{d\theta}\right)^2\right]^{\frac{1}{2}} d\theta.$$

These relations between $\rho$ and the differentials $ds$, $d\rho$, and $d\theta$ are correctly represented by a right triangle whose hypotenuse is $ds$ and whose sides are $d\rho$ and $\rho d\theta$. Then

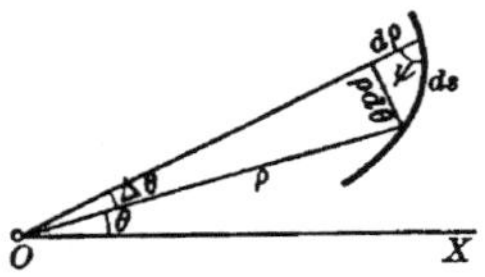

$$ds = \sqrt{(\rho d\theta)^2 + (d\rho)^2},$$

and dividing by $d\theta$ gives (30).

Denoting by $\psi$ the angle between $d\rho$ and $ds$, we get at once

$$\tan\psi = \rho\frac{d\theta}{d\rho},$$

which is the same as (*A*), p. 84.

ILLUSTRATIVE EXAMPLE 1. Find the differential of the arc of the circle $x^2 + y^2 = r^2$.

**Solution.** Differentiating, $\qquad \dfrac{dy}{dx} = -\dfrac{x}{y}.$

To find $ds$ in terms of $x$ we substitute in (27), giving

$$ds = \left[1 + \frac{x^2}{y^2}\right]^{\frac{1}{2}} dx = \left[\frac{y^2 + x^2}{y^2}\right]^{\frac{1}{2}} dx = \left[\frac{r^2}{y^2}\right]^{\frac{1}{2}} dx = \frac{r\,dx}{\sqrt{r^2 - x^2}}.$$

To find $ds$ in terms of $y$ we substitute in (28), giving

$$ds = \left[1 + \frac{y^2}{x^2}\right]^{\frac{1}{2}} dy = \left[\frac{x^2 + y^2}{x^2}\right]^{\frac{1}{2}} dy = \left[\frac{r^2}{x^2}\right]^{\frac{1}{2}} = \frac{r\,dy}{\sqrt{r^2 - y^2}}.$$

ILLUSTRATIVE EXAMPLE 2. Find the differential of the arc of the cardioid $\rho = a(1 - \cos\theta)$ in terms of $\theta$.

**Solution.** Differentiating, $\qquad \dfrac{d\rho}{d\theta} = a\sin\theta.$

Substituting in (31), gives

$$ds = [a^2(1 - \cos\theta)^2 + a^2\sin^2\theta]^{\frac{1}{2}} d\theta = a[2 - 2\cos\theta]^{\frac{1}{2}} d\theta = a\left[4\sin^2\frac{\theta}{2}\right]^{\frac{1}{2}} d\theta = 2a\sin\frac{\theta}{2}\,d\theta.$$

* $\displaystyle \lim_{\Delta\theta=0}\frac{\text{chord } PQ}{\Delta\theta} = \lim_{\Delta\theta=0}\frac{\Delta s}{\Delta\theta} = \frac{ds}{d\theta}.$ By (*G*), p. 134

$\displaystyle \lim_{\Delta\theta=0}\frac{\sin\Delta\theta}{\Delta\theta} = 1.$ By § 22, p. 21

$\displaystyle \lim_{\Delta\theta=0}\frac{1 - \cos\Delta\theta}{\Delta\theta} = \lim_{\Delta\theta=0}\frac{2\sin^2\frac{\Delta\theta}{2}}{\Delta\theta} = \lim_{\Delta\theta=0}\sin\frac{\Delta\theta}{2}\cdot\frac{\sin\frac{\Delta\theta}{2}}{\frac{\Delta\theta}{2}} = 0\cdot 1 = 0.$ By 39, p. 2, and § 22, p. 21

**EXAMPLES**

Find the differential of arc in each of the following curves:

1. $y^2 = 4x$. *Ans.* $ds = \sqrt{\frac{1+x}{x}}\,dx$.

2. $y = ax^2$. $ds = \sqrt{1 + 4a^2x^2}\,dx$.

3. $y = x^3$. $ds = \sqrt{1 + 9x^4}\,dx$.

4. $y^3 = x^2$. $ds = \frac{1}{3}\sqrt{4 + 9y}\,dy$.

5. $x^{\frac{2}{3}} + y^{\frac{2}{3}} = a^{\frac{2}{3}}$. $ds = \sqrt[3]{\frac{a}{y}}\,dy$.

6. $b^2x^2 + a^2y^2 = a^2b^2$. $ds = \sqrt{\frac{a^2 - e^2x^2}{a^2 - x^2}}\,dx$.

HINT. $e^2 = \frac{a^2 - b^2}{a^2}$.

7. $e^y \cos x = 1$. $ds = \sec x\,dx$.

8. $\rho = a\cos\theta$. $ds = a\,d\theta$.

9. $\rho^2 = a^2\cos 2\theta$. $ds = a\sqrt{\sec 2\theta}\,d\theta$.

10. $\rho = ae^{\theta\cot\alpha}$. $ds = \rho\csc\alpha\,d\theta$.

11. $\rho = a^\theta$. $ds = a^\theta\sqrt{1 + \log^2 a}\,d\theta$.

12. $\rho = a\theta$. $ds = \frac{1}{a}\sqrt{a^2 + \rho^2}\,d\rho$.

13. (a) $x^2 - y^2 = a^2$.
(b) $x^2 = 4ay$.
(c) $y = e^x + e^{-x}$.
(d) $xy = a$.
(e) $y = \log\sec x$.
(f) $\rho = 2a\tan\theta\sin\theta$.
(g) $\rho = a\sec^3\frac{\theta}{3}$.
(h) $x^{\frac{1}{2}} + y^{\frac{1}{2}} = a^{\frac{1}{2}}$.
(i) $y^2 = ax^3$.
(j) $y = \log x$.
(k) $4x = y^3$.
(l) $\rho = a\sec^2\frac{\theta}{2}$.
(m) $\rho = 1 + \sin\theta$.
(n) $\rho\theta = a$.

**92. Formulas for finding the differentials of functions.** Since the differential of a function is its derivative multiplied by the differential of the independent variable, it follows at once that the formulas for finding differentials are the same as those for finding derivatives given in § 33, pp. 34–36, if we multiply each one by $dx$.

This gives us

I $\quad d(c) = 0.$

II $\quad d(x) = dx.$

III $\quad d(u + v - w) = du + dv - dw.$

IV $\quad d(cv) = c\,dv.$

V $\quad d(uv) = u\,dv + v\,du.$

VI $\quad d(v^n) = nv^{n-1}\,dv.$

VI a $$d(x^n) = nx^{n-1}dx.$$

VII $$d\left(\frac{u}{v}\right) = \frac{v\,du - u\,dv}{v^2}.$$

VII a $$d\left(\frac{u}{c}\right) = \frac{du}{c}.$$

VIII $$d(\log_a v) = \log_a e\frac{dv}{v}.$$

IX $$d(a^v) = a^v \log a\,dv.$$

IX a $$d(e^v) = e^v\,dv.$$

X $$d(u^v) = vu^{v-1}du + \log u \cdot u^v \cdot dv.$$

XI $$d(\sin v) = \cos v\,dv.$$

XII $$d(\cos v) = -\sin v\,dv.$$

XIII $$d(\tan v) = \sec^2 v\,dv, \text{ etc.}$$

XVIII $$d(\text{arc}\sin v) = \frac{dv}{\sqrt{1-v^2}}, \text{ etc.}$$

The term "differentiation" also includes the operation of finding differentials.

In finding differentials the easiest way is to find the derivative as usual, and then multiply the result by $dx$.

ILLUSTRATIVE EXAMPLE 1. Find the differential of

$$y = \frac{x+3}{x^2+3}.$$

**Solution.** $$dy = d\left(\frac{x+3}{x^2+3}\right) = \frac{(x^2+3)\,d(x+3) - (x+3)\,d(x^2+3)}{(x^2+3)^2}$$

$$= \frac{(x^2+3)\,dx - (x+3)\,2\,x dx}{(x^2+3)^2} = \frac{(3-6x-x^2)\,dx}{(x^2+3)^2}. \quad \textit{Ans.}$$

ILLUSTRATIVE EXAMPLE 2. Find $dy$ from

$$b^2x^2 - a^2y^2 = a^2b^2.$$

**Solution.** $$2\,b^2x dx - 2\,a^2y dy = 0.$$

$$\therefore dy = \frac{b^2x}{a^2y}dx. \quad \textit{Ans.}$$

ILLUSTRATIVE EXAMPLE 3. Find $d\rho$ from

$$\rho^2 = a^2\cos 2\theta.$$

**Solution.** $$2\,\rho d\rho = -a^2 \sin 2\theta \cdot 2\,d\theta.$$

$$\therefore d\rho = -\frac{a^2\sin 2\theta}{\rho}d\theta.$$

ILLUSTRATIVE EXAMPLE 4. Find $d[\text{arc}\sin(3t - 4t^3)]$.

**Solution.** $$d[\text{arc}\sin(3t-4t^3)] = \frac{d(3t-4t^3)}{\sqrt{1-(3t-4t^3)^2}} = \frac{3\,dt}{\sqrt{1-t^2}}. \quad \textit{Ans.}$$

**93. Successive differentials.** As the differential of a function is in general also a function of the independent variable, we may deal with its differential. Consider the function

$$y = f(x).$$

$d(dy)$ is called the *second differential of y* (or *of the function*) and is denoted by the symbol $$d^2y.$$

Similarly, the *third differential of y*, $d[d(dy)]$, is written

$$d^3y,$$

and so on, to the *n*th *differential of y*,

$$d^ny.$$

Since $dx$, the differential of the independent variable, is independent of $x$ (see footnote, p. 131), it must be treated as a constant when differentiating with respect to $x$. Bearing this in mind, we get very simple relations between *successive differentials* and *successive derivatives*. For

$$dy = f'(x)\,dx,$$

and

$$d^2y = f''(x)(dx)^2,$$

since $dx$ is regarded as a constant.

Also,

$$d^3y = f'''(x)(dx)^3,$$

and in general

$$d^ny = f^{(n)}(x)(dx)^n.$$

Dividing both sides of each expression by the power of $dx$ occurring on the right, we get our ordinary derivative notation

$$\frac{d^2y}{dx^2} = f''(x),\quad \frac{d^3y}{dx^3} = f'''(x),\quad \cdots,\quad \frac{d^ny}{dx^n} = f^{(n)}(x).$$

Powers of an infinitesimal are called *infinitesimals of a higher order*. More generally, if for the infinitesimals $\alpha$ and $\beta$,

$$\text{limit}\,\frac{\beta}{\alpha} = 0,$$

then $\beta$ is said to be an infinitesimal of a higher order than $\alpha$.

ILLUSTRATIVE EXAMPLE 1. Find the third differential of

$$y = x^5 - 2x^3 + 3x - 5.$$

**Solution.**

$$dy = (5x^4 - 6x^2 + 3)\,dx,$$

$$d^2y = (20x^3 - 12x)(dx)^2,$$

$$d^3y = (60x^2 - 12)(dx)^3.\quad \textit{Ans.}$$

NOTE. This is evidently the third derivative of the function multiplied by the cube of the differential of the independent variable. Dividing through by $(dx)^3$, we get the third derivative

$$\frac{d^3y}{dx^3} = 60x^2 - 12.$$

**EXAMPLES**

Differentiate the following, using differentials:

1. $y = ax^3 - bx^2 + cx + d.$ *Ans.* $dy = (3\,ax^2 - 2\,bx + c)\,dx.$

2. $y = 2\,x^{\frac{5}{2}} - 3\,x^{\frac{2}{3}} + 6\,x^{-1} + 5.$ $dy = (5\,x^{\frac{3}{2}} - 2\,x^{-\frac{1}{3}} - 6\,x^{-2})\,dx.$

3. $y = (a^2 - x^2)^5.$ $dy = -10\,x\,(a^2 - x^2)^4 dx.$

4. $y = \sqrt{1 + x^2}.$ $dy = \dfrac{x}{\sqrt{1 + x^2}}\,dx.$

5. $y = \dfrac{x^{2n}}{(1 + x^2)^n}.$ $dy = \dfrac{2\,nx^{2n-1}}{(1 + x^2)^{n+1}}\,dx.$

6. $y = \log \sqrt{1 - x^3}.$ $dy = \dfrac{3\,x^2 dx}{2\,(x^3 - 1)}.$

7. $y = (e^x + e^{-x})^2.$ $dy = 2\,(e^{2x} - e^{-2x})\,dx.$

8. $y = e^x \log x.$ $dy = e^x\left(\log x + \dfrac{1}{x}\right)dx.$

9. $s = t - \dfrac{e^t - e^{-t}}{e^t + e^{-t}}.$ $ds = \left(\dfrac{e^t - e^{-t}}{e^t + e^{-t}}\right)^2 dt.$

10. $\rho = \tan\phi + \sec\phi.$ $d\rho = \dfrac{1 + \sin\phi}{\cos^2\phi}\,d\phi.$

11. $r = \frac{1}{3}\tan^3\theta + \tan\theta.$ $dr = \sec^4\theta d\theta.$

12. $f(x) = (\log x)^3.$ $f'(x)\,dx = \dfrac{3\,(\log x)^2 dx}{x}.$

13. $\phi(t) = \dfrac{t^3}{(1 - t^2)^{\frac{3}{2}}}.$ $\phi'(t)\,dt = \dfrac{3\,t^2 dt}{(1 - t^2)^{\frac{5}{2}}}.$

14. $d\left[\dfrac{x\log x}{1 - x} + \log(1 - x)\right] = \dfrac{\log x\,dx}{(1 - x)^2}.$

15. $d\,[\text{arc tan}\log y] = \dfrac{dy}{y\,[1 + (\log y)^2]}.$

16. $d\left[r\,\text{arc vers}\dfrac{y}{r} - \sqrt{2\,ry - y^2}\right] = \dfrac{ydy}{\sqrt{2\,ry - y^2}}.$

17. $d\left[\dfrac{\cos\phi}{2\sin^2\phi} - \dfrac{1}{2}\log\tan\dfrac{\phi}{2}\right] = -\dfrac{d\phi}{\sin^3\phi}.$

# CHAPTER X

## RATES

**94. The derivative considered as the ratio of two rates.** Let

$$y = f(x)$$

be the equation of a curve generated by a moving point $P$. Its coördinates $x$ and $y$ may then be considered as functions of the time, as explained in § 71, p. 91. Differentiating with respect to $t$, by **XXV**, we have

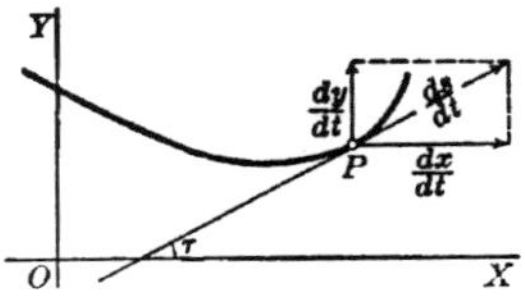


$$\frac{dy}{dt} = f'(x)\frac{dx}{dt}. \tag{32}$$

*At any instant the time rate of change of $y$ (or the function) equals its derivative multiplied by the time rate of change of the independent variable.*

Or, write (**32**) in the form

$$\frac{\frac{dy}{dt}}{\frac{dx}{dt}} = f'(x) = \frac{dy}{dx}. \tag{33}$$

*The derivative measures the ratio of the time rate of change of $y$ to that of $x$.*

$\frac{ds}{dt}$ being the time rate of change of length of arc, we have from (**12**), p. 92,

$$\frac{ds}{dt} = \sqrt{\left(\frac{dx}{dt}\right)^2 + \left(\frac{dy}{dt}\right)^2}, \tag{34}$$

which is the relation indicated by the above figure.

As a guide in solving rate problems use the following **rule**:

First Step. *Draw a figure illustrating the problem. Denote by $x$, $y$, $z$, etc., the quantities which vary with the time.*

Second Step. *Obtain a relation between the variables involved which will hold true at any instant.*

THIRD STEP. *Differentiate with respect to the time.*

FOURTH STEP. *Make a list of the given and required quantities.*

FIFTH STEP. *Substitute the known quantities in the result found by differentiating (third step), and solve for the unknown.*

## EXAMPLES

1. A man is walking at the rate of 5 miles per hour towards the foot of a tower 60 ft. high. At what rate is he approaching the top when he is 80 ft. from the foot of the tower?

**Solution.** Apply the above rule.

*First step.* Draw the figure. Let $x$ = distance of the man from the foot and $y$ = his distance from the top of the tower at any instant.

*Second step.* Since we have a right triangle,

$$y^2 = x^2 + 3600.$$

*Third step.* Differentiating, we get

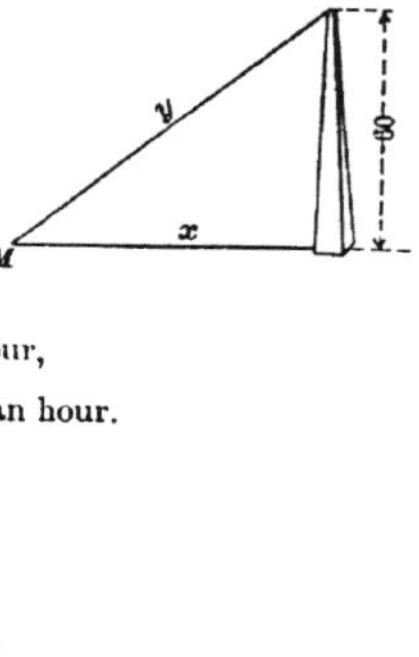


$$2y\frac{dy}{dt} = 2x\frac{dx}{dt}, \text{ or,}$$

$$(A) \quad \frac{dy}{dt} = \frac{x}{y}\frac{dx}{dt},$$ meaning that at *any* instant whatever

*(Rate of change of y)* $= \left(\frac{x}{y}\right)$ *(rate of change of x).*

*Fourth step.* $x = 80$, $\frac{dx}{dt} = 5$ miles an hour,

$= 5 \times 5280$ ft. an hour.

$y = \sqrt{x^2 + 3600}$ $= 100.$ $\frac{dy}{dt} = ?$

*Fifth step.* Substituting back in $(A)$,

$$\frac{dy}{dt} = \frac{80}{100} \times 5 \times 5280 \text{ ft. per hour}$$
$$= 4 \text{ miles per hour.} \quad \textit{Ans.}$$

2. A point moves on the parabola $6y = x^2$ in such a way that when $x = 6$, the abscissa is increasing at the rate of 2 ft. per second. At what rates are the ordinate and length of arc increasing at the same instant?

**Solution.** *First step.* Plot the parabola.

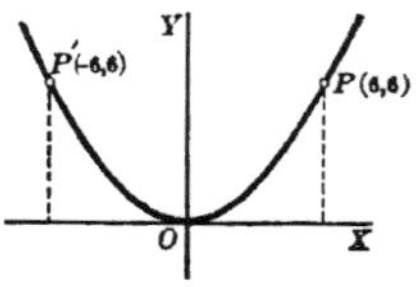


*Second step.* $6y = x^2.$

*Third step.* $$6\frac{dy}{dt} = 2x\frac{dx}{dt}, \text{ or,}$$

$$(B) \quad \frac{dy}{dt} = \frac{x}{3}\cdot\frac{dx}{dt}.$$

This means that at *any* point on the parabola

*(Rate of change of ordinate)* $= \left(\frac{x}{3}\right)$ *(rate of change of abscissa).*

*Fourth step.* $\frac{dx}{dt} = 2$ ft. per second.

$x = 6.$ $\frac{dy}{dt} = ?$

$y = \frac{x^2}{6} = 6.$ $\frac{ds}{dt} = ?$

*Fifth step.* Substituting back in $(B)$,

$$\frac{dy}{dt} = \frac{6}{3} \times 2 = 4 \text{ ft. per second. } \textit{Ans.}$$

Substituting in (34), p. 141,

$$\frac{ds}{dt} = \sqrt{(2)^2 + (4)^2} = 2\sqrt{5} \text{ ft. per second. } \textit{Ans.}$$

From the first result we note that at the point $P$ (6, 6) the ordinate changes twice as rapidly as the abscissa.

If we consider the point $P'$ (– 6, 6) instead, the result is $\frac{dy}{dt} = -4$ ft. per second, the minus sign indicating that the ordinate is decreasing as the abscissa increases.

3. A circular plate of metal expands by heat so that its radius increases uniformly at the rate of .01 inch per second. At what rate is the surface increasing when the radius is two inches?

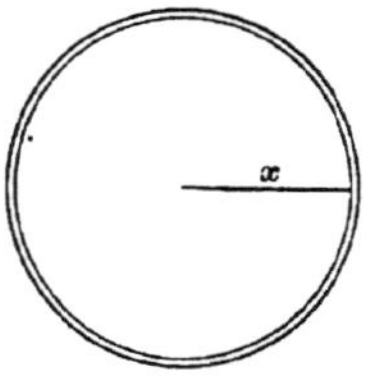


**Solution.** Let $x$ = radius and $y$ = area of plate. Then

$$y = \pi x^2.$$

$$(C) \qquad \frac{dy}{dt} = 2\pi x \frac{dx}{dt}.$$

That is, at any instant the area of the plate is increasing in square inches $2\pi x$ times as fast as the radius is increasing in linear inches.

$$x = 2, \frac{dx}{dt} = .01, \frac{dy}{dt} = ?$$

Substituting in $(C)$,

$$\frac{dy}{dt} = 2\pi \times 2 \times .01 = .04\pi \text{ sq. in. per sec. } \textit{Ans.}$$

4. An arc light is hung 12 ft. directly above a straight horizontal walk on which a boy 5 ft. in height is walking. How fast is the boy's shadow lengthening when he is walking away from the light at the rate of 168 ft. per minute?

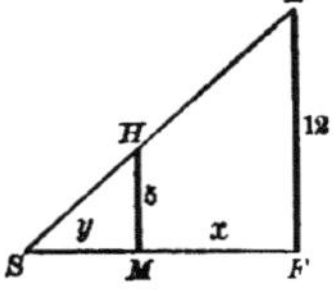


**Solution.** Let $x$ = distance of boy from a point directly under light $L$, and $y$ = length of boy's shadow. From the figure,

$$y : y + x :: 5 : 12,$$

or

$$y = \tfrac{5}{7}x.$$

Differentiating,

$$\frac{dy}{dt} = \frac{5}{7}\frac{dx}{dt};$$

i.e. the shadow is lengthening $\frac{5}{7}$ as fast as the boy is walking, or 120 ft. per minute.

5. In a parabola $y^2 = 12x$, if $x$ increases uniformly at the rate of 2 in. per second, at what rate is $y$ increasing when $x = 3$ in.? *Ans.* 2 in. per sec.

**6.** At what point on the parabola of the last example do the abscissa and ordinate increase at the same rate? *Ans.* (3, 6).

**7.** In the function $y = 2x^3 + 6$, what is the value of $x$ at the point where $y$ increases 24 times as fast as $x$? *Ans.* $x = \pm 2$.

**8.** The ordinate of a point describing the curve $x^2 + y^2 = 25$ is decreasing at the rate of $1\frac{1}{2}$ in. per second. How rapidly is the abscissa changing when the ordinate is 4 inches?

*Ans.* $\frac{dx}{dt} = 2$ in. per sec.

**9.** Find the values of $x$ at the points where the rate of change of

$$x^3 - 12x^2 + 45x - 13$$

is zero. *Ans.* $x = 3$ and 5.

**10.** At what points on the ellipse $16x^2 + 9y^2 = 400$ does $y$ decrease at the same rate that $x$ increases? *Ans.* $(3, \frac{16}{3})$, $(-3, -\frac{16}{3})$.

**11.** Where in the first quadrant, on $x^2 + y^2 = r^2$, does the arc increase twice as fast as the ordinate? *Ans.* At 60°.

A point generates each of the following curves. Find the rate at which the arc is increasing in each case:

**12.** $y^2 = 2x$; $\frac{dx}{dt} = 2$, $x = 2$. *Ans.* $\frac{ds}{dt} = \sqrt{5}$.

**13.** $xy = 6$; $\frac{dy}{dt} = 2$, $y = 3$. $\frac{ds}{dt} = \frac{2}{3}\sqrt{13}$.

**14.** $x^2 + 4y^2 = 20$; $\frac{dx}{dt} = -1$, $y = 1$. $\frac{ds}{dt} = \sqrt{2}$.

**15.** $y = x^3$; $\frac{dx}{dt} = 3$, $x = -3$.

**16.** $y^2 = x^3$; $\frac{dy}{dt} = 4$, $y = 8$.

**17.** The side of an equilateral triangle is 24 inches long, and is increasing at the rate of 3 inches per hour. How fast is the area increasing?

*Ans.* $36\sqrt{3}$ sq. in. per hour.

**18.** Find the rate of change of the area of a square when the side $b$ is increasing at the rate of $a$ units per second. *Ans.* $2ab$ sq. units per sec.

**19.** (a) The volume of a spherical soap bubble increases how many times as fast as the radius? (b) When its radius is 4 in. and increasing at the rate of $\frac{1}{2}$ in. per second, how fast is the volume increasing? *Ans.* (a) $4\pi r^2$ times as fast;
(b) $32\pi$ cu. in. per sec.

How fast is the surface increasing in the last case?

**20.** One end of a ladder 50 ft. long is leaning against a perpendicular wall standing on a horizontal plane. Supposing the foot of the ladder to be pulled away from the wall at the rate of 3 ft. per minute; (a) how fast is the top of the ladder descending when the foot is 14 ft. from the wall? (b) when will the top and bottom of the ladder move at the same rate? (c) when is the top of the ladder descending at the rate of 4 ft. per minute?

*Ans.* (a) $\frac{7}{8}$ ft. per min.;
(b) when $25\sqrt{2}$ ft. from wall;
(c) when 40 ft. from wall.

**21.** A barge whose deck is 12 ft. below the level of a dock is drawn up to it by means of a cable attached to a ring in the floor of the dock, the cable being hauled in by a windlass on deck at the rate of 8 ft. per minute. How fast is the barge moving towards the dock when 16 ft. away? *Ans.* 10 ft. per minute.

**22.** An elevated car is 40 ft. immediately above a surface car, their tracks intersecting at right angles. If the speed of the elevated car is 16 miles per hour and of the surface car 8 miles per hour, at what rate are the cars separating 5 minutes after they meet? *Ans.* 17.9 miles per hour.

**23.** One ship was sailing south at the rate of 6 miles per hour; another east at the rate of 8 miles per hour. At 4 P.M. the second crossed the track of the first where the first was two hours before; (a) how was the distance between the ships changing at 3 P.M.? (b) how at 5 P.M.? (c) when was the distance between them not changing?

*Ans.* (a) Diminishing 2.8 miles per hour;
(b) increasing 8.73 miles per hour;
(c) 3 : 17 P.M.

**24.** Assuming the volume of the wood in a tree to be proportional to the cube of its diameter, and that the latter increases uniformly year by year when growing, show that the rate of growth when the diameter is 3 ft. is 36 times as great as when the diameter is 6 inches.

**25.** A railroad train is running 15 miles an hour past a station 800 ft. long, the track having the form of the parabola

$$y^2 = 600\,x,$$

and situated as shown in the figure. If the sun is just rising in the east, find how fast the shadow $S$ of the locomotive $L$ is moving along the wall of the station at the instant it reaches the end of the wall.

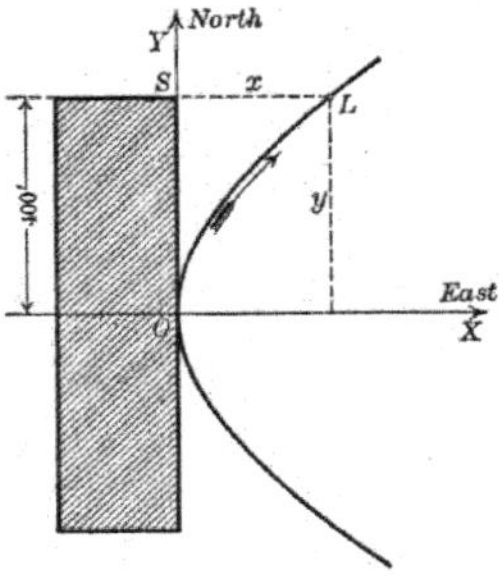


**Solution.**

$$y^2 = 600\,x.$$

$$2y\frac{dy}{dt} = 600\frac{dx}{dt},$$

or,

$$\frac{dx}{dt} = \frac{y}{300}\frac{dy}{dt}.$$

Substituting this value of $\frac{dx}{dt}$ in

$$\frac{ds}{dt} = \sqrt{\left(\frac{dx}{dt}\right)^2 + \left(\frac{dy}{dt}\right)^2},\text{ we get}$$

$$(D)\qquad \left(\frac{ds}{dt}\right)^2 = \left(\frac{y}{300}\frac{dy}{dt}\right)^2 + \left(\frac{dy}{dt}\right)^2.$$

Now

$$\frac{ds}{dt} = 15 \text{ miles per hour}$$
$$= 22 \text{ ft. per sec.}$$

$$y = 400 \text{ and } \frac{dy}{dt} = ?$$

Substituting back in $(D)$, we get

$$(22)^2 = \left(\frac{16}{9} + 1\right)\left(\frac{dy}{dt}\right)^2,$$

or,

$$\frac{dy}{dt} = 13\tfrac{1}{5} \text{ ft. per second. } \textit{Ans.}$$

**26.** An express train and a balloon start from the same point at the same instant. The former travels 50 miles an hour and the latter rises at the rate of 10 miles an hour. How fast are they separating? *Ans.* 51 miles an hour.

27. A man 6 ft. tall walks away from a lamp-post 10 ft. high at the rate of 4 miles an hour. How fast does the shadow of his head move? *Ans.* 10 miles an hour.

28. The rays of the sun make an angle of 30° with the horizon. A ball is thrown vertically upward to a height of 64 ft. How fast is the shadow of the ball moving along the ground just before it strikes the ground? *Ans.* 110.8 ft. per sec.

29. A ship is anchored in 18 ft. of water. The cable passes over a sheave on the bow 6 ft. above the surface of the water. If the cable is taken in at the rate of 1 ft. a second, how fast is the ship moving when there are 30 ft. of cable out?

*Ans.* $1\frac{2}{3}$ ft. per sec.

30. A man is hoisting a chest to a window 50 ft. up by means of a block and tackle. If he pulls in the rope at the rate of 10 ft. a minute while walking away from the building at the rate of 5 ft. a minute, how fast is the chest rising at the end of the second minute? *Ans.* 10.98 ft. per min.

31. Water flows from a faucet into a hemispherical basin of diameter 14 inches at the rate of 2 cu. in. per second. How fast is the water rising (a) when the water is halfway to the top? (b) just as it runs over? (The volume of a spherical segment $= \frac{1}{2}\pi r^2 h + \frac{1}{6}\pi h^3$, where $h =$ altitude of segment.)

32. Sand is being poured on the ground from the orifice of an elevated pipe, and forms a pile which has always the shape of a right circular cone whose height is equal to the radius of the base. If sand is falling at the rate of 6 cu. ft. per sec., how fast is the height of the pile increasing when the height is 5 ft.?

33. An aëroplane is 528 ft. directly above an automobile and starts east at the rate of 20 miles an hour at the same instant the automobile starts east at the rate of 40 miles an hour. How fast are they separating?

34. A revolving light sending out a bundle of parallel rays is at a distance of $\frac{1}{2}$ a mile from the shore and makes 1 revolution a minute. Find how fast the light is traveling along the straight beach when at a distance of 1 mile from the nearest point of the shore. *Ans.* 15.7 miles per min.

35. A kite is 150 ft. high and 200 ft. of string are out. If the kite starts drifting away horizontally at the rate of 4 miles an hour, how fast is the string being paid out at the start? *Ans.* 2.64 miles an hour.

36. A solution is poured into a conical filter of base radius 6 cm. and height 24 cm. at the rate of 2 cu. cm. a second, and filters out at the rate of 1 cu. cm. a second. How fast is the level of the solution rising when (a) one third of the way up? (b) at the top? *Ans.* (a) .079 cm. per sec.;
(b) .009 cm. per sec.

37. A horse runs 10 miles per hour on a circular track in the center of which is an arc light. How fast will his shadow move along a straight board fence (tangent to the track at the starting point) when he has completed one eighth of the circuit?

*Ans.* 20 miles per hour.

38. The edges of a cube are 24 inches and are increasing at the rate of .02 in. per minute. At what rate is (a) the volume increasing? (b) the area increasing?

39. The edges of a regular tetrahedron are 10 inches and are increasing at the rate of .3 in. per hour. At what rate is (a) the volume increasing? (b) the area increasing?

40. An electric light hangs 40 ft. from a stone wall. A man is walking 12 ft. per second on a straight path 10 ft. from the light and perpendicular to the wall. How fast is the man's shadow moving when he is 30 ft. from the wall? *Ans.* 48 ft. per sec.

**41.** The approach to a drawbridge has a gate whose two arms rotate about the same axis as shown in the figure. The arm over the driveway is 4 yards long and the arm over the footwalk is 3 yards long. Both arms rotate at the rate of 5 radians per minute. At what rate is the distance between the extremities of the arms changing when they make an angle of 45° with the horizontal?

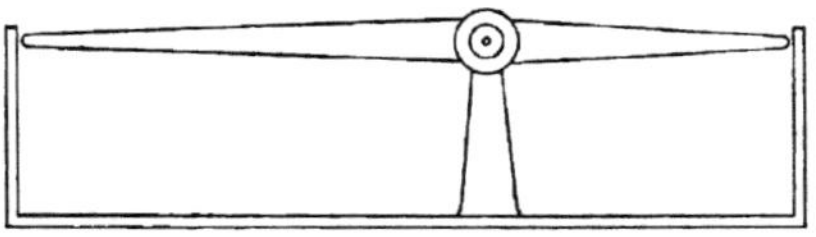

*Ans.* 24 yd. per min.

**42.** A conical funnel of radius 3 inches and of the same depth is filled with a solution which filters at the rate of 1 cu. in. per minute. How fast is the surface falling when it is 1 inch from the top of the funnel?

*Ans.* $\frac{1}{4\pi}$ in. per min.

**43.** An angle is increasing at a constant rate. Show that the tangent and sine are increasing at the same rate when the angle is zero, and that the tangent increases eight times as fast as the sine when the angle is 60°.

# CHAPTER XI

## CHANGE OF VARIABLE

**95. Interchange of dependent and independent variables.** It is sometimes desirable to transform an expression involving derivatives of $y$ with respect to $x$ into an equivalent expression involving instead derivatives of $x$ with respect to $y$. Our examples will show that in many cases such a change transforms the given expression into a much simpler one. Or perhaps $x$ is given as an explicit function of $y$ in a problem, and it is found more convenient to use a formula involving $\frac{dx}{dy}, \frac{d^2x}{dy^2}$, etc., than one involving $\frac{dy}{dx}, \frac{d^2y}{dx^2}$, etc. We shall now proceed to find the formulas necessary for making such transformations.

Given $y=f(x)$, then from **XXVI** we have

$$\frac{dy}{dx}=\frac{1}{\frac{dx}{dy}}, \qquad \frac{dx}{dy}\neq 0 \tag{35}$$

giving $\frac{dy}{dx}$ in terms of $\frac{dx}{dy}$. Also, by **XXV**,

$$\frac{d^2y}{dx^2}=\frac{d}{dx}\left(\frac{dy}{dx}\right)=\frac{d}{dy}\left(\frac{dy}{dx}\right)\frac{dy}{dx},$$

or

$$\frac{d^2y}{dx^2}=\frac{d}{dy}\left(\frac{1}{\frac{dx}{dy}}\right)\frac{dy}{dx}. \tag{A}$$

But $$\frac{d}{dy}\left(\frac{1}{\frac{dx}{dy}}\right)=-\frac{\frac{d^2x}{dy^2}}{\left(\frac{dx}{dy}\right)^2}; \text{ and } \frac{dy}{dx}=\frac{1}{\frac{dx}{dy}} \text{ from (35).}$$

Substituting these in $(A)$, we get

$$\frac{d^2y}{dx^2}=-\frac{\frac{d^2x}{dy^2}}{\left(\frac{dx}{dy}\right)^3}, \tag{36}$$

giving $\frac{d^2y}{dx^2}$ in terms of $\frac{dx}{dy}$ and $\frac{d^2x}{dy^2}$. Similarly,

$$(37) \qquad \frac{d^3y}{dx^3} = -\frac{\frac{d^3x}{dy^3}\frac{dx}{dy} - 3\left(\frac{d^2x}{dy^2}\right)^2}{\left(\frac{dx}{dy}\right)^5};$$

and so on for higher derivatives. This transformation is called *changing the independent variable from x to y.*

ILLUSTRATIVE EXAMPLE 1. Change the independent variable from $x$ to $y$ in the equation

$$3\left(\frac{d^2y}{dx^2}\right)^2 - \frac{dy}{dx}\frac{d^3y}{dx^3} - \frac{d^2y}{dx^2}\left(\frac{dy}{dx}\right)^2 = 0.$$

**Solution.** Substituting from (35), (36), (37),

$$3\left(-\frac{\frac{d^2x}{dy^2}}{\left(\frac{dx}{dy}\right)^3}\right)^2 - \left(\frac{1}{\frac{dx}{dy}}\right)\left(-\frac{\frac{d^3x}{dy^3}\frac{dx}{dy} - 3\left(\frac{d^2x}{dy^2}\right)^2}{\left(\frac{dx}{dy}\right)^5}\right) - \left(-\frac{\frac{d^2x}{dy^2}}{\left(\frac{dx}{dy}\right)^3}\right)\left(\frac{1}{\frac{dx}{dy}}\right)^2 = 0.$$

Reducing, we get

$$\frac{d^3x}{dy^3} + \frac{d^2x}{dy^2} = 0,$$

a much simpler equation.

**96. Change of the dependent variable.** Let

$$(A) \qquad y = f(x),$$

and suppose at the same time $y$ is a function of $z$, say

$$(B) \qquad y = \phi(z).$$

We may then express $\frac{dy}{dx}$, $\frac{d^2y}{dx^2}$, etc., in terms of $\frac{dz}{dx}$, $\frac{d^2z}{dx^2}$, etc., as follows.

In general, $z$ is a function of $y$ by (*B*), p. 45; and since $y$ is a function of $x$ by (*A*), it is evident that $z$ is a function of $x$. Hence by **XXV** we have

$$(C) \qquad \frac{dy}{dx} = \frac{dy}{dz}\frac{dz}{dx} = \phi'(z)\frac{dz}{dx}.$$

Also $$\frac{d^2y}{dx^2} = \frac{d}{dx}\left(\phi'(z)\frac{dz}{dx}\right) = \frac{dz}{dx}\frac{d}{dx}\phi'(z) + \phi'(z)\frac{d^2z}{dx^2}.$$ By **V**

But $$\frac{d}{dx}\phi'(z) = \frac{d}{dz}\phi'(z)\frac{dz}{dx} = \phi''(z)\frac{dz}{dx}.$$ By **XXV**

$$(D) \qquad \therefore \frac{d^2y}{dx^2} = \phi''(z)\left(\frac{dz}{dx}\right)^2 + \phi'(z)\frac{d^2z}{dx^2}.$$

Similarly for higher derivatives. This transformation is called *changing the dependent variable from y to z*, the independent variable remaining $x$ throughout. We will now illustrate this process by means of an example.

ILLUSTRATIVE EXAMPLE 1. Having given the equation

$$(E) \qquad \frac{d^2y}{dx^2} = 1 + \frac{2(1+y)}{1+y^2}\left(\frac{dy}{dx}\right)^2,$$

change the dependent variable from $y$ to $z$ by means of the relation

$$(F) \qquad y = \tan z.$$

**Solution.** From $(F)$,

$$\frac{dy}{dx} = \sec^2 z\frac{dz}{dx}, \quad \frac{d^2y}{dx^2} = \sec^2 z\frac{d^2z}{dx^2} + 2\sec^2 z\tan z\left(\frac{dz}{dx}\right)^2.$$

Substituting in $(E)$,

$$\sec^2 z\frac{d^2z}{dx^2} + 2\sec^2 z\tan z\left(\frac{dz}{dx}\right)^2 = 1 + \frac{2(1+\tan z)}{1+\tan^2 z}\left(\sec^2 z\frac{dz}{dx}\right)^2,$$

and reducing, we get $\dfrac{d^2z}{dx^2} - 2\left(\dfrac{dz}{dx}\right)^2 = \cos^2 z.$ *Ans.*

**97. Change of the independent variable.** Let $y$ be a function of $x$, and at the same time let $x$ (and hence also $y$) be a function of a new variable $t$. It is required to express

$$\frac{dy}{dx}, \frac{d^2y}{dx^2}, \text{ etc.,}$$

in terms of new derivatives having $t$ as the independent variable. By **XXV**

$$\frac{dy}{dt} = \frac{dy}{dx}\frac{dx}{dt}, \text{ or}$$

$$(A) \qquad \frac{dy}{dx} = \frac{\dfrac{dy}{dt}}{\dfrac{dx}{dt}}.$$

Also

$$\frac{d^2y}{dx^2} = \frac{d}{dx}\left(\frac{dy}{dx}\right) = \frac{d}{dt}\left(\frac{dy}{dx}\right)\frac{dt}{dx} = \frac{\dfrac{d}{dt}\left(\dfrac{dy}{dx}\right)}{\dfrac{dx}{dt}}.$$

But differentiating $(A)$ with respect to $t$,

$$\frac{d}{dt}\left(\frac{dy}{dx}\right) = \frac{d}{dt}\left(\frac{\dfrac{dy}{dt}}{\dfrac{dx}{dt}}\right) = \frac{\dfrac{dx}{dt}\dfrac{d^2y}{dt^2} - \dfrac{dy}{dt}\dfrac{d^2x}{dt^2}}{\left(\dfrac{dx}{dt}\right)^2}.$$

Therefore

$$(B)\qquad \frac{d^2y}{dx^2}=\frac{\frac{dx}{dt}\frac{d^2y}{dt^2}-\frac{dy}{dt}\frac{d^2x}{dt^2}}{\left(\frac{dx}{dt}\right)^3};$$

and so on for higher derivatives. This transformation is called *changing the independent variable from x to t.* It is usually better to work out examples by the methods illustrated above rather than by using the formulas deduced.

ILLUSTRATIVE EXAMPLE 1. Change the independent variable from $x$ to $t$ in the equation.

$$(C)\qquad x^2\frac{d^2y}{dx^2}+x\frac{dy}{dx}+y=0,$$

by means of the relation

$$(D)\qquad x=e^t.$$

Solution. $\frac{dx}{dt}=e^t$; therefore

$$(E)\qquad \frac{dt}{dx}=e^{-t}.$$

Also $\frac{dy}{dx}=\frac{dy}{dt}\frac{dt}{dx}$; therefore

$$(F)\qquad \frac{dy}{dx}=e^{-t}\frac{dy}{dt}.$$

Also
$$\frac{d^2y}{dx^2}=e^{-t}\frac{d}{dx}\left(\frac{dy}{dt}\right)-\frac{dy}{dt}e^{-t}\frac{dt}{dx}=e^{-t}\frac{d}{dt}\left(\frac{dy}{dt}\right)\frac{dt}{dx}-\frac{dy}{dt}e^{-t}\frac{dt}{dx}.$$

Substituting in the last result from $(E)$,

$$(G)\qquad \frac{d^2y}{dx^2}=e^{-2t}\frac{d^2y}{dt^2}-\frac{dy}{dt}e^{-2t}.$$

Substituting $(D)$, $(F)$, $(G)$ in $(C)$,

$$e^{2t}\left(e^{-2t}\frac{d^2y}{dt^2}-\frac{dy}{dt}e^{-2t}\right)+e^t\left(e^{-t}\frac{dy}{dt}\right)+y=0;$$

and reducing, we get $\frac{d^2y}{dt^2}+y=0.$ *Ans.*

Since the formulas deduced in the Differential Calculus generally involve derivatives of $y$ with respect to $x$, such formulas as $(A)$ and $(B)$ are especially useful when the parametric equations of a curve are given. Such examples were given on pp. 82, 83, and many others will be employed in what follows.

**98. Simultaneous change of both independent and dependent variables.** It is often desirable to change both variables simultaneously. An important case is that arising in the transformation from rectangular to polar coördinates. Since

$$x = \rho \cos\theta \text{ and } y = \rho \sin\theta,$$

the equation

$$f(x, y) = 0$$

becomes by substitution an equation between $\rho$ and $\theta$, defining $\rho$ as a function of $\theta$. Hence $\rho$, $x$, $y$ are all functions of $\theta$.

ILLUSTRATIVE EXAMPLE 1. Transform the formula for the radius of curvature (42), p. 159,

$$(A) \qquad R = \frac{\left[1 + \left(\frac{dy}{dx}\right)^2\right]^{\frac{3}{2}}}{\frac{d^2y}{dx^2}},$$

into polar coördinates.

**Solution.** Since in $(A)$ and $(B)$, pp. 150, 151, $t$ is any variable on which $x$ and $y$ depend, we may in this case let $t = \theta$, giving

$$(B) \qquad \frac{dy}{dx} = \frac{\frac{dy}{d\theta}}{\frac{dx}{d\theta}}, \text{ and}$$

$$(C) \qquad \frac{d^2y}{dx^2} = \frac{\frac{dx}{d\theta}\frac{d^2y}{d\theta^2} - \frac{dy}{d\theta}\frac{d^2x}{d\theta^2}}{\left(\frac{dx}{d\theta}\right)^3}.$$

Substituting $(B)$ and $(C)$ in $(A)$, we get

$$R = \left[\frac{\left(\frac{dx}{d\theta}\right)^2 + \left(\frac{dy}{d\theta}\right)^2}{\left(\frac{dx}{d\theta}\right)^2}\right]^{\frac{3}{2}} \div \frac{\frac{dx}{d\theta}\frac{d^2y}{d\theta^2} - \frac{dy}{d\theta}\frac{d^2x}{d\theta^2}}{\left(\frac{dx}{d\theta}\right)^3}, \text{ or}$$

$$(D) \qquad R = \frac{\left[\left(\frac{dx}{d\theta}\right)^2 + \left(\frac{dy}{d\theta}\right)^2\right]^{\frac{3}{2}}}{\frac{dx}{d\theta}\frac{d^2y}{d\theta^2} - \frac{dy}{d\theta}\frac{d^2x}{d\theta^2}}.$$

But since $x = \rho\cos\theta$ and $y = \rho\sin\theta$, we have

$$\frac{dx}{d\theta} = -\rho\sin\theta + \cos\theta\frac{d\rho}{d\theta}; \quad \frac{dy}{d\theta} = \rho\cos\theta + \sin\theta\frac{d\rho}{d\theta};$$

$$\frac{d^2x}{d\theta^2} = -\rho\cos\theta - 2\sin\theta\frac{d\rho}{d\theta} + \cos\theta\frac{d^2\rho}{d\theta^2}; \quad \frac{d^2y}{d\theta^2} = -\rho\sin\theta + 2\cos\theta\frac{d\rho}{d\theta} + \sin\theta\frac{d^2\rho}{d\theta^2}.$$

Substituting these in $(D)$ and reducing,

$$R = \frac{\left[\rho^2 + \left(\frac{d\rho}{d\theta}\right)^2\right]^{\frac{3}{2}}}{\rho^2 + 2\left(\frac{d\rho}{d\theta}\right)^2 - \rho\frac{d^2\rho}{d\theta^2}}. \quad \textit{Ans.}$$

## EXAMPLES

Change the independent variable from $x$ to $y$ in the four following equations:

**1.** $R = \dfrac{\left[1+\left(\dfrac{dy}{dx}\right)^2\right]^{\frac{3}{2}}}{\dfrac{d^2y}{dx^2}}.$ *Ans.* $R = -\dfrac{\left[1+\left(\dfrac{dx}{dy}\right)^2\right]^{\frac{3}{2}}}{\dfrac{d^2x}{dy^2}}.$

**2.** $\dfrac{d^2y}{dx^2} + 2y\left(\dfrac{dy}{dx}\right)^2 = 0.$ $\dfrac{d^2x}{dy^2} - 2y\dfrac{dx}{dy} = 0.$

**3.** $x\dfrac{d^2y}{dx^2} + \left(\dfrac{dy}{dx}\right)^3 - \dfrac{dy}{dx} = 0.$ $x\dfrac{d^2x}{dy^2} - 1 + \left(\dfrac{dx}{dy}\right)^2 = 0.$

**4.** $\left(3a\dfrac{dy}{dx} + 2\right)\left(\dfrac{d^2y}{dx^2}\right)^2 = \left(a\dfrac{dy}{dx} + 1\right)\dfrac{dy}{dx}\dfrac{d^3y}{dx^3}.$ $\left(\dfrac{d^2x}{dy^2}\right)^2 = \left(\dfrac{dx}{dy} + a\right)\dfrac{d^3x}{dy^3}.$

Change the dependent variable from $y$ to $z$ in the following equations:

**5.** $(1+y)^2\left(\dfrac{d^3y}{dx^3} - 2y\right) + \left(\dfrac{dy}{dx}\right)^3 = 2\left(1+y\right)\dfrac{dy}{dx}\dfrac{d^2y}{dx^2},\ y = z^2 + 2z.$

*Ans.* $(z+1)\dfrac{d^3z}{dx^3} = \dfrac{dz}{dx}\dfrac{d^2z}{dx^2} + z^2 + 2z.$

**6.** $\dfrac{d^2y}{dx^2} = 1 + \dfrac{2(1+y)}{1+y^2}\left(\dfrac{dy}{dx}\right)^2,\ y = \tan z.$ *Ans.* $\dfrac{d^2z}{dx^2} - 2\left(\dfrac{dz}{dx}\right)^2 = \cos^2 z.$

**7.** $y^2\dfrac{d^3y}{dx^3} - \left(3y\dfrac{dy}{dx} + 2xy^2\right)\dfrac{d^2y}{dx^2} + \left\{2\left(\dfrac{dy}{dx}\right)^2 + 2xy\dfrac{dy}{dx} + 3x^2y^2\right\}\dfrac{dy}{dx} + x^3y^3 = 0,\ y = e^z.$

*Ans.* $\dfrac{d^3z}{dx^3} - 2x\dfrac{d^2z}{dx^2} + 3x^2\dfrac{dz}{dx} + x^3 = 0.$

Change the independent variable in the following eight equations:

**8.** $\dfrac{d^2y}{dx^2} - \dfrac{x}{1-x^2}\dfrac{dy}{dx} + \dfrac{y}{1-x^2} = 0,$ $x = \cos t.$ *Ans.* $\dfrac{d^2y}{dt^2} + y = 0.$

**9.** $(1-x^2)\dfrac{d^2y}{dx^2} - x\dfrac{dy}{dx} = 0,$ $x = \cos z.$ $\dfrac{d^2y}{dz^2} = 0.$

**10.** $(1-y^2)\dfrac{d^2u}{dy^2} - y\dfrac{du}{dy} + a^2u = 0,$ $y = \sin x.$ $\dfrac{d^2u}{dx^2} + a^2u = 0.$

**11.** $x^2\dfrac{d^2y}{dx^2} + 2x\dfrac{dy}{dx} + \dfrac{a^2}{x^2}y = 0,$ $x = \dfrac{1}{z}.$ $\dfrac{d^2y}{dz^2} + a^2y = 0.$

**12.** $x^3\dfrac{d^3v}{dx^3} + 3x^2\dfrac{d^2v}{dx^2} + x\dfrac{dv}{dx} + v = 0,$ $x = e^t.$ $\dfrac{d^3v}{dt^3} + v = 0.$

**13.** $\dfrac{d^2y}{dx^2} + \dfrac{2x}{1+x^2}\dfrac{dy}{dx} + \dfrac{y}{(1+x^2)^2} = 0,$ $x = \tan\theta.$ $\dfrac{d^2y}{d\theta^2} + y = 0.$

**14.** $\dfrac{d^2u}{ds^2} + su\dfrac{du}{ds} + \sec^2 s = 0,$ $s = \operatorname{arc}\tan t.$

*Ans.* $(1+t^2)\dfrac{d^2u}{dt^2} + (2t + u\operatorname{arc}\tan t)\dfrac{du}{dt} + 1 = 0.$

**15.** $x^4\dfrac{d^2y}{dx^2} + a^2y = 0,$ $x = \dfrac{1}{z}.$ *Ans.* $\dfrac{d^2y}{dz^2} + \dfrac{2}{z}\dfrac{dy}{dz} + a^2y = 0.$

In the following seven examples the equations are given in parametric form. Find $\frac{dy}{dx}$ and $\frac{d^2y}{dx^2}$ in each case:

**16.** $x = 7 + t^2,\ y = 3 + t^2 - 3t^4.$ *Ans.* $\frac{dy}{dx} = 1 - 6t^2,\ \frac{d^2y}{dx^2} = -6.$

**17.** $x = \cot t,\ y = \sin^3 t.$ *Ans.* $\frac{dy}{dx} = -3\sin^4 t\cos t,\ \frac{d^2y}{dx^2} = 3\sin^5 t(4 - 5\sin^2 t).$

**18.** $x = a(\cos t + t\sin t),\ y = a(\sin t - t\cos t).$ *Ans.* $\frac{dy}{dx} = \tan t,\ \frac{d^2y}{dx^2} = \frac{1}{at\cos^3 t}.$

**19.** $x = \frac{1-t}{1+t},\ y = \frac{2t}{1+t}.$

**20.** $x = 2t,\ y = 2 - t^2.$

**21.** $x = 1 - t^2,\ y = t^3.$

**22.** $x = a\cos t,\ y = b\sin t.$

**23.** Transform $\dfrac{x\frac{dy}{dx} - y}{\sqrt{1 + \left(\frac{dy}{dx}\right)^2}}$ by assuming $x = \rho\cos\theta,\ y = \rho\sin\theta.$

*Ans.* $\dfrac{\rho^2}{\sqrt{\rho^2 + \left(\frac{d\rho}{d\theta}\right)^2}}.$

**24.** Let $f(x, y) = 0$ be the equation of a curve. Find an expression for its slope $\left(\frac{dy}{dx}\right)$ in terms of polar coördinates.

*Ans.* $\dfrac{dy}{dx} = \dfrac{\rho\cos\theta + \sin\theta\frac{d\rho}{d\theta}}{-\rho\sin\theta + \cos\theta\frac{d\rho}{d\theta}}.$

# CHAPTER XII

## CURVATURE. RADIUS OF CURVATURE

**99. Curvature.** The shape of a curve depends very largely upon the rate at which the direction of the tangent changes as the point of contact describes the curve. This rate of change of direction is called *curvature* and is denoted by $K$. We now proceed to find its analytical expression, first for the simple case of the circle, and then for curves in general.

**100. Curvature of a circle.** Consider a circle of radius $R$. Let

$\tau =$ angle that the tangent at $P$ makes with $OX$, and

$\tau + \Delta\tau =$ angle made by the tangent at a neighboring point $P'$.

Then we say

$\Delta\tau =$ *total curvature of arc* $PP'$.

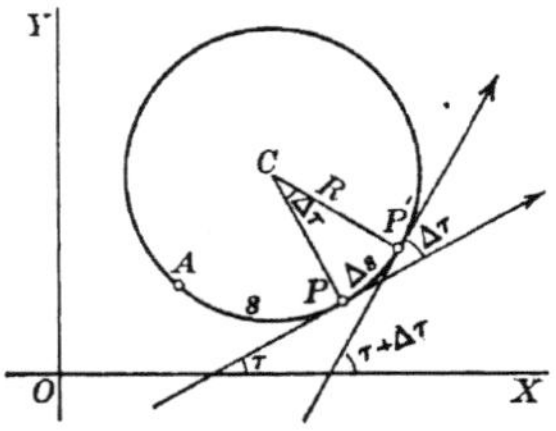


If the point $P$ with its tangent be supposed to move along the curve to $P'$, the total curvature $(= \Delta\tau)$ would measure the total change in direction, or rotation, of the tangent; or, what is the same thing, the total change in direction of the arc itself. Denoting by $s$ the length of the arc of the curve measured from some fixed point (as $A$) to $P$, and by $\Delta s$ the length of the arc $PP'$, then the ratio

$$\frac{\Delta\tau}{\Delta s}$$

measures the average change in direction per unit length of arc.* Since, from the figure,

$$\Delta s = R \cdot \Delta\tau,$$

or

$$\frac{\Delta\tau}{\Delta s} = \frac{1}{R},$$

* Thus, if $\Delta\tau = \frac{\pi}{6}$ radians (= 30°), and $\Delta s = 3$ centimeters, then $\frac{\Delta\tau}{\Delta s} = \frac{\pi}{18}$ radians per centimeter = 10° per centimeter = average rate of change of direction.

it is evident that this ratio is constant everywhere on the circle. This ratio is, by definition, the *curvature of the circle*, and we have

$$K = \frac{1}{R}. \tag{38}$$

*The curvature of a circle equals the reciprocal of its radius.*

**101. Curvature at a point.** Consider any curve. As in the last section,

$$\Delta\tau = \textit{total curvature of the arc } PP',$$

and

$$\frac{\Delta\tau}{\Delta s} = \textit{average curvature of the arc } PP'.$$

More important, however, than the notion of the average curvature of an arc is that of *curvature at a point.* This is obtained as follows. Imagine $P'$ to approach $P$ along the curve; then the limiting value of the average curvature $\left(= \frac{\Delta\tau}{\Delta s}\right)$ as $P'$ approaches $P$ along the curve is defined as the curvature at $P$, that is,

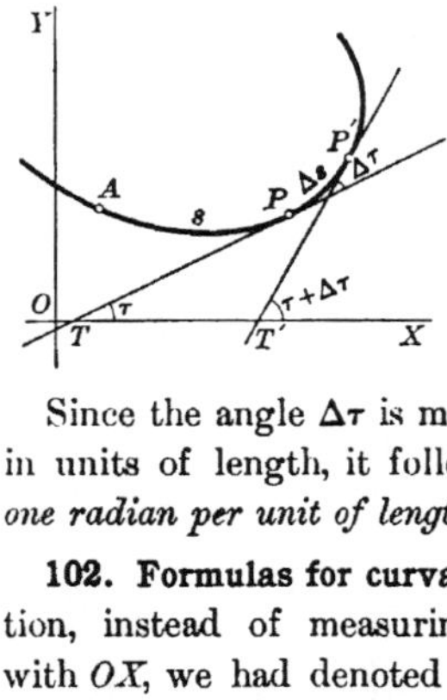


$$\textit{Curvature at a point} = \underset{\Delta s = 0}{\text{limit}} \left(\frac{\Delta\tau}{\Delta s}\right) = \frac{d\tau}{ds}.$$

$$\therefore K = \frac{d\tau}{ds} = \textbf{curvature.} \tag{39}$$

Since the angle $\Delta\tau$ is measured in radians and the length of arc $\Delta s$ in units of length, it follows that *the unit of curvature at a point is one radian per unit of length.*

**102. Formulas for curvature.** It is evident that if, in the last section, instead of measuring the angles which the tangents made with $OX$, we had denoted by $\tau$ and $\tau + \Delta\tau$ the angles made by the tangents with any arbitrarily fixed line, the different steps would in no wise have been changed, and consequently the results are entirely independent of the system of coördinates used. However, since the equations of the curves we shall consider are all given in either rectangular or polar coördinates, it is necessary to deduce formulas for $K$ in terms of both. We have

$$\tan\tau = \frac{dy}{dx}, \qquad \text{§ 32, p. 31}$$

or

$$\tau = \text{arc} \tan \frac{dy}{dx}.$$

Differentiating with respect to $x$, using **XX**

$$(A) \qquad \frac{d\tau}{dx} = \frac{\frac{d^2y}{dx^2}}{1+\left(\frac{dy}{dx}\right)^2}. \quad \text{Also}$$

$$(B) \qquad \frac{ds}{dx} = \left[1+\left(\frac{dy}{dx}\right)^2\right]^{\frac{1}{2}}. \qquad \text{From (24), p. 134}$$

Dividing $(A)$ by $(B)$ gives

$$\frac{\frac{d\tau}{dx}}{\frac{ds}{dx}} = \frac{\frac{d^2y}{dx^2}}{\left[1+\left(\frac{dy}{dx}\right)^2\right]^{\frac{3}{2}}}.$$

But

$$\frac{\frac{d\tau}{dx}}{\frac{ds}{dx}} = \frac{d\tau}{ds} = K. \quad \text{Hence}$$

$$(40) \qquad K = \frac{\frac{d^2y}{dx^2}}{\left[1+\left(\frac{dy}{dx}\right)^2\right]^{\frac{3}{2}}}.$$

If the equation of the curve be given in polar coördinates, $K$ may be found as follows:

From (*B*), p. 84,

$$\tau = \theta + \psi. \quad \text{Differentiating,}$$

$$(C) \qquad \frac{d\tau}{d\theta} = 1 + \frac{d\psi}{d\theta}.$$

But

$$\tan\psi = \frac{\rho}{\frac{d\rho}{d\theta}}. \qquad \text{From } (A)\text{, p. 84}$$

$$\therefore \psi = \text{arc} \tan \frac{\rho}{\frac{d\rho}{d\theta}}.$$

Differentiating by **XX** with respect to $\theta$ and reducing,

$$(D) \qquad \frac{d\psi}{d\theta} = \frac{\left(\frac{d\rho}{d\theta}\right)^2 - \rho\frac{d^2\rho}{d\theta^2}}{\rho^2+\left(\frac{d\rho}{d\theta}\right)^2}.$$

Substituting ($D$) in ($C$), we get

$$(E)\qquad \frac{d\tau}{d\theta}=\frac{\rho^2-\rho\dfrac{d^2\rho}{d\theta^2}+2\left(\dfrac{d\rho}{d\theta}\right)^2}{\rho^2+\left(\dfrac{d\rho}{d\theta}\right)^2}.\quad \text{Also}$$

$$(F)\qquad \frac{ds}{d\theta}=\left[\rho^2+\left(\frac{d\rho}{d\theta}\right)^2\right]^{\frac{1}{2}}. \qquad \text{From (30), p. 136}$$

Dividing ($E$) by ($F$) gives

$$\frac{\dfrac{d\tau}{d\theta}}{\dfrac{ds}{d\theta}}=\frac{\rho^2-\rho\dfrac{d^2\rho}{d\theta^2}+2\left(\dfrac{d\rho}{d\theta}\right)^2}{\left[\rho^2+\left(\dfrac{d\rho}{d\theta}\right)^2\right]^{\frac{3}{2}}}.$$

But

$$\frac{\dfrac{d\tau}{d\theta}}{\dfrac{ds}{d\theta}}=\frac{d\tau}{ds}=K.\quad \text{Hence}$$

$$\textbf{(41)}\qquad K=\frac{\rho^2-\rho\dfrac{d^2\rho}{d\theta^2}+2\left(\dfrac{d\rho}{d\theta}\right)^2}{\left[\rho^2+\left(\dfrac{d\rho}{d\theta}\right)^2\right]^{\frac{3}{2}}}.$$

ILLUSTRATIVE EXAMPLE 1. Find the curvature of the parabola $y^2 = 4px$ at the upper end of the latus rectum.

**Solution.** $$\frac{dy}{dx}=\frac{2p}{y};\quad \frac{d^2y}{dx^2}=-\frac{2p}{y^2}\frac{dy}{dx}=-\frac{4p^2}{y^3}.$$

Substituting in (40), $$K=-\frac{4p^2}{(y^2+4p^2)^{\frac{3}{2}}},$$

giving the curvature at *any point*. At the upper end of the latus rectum $(p, 2p)$

$$K=-\frac{4p^2}{(4p^2+4p^2)^{\frac{3}{2}}}=-\frac{4p^2}{16\sqrt{2}p^3}=-\frac{1}{4\sqrt{2}p}.^{*}\quad \textit{Ans.}$$

ILLUSTRATIVE EXAMPLE 2. Find the curvature of the logarithmic spiral $\rho = e^{a\theta}$ at any point.

**Solution.** $$\frac{d\rho}{d\theta}=ae^{a\theta}=a\rho;\quad \frac{d^2\rho}{d\theta^2}=a^2e^{a\theta}=a^2\rho.$$

Substituting in (41), $$K=\frac{1}{\rho\sqrt{1+a^2}}.\quad \textit{Ans.}$$

* While in our work it is generally only the numerical value of $K$ that is of importance, yet we can give a geometric meaning to its sign. Throughout our work we have taken the positive sign of the radical $\sqrt{1+\left(\frac{dy}{dx}\right)^2}$. Therefore $K$ will be positive or negative at the same time as $\frac{d^2y}{dx^2}$, that is (§ 85, p. 125), according as the curve is concave upwards or concave downwards.

In laying out the curves on a railroad it will not do, on account of the high speed of trains, to pass abruptly from a straight stretch of track to a circular curve. In order to make the change of direction gradual, engineers make use of *transition curves* to connect the straight part of a track with a circular curve. Arcs of cubical parabolas are generally employed as transition curves.

ILLUSTRATIVE EXAMPLE 3. The transition curve on a railway track has the shape of an arc of the cubical parabola $y = \frac{1}{3}x^3$. At what rate is a car on this track changing its direction (1 mi. = unit of length) when it is passing through (a) the point (3, 9)? (b) the point $(2, \frac{8}{3})$? (c) the point $(1, \frac{1}{3})$?

**Solution.**
$$\frac{dy}{dx} = x^2, \quad \frac{d^2y}{dx^2} = 2x.$$

Substituting in (40),
$$K = \frac{2x}{(1 + x^4)^{\frac{3}{2}}}.$$

(a) At (3, 9), $K = \frac{6}{(82)^{\frac{3}{2}}}$ radians per mile $= 28'$ per mile.

(b) At $(2, \frac{8}{3})$, $K = \frac{4}{(17)^{\frac{3}{2}}}$ radians per mile $= 3° 16'$ per mile. *Ans.*

(c) At $(1, \frac{1}{3})$, $K = \frac{2}{(2)^{\frac{3}{2}}} = \frac{1}{\sqrt{2}}$ radians per mile $= 40° 30'$ per mile. *Ans.*

**103. Radius of curvature.** By analogy with the circle (see (**38**), p. 156), the *radius of curvature of a curve at a point* is defined as the reciprocal of the curvature of the curve at that point. Denoting the radius of curvature by $R$, we have

$$R = \frac{1}{K};\text{*}$$

or, substituting the values of $K$ from (**40**) and (**41**),

(**42**)
$$R = \frac{\left[1 + \left(\frac{dy}{dx}\right)^2\right]^{\frac{3}{2}}}{\frac{d^2y}{dx^2}};$$

(**43**)
$$R = \frac{\left[\rho^2 + \left(\frac{d\rho}{d\theta}\right)^2\right]^{\frac{3}{2}}}{\rho^2 - \rho\frac{d^2\rho}{d\theta^2} + 2\left(\frac{d\rho}{d\theta}\right)^2}.\text{†}$$

* Hence the radius of curvature will have the same sign as the curvature, that is, + or −, according as the curve is concave upwards or concave downwards.

† In § 98, p. 152, (43) is derived from (42) by transforming from rectangular to polar coördinates.

ILLUSTRATIVE EXAMPLE 1. Find the radius of curvature at any point of the catenary $y = \frac{a}{2}(e^{\frac{x}{a}} + e^{-\frac{x}{a}})$.

**Solution.**
$$\frac{dy}{dx} = \frac{1}{2}(e^{\frac{x}{a}} - e^{-\frac{x}{a}}); \quad \frac{d^2y}{dx^2} = \frac{1}{2a}(e^{\frac{x}{a}} + e^{-\frac{x}{a}}).$$

Substituting in (42),

$$R = \frac{\left[1 + \left(\frac{e^{\frac{x}{a}} - e^{-\frac{x}{a}}}{2}\right)^2\right]^{\frac{3}{2}}}{\frac{e^{\frac{x}{a}} + e^{-\frac{x}{a}}}{2a}} = \frac{\left(\frac{e^{\frac{x}{a}} + e^{-\frac{x}{a}}}{2}\right)^3}{\frac{e^{\frac{x}{a}} + e^{-\frac{x}{a}}}{2a}} = \frac{a\left(e^{\frac{x}{a}} + e^{-\frac{x}{a}}\right)^2}{4} = \frac{y^2}{a}. \quad \textit{Ans.}$$

If the equation of the curve is given in parametric form, find the first and second derivatives of $y$ with respect to $x$ from (*A*) and (*B*), pp. 150, 151, namely:

$$(G) \qquad \frac{dy}{dx} = \frac{\frac{dy}{dt}}{\frac{dx}{dt}}, \text{ and}$$

$$(H) \qquad \frac{d^2y}{dx^2} = \frac{\frac{dx}{dt}\frac{d^2y}{dt^2} - \frac{dy}{dt}\frac{d^2x}{dt^2}}{\left(\frac{dx}{dt}\right)^3};$$

and then substitute the results in (**42**).*

ILLUSTRATIVE EXAMPLE 2. Find the radius of curvature of the cycloid

$$x = a(t - \sin t),$$
$$y = a(1 - \cos t).$$

**Solution.**
$$\frac{dx}{dt} = a(1 - \cos t), \quad \frac{dy}{dt} = a \sin t;$$
$$\frac{d^2x}{dt^2} = a \sin t, \quad \frac{d^2y}{dt^2} = a \cos t.$$

Substituting in (*G*) and (*H*), and then in (42), p. 159, we get

$$\frac{dy}{dx} = \frac{\sin t}{1 - \cos t}, \quad \frac{d^2y}{dx^2} = \frac{a(1 - \cos t)a \cos t - a \sin t \, a \sin t}{a^3(1 - \cos t)^3} = -\frac{1}{a(1 - \cos t)^2}.$$

$$R = \frac{\left[1 + \left(\frac{\sin t}{1 - \cos t}\right)^2\right]^{\frac{3}{2}}}{-\frac{1}{a(1 - \cos t)^2}} = -2a\sqrt{2 - 2\cos t}. \quad \textit{Ans.}$$

* Substituting (*G*) and (*H*) in (42) gives $R = \frac{\left[\left(\frac{dx}{dt}\right)^2 + \left(\frac{dy}{dt}\right)^2\right]^{\frac{3}{2}}}{\frac{dx}{dt}\frac{d^2y}{dt^2} - \frac{dy}{dt}\frac{d^2x}{dt^2}}$.

**104. Circle of curvature.** Consider any point $P$ on the curve $C$. The tangent drawn to the curve at $P$ has the same slope as the curve itself at $P$ (§ 64, p. 73). In an analogous manner we may construct for each point of the curve a circle whose curvature is the same as the curvature of the curve itself at that point. To do this, proceed as follows. Draw the normal to the curve at $P$ on the concave side of the curve. Lay off on this normal the distance $PC$ = radius of curvature (= $R$) at $P$. With $C$ as a center draw the circle passing through $P$. The curvature of this circle is then

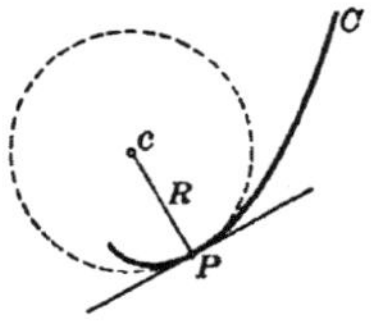


$$K = \frac{1}{R},$$

which also equals the curvature of the curve itself at $P$. The circle so constructed is called the *circle of curvature* for the point $P$ on the curve.

In general, the circle of curvature of a curve at a point will cross the curve at that point. This is illustrated in the above figure.

Just as the tangent at $P$ shows the direction of the curve at $P$, so the circle of curvature at $P$ aids us very materially in forming a geometric concept of the curvature of the curve at $P$, the rate of change of direction of the curve and of the circle being the same at $P$.

In a subsequent section (§ 116) the circle of curvature will be defined as the limiting position of a secant circle, a definition analogous to that of the tangent given in § 32, p. 31.

ILLUSTRATIVE EXAMPLE 4. Find the radius of curvature at the point (3, 4) on the equilateral hyperbola $xy = 12$, and draw the corresponding circle of curvature.

**Solution.** $\frac{dy}{dx} = -\frac{y}{x}, \quad \frac{d^2y}{dx^2} = \frac{2y}{x^2}.$

For (3, 4), $\frac{dy}{dx} = -\frac{4}{3}, \quad \frac{d^2y}{dx^2} = \frac{8}{9}.$

$$\therefore R = \frac{[1 + \frac{16}{9}]^{\frac{3}{2}}}{\frac{8}{9}} = \frac{125}{24} = 5\tfrac{5}{24}.$$

The circle of curvature crosses the curve at two points.

**EXAMPLES**

1. Find the radius of curvature for each of the following curves, at the point indicated; draw the curve and the corresponding circle of curvature:

(a) $b^2x^2 + a^2y^2 = a^2b^2$, $(a, 0)$. *Ans.* $R = \dfrac{b^2}{a}$.

(b) $b^2x^2 + a^2y^2 = a^2b^2$, $(0, b)$. $R = \dfrac{a^2}{b}$.

(c) $y = x^4 - 4x^3 - 18x^2$, $(0, 0)$. $R = \frac{1}{36}$.

(d) $16y^2 = 4x^4 - x^6$, $(2, 0)$. $R = 2$.

(e) $y = x^3$, $(x_1, y_1)$. $R = \dfrac{(1 + 9x_1^4)^{\frac{3}{2}}}{6x_1}$.

(f) $y^2 = x^3$, $(4, 8)$. $R = \frac{1}{3}(40)^{\frac{3}{2}}$.

(g) $y^2 = 8x$, $(\frac{9}{8}, 3)$. $R = 7\frac{13}{16}$.

(h) $\left(\dfrac{x}{a}\right)^2 + \left(\dfrac{y}{b}\right)^{\frac{2}{3}} = 1$, $(0, b)$. $R = \dfrac{a^2}{3b}$.

(i) $x^2 = 4ay$, $(0, 0)$. $R = 2a$.

(j) $(y - x^2)^2 = x^5$, $(0, 0)$. $R = \frac{1}{2}$.

(k) $b^2x^2 - a^2y^2 = a^2b^2$, $(x_1, y_1)$. $R = \dfrac{(b^4x_1^2 + a^4y_1^2)^{\frac{3}{2}}}{a^4b^4}$.

(l) $e^x = \sin y$, $(x_1, y_1)$.

(m) $y = \sin x$, $\left(\dfrac{\pi}{2}, 1\right)$.

(n) $y = \cos x$, $\left(\dfrac{\pi}{4}, \sqrt{2}\right)$.

(o) $y = \log x$, $x = e$.

(p) $9y = x^3$, $x = 3$.

(q) $4y^2 = x^3$, $x = 4$.

(r) $x^2 - y^2 = a^2$, $y = 0$.

(s) $x^2 + 2y^2 = 9$, $(1, -2)$.

2. Determine the radius of curvature of the curve $a^2y = bx^2 + cx^2y$ at the origin.
*Ans.* $R = \dfrac{a^2}{2b}$.

3. Show that the radius of curvature of the witch $y^2 = \dfrac{a^2(a - x)}{x}$ at the vertex is $\dfrac{a}{2}$.

4. Find the radius of curvature of the curve $y = \log \sec x$ at the point $(x_1, y_1)$.
*Ans.* $R = \sec x_1$.

5. Find $K$ at any point on the parabola $x^{\frac{1}{2}} + y^{\frac{1}{2}} = a^{\frac{1}{2}}$. *Ans.* $K = \dfrac{a^{\frac{1}{2}}}{2(x + y)^{\frac{3}{2}}}$.

6. Find $R$ at any point on the hypocycloid $x^{\frac{2}{3}} + y^{\frac{2}{3}} = a^{\frac{2}{3}}$. *Ans.* $R = 3(axy)^{\frac{1}{3}}$.

7. Find $R$ at any point on the cycloid $x = r \text{ arc vers } \dfrac{y}{r} - \sqrt{2ry - y^2}$.
*Ans.* $R = 2\sqrt{2ry}$.

Find the radius of curvature of the following curves at any point:

8. The circle $\rho = a \sin \theta$. *Ans.* $R = \frac{a}{2}$.

9. The spiral of Archimedes $\rho = a\theta$. $R = \dfrac{(\rho^2 + a^2)^{\frac{3}{2}}}{\rho^2 + 2a^2}$.

10. The cardioid $\rho = a(1 - \cos \theta)$. $R = \frac{2}{3}\sqrt{2a\rho}$.

11. The lemniscate $\rho^2 = a^2 \cos 2\theta$. $R = \frac{a^2}{3\rho}$.

12. The parabola $\rho = a \sec^2 \frac{\theta}{2}$. $R = 2a \sec^3 \frac{\theta}{2}$.

13. The curve $\rho = a \sin^3 \frac{\theta}{3}$. $R = \frac{3}{4} a \sin^2 \frac{\theta}{3}$.

**14.** The trisectrix $\rho = 2a\cos\theta - a$. *Ans.* $R = \dfrac{a(5-4\cos\theta)^{\frac{3}{2}}}{9-6\cos\theta}$.

**15.** The equilateral hyperbola $\rho^2\cos 2\theta = a^2$. *Ans.* $R = \dfrac{\rho^3}{a^2}$.

**16.** The conic $\rho = \dfrac{a(1-e^2)}{1-e\cos\theta}$. *Ans.* $R = \dfrac{a(1-e^2)(1-2e\cos\theta+e^2)^{\frac{3}{2}}}{(1-e\cos\theta)^3}$.

**17.** The curve $\begin{cases} x = 3t^2, \\ y = 3t - t^3. \end{cases}$ $t = 1$. *Ans.* $R = 6$.

**18.** The hypocycloid $\begin{cases} x = a\cos^3 t, \\ y = a\sin^3 t. \end{cases}$ $t = t_1$. *Ans.* $R = 3a\sin t_1 \cos t_1$.

**19.** The curve $\begin{cases} x = a(\cos t + t\sin t), \\ y = a(\sin t - t\cos t). \end{cases}$ $t = \dfrac{\pi}{2}$. *Ans.* $R = \dfrac{\pi a}{2}$.

**20.** The curve $\begin{cases} x = a(m\cos t + \cos mt), \\ y = a(m\sin t - \sin mt). \end{cases}$ $t = t_0$.
*Ans.* $R = \dfrac{4ma}{m-1}\sin\left(\dfrac{m+1}{2}\right)t_0$.

**21.** Find the radius of curvature for each of the following curves at the point indicated; draw the curve and the corresponding circle of curvature:

(a) $x = t^2,\ 2y = t;\ t = 1$.
(b) $x = t^2,\ y = t^3;\ t = 1$.
(c) $x = \sin t,\ y = \cos 2t;\ t = \dfrac{\pi}{6}$.
(d) $x = 1 - t,\ y = t^3;\ t = 3$.
(e) $x = t,\ y = 6t^{-1};\ t = 2$.
(f) $x = 2e^t,\ y = e^{-t};\ t = 0$.
(g) $x = \sin t,\ y = 2\cos t;\ t = \dfrac{\pi}{4}$.
(h) $x = t^3,\ y = t^2 + 2t;\ t = 1$.

**22.** An automobile race track has the form of the ellipse $x^2 + 16y^2 = 16$, the unit being one mile. At what rate is a car on this track changing its direction

(a) when passing through one end of the major axis?
(b) when passing through one end of the minor axis?
(c) when two miles from the minor axis?
(d) when equidistant from the minor and major axes?

*Ans.* (a) 4 radians per mile; (b) $\frac{1}{16}$ radian per mile.

**23.** On leaving her dock a steamship moves on an arc of the semicubical parabola $4y^2 = x^3$. If the shore line coincides with the axis of $y$, and the unit of length is one mile, how fast is the ship changing its direction when one mile from the shore?

*Ans.* $\frac{24}{125}$ radians per mile.

**24.** A battleship 400 ft. long has changed its direction 30° while moving through a distance equal to its own length. What is the radius of the circle in which it is moving? *Ans.* 764 ft.

**25.** At what rate is a bicycle rider on a circular track of half a mile diameter changing his direction? *Ans.* 4 rad. per mile = 43′ per rod.

**26.** The origin being directly above the starting point, an aëroplane follows approximately the spiral $\rho = \theta$, the unit of length being one mile. How rapidly is the aëroplane turning at the instant it has circled the starting point once?

**27.** A railway track has curves of approximately the form of arcs from the following curves. At what rate will an engine change its direction when passing through the points indicated (1 mi. = unit of length):

(a) $y = x^3$, (2, 8)?
(b) $y = x^2$, (3, 9)?
(c) $x^2 - y^2 = 8$, (3, 1)?
(d) $y = e^x$, $x = 0$?
(e) $y = \cos x$, $x = \dfrac{\pi}{4}$?
(f) $\rho\theta = 4$, $\theta = 1$?

# CHAPTER XIII

## THEOREM OF MEAN VALUE. INDETERMINATE FORMS

**105. Rolle's Theorem.** Let $y = f(x)$ be a continuous single-valued function of $x$, vanishing for $x = a$ and $x = b$, and suppose that $f'(x)$ changes continuously when $x$ varies from $a$ to $b$. The function will then be represented graphically by a continuous curve as in the figure. Geometric intuition shows us at once that for *at least one value of $x$ between $a$ and $b$* the tangent is parallel to the axis of $X$ (as at $P$); that is, the slope is zero. This illustrates **Rolle's Theorem:**

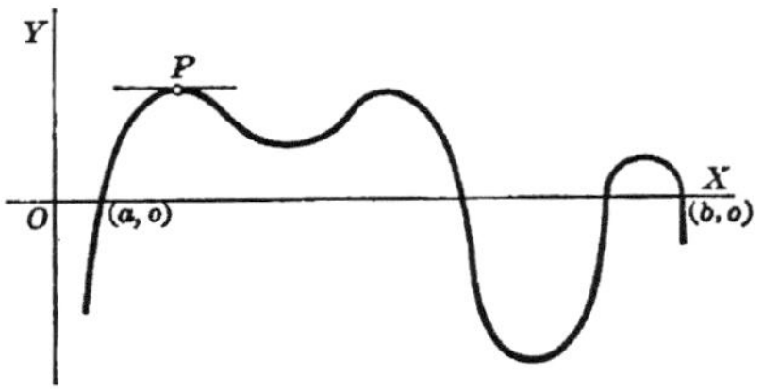


*If $f(x)$ vanishes when $x = a$ and $x = b$, and $f(x)$ and $f'(x)$ are continuous for all values of $x$ from $x = a$ to $x = b$, then $f'(x)$ will be zero for at least one value of $x$ between $a$ and $b$.*

This theorem is obviously true, because as $x$ increases from $a$ to $b$, $f(x)$ cannot always increase or always decrease as $x$ increases, since $f(a) = 0$ and $f(b) = 0$. Hence for at least one value of $x$ between $a$ and $b$, $f(x)$ must cease to increase and begin to decrease, or else cease to decrease and begin to increase; and for that particular value of $x$ the first derivative must be zero (§ 81, p. 108).

That Rolle's Theorem does not apply when $f(x)$ or $f'(x)$ are discontinuous is illustrated as follows:

Fig. $a$ shows the graph of a function which is discontinuous ($= \infty$) for $x = c$, a value lying between $a$ and $b$. Fig. $b$ shows a continuous function whose first derivative is discontinuous ($= \infty$) for such an intermediate value $x = c$. In either case it is seen that at no point on the graph between $x = a$ and $x = b$ does the tangent (or curve) become parallel to $OX$.

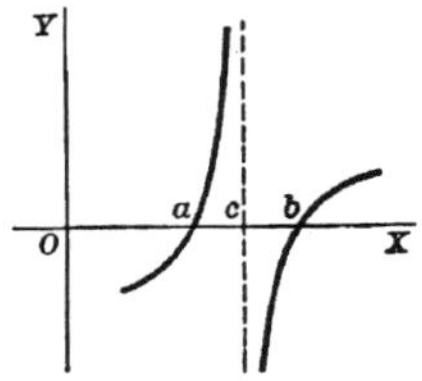


FIG. $a$

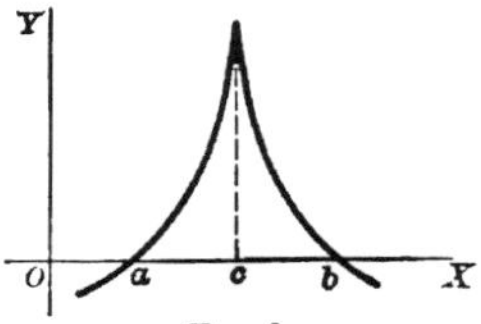


FIG. $b$

**106. The Theorem of Mean Value.*** Consider the quantity $Q$ defined by the equation

$$(A) \qquad \frac{f(b)-f(a)}{b-a}=Q, \text{ or}$$

$$(B) \qquad f(b)-f(a)-(b-a)Q=0.$$

Let $F(x)$ be a function formed by replacing $b$ by $x$ in the left-hand member of $(B)$; that is,

$$(C) \qquad F(x)=f(x)-f(a)-(x-a)Q.$$

From $(B)$, $\quad F(b)=0$, and from $(C)$, $F(a)=0$;

therefore, by Rolle's Theorem (p. 164) $F'(x)$ must be zero for at least one value of $x$ *between* $a$ and $b$, say for $x_1$. But by differentiating $(C)$ we get

$$F'(x)=f'(x)-Q.$$

Therefore, since $\quad F'(x_1)=0$, then also $f'(x_1)-Q=0$,

and

$$Q=f'(x_1).$$

Substituting this value of $Q$ in $(A)$, we get the **Theorem of Mean Value,**

$$\textbf{(44)} \qquad \frac{f(b)-f(a)}{b-a}=f'(x_1), \qquad a<x_1<b$$

where in general all we know about $x_1$ is that it lies *between* $a$ and $b$.

**The Theorem of Mean Value interpreted Geometrically.** Let the curve in the figure be the locus of

$$y=f(x).$$

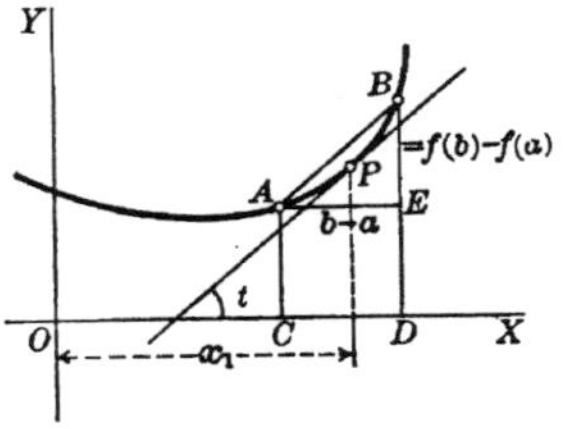


Take $OC=a$ and $OD=b$; then $f(a)=CA$ and $f(b)=DB$, giving $AE=b-a$ and $EB=f(b)-f(a)$.

Therefore the slope of the chord $AB$ is

$$(D) \qquad \tan EAB=\frac{EB}{AE}=\frac{f(b)-f(a)}{b-a}.$$

There is at least one point on the curve between $A$ and $B$ (as $P$) where the tangent (or curve) is parallel to the chord $AB$. If the abscissa of $P$ is $x_1$, the slope at $P$ is

$$(E) \qquad \tan t=f'(x_1)=\tan EAB.$$

* Also called the *Law of the Mean.*

Equating $(D)$ and $(E)$, we get

$$\frac{f(b)-f(a)}{b-a}=f'(x_1),$$

which is the Theorem of Mean Value.

The student should draw curves (as the one on p. 164) to show that there may be more than one such point in the interval; and curves to illustrate, on the other hand, that the theorem may not be true if $f(x)$ becomes discontinuous for any value of $x$ between $a$ and $b$ (Fig. $a$, p. 164), or if $f'(x)$ becomes discontinuous (Fig. $b$, p. 164).

Clearing (**44**) of fractions, we may also write the theorem in the form

$$\textbf{(45)} \qquad f(b)=f(a)+(b-a)f'(x_1).$$

Let $b=a+\Delta a$; then $b-a=\Delta a$, and since $x_1$ is a number lying between $a$ and $b$, we may write

$$x_1=a+\theta\cdot\Delta a,$$

where $\theta$ is a positive proper fraction. Substituting in (**45**), we get *another form of the* **Theorem of Mean Value.**

$$\textbf{(46)} \qquad f(a+\Delta a)-f(a)=\Delta a f'(a+\theta\cdot\Delta a). \qquad 0<\theta<1$$

**107. The Extended Theorem of Mean Value.*** Following the method of the last section, let $R$ be defined by the equation

$$(A) \qquad f(b)-f(a)-(b-a)f'(a)-\tfrac{1}{2}(b-a)^2R=0.$$

Let $F(x)$ be a function formed by replacing $b$ by $x$ in the left-hand member of $(A)$; that is,

$$(B) \qquad F(x)=f(x)-f(a)-(x-a)f'(a)-\tfrac{1}{2}(x-a)^2R.$$

From $(A)$, $F(b)=0$; and from $(B)$, $F(a)=0$; therefore, by Rolle's Theorem (p. 164), at least one value of $x$ between $a$ and $b$, say $x_1$, will cause $F'(x)$ to vanish. Hence, since

$$F'(x)=f'(x)-f'(a)-(x-a)R, \text{ we get}$$
$$F'(x_1)=f'(x_1)-f'(a)-(x_1-a)R=0.$$

Since $F'(x_1)=0$ and $F'(a)=0$, it is evident that $F'(x)$ also satisfies the conditions of Rolle's Theorem, so that *its derivative*, namely $F''(x)$, must vanish for at least one value of $x$ between $a$ and $x_1$, say $x_2$, and therefore $x_2$ also lies between $a$ and $b$. But

$$F''(x)=f''(x)-R; \text{ therefore } F''(x_2)=f''(x_2)-R=0,$$

and

$$R=f''(x_2).$$

* Also called the *Extended Law of the Mean.*

Substituting this result in $(A)$, we get

$$(C) \qquad f(b)=f(a)+(b-a)f'(a)+\frac{1}{\underline{|2}}(b-a)^2 f''(x_2). \qquad a<x_2<b$$

In the same manner, if we define $S$ by means of the equation

$$f(b)-f(a)-(b-a)f'(a)-\frac{1}{\underline{|2}}(b-a)^2 f''(a)-\frac{1}{\underline{|3}}(b-a)^3 S=0,$$

we can derive the equation

$$(D) \qquad f(b)=f(a)+(b-a)f'(a)+\frac{1}{\underline{|2}}(b-a)^2 f''(a) + \frac{1}{\underline{|3}}(b-a)^3 f'''(x_3), \qquad a<x_3<b$$

where $x_3$ lies between $a$ and $b$.

By continuing this process we get the general result,

$$(E) \qquad f(b)=f(a)+\frac{(b-a)}{\underline{|1}}f'(a)+\frac{(b-a)^2}{\underline{|2}}f''(a) + \frac{(b-a)^3}{\underline{|3}}f'''(a)+\cdots+\frac{(b-a)^{n-1}}{\underline{|n-1}}f^{(n-1)}(a) + \frac{(b-a)^n}{\underline{|n}}f^{(n)}(x_1), \qquad a<x_1<b$$

where $x_1$ lies between $a$ and $b$. $(E)$ is called the *Extended Theorem of Mean Value.*

**108. Maxima and minima treated analytically.** By making use of the results of the last two sections we can now give a general discussion of *maxima and minima of functions of a single independent variable.*

Given the function $f(x)$. Let $h$ be a positive number as small as we please; then the definitions given in § 82, p. 109, may be stated as follows:

If, for all values of $x$ different from $a$ in the interval $[a-h, a+h]$,

$$(A) \qquad f(x)-f(a)=\textit{a negative number},$$

then $f(x)$ is said to be a *maximum when* $x=a$.

If, on the other hand,

$$(B) \qquad f(x)-f(a)=\textit{a positive number},$$

then $f(x)$ is said to be a *minimum when* $x=a$.

Consider the following cases:

I. *Let* $f'(a)\neq 0$.

From (**45**), p. 166, replacing $b$ by $x$ and transposing $f(a)$,

$$(C) \qquad f(x)-f(a)=(x-a)f'(x_1). \qquad a<x_1<x$$

Since $f'(a) \neq 0$, and $f'(x)$ is assumed as continuous, $h$ may be chosen so small that $f'(x)$ will have the same sign as $f'(a)$ for all values of $x$ in the interval $[a-h, a+h]$. Therefore $f'(x_1)$ has the same sign as $f'(a)$ (Chap. III). But $x-a$ changes sign according as $x$ is less or greater than $a$. Therefore, from $(C)$, the difference

$$f(x)-f(a)$$

will also change sign, and, by $(A)$ and $(B)$, $f(a)$ will be neither a maximum nor a minimum. This result agrees with the discussion in § 82, where it was shown that *for all values of $x$ for which $f(x)$ is a maximum or a minimum, the first derivative $f'(x)$ must vanish.*

II. *Let $f'(a)=0$, and $f''(a)\neq 0$.*

From $(C)$, p. 167, replacing $b$ by $x$ and transposing $f(a)$,

$$(D) \qquad f(x)-f(a)=\frac{(x-a)^2}{\underline{|2}}f''(x_2). \qquad a<x_2<x$$

Since $f''(a)\neq 0$, and $f''(x)$ is assumed as continuous, we may choose our interval $[a-h, a+h]$ so small that $f''(x_2)$ will have the same sign as $f''(a)$ (Chap. III). Also $(x-a)^2$ does not change sign. Therefore the second member of $(D)$ will not change sign, and the difference

$$f(x)-f(a)$$

will have the same sign for all values of $x$ in the interval $[a-h, a+h]$, and, moreover, *this sign will be the same as the sign of $f''(a)$.* It therefore follows from our definitions $(A)$ and $(B)$ that

$(E)$ *$f(a)$ is a maximum if $f'(a)=0$ and $f''(a)=$ a negative number;*

$(F)$ *$f(a)$ is a minimum if $f'(a)=0$ and $f''(a)=$ a positive number.*

These conditions are the same as (**21**) and (**22**), p. 113.

III. *Let $f'(a)=f''(a)=0$, and $f'''(a)\neq 0$.*

From $(D)$, p. 167, replacing $b$ by $x$ and transposing $f(a)$,

$$(G) \qquad f(x)-f(a)=\frac{1}{\underline{|3}}(x-a)^3 f'''(x_3). \qquad a<x_3<x$$

As before, $f'''(x_3)$ will have the same sign as $f'''(a)$. But $(x-a)^3$ changes its sign from $-$ to $+$ as $x$ increases through $a$. Therefore the difference

$$f(x)-f(a)$$

must change sign, and $f(a)$ is neither a maximum nor a minimum.

IV. *Let* $f'(a)=f''(a)=\cdots=f^{(n-1)}(a)=0$, *and* $f^{(n)}(a)\neq 0$.

By continuing the process as illustrated in I, II, and III, it is seen that if the first derivative of $f(x)$ which does not vanish for $x=a$ is of **even order** ($=n$), then

(47) **$f(a)$ is a maximum if $f^{(n)}(a)=$ a negative number;**

(48) **$f(a)$ is a minimum if $f^{(n)}(a)=$ a positive number.***

If the first derivative of $f(x)$ which does not vanish for $x=a$ is of **odd order, then $f(a)$ will be neither a maximum nor a minimum.**

ILLUSTRATIVE EXAMPLE 1. Examine $x^3-9x^2+24x-7$ for maximum and minimum values.

**Solution.** $f(x)=x^3-9x^2+24x-7$.

$f'(x)=3x^2-18x+24$.

Solving $3x^2-18x+24=0$

gives the critical values $x=2$ and $x=4$. $\therefore f'(2)=0$, and $f'(4)=0$.

Differentiating again, $f''(x)=6x-18$.

Since $f''(2)=-6$, we know from (47) that $f(2)=13$ is a maximum.

Since $f''(4)=+6$, we know from (48) that $f(4)=9$ is a minimum.

ILLUSTRATIVE EXAMPLE 2. Examine $e^x+2\cos x+e^{-x}$ for maximum and minimum values.

**Solution.**

$$f(x)=e^x+2\cos x+e^{-x},$$
$$f'(x)=e^x-2\sin x-e^{-x}=0, \text{ for } x=0,\dagger$$
$$f''(x)=e^x-2\cos x+e^{-x}=0, \text{ for } x=0,$$
$$f'''(x)=e^x+2\sin x-e^{-x}=0, \text{ for } x=0,$$
$$f^{\text{iv}}(x)=e^x+2\cos x+e^{-x}=4, \text{ for } x=0.$$

Hence, from (48), $f(0)=4$ is a minimum.

## EXAMPLES

Examine the following functions for maximum and minimum values, using the method of the last section:

1. $3x^4-4x^3+1$. *Ans.* $x=1$ gives min. $=0$; $x=0$ gives neither.

2. $x^3-6x^2+12x+48$. $x=2$ gives neither.

3. $(x-1)^2(x+1)^3$. $x=1$ gives min. $=0$; $x=\frac{1}{5}$ gives max.; $x=-1$ gives neither.

4. Investigate $x^5-5x^4+5x^3-1$, at $x=1$ and $x=3$.

5. Investigate $x^3-3x^2+3x+7$, at $x=1$.

6. Show that if the first derivative of $f(x)$ which does not vanish for $x=a$ is of odd order ($=n$), then $f(x)$ is an increasing or decreasing function when $x=a$, according as $f^{(n)}(a)$ is positive or negative.

* As in § 82, a critical value $x=a$ is found by placing the first derivative equal to zero and solving the resulting equation for real roots.

† $x=0$ is the only root of the equation $e^x-2\sin x-e^{-x}=0$.

**109. Indeterminate forms.** When, for a particular value of the independent variable, a function takes on one of the forms

$$\frac{0}{0},\quad \frac{\infty}{\infty},\quad 0\cdot\infty,\quad \infty-\infty,\quad 0^0,\quad \infty^0,\quad 1^\infty,$$

it is said to be *indeterminate*, and the function is *not* defined for that value of the independent variable by the given analytical expression. For example, suppose we have

$$y=\frac{f(x)}{F(x)},$$

where for some value of the variable, as $x=a$,

$$f(a)=0,\qquad F(a)=0.$$

For this value of $x$ our function is *not* defined and we may therefore assign to it any value we please. It is evident from what has gone before (Case II, p. 15) that it is desirable to assign to the function a value that will make it continuous when $x=a$ whenever it is possible to do so.

**110. Evaluation of a function taking on an indeterminate form.** If when $x=a$ the function $f(x)$ assumes an indeterminate form, then

$$\underset{x=a}{\text{limit}} f(x)^*$$

*is taken as the value of $f(x)$ for $x=a$.*

The assumption of this limiting value makes $f(x)$ continuous for $x=a$. This agrees with the theorem under Case II, p. 15, and also with our practice in Chapter III, where several functions assuming the indeterminate form $\frac{0}{0}$ were evaluated. Thus, for $x=2$ the function $\frac{x^2-4}{x-2}$ assumes the form $\frac{0}{0}$, but

$$\underset{x=2}{\text{limit}}\frac{x^2-4}{x-2}=4.$$

Hence 4 is taken as the value of the function for $x=2$. Let us now illustrate graphically the fact that if we assume 4 as the value of the function for $x=2$, then the function is continuous for $x=2$.

Let
$$y=\frac{x^2-4}{x-2}.$$

This equation may also be written in the form

$$y(x-2)=(x-2)(x+2);$$

or,
$$(x-2)(y-x-2)=0.$$

* The calculation of this limiting value is called *evaluating the indeterminate form*.

Placing each factor separately equal to zero, we have

$$x = 2, \text{ and } y = x + 2.$$

In plotting, the loci of these equations are found to be the two lines $AB$ and $CD$ respectively. Since there are infinitely many points on the line $AB$ having the abscissa 2, it is clear that when $x = 2$ $(= OM)$, the value of $y$ (or the function) may be taken as any number whatever; but when $x$ is different from 2, it is seen from the graph of the function that the corresponding value of $y$ (or the function) is always found from

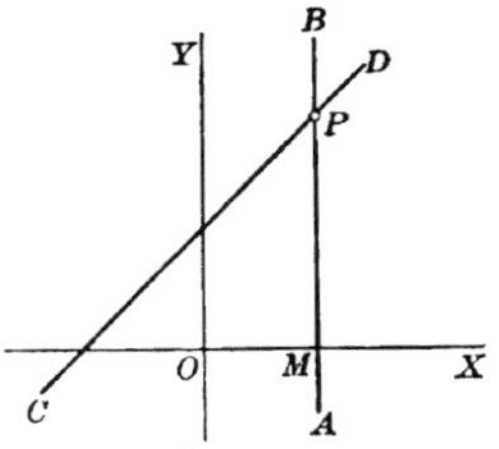


$$y = x + 2,$$

the equation of the line $CD$. Also, on $CD$, when $x = 2$, we get

$$y = MP = 4,$$

which we saw was also the limiting value of $y$ (or the function) for $x = 2$; and it is evident from geometrical considerations that if we assume 4 as the value of the function for $x = 2$, then the function is continuous for $x = 2$.

Similarly, several of the examples given in Chapter III illustrate how the limiting values of many functions assuming indeterminate forms may be found by employing suitable algebraic or trigonometric transformations, and how in general these limiting values make the corresponding functions continuous at the points in question. The most general methods, however, for evaluating indeterminate forms depend on differentiation.

**111. Evaluation of the indeterminate form $\frac{0}{0}$.** Given a function of the form $\frac{f(x)}{F(x)}$ such that $f(a) = 0$ and $F(a) = 0$; that is, the function takes on the indeterminate form $\frac{0}{0}$ when $a$ is substituted for $x$. It is then required to find

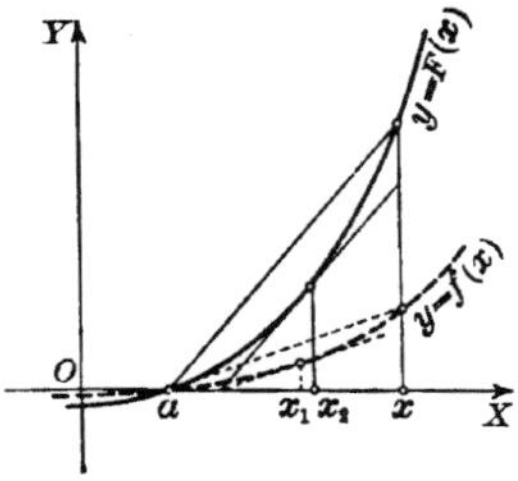


$$\lim_{x=a} \frac{f(x)}{F(x)}.$$

Draw the graphs of the functions $f(x)$ and $F(x)$. Since, by hypothesis, $f(a) = 0$ and $F(a) = 0$, these graphs intersect at $(a, 0)$.

Applying the Theorem of Mean Value to each of these functions (replacing $b$ by $x$), we get

$$f(x)=f(a)+(x-a)f'(x_1), \qquad a<x_1<x$$

$$F(x)=F(a)+(x-a)F'(x_2). \qquad a<x_2<x$$

Since $f(a)=0$ and $F(a)=0$, we get, after canceling out $(x-a)$,

$$\frac{f(x)}{F(x)}=\frac{f'(x_1)}{F'(x_2)}.$$

Now let $x \doteq a$; then $x_1 \doteq a$, $x_2 \doteq a$, and

$$\underset{x=a}{\text{limit}} f'(x_1)=f'(a), \quad \underset{x=a}{\text{limit}} F'(x_2)=F'(a).$$

(49) $$\therefore \underset{x=a}{\textbf{limit}} \frac{f(x)}{F(x)}=\frac{f'(a)}{F'(a)}. \qquad F'(a)\neq 0$$

**Rule for evaluating the indeterminate form $\frac{0}{0}$.** *Differentiate the numerator for a new numerator and the denominator for a new denominator.* The value of this new fraction for the assigned value † of the variable will be the limiting value of the original fraction.*

In case it so happens that

$$f'(a)=0 \text{ and } F'(a)=0,$$

that is, the first derivatives also vanish for $x=a$, then we still have the indeterminate form $\frac{0}{0}$, and the theorem can be applied anew to the ratio

$$\frac{f'(x)}{F'(x)},$$

giving us

$$\underset{x=a}{\text{limit}} \frac{f(x)}{F(x)}=\frac{f''(a)}{F''(a)}.$$

When also $f''(a)=0$ and $F''(a)=0$, we get in the same manner

$$\underset{x=a}{\text{limit}} \frac{f(x)}{F(x)}=\frac{f'''(a)}{F'''(a)},$$

and so on.

It may be necessary to repeat this process several times.

* The student is warned against the very careless but common mistake of differentiating the whole expression as a fraction by VII.

† If $a=\infty$, the substitution $x=\frac{1}{z}$ reduces the problem to the evaluation of the limit for $z=0$.

Thus $$\underset{x=\infty}{\text{limit}} \frac{f(x)}{F(x)}=\underset{z=0}{\text{limit}} \frac{-f'\left(\frac{1}{z}\right)\frac{1}{z^2}}{-F'\left(\frac{1}{z}\right)\frac{1}{z^2}}=\underset{z=0}{\text{limit}} \frac{f'\left(\frac{1}{z}\right)}{F'\left(\frac{1}{z}\right)}=\underset{x=\infty}{\text{limit}} \frac{f'(x)}{F'(x)}.$$

Therefore the rule holds in this case also.

ILLUSTRATIVE EXAMPLE 1. Evaluate $\frac{f(x)}{F(x)} = \frac{x^3 - 3x + 2}{x^3 - x^2 - x + 1}$ when $x = 1$.

**Solution.** $\frac{f(1)}{F(1)} = \frac{x^3 - 3x + 2}{x^3 - x^2 - x + 1}\Big]_{x=1} = \frac{1 - 3 + 2}{1 - 1 - 1 + 1} = \frac{0}{0}$. $\therefore$ indeterminate.

$$\frac{f'(1)}{F'(1)} = \frac{3x^2 - 3}{3x^2 - 2x - 1}\Big]_{x=1} = \frac{3 - 3}{3 - 2 - 1} = \frac{0}{0}. \quad \therefore \text{ indeterminate.}$$

$$\frac{f''(1)}{F''(1)} = \frac{6x}{6x - 2}\Big]_{x=1} = \frac{6}{6 - 2} = \frac{3}{2}. \quad \textit{Ans.}$$

ILLUSTRATIVE EXAMPLE 2. Evaluate $\underset{x=0}{\text{limit}} \frac{e^x - e^{-x} - 2x}{x - \sin x}$.

**Solution.** $\frac{f(0)}{F(0)} = \frac{e^x - e^{-x} - 2x}{x - \sin x}\Big]_{x=0} = \frac{1 - 1 - 0}{0 - 0} = \frac{0}{0}$. $\therefore$ indeterminate.

$$\frac{f'(0)}{F'(0)} = \frac{e^x + e^{-x} - 2}{1 - \cos x}\Big]_{x=0} = \frac{1 + 1 - 2}{1 - 1} = \frac{0}{0}. \quad \therefore \text{ indeterminate.}$$

$$\frac{f''(0)}{F''(0)} = \frac{e^x - e^{-x}}{\sin x}\Big]_{x=0} = \frac{1 - 1}{0} = \frac{0}{0}. \quad \therefore \text{ indeterminate.}$$

$$\frac{f'''(0)}{F'''(0)} = \frac{e^x + e^{-x}}{\cos x}\Big]_{x=0} = \frac{1 + 1}{1} = 2. \quad \textit{Ans.}$$

**EXAMPLES**

Evaluate the following by differentiation:*

1. $\underset{x=4}{\text{limit}} \frac{x^2 - 16}{x^2 + x - 20}$. *Ans.* $\frac{8}{9}$.

2. $\underset{x=1}{\text{limit}} \frac{x - 1}{x^n - 1}$. $\frac{1}{n}$.

3. $\underset{x=1}{\text{limit}} \frac{\log x}{x - 1}$. 1.

4. $\underset{x=0}{\text{limit}} \frac{e^x - e^{-x}}{\sin x}$. 2.

5. $\underset{x=0}{\text{limit}} \frac{\tan x - x}{x - \sin x}$. 2.

6. $\underset{x=\frac{\pi}{2}}{\text{limit}} \frac{\log \sin x}{(\pi - 2x)^2}$. $-\frac{1}{8}$.

7. $\underset{x=0}{\text{limit}} \frac{a^x - b^x}{x}$. $\log \frac{a}{b}$.

8. $\underset{r=a}{\text{limit}} \frac{r^3 - ar^2 - a^2r + a^3}{r^2 - a^2}$. 0.

9. $\underset{\theta=0}{\text{limit}} \frac{\theta - \arcsin \theta}{\sin^3 \theta}$. *Ans.* $-\frac{1}{6}$.

10. $\underset{x=\phi}{\text{limit}} \frac{\sin x - \sin \phi}{x - \phi}$. $\cos \phi$.

11. $\underset{y=0}{\text{limit}} \frac{e^y + \sin y - 1}{\log(1 + y)}$. 2.

12. $\underset{\theta=0}{\text{limit}} \frac{\tan \theta + \sec \theta - 1}{\tan \theta - \sec \theta + 1}$. 1.

13. $\underset{\phi=\frac{\pi}{4}}{\text{limit}} \frac{\sec^2 \phi - 2 \tan \phi}{1 + \cos 4\phi}$. $\frac{1}{2}$.

14. $\underset{z=a}{\text{limit}} \frac{az - z^2}{a^4 - 2a^3z + 2az^3 - z^4}$. $+\infty$.

15. $\underset{x=2}{\text{limit}} \frac{(e^x - e^2)^3}{(x - 4)e^x + e^2x}$. $6e^4$.

16. $\underset{x=1}{\text{limit}} \frac{x^2 + x - 2}{x^2 - 1}$.

17. $\underset{x=-2}{\text{limit}} \frac{x^3 + 8}{x^5 + 32}$.

18. $\underset{x=0}{\text{limit}} \frac{\sin 2x}{x}$.

19. $\underset{x=0}{\text{limit}} \frac{x - \sin x}{x^3}$.

20. $\underset{x=1}{\text{limit}} \frac{\log \cos(x - 1)}{1 - \sin \frac{\pi x}{2}}$.

21. $\underset{x=0}{\text{limit}} \frac{\tan x - \sin x}{\sin^3 x}$.

* After differentiating, the student should in every case reduce the resulting expression to its simplest possible form before substituting the value of the variable.

**112. Evaluation of the indeterminate form $\frac{\infty}{\infty}$.** In order to find

$$\underset{x=a}{\text{limit}} \frac{f(x)}{F(x)}$$

when
$$\underset{x=a}{\text{limit}} f(x) = \infty \text{ and } \underset{x=a}{\text{limit}} F(x) = \infty,$$

that is, when for $x = a$ the function

$$\frac{f(x)}{F(x)}$$

assumes the indeterminate form

$$\frac{\infty}{\infty},$$

we follow the same rule as that given on p. 172 for evaluating the indeterminate form $\frac{0}{0}\cdot$ Hence

**Rule for evaluating the indeterminate form $\frac{\infty}{\infty}$.** *Differentiate the numerator for a new numerator and the denominator for a new denominator. The value of this new fraction for the assigned value of the variable will be the limiting value of the original fraction.*

A rigorous proof of this rule is beyond the scope of this book and is left for more advanced treatises.

ILLUSTRATIVE EXAMPLE 1. Evaluate $\frac{\log x}{\csc x}$ for $x = 0$.

**Solution.** $\frac{f(0)}{F(0)} = \frac{\log x}{\csc x}\Big]_{x=0} = \frac{-\infty}{\infty}\cdot \quad \therefore$ indeterminate.

$$\frac{f'(0)}{F'(0)} = \frac{\frac{1}{x}}{-\csc x \cot x}\Bigg]_{x=0} = -\frac{\sin^2 x}{x \cos x}\Bigg]_{x=0} = \frac{0}{0}\cdot \quad \therefore \text{ indeterminate.}$$

$$\frac{f''(0)}{F''(0)} = -\frac{2 \sin x \cos x}{\cos x - x \sin x}\Bigg]_{x=0} = -\frac{0}{1} = 0. \quad \textit{Ans.}$$

**113. Evaluation of the indeterminate form $0 \cdot \infty$.** If a function $f(x) \cdot \phi(x)$ takes on the indeterminate form $0 \cdot \infty$ for $x = a$, we write the given function

$$f(x) \cdot \phi(x) = \frac{f(x)}{\frac{1}{\phi(x)}} \left( \text{or } = \frac{\phi(x)}{\frac{1}{f(x)}} \right)$$

so as to cause it to take on one of the forms $\frac{0}{0}$ or $\frac{\infty}{\infty}$, thus bringing it under § 111 or § 112.

ILLUSTRATIVE EXAMPLE 1. Evaluate $\sec 3x \cos 5x$ for $x = \frac{\pi}{2}$.

**Solution.** $\sec 3x \cos 5x\Big]_{x=\frac{\pi}{2}} = \infty \cdot 0. \quad \therefore$ indeterminate.

Substituting $\frac{1}{\cos 3x}$ for $\sec 3x$, the function becomes $\frac{\cos 5x}{\cos 3x} = \frac{f(x)}{F(x)}$.

$$\frac{f\left(\frac{\pi}{2}\right)}{F\left(\frac{\pi}{2}\right)} = \frac{\cos 5x}{\cos 3x}\Big]_{x=\frac{\pi}{2}} = \frac{0}{0}. \quad \therefore \text{ indeterminate.}$$

$$\frac{f'\left(\frac{\pi}{2}\right)}{F'\left(\frac{\pi}{2}\right)} = \frac{-\sin 5x \cdot 5}{-\sin 3x \cdot 3}\Big]_{x=\frac{\pi}{2}} = -\frac{5}{3}. \quad \textit{Ans.}$$

**114. Evaluation of the indeterminate form $\infty - \infty$.** It is possible in general to transform the expression into a fraction which will assume either the form $\frac{0}{0}$ or $\frac{\infty}{\infty}$.

ILLUSTRATIVE EXAMPLE 1. Evaluate $\sec x - \tan x$ for $x = \frac{\pi}{2}$.

**Solution.** $\sec x - \tan x\Big]_{x=\frac{\pi}{2}} = \infty - \infty. \quad \therefore$ indeterminate.

By Trigonometry, $\sec x - \tan x = \frac{1}{\cos x} - \frac{\sin x}{\cos x} = \frac{1 - \sin x}{\cos x} = \frac{f(x)}{F(x)}$.

$$\frac{f\left(\frac{\pi}{2}\right)}{F\left(\frac{\pi}{2}\right)} = \frac{1 - \sin x}{\cos x}\Big]_{x=\frac{\pi}{2}} = \frac{1-1}{0} = \frac{0}{0}. \quad \therefore \text{ indeterminate.}$$

$$\frac{f'\left(\frac{\pi}{2}\right)}{F'\left(\frac{\pi}{2}\right)} = \frac{-\cos x}{-\sin x}\Big]_{x=\frac{\pi}{2}} = \frac{0}{-1} = 0. \quad \textit{Ans.}$$

**EXAMPLES**

Evaluate the following expressions by differentiation : *

1. $\underset{x=\infty}{\text{limit}} \frac{ax^2 + b}{cx^2 + d}$. *Ans.* $\frac{a}{c}$.

2. $\underset{x=0}{\text{limit}} \frac{\cot x}{\log x}$. $-\infty$.

3. $\underset{x=\infty}{\text{limit}} \frac{\log x}{x^n}$. 0.

4. $\underset{x=\infty}{\text{limit}} \frac{x^2}{e^x}$. 0.

5. $\underset{x=\infty}{\text{limit}} \frac{e^x}{\log x}$. $\infty$.

6. $\underset{x=0}{\text{limit}} \frac{\log \sin 2x}{\log \sin x}$. *Ans.* 1.

7. $\underset{\theta=\frac{\pi}{2}}{\text{limit}} \frac{\tan \theta}{\tan 3\theta}$. 3.

8. $\underset{\phi=\frac{\pi}{2}}{\text{limit}} \frac{\log\left(\phi - \frac{\pi}{2}\right)}{\tan \phi}$. 0.

9. $\underset{x=0}{\text{limit}} \frac{\log x}{\cot x}$. 0.

10. $\underset{x=0}{\text{limit}}\, x \log \sin x$. 0.

* In solving the remaining examples in this chapter it may be of assistance to the student to refer to § 24, pp. 23, 24, where many special forms *not indeterminate* are evaluated.

11. $\lim_{x=0} x \cot \pi x.$ *Ans.* $\frac{1}{\pi}.$

12. $\lim_{y=\infty} \frac{y}{e^{ay}}.$ 0.

13. $\lim_{x=\frac{\pi}{2}} (\pi - 2x) \tan x.$ 2.

14. $\lim_{x=\infty} x \sin \frac{a}{x}.$ $a.$

15. $\lim_{x=0} x^n \log x.$ [$n$ positive.] 0.

16. $\lim_{\theta=\frac{\pi}{4}} (1 - \tan \theta) \sec 2\theta.$ 1.

17. $\lim_{\phi=a} (a^2 - \phi^2) \tan \frac{\pi\phi}{2a}.$ $\frac{4a^2}{\pi}.$

18. $\lim_{x=1} \left[\frac{2}{x^2-1} - \frac{1}{x-1}\right].$ *Ans.* $-\frac{1}{2}$

19. $\lim_{x=1} \left[\frac{1}{\log x} - \frac{x}{\log x}\right].$ $-1.$

20. $\lim_{\theta=\frac{\pi}{2}} [\sec \theta - \tan \theta].$ 0.

21. $\lim_{\phi=0} \left[\frac{2}{\sin^2 \phi} - \frac{1}{1-\cos \phi}\right].$ $\frac{1}{2}.$

22. $\lim_{y=1} \left[\frac{y}{y-1} - \frac{1}{\log y}\right].$ $\frac{1}{2}.$

23. $\lim_{z=0} \left[\frac{\pi}{4z} - \frac{\pi}{2z(e^{\pi z}+1)}\right].$ $\frac{\pi^2}{8}.$

**115. Evaluation of the indeterminate forms $0^0$, $1^\infty$, $\infty^0$.** Given a function of the form

$$f(x)^{\phi(x)}.$$

In order that the function shall take on one of the above three forms, we must have for a certain value of $x$

$$f(x) = 0, \quad \phi(x) = 0, \text{ giving } 0^0;$$

or, $$f(x) = 1, \quad \phi(x) = \infty, \text{ giving } 1^\infty;$$

or, $$f(x) = \infty, \quad \phi(x) = 0, \text{ giving } \infty^0.$$

Let $$y = f(x)^{\phi(x)};$$

taking the logarithm of both sides,

$$\log y = \phi(x) \log f(x).$$

In any of the above cases the logarithm of $y$ (the function) will take on the indeterminate form

$$0 \cdot \infty.$$

Evaluating this by the process illustrated in § 113 gives the limit of the logarithm of the function. This being equal to the logarithm of the limit of the function, the limit of the function is known.*

ILLUSTRATIVE EXAMPLE 1. Evaluate $x^x$ when $x = 0$.

**Solution.** This function assumes the indeterminate form $0^0$ for $x = 0$.

Let $$y = x^x;$$

then $$\log y = x \log x = 0 \cdot -\infty, \quad \text{when } x = 0.$$

By § 113, p. 174, $$\log y = \frac{\log x}{\frac{1}{x}} = \frac{-\infty}{\infty}, \quad \text{when } x = 0.$$

* Thus, if limit $\log_e y = a$, then $y = e^a$.

By § 112, p. 174, $\log y = \dfrac{\frac{1}{x}}{-\frac{1}{x^2}} = -x = 0,$ when $x = 0$.

Since $y = x^x$, this gives $\log_e x^x = 0$; i.e., $x^x = 1$. *Ans.*

ILLUSTRATIVE EXAMPLE 2. Evaluate $(1 + x)^{\frac{1}{x}}$ when $x = 0$.

**Solution.** This function assumes the indeterminate form $1^\infty$ for $x = 0$.

Let $y = (1 + x)^{\frac{1}{x}}$;

then $\log y = \frac{1}{x} \log(1 + x) = \infty \cdot 0,$ when $x = 0$.

By § 113, p. 174, $\log y = \dfrac{\log(1 + x)}{x} = \dfrac{0}{0},$ when $x = 0$.

By § 111, p. 171, $\log y = \dfrac{\frac{1}{1 + x}}{1} = \dfrac{1}{1 + x} = 1,$ when $x = 0$.

Since $y = (1 + x)^{\frac{1}{x}}$, this gives $\log_e(1 + x)^{\frac{1}{x}} = 1$; i.e. $(1 + x)^{\frac{1}{x}} = e$. *Ans.*

ILLUSTRATIVE EXAMPLE 3. Evaluate $(\cot x)^{\sin x}$ for $x = 0$.

**Solution.** This function assumes the indeterminate form $\infty^0$ for $x = 0$.

Let $y = (\cot x)^{\sin x}$;

then $\log y = \sin x \log \cot x = 0 \cdot \infty,$ when $x = 0$.

By § 113, p. 174, $\log y = \dfrac{\log \cot x}{\csc x} = \dfrac{\infty}{\infty},$ when $x = 0$.

By § 112, p. 174, $\log y = \dfrac{\frac{-\csc^2 x}{\cot x}}{-\csc x \cot x} = \dfrac{\sin x}{\cos^2 x} = 0,$ when $x = 0$.

Since $y = (\cot x)^{\sin x}$, this gives $\log_e(\cot x)^{\sin x} = 0$; i.e. $(\cot x)^{\sin x} = 1$. *Ans.*

**EXAMPLES**

Evaluate the following expressions by differentiation :

1. $\underset{x=1}{\text{limit}}\, x^{\frac{1}{1-x}}$. *Ans.* $\frac{1}{e}$.

2. $\underset{x=0}{\text{limit}} \left(\frac{1}{x}\right)^{\tan x}$. 1.

3. $\underset{\theta=\frac{\pi}{2}}{\text{limit}} (\sin\theta)^{\tan\theta}$. 1.

4. $\underset{y=\infty}{\text{limit}} \left(1 + \frac{a}{y}\right)^y$. $e^a$.

5. $\underset{x=0}{\text{limit}} (1 + \sin x)^{\cot x}$. $e$.

6. $\underset{x=\infty}{\text{limit}} \left(\frac{2}{x} + 1\right)^x$. $e^2$.

7. $\underset{x=0}{\text{limit}} (e^x + x)^{\frac{1}{x}}$. *Ans.* $e^2$.

8. $\underset{x=0}{\text{limit}} (\cot x)^{\frac{1}{\log x}}$. $\frac{1}{e}$.

9. $\underset{z=0}{\text{limit}} (1 + nz)^{\frac{1}{z}}$. $e^n$.

10. $\underset{\phi=1}{\text{limit}} \left(\tan\frac{\pi\phi}{4}\right)^{\tan\frac{\pi\phi}{2}}$. $\frac{1}{e}$.

11. $\underset{\theta=0}{\text{limit}} (\cos m\theta)^{\frac{n}{\theta^2}}$. $e^{-\frac{1}{2}nm^2}$.

12. $\underset{x=0}{\text{limit}} (\cot x)^x$. 1.

13. $\underset{x=a}{\text{limit}} \left(2 - \frac{x}{a}\right)^{\tan\frac{\pi x}{2a}}$. $e^{\frac{2}{\pi}}$.

# CHAPTER XIV

## CIRCLE OF CURVATURE. CENTER OF CURVATURE

**116. Circle of curvature.* Center of curvature.** If a circle be drawn through three points $P_0$, $P_1$, $P_2$ on a plane curve, and if $P_1$ and $P_2$ be made to approach $P_0$ along the curve as a limiting position, then the circle will in general approach in magnitude and position a limiting circle called the *circle of curvature of the curve at the point* $P_0$. The center of this circle is called the *center of curvature.*

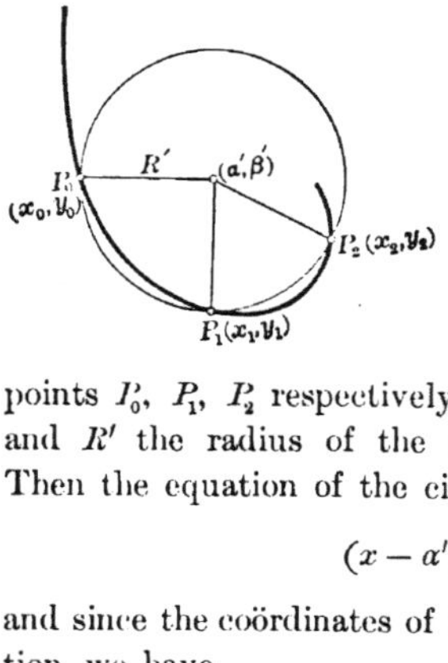


Let the equation of the curve be

$$(1) \qquad y = f(x);$$

and let $x_0$, $x_1$, $x_2$ be the abscissas of the points $P_0$, $P_1$, $P_2$ respectively, $(\alpha', \beta')$ the coördinates of the center, and $R'$ the radius of the circle passing through the three points. Then the equation of the circle is

$$(x-\alpha')^2 + (y-\beta')^2 = R'^2;$$

and since the coördinates of the points $P_0$, $P_1$, $P_2$ must satisfy this equation, we have

$$(2) \qquad \begin{cases} (x_0-\alpha')^2 + (y_0-\beta')^2 - R'^2 = 0, \\ (x_1-\alpha')^2 + (y_1-\beta')^2 - R'^2 = 0, \\ (x_2-\alpha')^2 + (y_2-\beta')^2 - R'^2 = 0. \end{cases}$$

Now consider the *function of x* defined by

$$F(x) = (x-\alpha')^2 + (y-\beta')^2 - R'^2,$$

in which $y$ has been replaced by $f(x)$ from (1).

Then from equations (2) we get

$$F(x_0) = 0, \quad F(x_1) = 0, \quad F(x_2) = 0.$$

* Sometimes called the *osculating circle.* The circle of curvature was defined from another point of view on p. 161.

Hence, by Rolle's Theorem (p. 164), $F'(x)$ must vanish for at least two values of $x$, one lying between $x_0$ and $x_1$, say $x'$, and the other lying between $x_1$ and $x_2$, say $x''$; that is,

$$F'(x') = 0, \quad F'(x'') = 0.$$

Again, for the same reason, $F''(x)$ must vanish for some value of $x$ between $x'$ and $x''$, say $x_3$; hence

$$F''(x_3) = 0.$$

Therefore the elements $\alpha'$, $\beta'$, $R'$ of the circle passing through the points $P_0$, $P_1$, $P_2$ must satisfy the three equations

$$F(x_0) = 0, \quad F'(x') = 0, \quad F''(x_3) = 0.$$

Now let the points $P_1$ and $P_2$ approach $P_0$ as a limiting position; then $x_1$, $x_2$, $x'$, $x''$, $x_3$ will all approach $x_0$ as a limit, and the elements $\alpha$, $\beta$, $R$ of the osculating circle are therefore determined by the three equations

$$F(x_0) = 0, \quad F'(x_0) = 0, \quad F''(x_0) = 0;$$

or, dropping the subscripts, which is the same thing,

$$(A) \qquad (x-\alpha)^2 + (y-\beta)^2 = R^2,$$

$$(B) \qquad (x-\alpha) + (y-\beta)\frac{dy}{dx} = 0, \text{ differentiating } (A).$$

$$(C) \qquad 1 + \left(\frac{dy}{dx}\right)^2 + (y-\beta)\frac{d^2y}{dx^2} = 0, \text{ differentiating } (B).$$

Solving $(B)$ and $(C)$ for $x-\alpha$ and $y-\beta$, we get $\left(\frac{d^2y}{dx^2} \neq 0\right)$,

$$(D) \qquad \begin{cases} x - \alpha = \dfrac{\dfrac{dy}{dx}\left[1 + \left(\dfrac{dy}{dx}\right)^2\right]}{\dfrac{d^2y}{dx^2}}, \\[2ex] y - \beta = -\dfrac{1 + \left(\dfrac{dy}{dx}\right)^2}{\dfrac{d^2y}{dx^2}}; \end{cases}$$

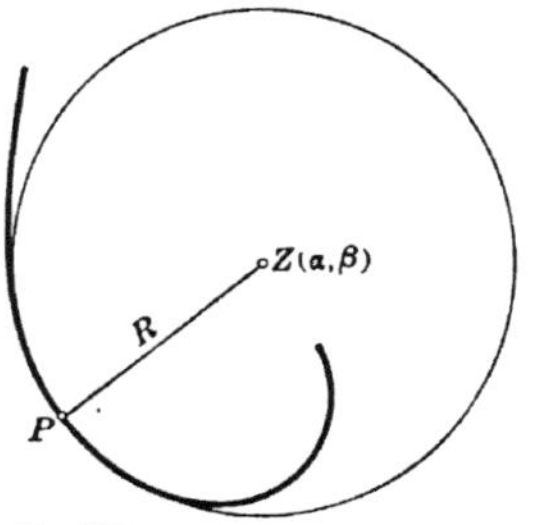


hence the *coördinates of the center of curvature are*

$$(E) \qquad \alpha = x - \frac{\dfrac{dy}{dx}\left[1 + \left(\dfrac{dy}{dx}\right)^2\right]}{\dfrac{d^2y}{dx^2}}; \quad \beta = y + \frac{1 + \left(\dfrac{dy}{dx}\right)^2}{\dfrac{d^2y}{dx^2}}. \qquad \left(\frac{d^2y}{dx^2} \neq 0\right)$$

Substituting the values of $x-\alpha$ and $y-\beta$ from $(D)$ in $(A)$, and solving for $R$, we get

$$R = \pm \frac{\left[1+\left(\frac{dy}{dx}\right)^2\right]^{\frac{3}{2}}}{\frac{d^2y}{dx^2}},$$

which is identical with (**42**), p. 159. Hence

**Theorem.** *The radius of the circle of curvature equals the radius of curvature.*

**117. Second method for finding center of curvature.** Here we shall make use of the definition of circle of curvature given on p. 161. Draw a figure showing the tangent line, circle of curvature, radius of curvature, and center of curvature $(\alpha, \beta)$ corresponding to the point $P(x, y)$ on the curve. Then

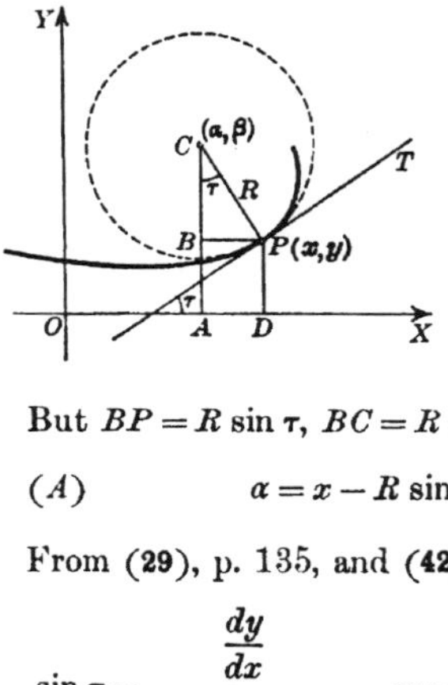


$$\alpha = OA = OD - AD = OD - BP = x - BP,$$
$$\beta = AC = AB + BC = DP + BC = y + BC.$$

But $BP = R\sin\tau$, $BC = R\cos\tau$. Hence

$(A)$ $$\alpha = x - R\sin\tau, \qquad \beta = y + R\cos\tau.$$

From (**29**), p. 135, and (**42**), p. 159,

$$\sin\tau = \frac{\frac{dy}{dx}}{\left[1+\left(\frac{dy}{dx}\right)^2\right]^{\frac{1}{2}}}, \quad \cos\tau = \frac{1}{\left[1+\left(\frac{dy}{dx}\right)^2\right]^{\frac{1}{2}}}, \quad R = \frac{\left[1+\left(\frac{dy}{dx}\right)^2\right]^{\frac{3}{2}}}{\frac{d^2y}{dx^2}}.$$

Substituting these back in $(A)$, we get

**(50)** $$\alpha = x - \frac{\frac{dy}{dx}\left[1+\left(\frac{dy}{dx}\right)^2\right]}{\frac{d^2y}{dx^2}}; \quad \beta = y + \frac{1+\left(\frac{dy}{dx}\right)^2}{\frac{d^2y}{dx^2}}.$$

From (**23**), p. 126, we know that at a point of inflection (as $Q$ in the next figure)

$$\frac{d^2y}{dx^2} = 0.$$

Therefore, by (**40**), p. 157, the curvature $K=0$; and from (**42**), p. 159, and (**50**), p. 180, we see that in general $\alpha$, $\beta$, $R$ increase without limit as the second derivative approaches zero. That is, if we suppose $P$ with its tangent to move along the curve to $P'$, at the point of inflection $Q$ the curvature is zero, the rotation of the tangent is momentarily arrested, and as the direction of rotation changes, the center of curvature moves out indefinitely and the radius of curvature becomes infinite.

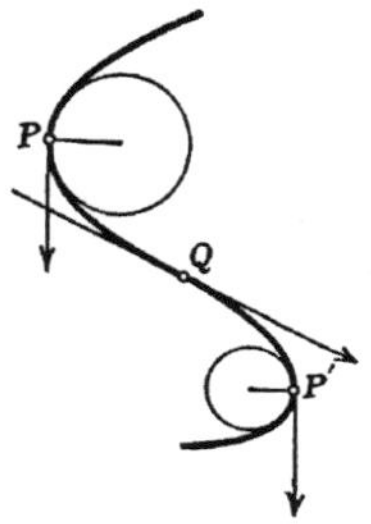


ILLUSTRATIVE EXAMPLE 1. Find the coördinates of the center of curvature of the parabola $y^2 = 4px$ corresponding (a) to any point on the curve; (b) to the vertex.

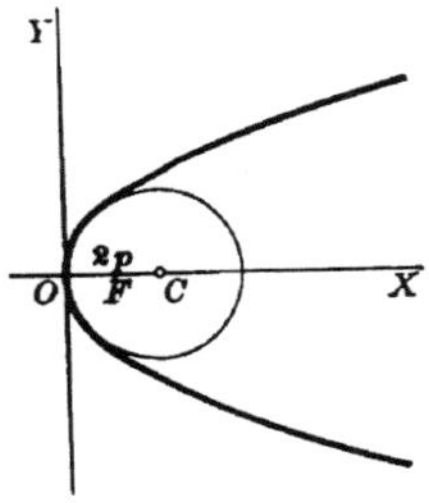


**Solution.** $\dfrac{dy}{dx}=\dfrac{2p}{y}$; $\dfrac{d^2y}{dx^2}=-\dfrac{4p^2}{y^3}$.

(a) Substituting in (*E*), p. 179,

$$\alpha = x+\frac{y^2+4p^2}{y^2}\cdot\frac{2p}{y}\cdot\frac{y^3}{4p^2}=3x+2p.$$

$$\beta = y-\frac{y^2+4p^2}{y^2}\cdot\frac{y^3}{4p^2}=-\frac{y^3}{4p^2}.$$

Therefore $\left(3x+2p,\ -\dfrac{y^3}{4p^2}\right)$ is the center of curvature corresponding to any point on the curve.

(b) $(2p, 0)$ is the center of curvature corresponding to the vertex $(0, 0)$.

**118. Center of curvature the limiting position of the intersection of normals at neighboring points.** Let the equation of a curve be

$$(A)\qquad y=f(x).$$

The equations of the normals to the curve at two neighboring points $P_0$ and $P_1$ are*

$$(x_0-X)+(y_0-Y)\frac{dy_0}{dx_0}=0,$$

$$(x_1-X)+(y_1-Y)\frac{dy_1}{dx_1}=0.$$

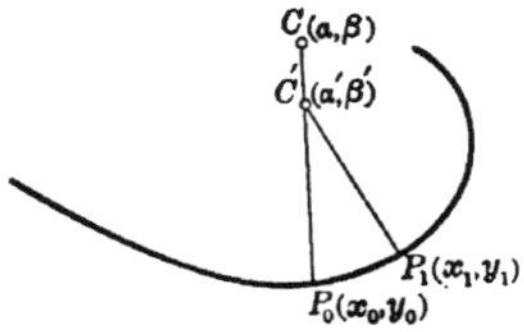


If the normals intersect at $C'(\alpha', \beta')$, the coördinates of this point must satisfy both equations, giving

$$(B)\qquad \begin{cases}(x_0-\alpha')+(y_0-\beta')\dfrac{dy_0}{dx_0}=0,\\[2mm] (x_1-\alpha')+(y_1-\beta')\dfrac{dy_1}{dx_1}=0.\end{cases}$$

* From (2), p. 77, $X$ and $Y$ being the variable coördinates.

Now consider the *function of $x$* defined by

$$\phi(x) = (x - \alpha') + (y - \beta')\frac{dy}{dx},$$

in which $y$ has been replaced by $f(x)$ from $(A)$.

Then equations $(B)$ show that

$$\phi(x_0) = 0, \quad \phi(x_1) = 0.$$

But then, by Rolle's Theorem (p. 164), $\phi'(x)$ must vanish for some value of $x$ between $x_0$ and $x_1$, say $x'$. Therefore $\alpha'$ and $\beta'$ are determined by the two equations

$$\phi(x_0) = 0, \quad \phi'(x') = 0.$$

If now $P_1$ approaches $P_0$ as a limiting position, then $x'$ approaches $x_0$, giving

$$\phi(x_0) = 0, \quad \phi'(x_0) = 0;$$

and $C'(\alpha', \beta')$ will approach as a limiting position the center of curvature $C(\alpha, \beta)$ corresponding to $P_0$ on the curve. For if we drop the subscripts and write the last two equations in the form

$$(x - \alpha') + (y - \beta')\frac{dy}{dx} = 0,$$

$$1 + \left(\frac{dy}{dx}\right)^2 + (y - \beta')\frac{d^2y}{dx^2} = 0,$$

it is evident that solving for $\alpha'$ and $\beta'$ will give the same results as solving $(B)$ and $(C)$, p. 179, for $\alpha$ and $\beta$. Hence

**Theorem.** *The center of curvature C corresponding to a point P on a curve is the limiting position of the intersection of the normal to the curve at P with a neighboring normal.*

**119. Evolutes.** The locus of the centers of curvature of a given curve is called the *evolute* of that curve. Consider the circle of curvature corresponding to a point $P$ on a curve. If $P$ moves along the given curve, we may suppose the corresponding circle of curvature to roll along the curve with it, its radius varying so as to be always equal to the radius of curvature of the curve at the point $P$. The curve $CC_7$ described by the center of the circle is the evolute of $PP_7$.

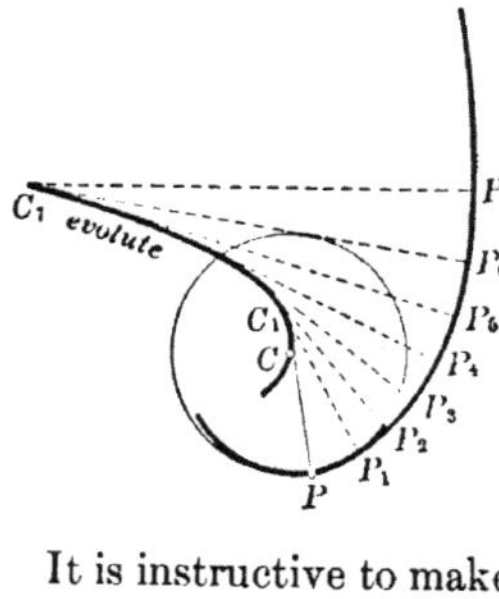


It is instructive to make an approximate construction of the evolute of a curve by estimating (from the shape of the curve) the lengths

of the radii of curvature at different points on the curve and then drawing them in and drawing the locus of the centers of curvature.

Formula (*E*), p. 179, gives the coördinates of any point $(\alpha, \beta)$ on the evolute expressed in terms of the coördinates of the corresponding point $(x, y)$ of the given curve. But $y$ is a function of $x$; therefore

$$\alpha = x - \frac{\left[1+\left(\frac{dy}{dx}\right)^2\right]\frac{dy}{dx}}{\frac{d^2y}{dx^2}}, \quad \beta = y + \frac{1+\left(\frac{dy}{dx}\right)^2}{\frac{d^2y}{dx^2}}$$

give us at once *the parametric equations of the evolute in terms of the parameter* $x$.

To find the ordinary rectangular equation of the evolute we eliminate $x$ between the two expressions. No general process of elimination can be given that will apply in all cases, the method to be adopted depending on the form of the given equation. In a large number of cases, however, the student can find the rectangular equation of the evolute by taking the following steps:

**General directions for finding the equation of the evolute in rectangular coördinates.**

FIRST STEP. *Find* $\alpha$ *and* $\beta$ *from* (**50**), p. 180.

SECOND STEP. *Solve the two resulting equations for* $x$ *and* $y$ *in terms of* $\alpha$ *and* $\beta$.

THIRD STEP. *Substitute these values of* $x$ *and* $y$ *in the given equation. This gives a relation between the variables* $\alpha$ *and* $\beta$ *which is the equation of the evolute.*

ILLUSTRATIVE EXAMPLE 1. Find the equation of the evolute of the parabola $y^2 = 4px$.

**Solution.** $\frac{dy}{dx} = \frac{2p}{y}, \quad \frac{d^2y}{dx^2} = -\frac{4p^2}{y^3}.$

*First step.* $\alpha = 3x + 2p, \quad \beta = -\frac{y^3}{4p^2}.$

*Second step.* $x = \frac{\alpha - 2p}{3}, \quad y = -(4p^2\beta)^{\frac{1}{3}}.$

*Third step* $(4p^2\beta)^{\frac{2}{3}} = 4p\left(\frac{\alpha - 2p}{3}\right);$

or, $p\beta^2 = \frac{4}{27}(\alpha - 2p)^3.$

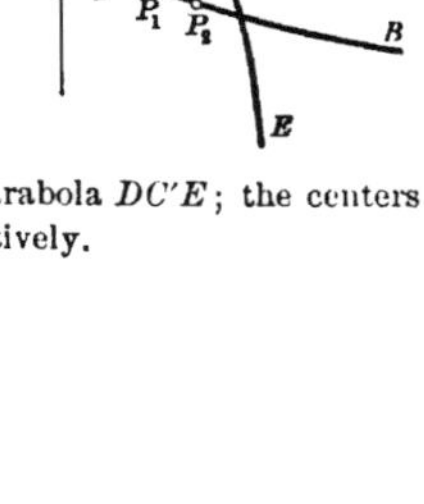


Remembering that $\alpha$ denotes the abscissa and $\beta$ the ordinate of a rectangular system of coördinates, we see that the evolute of the parabola $AOB$ is the semicubical parabola $DC'E$; the centers of curvature for $O, P, P_1, P_2$ being at $C', C, C_1, C_2$ respectively.

ILLUSTRATIVE EXAMPLE 2. Find the equation of the evolute of the ellipse $b^2x^2 + a^2y^2 = a^2b^2$.

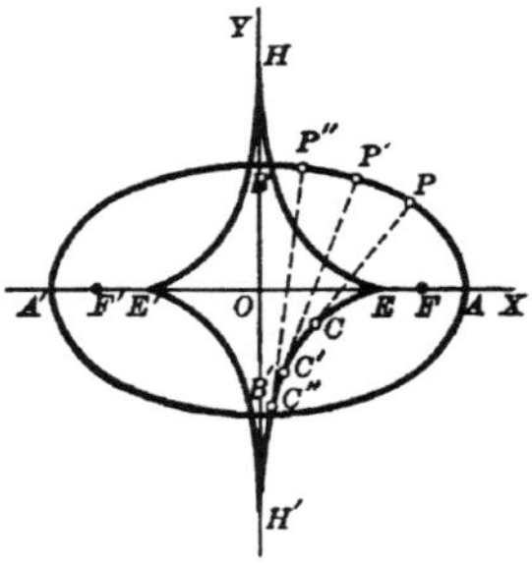


*Solution.* $$\frac{dy}{dx} = -\frac{b^2x}{a^2y},\ \frac{d^2y}{dx^2} = -\frac{b^4}{a^2y^3}.$$

*First step.* $$\alpha = \frac{(a^2 - b^2)x^3}{a^4},$$

$$\beta = -\frac{(a^2 - b^2)y^3}{b^4}.$$

*Second step.* $$x = \left(\frac{a^4\alpha}{a^2 - b^2}\right)^{\frac{1}{3}},$$

$$y = -\left(\frac{b^4\beta}{a^2 - b^2}\right)^{\frac{1}{3}}.$$

*Third step.* $(a\alpha)^{\frac{2}{3}} + (b\beta)^{\frac{2}{3}} = (a^2 - b^2)^{\frac{2}{3}}$, the equation of the evolute $EHE'H'$ of the ellipse $ABA'B'$. $E$, $E'$, $H'$, $H$ are the centers of curvature corresponding to the points $A$, $A'$, $B$, $B'$, on the curve, and $C$, $C'$, $C''$ correspond to the points $P$, $P'$, $P''$.

**When the equations of the curve are given in parametric form,** we proceed to find $\frac{dy}{dx}$ and $\frac{d^2y}{dx^2}$, as on p. 160, from

$$(A)\qquad \frac{dy}{dx} = \frac{\frac{dy}{dt}}{\frac{dx}{dt}},\quad \frac{d^2y}{dx^2} = \frac{\frac{dx}{dt}\frac{d^2y}{dt^2} - \frac{dy}{dt}\frac{d^2x}{dt^2}}{\left(\frac{dx}{dt}\right)^3};$$

and then substitute the results in formulas (**50**), p. 180. This gives the parametric equations of the evolute in terms of the same parameter that occurs in the given equations.

ILLUSTRATIVE EXAMPLE 3. The parametric equations of a curve are

$$(B)\qquad x = \frac{t^2 + 1}{4},\ y = \frac{t^3}{6}.$$

Find the equation of the evolute in parametric form, plot the curve and the evolute, find the radius of curvature at the point where $t = 1$, and draw the corresponding circle of curvature.

*Solution.* $$\frac{dx}{dt} = \frac{t}{2},\ \frac{d^2x}{dt^2} = \frac{1}{2},$$

$$\frac{dy}{dt} = \frac{t^2}{2},\ \frac{d^2y}{dt^2} = t.$$

Substituting in above formulas $(A)$ and then in (50), p. 180, gives

$$(C)\qquad \alpha = \frac{1 - t^2 - 2t^4}{4},\ \beta = \frac{4t^3 + 3t}{6},$$

the parametric equations of the evolute. Assuming values of the parameter $t$, we calculate $x, y$ ; $\alpha, \beta$ from ($B$) and ($C$) ; and tabulate the results as follows :

Now plot the curve and its evolute.

| $t$ | $x$ | $y$ | $\alpha$ | $\beta$ |
|---|---|---|---|---|
| 3 | $\frac{5}{2}$ | $\frac{9}{2}$ | | |
| 2 | $\frac{5}{4}$ | $\frac{4}{3}$ | $-\frac{35}{4}$ | $\frac{19}{3}$ |
| $\frac{3}{2}$ | $\frac{13}{16}$ | $\frac{9}{16}$ | $-\frac{91}{32}$ | 3 |
| 1 | $\frac{1}{2}$ | $\frac{1}{6}$ | $-\frac{1}{2}$ | $\frac{7}{6}$ |
| 0 | $\frac{1}{4}$ | 0 | $\frac{1}{4}$ | 0 |
| $-1$ | $\frac{1}{2}$ | $-\frac{1}{6}$ | $-\frac{1}{2}$ | $-\frac{7}{6}$ |
| $-\frac{3}{2}$ | $\frac{13}{16}$ | $-\frac{9}{16}$ | $-\frac{91}{32}$ | $-3$ |
| $-2$ | $\frac{5}{4}$ | $-\frac{4}{3}$ | $-\frac{35}{4}$ | $-\frac{19}{3}$ |
| $-3$ | $\frac{5}{2}$ | $-\frac{9}{2}$ | | |

The point $(\frac{1}{4}, 0)$ is common to the given curve and its evolute. The given curve (semicubical parabola) lies entirely to the right and the evolute entirely to the left of $x = \frac{1}{4}$.

The circle of curvature at $A$ $(\frac{1}{2}, \frac{1}{6})$, where $t = 1$, will have its center at $A'$ $(-\frac{1}{2}, \frac{7}{6})$ on the evolute and radius $= AA'$. To verify our work find radius of curvature at $A$. From (42), p. 159, we get

$$R = \frac{t(1+t^2)^{\frac{3}{2}}}{2} = \sqrt{2}, \text{ when } t = 1.$$

This should equal the distance

$$AA' = \sqrt{(\tfrac{1}{2} + \tfrac{1}{2})^2 + (\tfrac{1}{6} - \tfrac{7}{6})^2} = \sqrt{2}.$$

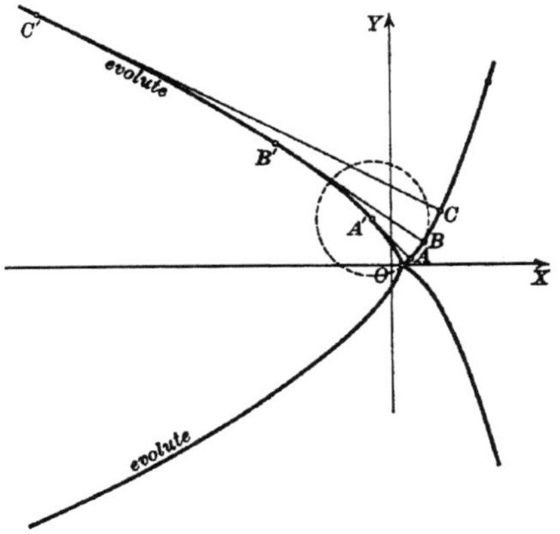


Illustrative Example 4. Find the parametric equations of the evolute of the cycloid,

$$(C) \qquad \begin{cases} x = a(t - \sin t), \\ y = a(1 - \cos t). \end{cases}$$

**Solution.** As in Illustrative Example 2, p. 160, we get

$$\frac{dy}{dx} = \frac{\sin t}{1 - \cos t}, \quad \frac{d^2y}{dx^2} = -\frac{1}{a(1 - \cos t)^2}.$$

Substituting these results in formulas (50), p. 180, we get

$$(D) \qquad \begin{cases} \alpha = a(t + \sin t), \\ \beta = -a(1 - \cos t). \quad \textit{Ans.} \end{cases}$$

**Note.** If we eliminate $t$ between equations $(D)$, there results the rectangular equation of the evolute $OO'Q^v$ referred to the axes $O'\alpha$ and $O'\beta$. The coördinates of $O$ with respect to these axes are $(-\pi a, -2a)$.

Let us transform equations $(D)$ to the new set of axes $OX$ and $OY$. Then

$$\alpha = x - \pi a, \qquad \beta = y - 2a,$$
$$t = t' - \pi.$$

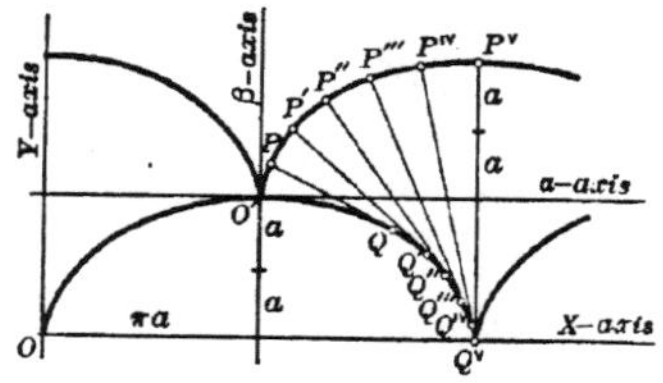


Substituting in $(D)$ and reducing, the equations of the evolute become

$$(E) \qquad \begin{cases} x = a(t' - \sin t'), \\ y = a(1 - \cos t'). \end{cases}$$

Since $(E)$ and $(C)$ are identical in form, we have:

*The evolute of a cycloid is itself a cycloid whose generating circle equals that of the given cycloid.*

**120. Properties of the evolute.** From $(A)$, p. 180,

$$(A) \qquad \alpha = x - R\sin\tau, \quad \beta = y + R\cos\tau.$$

Let us choose as independent variable the length $s$ of the arc on the given curve; then $x, y, R, \tau, \alpha, \beta$ are functions of $s$. Differentiating $(A)$ with respect to $s$ gives

$$(B) \qquad \frac{d\alpha}{ds} = \frac{dx}{ds} - R\cos\tau\frac{d\tau}{ds} - \sin\tau\frac{dR}{ds},$$

$$(C) \qquad \frac{d\beta}{ds} = \frac{dy}{ds} - R\sin\tau\frac{d\tau}{ds} + \cos\tau\frac{dR}{ds}.$$

But $\frac{dx}{ds} = \cos\tau$, $\frac{dy}{ds} = \sin\tau$, from **(26)**, p. 134; and $\frac{d\tau}{ds} = \frac{1}{R}$, from **(38)** and **(39)**, p. 156.

Substituting in $(B)$ and $(C)$, we obtain

$$(D) \qquad \frac{d\alpha}{ds} = \cos\tau - R\cos\tau\cdot\frac{1}{R} - \sin\tau\frac{dR}{ds} = -\sin\tau\frac{dR}{ds},$$

$$(E) \qquad \frac{d\beta}{ds} = \sin\tau - R\sin\tau\cdot\frac{1}{R} + \cos\tau\frac{dR}{ds} = \cos\tau\frac{dR}{ds}.$$

Dividing $(E)$ by $(D)$ gives

$$(F) \qquad \frac{d\beta}{d\alpha} = -\cot\tau = -\frac{1}{\tan\tau} = -\frac{1}{\frac{dy}{dx}}.$$

But $\frac{d\beta}{d\alpha} = \tan\tau' =$ slope of tangent to the evolute at $C$, and

$\frac{dy}{dx} = \tan\tau =$ slope of tangent to the given curve at the corresponding point $P(x, y)$.

Substituting the last two results in $(F)$, we get

$$\tan\tau' = -\frac{1}{\tan\tau}.$$

Since the slope of one tangent is the negative reciprocal of the slope of the other, they are perpendicular. But a line perpendicular to the tangent at $P$ is a normal to the curve. Hence

*A normal to the given curve is a tangent to its evolute.*

Again, squaring equations $(D)$ and $(E)$ and adding, we get

$$(G) \qquad \left(\frac{d\alpha}{ds}\right)^2 + \left(\frac{d\beta}{ds}\right)^2 = \left(\frac{dR}{ds}\right)^2.$$

But if $s'$ = length of arc of the evolute, the left-hand member of $(G)$ is precisely the square of $\frac{ds'}{ds}$ (from **(34)**, p. 141, where $t = s$, $s = s'$, $x = \alpha$, $y = \beta$). Hence $(G)$ asserts that

$$\left(\frac{ds'}{ds}\right)^2 = \left(\frac{dR}{ds}\right)^2, \text{ or } \frac{ds'}{ds} = \pm\frac{dR}{ds}.$$

That is, *the radius of curvature of the given curve increases or decreases as fast as the arc of the evolute increases.* In our figure this means that

$$P_1C_1 - PC = \text{arc } CC_1.$$

*The length of an arc of the evolute is equal to the difference between the radii of curvature of the given curve which are tangent to this arc at its extremities.*

Thus in Illustrative Example 4, p. 186, we observe that if we fold $Q^{\text{v}}P^{\text{v}}(= 4a)$ over to the left on the evolute, $P^{\text{v}}$ will reach to $O'$, and we have:

*The length of one arc of the cycloid (as $OO'Q^{\text{v}}$) is eight times the length of the radius of the generating circle.*

**121. Involutes and their mechanical construction.** Let a flexible ruler be bent in the form of the curve $C_1C_9$, the evolute of the curve $P_1P_9$, and suppose a string of length $R_9$, with one end fastened at $C_9$, to be wrapped around the ruler (or curve). It is clear from the results of the last section that when the string is unwound and kept taut, the free end will describe the curve $P_1P_9$. Hence the name *evolute.*

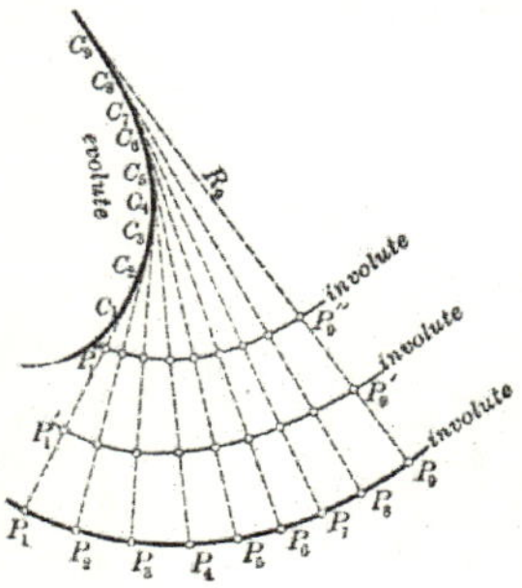


The curve $P_1P_9$ is said to be an *involute* of $C_1C_9$. Obviously any point on the string will describe an involute, so that a given curve has an infinite number of involutes but only one evolute.

The involutes $P_1P_9$, $P_1'P_9'$, $P_1''P_9''$ are called *parallel curves* since the distance between any two of them measured along their common normals is constant.

The student should observe how the parabola and ellipse on pp. 183, 184 may be constructed in this way from their evolutes.

## EXAMPLES

Find the coördinates of the center of curvature and the equation of the evolute of each of the following curves. Draw the curve and its evolute, and draw at least one circle of curvature.

1. The hyperbola $\dfrac{x^2}{a^2} - \dfrac{y^2}{b^2} = 1$. *Ans.* $\alpha = \dfrac{(a^2+b^2)x^3}{a^4}$, $\beta = -\dfrac{(a^2+b^2)y^3}{b^4}$;
evolute $(a\alpha)^{\frac{2}{3}} - (b\beta)^{\frac{2}{3}} = (a^2+b^2)^{\frac{2}{3}}$.

2. The hypocycloid $x^{\frac{2}{3}} + y^{\frac{2}{3}} = a^{\frac{2}{3}}$. *Ans.* $\alpha = x + 3x^{\frac{1}{3}}y^{\frac{2}{3}}$, $\beta = y + 3x^{\frac{2}{3}}y^{\frac{1}{3}}$;
evolute $(\alpha+\beta)^{\frac{2}{3}} + (\alpha-\beta)^{\frac{2}{3}} = 2a^{\frac{2}{3}}$.

3. Find the coördinates of the center of curvature of the cubical parabola $y^3 = a^2x$.
*Ans.* $\alpha = \dfrac{a^4+15y^4}{6a^2y}$, $\beta = \dfrac{a^4y - 9y^5}{2a^4}$.

4. Show that in the parabola $x^{\frac{1}{2}} + y^{\frac{1}{2}} = a^{\frac{1}{2}}$ we have the relation $\alpha + \beta = 3(x+y)$.

5. Given the equation of the equilateral hyperbola $2xy = a^2$; show that

$$\alpha + \beta = \frac{(y+x)^3}{a^2}, \quad \alpha - \beta = \frac{(y-x)^3}{a^2}.$$

From this derive the equation of the evolute $(\alpha+\beta)^{\frac{2}{3}} - (\alpha-\beta)^{\frac{2}{3}} = 2a^{\frac{2}{3}}$.

Find the parametric equations of the evolutes of the following curves in terms of the parameter $t$. Draw the curve and its evolute, and draw at least one circle of curvature.

6. The hypocycloid $\begin{cases} x = a\cos^3 t, \\ y = a\sin^3 t. \end{cases}$ *Ans.* $\begin{cases} \alpha = a\cos^3 t + 3a\cos t\sin^2 t, \\ \beta = 3a\cos^2 t\sin t + a\sin^3 t. \end{cases}$

7. The curve $\begin{cases} x = 3t^2, \\ y = 3t - t^3. \end{cases}$ $\begin{cases} \alpha = \frac{3}{2}(1 + 2t^2 - t^4), \\ \beta = -4t^3. \end{cases}$

8. The curve $\begin{cases} x = a(\cos t + t \sin t), \\ y = a(\sin t - t \cos t). \end{cases}$ *Ans.* $\begin{cases} \alpha = a \cos t, \\ \beta = a \sin t. \end{cases}$

9. The curve $\begin{cases} x = 3t, \\ y = t^2 - 6. \end{cases}$ $\begin{cases} \alpha = -\frac{4}{3}t^3, \\ \beta = 3t^2 - \frac{3}{2}. \end{cases}$

10. The curve $\begin{cases} x = 6 - t^2, \\ y = 2t. \end{cases}$ $\begin{cases} \alpha = 4 - 3t^2, \\ \beta = -2t^3. \end{cases}$

11. The curve $\begin{cases} x = 2t, \\ y = t^2 - 2. \end{cases}$ $\begin{cases} \alpha = -2t^3, \\ \beta = 3t^2. \end{cases}$

12. The curve $\begin{cases} x = 4t, \\ y = 3 + t^2. \end{cases}$ $\begin{cases} \alpha = -t^3, \\ \beta = 11 + 3t^2. \end{cases}$

13. The curve $\begin{cases} x = 9 - t^2, \\ y = 2t. \end{cases}$ $\begin{cases} \alpha = 7 - 3t^2, \\ \beta = -2t^3. \end{cases}$

14. The curve $\begin{cases} x = 2t, \\ y = \frac{1}{3}t^3. \end{cases}$ $\begin{cases} \alpha = \frac{4t - t^5}{4}, \\ \beta = \frac{12 + 5t^4}{6t}. \end{cases}$

15. The curve $\begin{cases} x = \frac{1}{3}t^3, \\ y = t^2. \end{cases}$ $\begin{cases} \alpha = \frac{4t^3 + 12t}{3}, \\ \beta = -\frac{2t^2 + t^4}{2}. \end{cases}$

16. The curve $\begin{cases} x = 2t, \\ y = \frac{3}{t}. \end{cases}$ $\begin{cases} \alpha = \frac{12t^4 + 9}{4t^3}, \\ \beta = \frac{27 + 4t^4}{6t}. \end{cases}$

17. $x = 4 - t^2,\ y = 2t.$

18. $x = 2t,\ y = 16 - t^2.$

19. $x = t,\ y = \sin t.$

20. $x = \frac{4}{t},\ y = 3t.$

21. $x = t^2,\ y = \frac{1}{3}t^3.$

22. $x = t,\ y = t^3.$

23. $x = \sin t,\ y = 3 \cos t.$

24. $x = 1 - \cos t,\ y = t - \sin t.$

25. $x = \cos^4 t,\ y = \sin^4 t.$

26. $x = a \sec t,\ y = b \tan t.$

# CHAPTER XV

## PARTIAL DIFFERENTIATION

**122. Continuous functions of two or more independent variables.** A function $f(x, y)$ of two independent variables $x$ and $y$ is defined as continuous for the values $(a, b)$ of $(x, y)$ when

$$\lim_{\substack{x=a\\y=b}} f(x, y) = f(a, b),$$

no matter in what way $x$ and $y$ approach their respective limits $a$ and $b$. This definition is sometimes roughly summed up in the statement that *a very small change in one or both of the independent variables shall produce a very small change in the value of the function.**

We may illustrate this geometrically by considering the surface represented by the equation

$$z = f(x, y).$$

Consider a fixed point $P$ on the surface where $x = a$ and $y = b$.

Denote by $\Delta x$ and $\Delta y$ the increments of the independent variables $x$ and $y$, and by $\Delta z$ the corresponding increment of the dependent variable $z$, the coördinates of $P'$ being

$$(x + \Delta x,\ y + \Delta y,\ z + \Delta z).$$

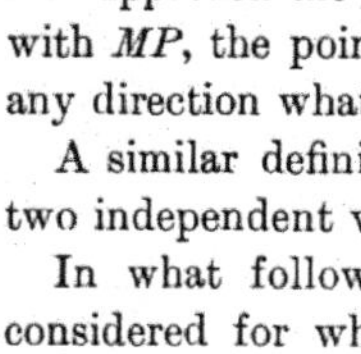


At $P$ the value of the function is

$$z = f(a, b) = MP.$$

If the function is continuous at $P$, then, however $\Delta x$ and $\Delta y$ may approach the limit zero, $\Delta z$ will also approach the limit zero. That is, $M'P'$ will approach coincidence with $MP$, the point $P'$ approaching the point $P$ on the surface from any direction whatever.

A similar definition holds for a continuous function of more than two independent variables.

In what follows, only values of the independent variables are considered for which a function is continuous.

* This will be better understood if the student again reads over § 18, p. 14, on continuous functions of a single variable.

**123. Partial derivatives.** Since $x$ and $y$ are independent in

$$z=f(x, y),$$

$x$ may be supposed to vary while $y$ remains constant, or the reverse.

The derivative of $z$ with respect to $x$ when $x$ varies and $y$ remains constant* is called the *partial derivative of $z$ with respect to $x$*, and is denoted by the symbol $\frac{\partial z}{\partial x}$. We may then write

$$(A)\qquad \frac{\partial z}{\partial x}=\underset{\Delta x=0}{\text{limit}}\left[\frac{f(x+\Delta x, y)-f(x, y)}{\Delta x}\right].$$

Similarly, when $x$ remains constant* and $y$ varies, *the partial derivative of $z$ with respect to $y$* is

$$(B)\qquad \frac{\partial z}{\partial y}=\underset{\Delta y=0}{\text{limit}}\left[\frac{f(x, y+\Delta y)-f(x, y)}{\Delta y}\right].$$

$\frac{\partial z}{\partial x}$ is also written $\frac{\partial}{\partial x}f(x, y)$, or $\frac{\partial f}{\partial x}$.

Similarly, $\frac{\partial z}{\partial y}$ is also written $\frac{\partial}{\partial y}f(x, y)$, or $\frac{\partial f}{\partial y}$.

In order to avoid confusion the round $\partial$† has been generally adopted to indicate partial differentiation. Other notations, however, which are in use are

$$\left(\frac{dz}{dx}\right), \left(\frac{dz}{dy}\right); f'_x(x, y), f'_y(x, y); f_x(x, y), f_y(x, y); D_x f, D_y f; z_x, z_y.$$

Our notation may be extended to a function of any number of independent variables. Thus, if

$$u=F(x, y, z),$$

then we have the three partial derivatives

$$\frac{\partial u}{\partial x}, \frac{\partial u}{\partial y}, \frac{\partial u}{\partial z}; \text{ or, } \frac{\partial F}{\partial x}, \frac{\partial F}{\partial y}, \frac{\partial F}{\partial z}.$$

ILLUSTRATIVE EXAMPLE 1. Find the partial derivatives of $z=ax^2+2bxy+cy^2$.

**Solution.** $\frac{\partial z}{\partial x}=2ax+2by$, treating $y$ as a constant,

$\frac{\partial z}{\partial y}=2bx+2cy$, treating $x$ as a constant.

* The constant values are substituted in the function before differentiating.
† Introduced by Jacobi (1804–1851).

ILLUSTRATIVE EXAMPLE 2. Find the partial derivatives of $u = \sin(ax + by + cz)$.

**Solution.** $\frac{\partial u}{\partial x} = a \cos(ax + by + cz)$, treating $y$ and $z$ as constants,

$\frac{\partial u}{\partial y} = b \cos(ax + by + cz)$, treating $x$ and $z$ as constants,

$\frac{\partial u}{\partial z} = c \cos(ax + by + cz)$, treating $y$ and $x$ as constants.

Again turning to the function

$$z = f(x, y),$$

we have, by (A), p. 191, defined $\frac{\partial z}{\partial x}$ as the limit of the ratio of the increment of the function ($y$ being constant) to the increment of $x$, as the increment of $x$ approaches the limit zero. Similarly, (B), p. 191, has defined $\frac{\partial z}{\partial y}$. It is evident, however, that if we look upon these partial derivatives from the point of view of § 94, p. 141, then

$$\frac{\partial z}{\partial x}$$

may be considered as the ratio of the time rates of change of $z$ and $x$ when $y$ is constant, and

$$\frac{\partial z}{\partial y}$$

as the ratio of the time rates of change of $z$ and $y$ when $x$ is constant.

**124. Partial derivatives interpreted geometrically.** Let the equation of the surface shown in the figure be

$$z = f(x, y).$$

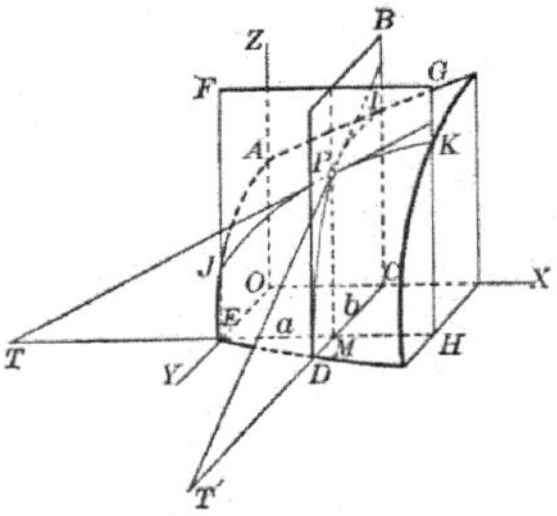


Pass a plane $EFGH$ through the point $P$ (where $x = a$ and $y = b$) on the surface parallel to the $XOZ$-plane. Since the equation of this plane is

$$y = b,$$

the equation of the section $JPK$ cut out of the surface is

$$z = f(x, b),$$

if we consider $EF$ as the axis of $Z$ and $EH$ as the axis of $X$. In this plane $\frac{\partial z}{\partial x}$ means the same as $\frac{dz}{dx}$, and we have

$$\frac{\partial z}{\partial x} = \tan MTP$$
$$= \textit{slope of section } JK \textit{ at } P.$$

Similarly, if we pass the plane $BCD$ through $P$ parallel to the $YOZ$-plane, its equation is

$$x = a,$$

and for the section $DPI$, $\frac{\partial z}{\partial y}$ means the same as $\frac{dz}{dy}$. Hence

$$\frac{\partial z}{\partial y} = \frac{dz}{dy} = -\tan MT'P = \textit{slope of section } DI \textit{ at } P.$$

ILLUSTRATIVE EXAMPLE 1. Given the ellipsoid $\frac{x^2}{24} + \frac{y^2}{12} + \frac{z^2}{6} = 1$; find the slope of the section of the ellipsoid made (a) by the plane $y = 1$ at the point where $x = 4$ and $z$ is positive; (b) by the plane $x = 2$ at the point where $y = 3$ and $z$ is positive.

**Solution.** Considering $y$ as constant,

$$\frac{2x}{24} + \frac{2z}{6}\frac{\partial z}{\partial x} = 0, \text{ or } \frac{\partial z}{\partial x} = -\frac{x}{4z}.$$

When $x$ is constant, $$\frac{2y}{12} + \frac{2z}{6}\frac{\partial z}{\partial y} = 0, \text{ or } \frac{\partial z}{\partial y} = -\frac{y}{2z}.$$

(a) When $y = 1$ and $x = 4$, $z = \sqrt{\frac{3}{2}}$. $\therefore \frac{\partial z}{\partial x} = -\sqrt{\frac{2}{3}}$. *Ans.*

(b) When $x = 2$ and $y = 3$, $z = \frac{1}{\sqrt{2}}$. $\therefore \frac{\partial z}{\partial y} = -\frac{3}{2}\sqrt{2}$. *Ans.*

## EXAMPLES

1. $u = x^3 + 3x^2y - y^3$. *Ans.* $\frac{\partial u}{\partial x} = 3x^2 + 6xy$; $\frac{\partial u}{\partial y} = 3x^2 - 3y^2$.

2. $u = Ax^2 + Bxy + Cy^2 + Dx + Ey + F$. $\frac{\partial u}{\partial x} = 2Ax + By + D$; $\frac{\partial u}{\partial y} = Bx + 2Cy + E$.

3. $u = (ax^2 + by^2 + cz^2)^n$. $\frac{\partial u}{\partial x} = \frac{2anxu}{ax^2 + by^2 + cz^2}$; $\frac{\partial u}{\partial y} = \frac{2bnyu}{ax^2 + by^2 + cz^2}$.

4. $u = \text{arc} \sin \frac{x}{y}$. $\frac{\partial u}{\partial x} = \frac{1}{\sqrt{y^2 - x^2}}$; $\frac{\partial u}{\partial y} = -\frac{x}{y\sqrt{y^2 - x^2}}$.

5. $u = x^y$. $\frac{\partial u}{\partial x} = yx^{y-1}$; $\frac{\partial u}{\partial y} = x^y \log x$.

6. $u = ax^3y^2z + bxy^3z^4 + cy^5 + dxz^3$. $\frac{\partial u}{\partial x} = 3ax^2y^2z + by^3z^4 + dz^3$; $\frac{\partial u}{\partial y} = 2ax^3yz + 3bxy^2z^4 + 5cy^4$; $\frac{\partial u}{\partial z} = ax^3y^2 + 4bxy^3z^3 + 3dxz^2$.

7. $u = x^3y^2 - 2xy^4 + 3x^2y^3$; show that $x\frac{\partial u}{\partial x} + y\frac{\partial u}{\partial y} = 5u$.

8. $u = \frac{xy}{x+y}$; show that $x\frac{\partial u}{\partial x} + y\frac{\partial u}{\partial y} = u$.

9. $u = (y-z)(z-x)(x-y)$; show that $\frac{\partial u}{\partial x} + \frac{\partial u}{\partial y} + \frac{\partial u}{\partial z} = 0$.

10. $u = \log(e^x + e^y)$; show that $\frac{\partial u}{\partial x} + \frac{\partial u}{\partial y} = 1$.

11. $u = \frac{e^{xy}}{e^x + e^y}$; show that $\frac{\partial u}{\partial x} + \frac{\partial u}{\partial y} = (x + y - 1)u$.

12. $u = x^y y^x$; show that $x\frac{\partial u}{\partial x} + y\frac{\partial u}{\partial y} = (x + y + \log u)u$.

13. $u = \log(x^3 + y^3 + z^3 - 3xyz)$; show that $\frac{\partial u}{\partial x} + \frac{\partial u}{\partial y} + \frac{\partial u}{\partial z} = \frac{3}{x+y+z}$.

14. $u = e^x \sin y + e^y \sin x$; show that

$$\left(\frac{\partial u}{\partial x}\right)^2 + \left(\frac{\partial u}{\partial y}\right)^2 = e^{2x} + e^{2y} + 2e^{x+y}\sin(x+y).$$

15. $u = \log(\tan x + \tan y + \tan z)$; show that

$$\sin 2x\frac{\partial u}{\partial x} + \sin 2y\frac{\partial u}{\partial y} + \sin 2z\frac{\partial u}{\partial z} = 2.$$

16. Let $y$ be the altitude of a right circular cone and $x$ the radius of its base. Show (a) that if the base remains constant, the volume changes $\frac{1}{3}\pi x^2$ times as fast as the altitude; (b) that if the altitude remains constant, the volume changes $\frac{2}{3}\pi xy$ times as fast as the radius of the base.

17. A point moves on the elliptic paraboloid $z = \frac{x^2}{9} + \frac{y^2}{4}$ and also in a plane parallel to the $XOZ$-plane. When $x = 3$ ft. and is increasing at the rate of 9 ft. per second, find (a) the time rate of change of $z$; (b) the magnitude of the velocity of the point; (c) the direction of its motion.

*Ans.* (a) $v_z = 6$ ft. per sec.; (b) $v = 3\sqrt{13}$ ft. per sec.;
(c) $\tau = \text{arc}\tan\frac{2}{3}$, the angle made with the $XOY$-plane.

18. If, on the surface of Ex. 17, the point moves in a plane parallel to the plane $YOZ$, find, when $y = 2$ and increases at the rate of 5 ft. per sec., (a) the time rate of change of $z$; (b) the magnitude of the velocity of the point; (c) the direction of its motion. *Ans.* (a) 5 ft. per sec.; (b) $5\sqrt{2}$ ft. per sec.;
(c) $\tau = \frac{\pi}{4}$, the angle made with the plane $XOY$.

**125. Total derivatives.** We have already considered the differentiation of a function of one function of a single independent variable. Thus, if

$$y = f(v) \text{ and } v = \phi(x),$$

it was shown that

$$\frac{dy}{dx} = \frac{dy}{dv} \cdot \frac{dv}{dx}.$$

We shall next consider a function of two variables, both of which depend on a single independent variable. Consider the function

$$u = f(x, y),$$

where $x$ and $y$ are functions of a third variable $t$.

Let $t$ take on the increment $\Delta t$, and let $\Delta x$, $\Delta y$, $\Delta u$ be the corresponding increments of $x$, $y$, $u$ respectively. Then the quantity

$$\Delta u = f(x + \Delta x, y + \Delta y) - f(x, y)$$

is called the *total increment* of $u$.

Adding and subtracting $f(x, y + \Delta y)$ in the second member,

$$(A)\quad \Delta u = [f(x+\Delta x, y+\Delta y) - f(x, y+\Delta y)] + [f(x, y+\Delta y) - f(x, y)].$$

Applying the Theorem of Mean Value (**46**), p. 166, to each of the two differences on the right-hand side of $(A)$, we get, for the first difference,

$$(B)\quad f(x + \Delta x, y + \Delta y) - f(x, y + \Delta y) = f'_x(x + \theta_1 \cdot \Delta x, y + \Delta y)\Delta x.$$

[$a = x$, $\Delta a = \Delta x$, and since $x$ varies while $y + \Delta y$ remains constant, we get the partial derivative with respect to $x$.]

For the second difference we get

$$(C)\qquad f(x, y + \Delta y) - f(x, y) = f'_y(x, y + \theta_2 \cdot \Delta y)\,\Delta y.$$

[$a = y$, $\Delta a = \Delta y$, and since $y$ varies while $x$ remains constant, we get the partial derivative with respect to $y$.]

Substituting $(B)$ and $(C)$ in $(A)$ gives

$$(D)\quad \Delta u = f'_x(x + \theta_1 \cdot \Delta x, y + \Delta y)\,\Delta x + f'_y(x, y + \theta_2 \cdot \Delta y)\,\Delta y,$$

where $\theta_1$ and $\theta_2$ are positive proper fractions. Dividing $(D)$ by $\Delta t$,

$$(E)\quad \frac{\Delta u}{\Delta t} = f'_x(x + \theta_1 \cdot \Delta x, y + \Delta y)\frac{\Delta x}{\Delta t} + f'_y(x, y + \theta_2 \cdot \Delta y)\frac{\Delta y}{\Delta t}.$$

Now let $\Delta t$ approach zero as a limit, then

$$(F)\qquad \frac{du}{dt} = f'_x(x, y)\frac{dx}{dt} + f'_y(x, y)\frac{dy}{dt}.$$

[Since $\Delta x$ and $\Delta y$ converge to zero with $\Delta t$, we get $\lim\limits_{\Delta t = 0} f'_x(x + \theta_1 \cdot \Delta x, y + \Delta y) = f'_x(x, y)$, and $\lim\limits_{\Delta t = 0} f'_y(x, y + \theta_2 \cdot \Delta y) = f'_y(x, y)$, $f'_x(x, y)$ and $f'_y(x, y)$ being assumed continuous.]

Replacing $f(x, y)$ by $u$ in $(F)$, we get the *total derivative*

$$(51)\qquad \frac{du}{dt} = \frac{\partial u}{\partial x}\frac{dx}{dt} + \frac{\partial u}{\partial y}\frac{dy}{dt}.$$

In the same way, if

$$u=f(x, y, z),$$

and $x$, $y$, $z$ are all functions of $t$, we get

$$\frac{du}{dt}=\frac{\partial u}{\partial x}\frac{dx}{dt}+\frac{\partial u}{\partial y}\frac{dy}{dt}+\frac{\partial u}{\partial z}\frac{dz}{dt}, \tag{52}$$

and so on for any number of variables.*

In (51) we may suppose $t=x$; then $y$ is a function of $x$, and $u$ is really a function of the one variable $x$, giving

$$\frac{du}{dx}=\frac{\partial u}{\partial x}+\frac{\partial u}{\partial y}\frac{dy}{dx}. \tag{53}$$

In the same way, from (52) we have

$$\frac{du}{dx}=\frac{\partial u}{\partial x}+\frac{\partial u}{\partial y}\cdot\frac{dy}{dx}+\frac{\partial u}{\partial z}\cdot\frac{dz}{dx}. \tag{54}$$

The student should observe that $\frac{\partial u}{\partial x}$ and $\frac{du}{dx}$ have quite different meanings. The partial derivative $\frac{\partial u}{\partial x}$ is formed on the supposition that the *particular variable x alone varies*, while

$$\frac{du}{dx}=\underset{\Delta x=0}{\text{limit}}\left(\frac{\Delta u}{\Delta x}\right),$$

where $\Delta u$ is the *total increment of u* caused by changes in *all the variables*, these increments being due to the change $\Delta x$ in the independent variable. In contradistinction to partial derivatives, $\frac{du}{dt}$, $\frac{du}{dx}$ are called *total derivatives* with respect to $t$ and $x$ respectively.†

* This is really only a special case of a general theorem which may be stated as follows: If $u$ is a function of the independent variables $x, y, z, \cdots$, each of these in turn being a function of the independent variables $r, s, t, \cdots$, then (with certain assumptions as to continuity)

$$\frac{\partial u}{\partial r}=\frac{\partial u}{\partial x}\frac{\partial x}{\partial r}+\frac{\partial u}{\partial y}\frac{\partial y}{\partial r}+\frac{\partial u}{\partial z}\frac{\partial z}{\partial r}+\cdots,$$

and similar expressions hold for $\frac{\partial u}{\partial s}$, $\frac{\partial u}{\partial t}$, etc.

† It should be observed that $\frac{\partial u}{\partial x}$ has a perfectly definite value for any point $(x, y)$, while $\frac{du}{dx}$ depends not only on the point $(x, y)$, but also on the particular direction chosen to reach that point. Hence

$\frac{\partial u}{\partial x}$ is called a point function; while

$\frac{du}{dx}$ is not called a point function unless it is agreed to approach the point from some particular direction.

ILLUSTRATIVE EXAMPLE 1. Given $u = \sin\frac{x}{y}$, $x = e^t$, $y = t^2$; find $\frac{du}{dt}$.

**Solution.** $\frac{\partial u}{\partial x} = \frac{1}{y}\cos\frac{x}{y}$, $\frac{\partial u}{\partial y} = -\frac{x}{y^2}\cos\frac{x}{y}$; $\frac{dx}{dt} = e^t$, $\frac{dy}{dt} = 2t$.

Substituting in (51), $\frac{du}{dt} = (t-2)\frac{e^t}{t^3}\cos\frac{e^t}{t^2}$. *Ans.*

ILLUSTRATIVE EXAMPLE 2. Given $u = e^{ax}(y - z)$, $y = a\sin x$, $z = \cos x$; find $\frac{du}{dx}$.

**Solution.** $\frac{\partial u}{\partial x} = ae^{ax}(y-z)$, $\frac{\partial u}{\partial y} = e^{ax}$, $\frac{\partial u}{\partial z} = -e^{ax}$; $\frac{dy}{dx} = a\cos x$, $\frac{dz}{dx} = -\sin x$.

Substituting in (54),

$$\frac{du}{dx} = ae^{ax}(y-z) + ae^{ax}\cos x + e^{ax}\sin x = e^{ax}(a^2+1)\sin x. \quad \textit{Ans.}$$

NOTE. In examples like the above, $u$ could, by substitution, be found explicitly in terms of the independent variable and then differentiated directly, but generally this process would be longer and in many cases could not be used at all.

Formulas (51) and (52) are very useful in all applications involving time rates of change of functions of two or more variables. The process is practically the same as that outlined in the rule given on p. 141, except that, instead of differentiating with respect to $t$ (Third Step), we find the partial derivatives and substitute in (51) or (52). Let us illustrate by an example.

ILLUSTRATIVE EXAMPLE 3. The altitude of a circular cone is 100 inches, and decreasing at the rate of 10 inches per second; and the radius of the base is 50 inches, and increasing at the rate of 5 inches per second. At what rate is the volume changing?

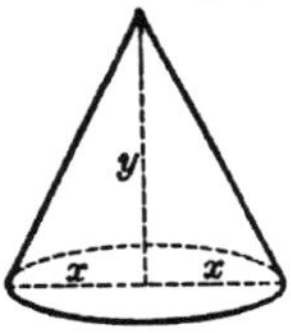


**Solution.** Let $x$ = radius of base, $y$ = altitude; then $u = \frac{1}{3}\pi x^2 y$ = volume, $\frac{\partial u}{\partial x} = \frac{2}{3}\pi xy$, $\frac{\partial u}{\partial y} = \frac{1}{3}\pi x^2$. Substitute in (51),

$$\frac{du}{dt} = \frac{2}{3}\pi xy\frac{dx}{dt} + \frac{1}{3}\pi x^2\frac{dy}{dt}.$$

But $x = 50$, $y = 100$, $\frac{dx}{dt} = 5$, $\frac{dy}{dt} = -10$.

$$\therefore \frac{du}{dt} = \frac{2}{3}\pi\cdot 5000\cdot 5 - \frac{1}{3}\pi\cdot 2500\cdot 10 = 15.15 \text{ cu. ft. per sec., increase.} \quad \textit{Ans.}$$

**126. Total differentials.** Multiplying (51) and (52) through by $dt$, we get

(55) $$du = \frac{\partial u}{\partial x}dx + \frac{\partial u}{\partial y}dy,$$

(56) $$du = \frac{\partial u}{\partial x}dx + \frac{\partial u}{\partial y}dy + \frac{\partial u}{\partial z}dz;$$

and so on.* Equations (55) and (56) define the quantity $du$, which is called a *total differential of u* or a *complete differential*,

and $$\frac{\partial u}{\partial x}dx,\ \frac{\partial u}{\partial y}dy,\ \frac{\partial u}{\partial z}dz$$

* A geometric interpretation of this result will be given on p. 266.

are called *partial differentials*. These partial differentials are sometimes denoted by $d_x u$, $d_y u$, $d_z u$, so that (**56**) is also written

$$du = d_x u + d_y u + d_z u.$$

Illustrative Example 1. Given $u = \text{arc} \tan \frac{y}{x}$, find $du$.

**Solution.**
$$\frac{\partial u}{\partial x} = -\frac{y}{x^2 + y^2}, \quad \frac{\partial u}{\partial y} = \frac{x}{x^2 + y^2}.$$

Substituting in (55),
$$du = \frac{xdy - ydx}{x^2 + y^2}. \quad Ans.$$

Illustrative Example 2. The base and altitude of a rectangle are 5 and 4 inches respectively. At a certain instant they are increasing continuously at the rate of 2 inches and 1 inch per second respectively. At what rate is the area of the rectangle increasing at that instant?

**Solution.** Let $x$ = base, $y$ = altitude; then $u = xy$ = area, $\frac{\partial u}{\partial x} = y$, $\frac{\partial u}{\partial y} = x$.

Substituting in (51),

$$(A) \qquad \frac{du}{dt} = y\frac{dx}{dt} + x\frac{dy}{dt}.$$

But $\quad x = 5$ in., $y = 4$ in., $\frac{dx}{dt} = 2$ in. per sec., $\frac{dy}{dt} = 1$ in. per sec.

$$\therefore \frac{du}{dt} = (8 + 5) \text{ sq. in. per sec.} = 13 \text{ sq. in. per sec.} \quad Ans.$$

Note. Considering $du$ as an infinitesimal increment of area due to the infinitesimal increments $dx$ and $dy$, $du$ is evidently the sum of two thin strips added on to the two sides. For, in $du = ydx + xdy$ (multiplying (A) by $dt$),

$ydx$ = area of vertical strip, and
$xdy$ = area of horizontal strip.

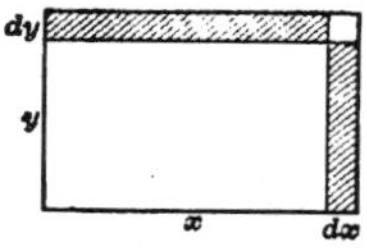


But the total increment $\Delta u$ due to the increments $dx$ and $dy$ is evidently $\quad \Delta u = ydx + xdy + dxdy.$

Hence the small rectangle in the upper right-hand corner ($= dxdy$) is evidently the difference between $\Delta u$ and $du$. This figure illustrates the fact that the total increment and the total differential of a function of several variables are not in general equal.

**127. Differentiation of implicit functions.** The equation

$$(A) \qquad f(x, y) = 0$$

defines either $x$ or $y$ as an implicit function of the other.* It represents any equation containing $x$ and $y$ when all its terms have been transposed to the first member. Let

$$(B) \qquad u = f(x, y);$$

then
$$\frac{du}{dx} = \frac{\partial u}{\partial x} + \frac{\partial u}{\partial y}\frac{dy}{dx}. \qquad (53), \text{ p. } 196$$

* We assume that a small change in the value of $x$ causes only a small change in the value of $y$.

But from $(A)$, $\quad f(x, y) = 0. \quad \therefore u = 0$ and $\dfrac{du}{dx} = 0$; that is,

$$(C) \qquad \frac{\partial u}{\partial x} + \frac{\partial u}{\partial y}\frac{dy}{dx} = 0.$$

Solving for $\dfrac{dy}{dx}$,* we get

$$(57) \qquad \frac{dy}{dx} = -\frac{\dfrac{\partial u}{\partial x}}{\dfrac{\partial u}{\partial y}}, \qquad \frac{\partial u}{\partial y} \neq 0$$

a formula for differentiating implicit functions. This formula in the form $(C)$ is equivalent to the process employed in § 62, p. 69, for differentiating implicit functions, and all the examples on p. 70 may be solved by using formula (57). Since

$$(D) \qquad f(x, y) = 0$$

for all admissible values of $x$ and $y$, we may say that (57) gives *the relative time rates of change of $x$ and $y$ which keep $f(x, y)$ from changing at all.* Geometrically this means that the point $(x, y)$ must move on the curve whose equation is $(D)$, and (57) determines the direction of its motion at any instant. Since

$$u = f(x, y),$$

we may write (57) in the form

$$(57\,a) \qquad \frac{dy}{dx} = -\frac{\dfrac{\partial f}{\partial x}}{\dfrac{\partial f}{\partial y}}. \qquad \frac{\partial f}{\partial y} \neq 0$$

Illustrative Example 1. Given $x^2y^4 + \sin y = 0$, find $\dfrac{dy}{dx}$.

**Solution.** Let $f(x, y) = x^2y^4 + \sin y$.

$$\frac{\partial f}{\partial x} = 2xy^4, \quad \frac{\partial f}{\partial y} = 4x^2y^3 + \cos y. \quad \therefore \text{from (57 } a\text{)}, \ \frac{dy}{dx} = -\frac{2xy^4}{4x^2y^3 + \cos y}. \quad \textit{Ans.}$$

Illustrative Example 2. If $x$ increases at the rate of 2 inches per second as it passes through the value $x = 3$ inches, at what rate must $y$ change when $y = 1$ inch, in order that the function $2xy^2 - 3x^2y$ shall remain constant?

**Solution.** Let $f(x, y) = 2xy^2 - 3x^2y$; then

$$\frac{\partial f}{\partial x} = 2y^2 - 6xy, \quad \frac{\partial f}{\partial y} = 4xy - 3x^2.$$

Substituting in (57 $a$),

$$\frac{dy}{dx} = -\frac{2y^2 - 6xy}{4xy - 3x^2}, \text{ or } \frac{\dfrac{dy}{dt}}{\dfrac{dx}{dt}} = -\frac{2y^2 - 6xy}{4xy - 3x^2}. \qquad \text{By (33), p. 141}$$

But $x = 3$, $y = 1$, $\dfrac{dx}{dt} = 2$. $\therefore \dfrac{dy}{dt} = -2\tfrac{2}{15}$ ft. per second. *Ans.*

* It is assumed that $\dfrac{\partial u}{\partial x}$ and $\dfrac{\partial u}{\partial y}$ exist.

Let $P$ be the point $(x, y, z)$ on the surface given by the equation

$$(E)\qquad u = F(x, y, z) = 0,$$

and let $PC$ and $AP$ be sections made by planes through $P$ parallel to the $YOZ$- and $XOZ$-planes respectively. Along the curve $AP$, $y$ is constant; therefore, from $(E)$, $z$ is an implicit function of $x$ alone, and we have, from (**57** $a$),

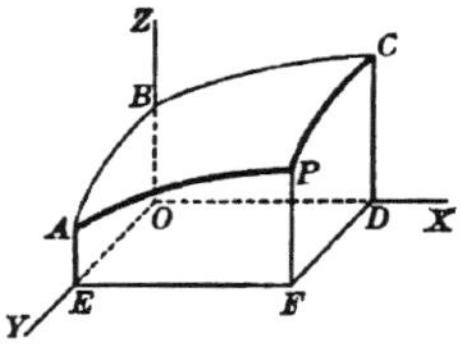


$$(58)\qquad \frac{\partial z}{\partial x} = -\frac{\frac{\partial F}{\partial x}}{\frac{\partial F}{\partial z}},$$

giving the slope at $P$ of the curve $AP$, § 122, p. 190.

$\frac{\partial z}{\partial x}$ is used instead of $\frac{dz}{dx}$ in the first member, since $z$ was originally, from $(E)$, an implicit function of $x$ and $y$; but (**58**) is deduced on the hypothesis that $y$ remains constant.

Similarly, the slope at $P$ of the curve $PC$ is

$$(59)\qquad \frac{\partial z}{\partial y} = -\frac{\frac{\partial F}{\partial y}}{\frac{\partial F}{\partial z}}.$$

## EXAMPLES

Find the total derivatives, using (51), (52), or (53), in the following six examples:

1. $u = z^2 + y^3 + zy,\ z = \sin x,\ y = e^x.$ *Ans.* $\frac{du}{dx} = 3e^{3x} + e^x(\sin x + \cos x) + \sin 2x.$

2. $u = \operatorname{arc}\tan(xy),\ y = e^x.$ *Ans.* $\frac{du}{dx} = \frac{e^x(1+x)}{1 + x^2e^{2x}}.$

3. $u = \log(a^2 - \rho^2),\ \rho = a\sin\theta.$ $\frac{du}{d\theta} = -2\tan\theta.$

4. $u = v^2 + vy,\ v = \log s,\ y = e^s.$ $\frac{du}{ds} = \frac{2v + y}{s} + ve^s.$

5. $u = \operatorname{arc}\sin(r - s),\ r = 3t,\ s = 4t^3.$ $\frac{du}{dt} = \frac{3}{\sqrt{1 - t^2}}.$

6. $u = \frac{e^{ax}(y - z)}{a^2 + 1},\ y = a\sin x,\ z = \cos x.$ $\frac{du}{dx} = e^{ax}\sin x.$

Using (55) or (56), find the total differentials in the next eight examples:

7. $u = by^2x + cx^2 + gy^3 + ex.$ *Ans.* $du = (by^2 + 2cx + e)\,dx + (2byx + 3gy^2)\,dy.$

8. $u = \log x^y.$ $du = \frac{y}{x}dx + \log x\,dy.$

9. $u = y^{\sin x}$. *Ans.* $du = y^{\sin x} \log y \cos x dx + \frac{\sin x}{y^{\operatorname{covers} x}} dy$.

10. $u = x^{\log y}$. $du = u\left(\frac{\log y}{x} dx + \frac{\log x}{y} dy\right)$.

11. $u = \frac{s+t}{s-t}$. $du = \frac{2(sdt - tds)}{(s-t)^2}$.

12. $u = \sin(pq)$. $du = \cos(pq)[qdp + pdq]$.

13. $u = x^{yz}$. $du = x^{yz-1}(yzdx + zx \log x dy + xy \log x dz)$.

14. $u = \tan^2\phi \tan^2\theta \tan^2\psi$. $du = 4u\left(\frac{d\phi}{\sin 2\phi} + \frac{d\theta}{\sin 2\theta} + \frac{d\psi}{\sin 2\psi}\right)$.

15. Assuming the characteristic equation of a perfect gas to be

$$vp = Rt,$$

where $v$ = volume, $p$ = pressure, $t$ = absolute temperature, and $R$ a constant, what is the relation between the differentials $dv$, $dp$, $dt$? *Ans.* $vdp + pdv = Rdt$.

16. Using the result in the last example as applied to air, suppose that in a given case we have found by actual experiment that

$$t = 300° \text{ C.}, \; p = 2000 \text{ lb. per sq. ft.}, \; v = 14.4 \text{ cubic feet.}$$

Find the change in $p$, assuming it to be uniform, when $t$ changes to 301° C., and $v$ to 14.5 cubic feet. $R = 96$. *Ans.* $-7.22$ lb. per sq. ft.

17. One side of a triangle is 8 ft. long, and increasing 4 inches per second; another side is 5 ft., and decreasing 2 inches per second. The included angle is 60°, and increasing 2° per second. At what rate is the area of the triangle changing?

*Ans.* Increasing 71.05 sq. in. per sec.

18. At what rate is the side opposite the given angle in the last example increasing?

*Ans.* 4.93 in. per sec.

19. One side of a rectangle is 10 in. and increasing 2 in. per sec. The other side is 15 in. and decreasing 1 in. per sec. At what rate is the area changing at the end of two seconds? *Ans.* Increasing 12 sq. in. per sec.

20. The three edges of a rectangular parallelepiped are 3, 4, 5 inches, and are each increasing at the rate of .02 in. per min. At what rate is the volume changing?

21. A boy starts flying a kite. If it moves horizontally at the rate of 2 ft. a sec. and rises at the rate of 5 ft. a sec., how fast is the string being paid out?

*Ans.* 5.38 ft. a sec.

22. A man standing on a dock is drawing in the painter of a boat at the rate of 2 ft. a sec. His hands are 6 ft. above the bow of the boat. How fast is the boat moving when it is 8 ft. from the dock? *Ans.* $\frac{5}{2}$ ft. a sec.

23. The volume and the radius of a cylindrical boiler are expanding at the rate of 1 cu. ft. and .001 ft. per min. respectively. How fast is the length of the boiler changing when the boiler contains 60 cu. ft. and has a radius of 2 ft.?

*Ans.* .078 ft. a min.

24. Water is running out of an opening in the vertex of a conical filtering glass, 8 inches high and 6 inches across the top, at the rate of .005 cu. in. per hour. How fast is the surface of the water falling when the depth of the water is 4 inches?

25. A covered water tank is made of sheet iron in the form of an inverted cone of altitude 8 ft. surmounted by a cylinder of altitude 5 ft. The diameter is 6 ft. If the sun's heat is increasing the diameter at the rate of .002 ft. per min., the altitude of the cylinder at the rate of .003 ft. per min., and the altitude of the cone at the rate of .0025 ft. per minute, at what rate is (a) the volume increasing; (b) the total area increasing?

In the remaining examples find $\frac{dy}{dx}$, using formula (57 a):

26. $(x^2+y^2)^2 - a^2(x^2-y^2) = 0.$ *Ans.* $\frac{dy}{dx} = -\frac{x}{y}\cdot\frac{2(x^2+y^2)-a^2}{2(x^2+y^2)+a^2}.$

27. $e^y - e^x + xy = 0.$ $\frac{dy}{dx} = \frac{e^x - y}{e^y + x}.$

28. $\sin(xy) - e^{xy} - x^2y = 0.$ $\frac{dy}{dx} = \frac{y[\cos(xy) - e^{xy} - 2x]}{x[x + e^{xy} - \cos(xy)]}.$

**128. Successive partial derivatives.** Consider the function

$$u = f(x, y);$$

then, in general,

$$\frac{\partial u}{\partial x} \text{ and } \frac{\partial u}{\partial y}$$

are functions of both $x$ and $y$, and may be differentiated again with respect to either independent variable, giving *successive partial derivatives*. Regarding $x$ alone as varying, we denote the results by

$$\frac{\partial^2 u}{\partial x^2}, \frac{\partial^3 u}{\partial x^3}, \frac{\partial^4 u}{\partial x^4}, \cdots, \frac{\partial^n u}{\partial x^n},$$

or, when $y$ alone varies,

$$\frac{\partial^2 u}{\partial y^2}, \frac{\partial^3 u}{\partial y^3}, \frac{\partial^4 u}{\partial y^4}, \cdots, \frac{\partial^n u}{\partial y^n},$$

the notation being similar to that employed for functions of a single variable.

If we differentiate $u$ with respect to $x$, regarding $y$ as constant, and then this result with respect to $y$, regarding $x$ as constant, we obtain

$$\frac{\partial}{\partial y}\left(\frac{\partial u}{\partial x}\right), \text{ which we denote by } \frac{\partial^2 u}{\partial y \partial x}.$$

Similarly, if we differentiate twice with respect to $x$ and then once with respect to $y$, the result is denoted by the symbol

$$\frac{\partial^3 u}{\partial y \partial x^2}.$$

**129. Order of differentiation immaterial.** Consider the function $f(x, y)$. Changing $x$ into $x+\Delta x$ and keeping $y$ constant, we get from the Theorem of Mean Value, (**46**), p. 166,

$$(A)\quad f(x+\Delta x, y)-f(x, y)=\Delta x \cdot f'_x(x+\theta\cdot\Delta x, y). \qquad 0<\theta<1$$

[$a=x$, $\Delta a=\Delta x$, and since $x$ varies while $y$ remains constant, we get the partial derivative with respect to $x$.]

If we now change $y$ to $y+\Delta y$ and keep $x$ and $\Delta x$ constant, the total increment of the left-hand member of $(A)$ is

$$(B)\quad [f(x+\Delta x, y+\Delta y)-f(x, y+\Delta y)]-[f(x+\Delta x, y)-f(x, y)].$$

The total increment of the right-hand member of $(A)$ found by the Theorem of Mean Value, (**46**), p. 166, is

$$(C)\quad \Delta x f'_x(x+\theta\cdot\Delta x, y+\Delta y)-\Delta x f'_x(x+\theta\cdot\Delta x, y) \qquad 0<\theta_1<1$$
$$=\Delta y\Delta x f''_{yx}(x+\theta_1\cdot\Delta x, y+\theta_2\cdot\Delta y). \qquad 0<\theta_2<1$$

[$a=y$, $\Delta a=\Delta y$, and since $y$ varies while $x$ and $\Delta x$ remain constant, we get the partial derivative with respect to $y$.]

Since the increments $(B)$ and $(C)$ must be equal,

$$(D)\quad [f(x+\Delta x, y+\Delta y)-f(x, y+\Delta y)]-[f(x+\Delta x, y)-f(x, y)]$$
$$=\Delta y\Delta x f''_{yx}(x+\theta_1\cdot\Delta x, y+\theta_2\cdot\Delta y).$$

In the same manner, if we take the increments in the reverse order,

$$(E)\quad [f(x+\Delta x, y+\Delta y)-f(x+\Delta x, y)]-[f(x, y+\Delta y)-f(x, y)]$$
$$=\Delta x\Delta y f''_{xy}(x+\theta_3\cdot\Delta x, y+\theta_4\cdot\Delta y),$$

$\theta_3$ and $\theta_4$ also lying between zero and unity.

The left-hand members of $(D)$ and $(E)$ being identical, we have

$$(F)\quad f''_{yx}(x+\theta_1\cdot\Delta x, y+\theta_2\cdot\Delta y)=f''_{xy}(x+\theta_3\cdot\Delta x, y+\theta_4\cdot\Delta y).$$

Taking the limit of both sides as $\Delta x$ and $\Delta y$ approach zero as limits, we have

$$(G)\qquad f''_{yx}(x, y)=f''_{xy}(x, y),$$

since these functions are assumed continuous. Placing

$$u=f(x, y),$$

$(G)$ may be written

$$(\mathbf{60})\qquad \frac{\partial^2 u}{\partial y\partial x}=\frac{\partial^2 u}{\partial x\partial y}.$$

That is, *the operations of differentiating with respect to $x$ and with respect to $y$ are commutative.*

This may be easily extended to higher derivatives. For instance, since (58) is true,

$$\frac{\partial^3 u}{\partial x^2 \partial y} = \frac{\partial}{\partial x}\left(\frac{\partial^2 u}{\partial x \partial y}\right) = \frac{\partial^3 u}{\partial x \partial y \partial x} = \frac{\partial^2}{\partial x \partial y}\left(\frac{\partial u}{\partial x}\right) = \frac{\partial^2}{\partial y \partial x}\left(\frac{\partial u}{\partial x}\right) = \frac{\partial^3 u}{\partial y \partial x^2}.$$

Similarly for functions of three or more variables.

ILLUSTRATIVE EXAMPLE 1. Given $u = x^3 y - 3x^2 y^3$; verify $\frac{\partial^2 u}{\partial y \partial x} = \frac{\partial^2 u}{\partial x \partial y}$.

**Solution.** $\frac{\partial u}{\partial x} = 3x^2 y - 6xy^3$, $\frac{\partial^2 u}{\partial y \partial x} = 3x^2 - 18xy^2$,

$\frac{\partial u}{\partial y} = x^3 - 9x^2 y^2$, $\frac{\partial^2 u}{\partial x \partial y} = 3x^2 - 18xy^2$; hence verified.

### EXAMPLES

1. $u = \cos(x + y)$; verify $\frac{\partial^2 u}{\partial y \partial x} = \frac{\partial^2 u}{\partial x \partial y}$.

2. $u = \frac{y^2 + x^2}{y^2 - x^2}$; verify $\frac{\partial^2 u}{\partial y \partial x} = \frac{\partial^2 u}{\partial x \partial y}$.

3. $u = y \log(1 + xy)$; verify $\frac{\partial^2 u}{\partial y \partial x} = \frac{\partial^2 u}{\partial x \partial y}$.

4. $u = \text{arc} \tan \frac{r}{s}$; verify $\frac{\partial^3 u}{\partial r^2 \partial s} = \frac{\partial^3 u}{\partial s \partial r^2}$.

5. $u = \sin(\theta^2 \phi)$; verify $\frac{\partial^3 u}{\partial \theta \partial \phi^2} = \frac{\partial^3 u}{\partial \phi^2 \partial \theta}$.

6. $u = 6e^x y^2 z + 3e^y x^2 z^2 + 2e^z x^3 y - xyz$; show that $\frac{\partial^4 u}{\partial x^2 \partial y \partial z} = 12(e^x y + e^y z + e^z x)$.

7. $u = e^{xyz}$; show that $\frac{\partial^3 u}{\partial x \partial y \partial z} = (1 + 3xyz + x^2 y^2 z^2)u$.

8. $u = \frac{x^2 y^2}{x + y}$; show that $x\frac{\partial^2 u}{\partial x^2} + y\frac{\partial^2 u}{\partial x \partial y} = 2\frac{\partial u}{\partial x}$.

9. $u = (x^2 + y^2)^{\frac{1}{3}}$; show that $3x\frac{\partial^2 u}{\partial x \partial y} + 3y\frac{\partial^2 u}{\partial y^2} + \frac{\partial u}{\partial y} = 0$.

10. $u = y^2 z^2 e^{\frac{x}{2}} + z^2 x^2 e^{\frac{y}{2}} + x^2 y^2 e^{\frac{z}{2}}$; show that $\frac{\partial^6 u}{\partial x^2 \partial y^2 \partial z^2} = e^{\frac{x}{2}} + e^{\frac{y}{2}} + e^{\frac{z}{2}}$.

11. $u = (x^2 + y^2 + z^2)^{-\frac{1}{2}}$; show that $\frac{\partial^2 u}{\partial x^2} + \frac{\partial^2 u}{\partial y^2} + \frac{\partial^2 u}{\partial z^2} = 0$.

# CHAPTER XVI

## ENVELOPES

**130. Family of curves. Variable parameter.** The equation of a curve generally involves, besides the variables $x$ and $y$, certain constants upon which the size, shape, and position of that particular curve depend. For example, the locus of the equation

$$(A) \qquad (x-\alpha)^2+y^2=r^2$$

is a circle whose center lies on the axis of $X$ at a distance of $\alpha$ from the origin, its size depending on the radius $r$. Suppose $\alpha$ to take on a series of values; then we shall have a corresponding series of circles differing in their distances from the origin, as shown in the figure.

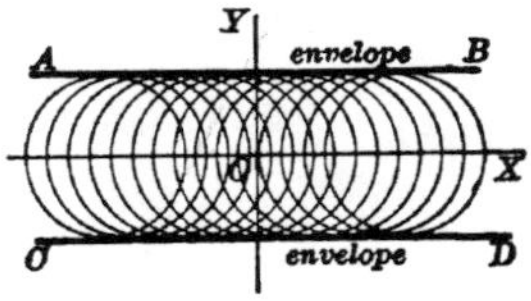


Any system of curves formed in this way is called a *family of curves*, and the quantity $\alpha$, which is constant for any one curve, but changes in passing from one curve to another, is called a *variable parameter*.

As will appear later on, problems occur which involve two or more parameters. The above series of circles is said to be a *family depending on one parameter*. To indicate that $\alpha$ enters as a variable parameter it is usual to insert it in the functional symbol, thus:

$$f(x, y, \alpha)=0.$$

**131. Envelope of a family of curves depending on one parameter.** The curves of a family may be tangent to the same curve or groups of curves, as in the above figure. In that case the name *envelope* of the family is applied to the curve or group of curves. We shall now explain a method for finding the equation of the envelope of a family of curves. Suppose that the curve whose parametric equations are

$$(A) \qquad x=\phi(\alpha), \quad y=\psi(\alpha)$$

touches (i.e. has a common tangent with) each curve of the family

$$(B) \qquad f(x, y, \alpha)=0,$$

the parameter $\alpha$ being the same in both cases. The slope of $(A)$ at any point is

$$(C) \qquad \frac{dy}{dx}=\frac{\psi'(\alpha)}{\phi'(\alpha)}, \qquad (D), \text{ p. } 80$$

and the slope of $(B)$ at any point is

$$(D) \qquad \frac{dy}{dx}=-\frac{f'_x(x,\ y,\ \alpha)}{f'_y(x,\ y,\ \alpha)}. \qquad (57\text{ a}), \text{ p. } 199$$

Hence if the curves $(A)$ and $(B)$ are tangent, the slopes $(C)$ and $(D)$ will be equal (for the same value of $\alpha$), giving

$$\frac{\psi'(\alpha)}{\phi'(\alpha)}=-\frac{f'_x(x,\ y,\ \alpha)}{f'_y(x,\ y,\ \alpha)}, \text{ or}$$

$$(E) \qquad f'_x(x,\ y,\ \alpha)\,\phi'(\alpha)+f'_y(x,\ y,\ \alpha)\,\psi'(\alpha)=0.$$

By hypothesis $(A)$ and $(B)$ are tangent for every value of $\alpha$; hence for all values of $\alpha$ the point $(x,\ y)$ given by $(A)$ must lie on a curve of the family $(B)$. If we then substitute the values of $x$ and $y$ from $(A)$ in $(B)$, the result will hold true for all values of $\alpha$; that is,

$$(F) \qquad f[\phi(\alpha),\ \psi(\alpha),\ \alpha]=0.$$

The total derivative of $(F)$ with respect to $\alpha$ must therefore vanish, and we get

$$(G) \qquad f'_x(x,\ y,\ \alpha)\,\phi'(\alpha)+f'_y(x,\ y,\ \alpha)\,\psi'(\alpha)+f'_\alpha(x,\ y,\ \alpha)=0,$$

where $x=\phi(\alpha)$, $y=\psi(\alpha)$.

Comparing $(E)$ and $(G)$ gives

$$(H) \qquad f'_\alpha(x,\ y,\ \alpha)=0.$$

Therefore the equations of the envelope satisfy the two equations $(B)$ and $(H)$, namely,

$$(I) \qquad f(x,\ y,\ \alpha)=0 \text{ and } f'_\alpha(x,\ y,\ \alpha)=0;$$

that is, the parametric equations of the envelope may be found by solving the two equations $(I)$ for $x$ and $y$ in terms of the parameter $\alpha$.

**General directions for finding the envelope.**

FIRST STEP. *Differentiate with respect to the variable parameter, considering all other quantities involved in the given equation as constants.*

SECOND STEP. *Solve the result and the given equation of the family of curves for $x$ and $y$ in terms of the parameter. These solutions will be the parametric equations of the envelope.*

NOTE. *In case the rectangular equation of the envelope is required we may either eliminate the parameter from the parametric equations of the envelope, or else eliminate the parameter from the given equation (B) of the family and the partial derivative (H).*

ILLUSTRATIVE EXAMPLE 1. Find the envelope of the family of straight lines $x \cos \alpha + y \sin \alpha = p$, $\alpha$ being the variable parameter.

**Solution.** (A) $$x \cos \alpha + y \sin \alpha = p.$$

*First step.* Differentiating (A) with respect to $\alpha$,

(B) $$-x \sin \alpha + y \cos \alpha = 0.$$

*Second step.* Multiplying (A) by $\cos \alpha$ and (B) by $\sin \alpha$ and subtracting, we get
$$x = p \cos \alpha.$$

Similarly, eliminating $x$ between (A) and (B), we get
$$y = p \sin \alpha.$$

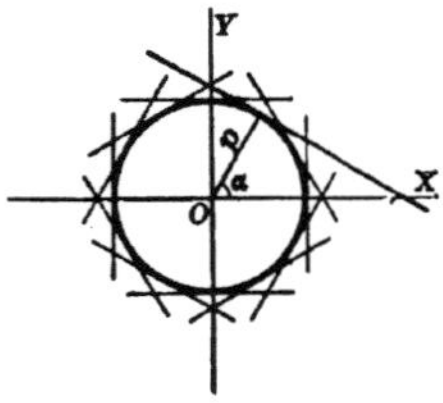


The *parametric equations of the envelope* are therefore

(C) $$\begin{cases} x = p \cos \alpha, \\ y = p \sin \alpha, \end{cases}$$

$\alpha$ being the parameter. Squaring equations (C) and adding, we get $$x^2 + y^2 = p^2,$$
the rectangular equation of the envelope, which is a circle.

ILLUSTRATIVE EXAMPLE 2. Find the envelope of a line of constant length $a$, whose extremities move along two fixed rectangular axes.

**Solution.** Let $AB = a$ in length, and let

(A) $$x \cos \alpha + y \sin \alpha - p = 0$$

be its equation. Now as $AB$ moves always touching the two axes, both $\alpha$ and $p$ will vary. But $p$ may be found in terms of $\alpha$. For $AO = AB \cos \alpha = a \cos \alpha$, and $p = AO \sin \alpha = a \sin \alpha \cos \alpha$. Substituting in (A),

(B) $$x \cos \alpha + y \sin \alpha - a \sin \alpha \cos \alpha = 0,$$

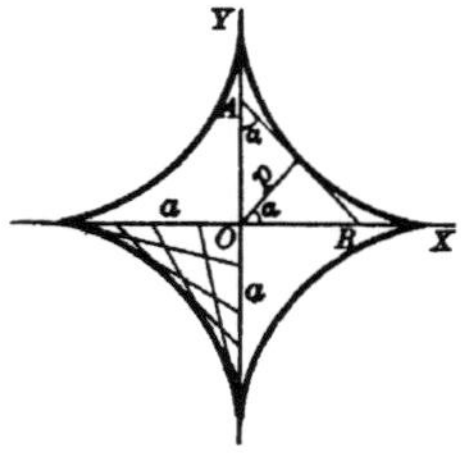


where $\alpha$ is the variable parameter. Differentiating (B) with respect to $\alpha$,

(C) $$-x \sin \alpha + y \cos \alpha + a \sin^2 \alpha - a \cos^2 \alpha = 0.$$

Solving (B) and (C) for $x$ and $y$ in terms of $\alpha$, we get

(D) $$\begin{cases} x = a \sin^3 \alpha, \\ y = a \cos^3 \alpha, \end{cases}$$

the parametric equations of the envelope, a hypocycloid.

The corresponding rectangular equation is found from equations (D) by eliminating $\alpha$ as follows:
$$x^{\frac{2}{3}} = a^{\frac{2}{3}} \sin^2 \alpha.$$
$$y^{\frac{2}{3}} = a^{\frac{2}{3}} \cos^2 \alpha.$$

Adding, $$x^{\frac{2}{3}} + y^{\frac{2}{3}} = a^{\frac{2}{3}},$$

the rectangular equation of the hypocycloid.

ILLUSTRATIVE EXAMPLE 1. Find the rectangular equation of the envelope of the straight line $y = mx + \frac{p}{m}$, where the slope $m$ is the variable parameter.

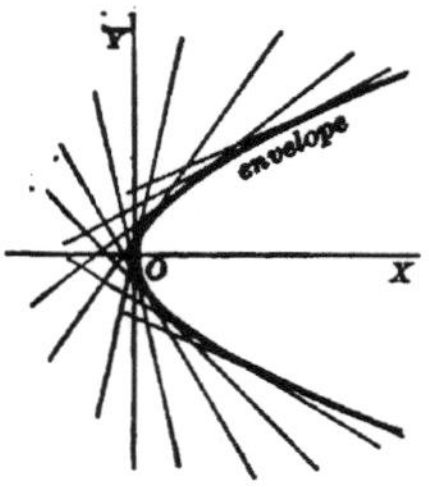


Solution. $y = mx + \frac{p}{m}$.

*First step.* $0 = x - \frac{p}{m^2}$.

Solving, $m = \pm\sqrt{\frac{p}{x}}$.

Substitute in the given equation,

$$y = \pm\sqrt{\frac{p}{x}}\cdot x \pm \sqrt{\frac{x}{p}}\cdot p = \pm 2\sqrt{px},$$

and squaring, $y^2 = 4px$, a parabola, is the equation of the envelope. The family of straight lines formed by varying the slope $m$ is shown in the figure, each line being tangent to the envelope, for we know from Analytic Geometry that $y = mx + \frac{p}{m}$ is the tangent to the parabola $y^2 = 4px$ expressed in terms of its own slope $m$.

**132. The evolute of a given curve considered as the envelope of its normals.** Since the normals to a curve are all tangent to the evolute, § 118, p. 181, it is evident that *the evolute of a curve may also be defined as the envelope of its normals;* that is, as the locus of the ultimate intersections of neighboring normals. It is also interesting to notice that if we find the parametric equations of the envelope by the method of the previous section, we get the coördinates $x$ and $y$ of the center of curvature; so that we have here *a second method for finding the coördinates of the center of curvature.* If we then eliminate the variable parameter, we have a relation between $x$ and $y$ which is the rectangular equation of the evolute (envelope of the normals).

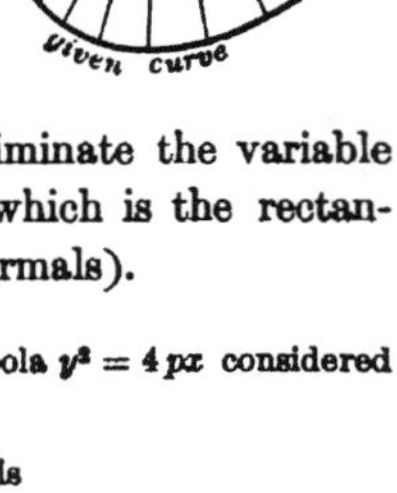


ILLUSTRATIVE EXAMPLE 1. Find the evolute of the parabola $y^2 = 4px$ considered as the envelope of its normals.

Solution. The equation of the normal at any point $(x', y')$ is

$$y - y' = -\frac{y'}{2p}(x - x')$$

from (2), p. 77. As we are considering the normals all along the curve, both $x'$ and $y'$ will vary. Eliminating $x'$ by means of $y'^2 = 4px'$, we get the equation of the normal to be

$$(A) \qquad y - y' = \frac{y'^3}{8p^2} - \frac{xy'}{2p}.$$

Considering $y'$ as the variable parameter, we wish to find the envelope of this family of normals. Differentiating $(A)$ with respect to $y'$,

$$-1 = \frac{3y'^2}{8p^2} - \frac{x}{2p},$$

and solving for $x$,

$$(B) \qquad x = \frac{3y'^2 + 8p^2}{4p}.$$

Substituting this value of $x$ in $(A)$ and solving for $y$,

$$(C) \qquad y = -\frac{y'^3}{4p^2}.$$

$(B)$ and $(C)$ are then the coördinates of the center of curvature of the parabola. Taken together, $(B)$ and $(C)$ are the parametric equations of the evolute in terms of the parameter $y'$. Eliminating $y'$ between $(B)$ and $(C)$ gives

$$27py^2 = 4(x - 2p)^3,$$

the rectangular equation of the evolute of the parabola. This is the same result we obtained in Illustrative Example 1, p. 183, by the first method.

**133. Two parameters connected by one equation of condition.** Many problems occur where it is convenient to use two parameters connected by an equation of condition. For instance, the example given in the last section involves the two parameters $x'$ and $y'$ which are connected by the equation of the curve. In this case we eliminated $x'$, leaving only the one parameter $y'$.

However, when the elimination is difficult to perform, both the given equation and the equation of condition between the two parameters may be differentiated with respect to one of the parameters, regarding either parameter as a function of the other. By studying the solution of the following problem the process will be made clear.

ILLUSTRATIVE EXAMPLE 1. Find the envelope of the family of ellipses whose axes coincide and whose area is constant.

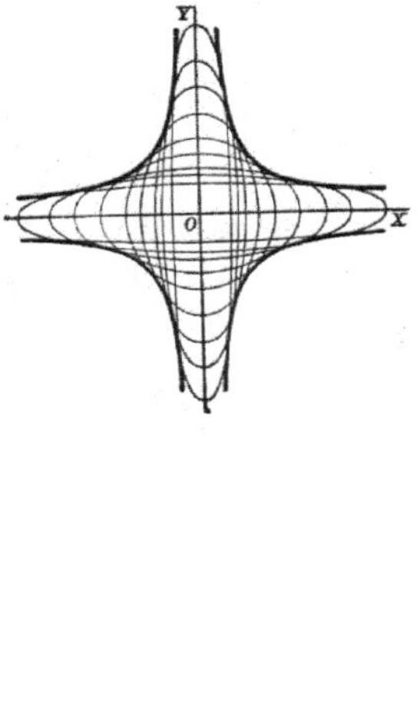


**Solution.** $(A)$ $\dfrac{x^2}{a^2} + \dfrac{y^2}{b^2} = 1$

is the equation of the ellipse where $a$ and $b$ are the variable parameters connected by the equation

$$(B) \qquad \pi ab = k,$$

$\pi ab$ being the area of an ellipse whose semi-axes are $a$ and $b$. Differentiating $(A)$ and $(B)$, regarding $a$ and $b$ as variables and $x$ and $y$ as constants, we have, using differentials,

$$\frac{x^2 da}{a^3} + \frac{y^2 db}{b^3} = 0, \text{ from } (A),$$

and $\qquad bda + adb = 0$, from $(B)$.

Transposing one term in each to the second member and dividing, we get

$$\frac{x^2}{a^2}=\frac{y^2}{b^2}.$$

Therefore, from (*A*), $\frac{x^2}{a^2}=\frac{1}{2}$ and $\frac{y^2}{b^2}=\frac{1}{2}$,

giving $a=\pm x\sqrt{2}$ and $b=\pm y\sqrt{2}$.

Substituting these values in (*B*), we get the envelope

$$xy=\pm\frac{k}{2\pi},$$

a pair of conjugate rectangular hyperbolas (see last figure).

## EXAMPLES

1. Find the envelope of the family of straight lines $y=2mx+m^4$, $m$ being the variable parameter. *Ans.* $x=-2m^3,\ y=-3m^4$; or $16y^3+27x^4=0$.*

2. Find the envelope of the family of parabolas $y^2=a(x-a)$, $a$ being the variable parameter. *Ans.* $x=2a,\ y=\pm a$; or $y=\pm\frac{1}{2}x$.

3. Find the envelope of the family of circles $x^2+(y-\beta)^2=r^2$, $\beta$ being the variable parameter. *Ans.* $x=\pm r$.

4. Find the equation of the curve having as tangents the family of straight lines $y=mx\pm\sqrt{a^2m^2+b^2}$, the slope $m$ being the variable parameter.
*Ans.* The ellipse $b^2x^2+a^2y^2=a^2b^2$.

5. Find the envelope of the family of circles whose diameters are double ordinates of the parabola $y^2=4px$. *Ans.* The parabola $y^2=4p(p+x)$.

6. Find the envelope of the family of circles whose diameters are double ordinates of the ellipse $b^2x^2+a^2y^2=a^2b^2$.
*Ans.* The ellipse $\frac{x^2}{a^2+b^2}+\frac{y^2}{b^2}=1$.

7. A circle moves with its center on the parabola $y^2=4ax$, and its circumference passes through the vertex of the parabola. Find the equation of the envelope of the circles. *Ans.* The cissoid $y^2(x+2a)+x^3=0$.

8. Find the curve whose tangents are $y=lx\pm\sqrt{al^2+bl+c}$, the slope $l$ being supposed to vary. *Ans.* $4(ay^2+bxy+cx^2)=4ac-b^2$.

9. Find the evolute of the ellipse $b^2x^2+a^2y^2=a^2b^2$, taking the equation of normal in the form

$$by=ax\tan\phi-(a^2-b^2)\sin\phi,$$

the eccentric angle $\phi$ being the parameter.

*Ans.* $x=\frac{a^2-b^2}{a}\cos^3\phi,\ y=\frac{b^2-a^2}{b}\sin^3\phi$; or $(ax)^{\frac{2}{3}}+(by)^{\frac{2}{3}}=(a^2-b^2)^{\frac{2}{3}}$.

10. Find the evolute of the hypocycloid $x^{\frac{2}{3}}+y^{\frac{2}{3}}=a^{\frac{2}{3}}$, the equation of whose normal is

$$y\cos\tau-x\sin\tau=a\cos 2\tau,$$

$\tau$ being the parameter. *Ans.* $(x+y)^{\frac{2}{3}}+(x-y)^{\frac{2}{3}}=2a^{\frac{2}{3}}$.

* When two answers are given, the first is in parametric form and the second in rectangular form.

**11.** Find the envelope of the circles which pass through the origin and have their centers on the hyperbola $x^2 - y^2 = c^2$.

*Ans.* The lemniscate $(x^2 + y^2)^2 = 4c^2(x^2 - y^2)$.

**12.** Find the envelope of a line such that the sum of its intercepts on the axes equals $c$.

*Ans.* The parabola $x^{\frac{1}{2}} + y^{\frac{1}{2}} = c^{\frac{1}{2}}$.

**13.** Find the equation of the envelope of the system of circles $x^2 + y^2 - 2(a + 2)x + a^2 = 0$, where $a$ is the parameter. Draw a figure illustrating the problem.

*Ans.* $y^2 = 4x$.

**14.** Find the envelope of the family of ellipses $b^2x^2 + a^2y^2 = a^2b^2$, when the sum of its semiaxes equals $c$.

*Ans.* The hypocycloid $x^{\frac{2}{3}} + y^{\frac{2}{3}} = c^{\frac{2}{3}}$.

**15.** Find the envelope of the ellipses whose axes coincide, and such that the distance between the extremities of the major and minor axes is constant and equal to $l$.

*Ans.* A square whose sides are $(x \pm y)^2 = l^2$.

**16.** Projectiles are fired from a gun with an initial velocity $v_0$. Supposing the gun can be given any elevation and is kept always in the same vertical plane, what is the envelope of all possible trajectories, the resistance of the air being neglected?

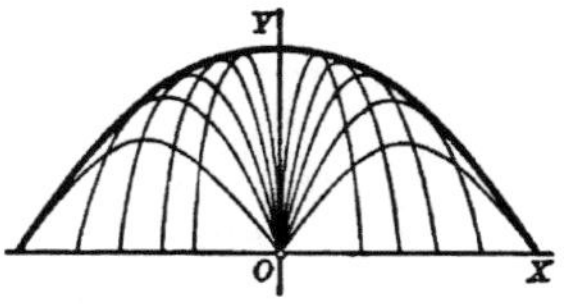


HINT. The equation of any trajectory is

$$y = x \tan \alpha - \frac{gx^2}{2v_0^2 \cos^2 \alpha},$$

$\alpha$ being the variable parameter.

*Ans.* The parabola $y = \dfrac{v_0^2}{2g} - \dfrac{gx^2}{2v_0^2}$.

**17.** Find the equation of the envelope of each of the following family of curves, $t$ being the parameter; draw the family and the envelope:

(a) $(x - t)^2 + y^2 = 1 - t^2$.
(b) $x^2 + (y - t)^2 = 2t$.
(c) $(x - t)^2 + y^2 = \frac{1}{2}t^2 - 1$.
(d) $x^2 + (y - t)^2 = \frac{1}{4}t^2$.
(e) $y = tx + t^2$.
(f) $x = 2ty + t^4$.
(g) $y = tx + \dfrac{1}{t}$.
(h) $y^2 = t(x + 2t)$.
(i) $(x - t)^2 + y^2 = 4t$.
(j) $x^2 + (y - t)^2 = 4 - t^2$.
(k) $(x - t)^2 + (y - t)^2 = t^2$.
(l) $(x - t)^2 + (y + t)^2 = t^2$.
(m) $y = t^2x + t$.
(n) $y = t(x - 2t)$.
(o) $x = \dfrac{y}{t} + t$.
(p) $(x - t)^2 + 4y^2 = t$.

# CHAPTER XVII

## SERIES

**134. Introduction.** A *series* is a succession of separate numbers which is formed according to some rule or law. Each number is called a term of the series. Thus

$$1, \quad 2, \quad 4, \quad 8, \quad \cdots, \quad 2^{n-1}$$

is a series whose law of formation is that each term after the first is found by multiplying the preceding term by 2; hence we may write down as many more terms of the series as we please, and any particular term of the series may be found by substituting *the number of that term in the series* for $n$ in the expression $2^{n-1}$, which is called the *general* or $n$th *term* of the series.

### EXAMPLES

In the following six series:

(a) Discover by inspection a law of formation;
(b) write down several terms more in each;
(c) find the $n$th or *general term*.

| *Series* | *nth term* |
|---|---|
| 1. $1, 3, 9, 27, \cdots$. | $3^{n-1}$. |
| 2. $-a, +a^2, -a^3, +a^4, \cdots$. | $(-a)^n$. |
| 3. $1, 4, 9, 16, \cdots$. | $n^2$. |
| 4. $x, \frac{x^2}{2}, \frac{x^3}{3}, \frac{x^4}{4}, \cdots$. | $\frac{x^n}{n}$. |
| 5. $4, -2, +1, -\frac{1}{2}, \cdots$. | $4(-\frac{1}{2})^{n-1}$. |
| 6. $\frac{3y}{2}, \frac{5y^2}{5}, \frac{7y^3}{10}, \cdots$. | $\frac{2n+1}{n^2+1}y^n$. |

Write down the first four terms of each series whose $n$th or *general term* is given below:

| *nth term* | *Series* |
|---|---|
| 7. $n^2x^n$. | $x, 4x^2, 9x^3, 16x^4$. |
| 8. $\frac{x^n}{1+\sqrt{n}}$. | $\frac{x}{2}, \frac{x^2}{1+\sqrt{2}}, \frac{x^3}{1+\sqrt{3}}, \frac{x^4}{1+\sqrt{4}}$. |
| 9. $\frac{n+2}{n^3+1}$. | $\frac{3}{2}, \frac{4}{9}, \frac{5}{28}, \frac{6}{65}$. |

10. $\dfrac{n}{2^n}$. $\qquad \dfrac{1}{2}, \dfrac{2}{4}, \dfrac{3}{8}, \dfrac{4}{16}$.

11. $\dfrac{(\log a)^n x^n}{\lfloor n}$. $\qquad \dfrac{\log a \cdot x}{1}, \dfrac{\log^2 a \cdot x^2}{2}, \dfrac{\log^3 a \cdot x^3}{6}, \dfrac{\log^4 a \cdot x^4}{24}$.

12. $\dfrac{(-1)^{n-1}x^{2n-2}}{\lfloor 2n-1}$. $\qquad \dfrac{1}{1}, -\dfrac{x^2}{\lfloor 3}, \dfrac{x^4}{\lfloor 5}, -\dfrac{x^6}{\lfloor 7}$.

**135. Infinite series.** Consider the series of $n$ terms

$$(A) \qquad 1, \frac{1}{2}, \frac{1}{4}, \frac{1}{8}, \ldots, \frac{1}{2^{n-1}};$$

and let $S_n$ denote the sum of the series. Then

$$(B) \qquad S_n = 1 + \frac{1}{2} + \frac{1}{4} + \frac{1}{8} + \cdots + \frac{1}{2^{n-1}}.$$

Evidently $S_n$ is a function of $n$, for

when $n=1$, $S_1 = 1 \qquad = 1,$

when $n=2$, $S_2 = 1 + \frac{1}{2} \qquad = 1\frac{1}{2},$

when $n=3$, $S_3 = 1 + \frac{1}{2} + \frac{1}{4} \qquad = 1\frac{3}{4},$

when $n=4$, $S_4 = 1 + \frac{1}{2} + \frac{1}{4} + \frac{1}{8} \qquad = 1\frac{7}{8},$

. . . . . . . . . . . . .

when $n=n$, $S_n = 1 + \frac{1}{2} + \frac{1}{4} + \frac{1}{8} + \cdots + \frac{1}{2^{n-1}} \qquad = 2 - \frac{1}{2^{n-1}}.$*

Mark off points on a straight line whose distances from a fixed point 0 correspond to these different sums. It is seen that the point

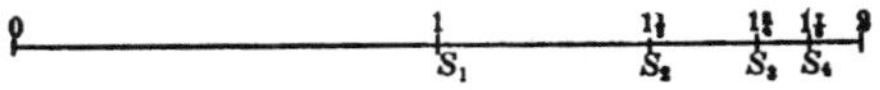


corresponding to any sum bisects the distance between the preceding point and 2. Hence it appears geometrically that when $n$ increases without limit

$$\text{limit } S_n = 2.$$

We also see that this is so from arithmetical considerations, for

$$\underset{n=\infty}{\text{limit}} S_n = \underset{n=\infty}{\text{limit}} \left(2 - \frac{1}{2^{n-1}}\right) = 2.\dagger$$

[Since when $n$ increases without limit $\frac{1}{2^{n-1}}$ approaches zero as a limit.]

* Found by 6, p. 1, for the sum of a geometric series.

† Such a result is sometimes, for the sake of brevity, called the *sum* of the series; but the student must not forget that 2 is *not* the sum but the *limit of the sum*, as the number of terms increases without limit.

We have so far discussed only a particular series ($A$) when the number of terms increases without limit. Let us now consider the general problem, using the series

$$(C) \qquad u_1, \quad u_2, \quad u_3, \quad u_4, \quad \cdots,$$

whose terms may be either positive or negative. Denoting by $S_n$ the sum of the first $n$ terms, we have

$$S_n = u_1 + u_2 + u_3 + \cdots + u_n,$$

and $S_n$ is a function of $n$. If we now let the number of terms ($= n$) increase without limit, one of two things may happen: either

CASE I. **$S_n$ approaches a limit,** say $u$, indicated by

$$\underset{n=\infty}{\text{limit}} S_n = u; \text{ or}$$

CASE II. **$S_n$ approaches no limit.**

In either case ($C$) is called an *infinite series*. In Case I the infinite series is said to be *convergent* and *to converge to the value u*, or *to have the value u*, or *to have the sum u*. The infinite geometric series discussed at the beginning of this section is an example of a convergent series, and it converges to the value 2. In fact, the simplest example of a convergent series is the infinite geometric series

$$a, \quad ar, \quad ar^2, \quad ar^3, \quad ar^4, \quad \cdots,$$

where $r$ is numerically less than unity. The sum of the first $n$ terms of this series is, by 6, p. 1,

$$S_n = \frac{a(1-r^n)}{1-r} = \frac{a}{1-r} - \frac{ar^n}{1-r}.$$

If we now suppose $n$ to increase without limit, the first fraction on the right-hand side remains unchanged, while the second approaches zero as a limit. Hence

$$\underset{n=\infty}{\text{limit}} S_n = \frac{a}{1-r},$$

a *perfectly definite number* in any given case.

In Case II the infinite series is said to be *nonconvergent*.* Series under this head may be divided into two classes.

FIRST CLASS. *Divergent series*, in which the sum of $n$ terms increases indefinitely in numerical value as $n$ increases without limit; take for example the series

$$S_n = 1 + 2 + 3 + \cdots + n.$$

* Some writers use *divergent* as equivalent to *nonconvergent*.

As $n$ increases without limit, $S_n$ increases without limit and therefore the series is *divergent.*

SECOND CLASS. *Oscillating series,* of which

$$S_n = 1 - 1 + 1 - 1 + \cdots + (-1)^{n-1}$$

is an example. Here $S_n$ is zero or unity according as $n$ is even or odd, and although $S_n$ does not become infinite as $n$ increases without limit, it does not tend to a limit, but oscillates. It is evident that if all the terms of a series have the same sign, the series cannot oscillate.

Since the sum of a converging series is a perfectly definite number, while such a thing as the sum of a nonconvergent series does not exist, it follows at once that it is absolutely essential in any given problem involving infinite series to determine whether or not the series is convergent. This is often a problem of great difficulty, and we shall consider only the simplest cases.

**136. Existence of a limit.** When a series is given we cannot in general, as in the case of a geometric series, actually find the number which is the limit of $S_n$. But although we may not know how to compute the numerical value of that limit, it is of prime importance to know that a *limit does exist,* for otherwise the series may be nonconvergent. When examining a series to determine whether or not it is convergent, the following theorems, which we state without proofs, are found to be of fundamental importance.*

**Theorem I.** *If $S_n$ is a variable that always increases as n increases, but always remains less than some definite fixed number A, then as n increases without limit, $S_n$ will approach a definite limit which is not greater than A.*

**Theorem II.** *If $S_n$ is a variable that always decreases as n increases, but always remains greater than some definite fixed number B, then as n increases without limit, $S_n$ will approach a definite limit which is not less than B.*

**Theorem III.** *The necessary and sufficient condition that $S_n$ shall approach some definite fixed number as a limit as n increases without limit is that*

$$\lim_{n=\infty} (S_{n+p} - S_n) = 0$$

*for all values of the integer p.*

* See Osgood's *Introduction to Infinite Series*, pp. 4, 14, 64.

**137. Fundamental test for convergence.** Summing up first $n$ and then $n+p$ terms of a series, we have

$$(A) \qquad S_n = u_1 + u_2 + u_3 + \cdots + u_n.$$

$$(B) \qquad S_{n+p} = u_1 + u_2 + u_3 + \cdots + u_n + u_{n+1} + \cdots + u_{n+p}.$$

Subtracting (*A*) from (*B*),

$$(C) \qquad S_{n+p} - S_n = u_{n+1} + u_{n+2} + \cdots + u_{n+p}.$$

From Theorem III we know that the *necessary* and *sufficient* condition that the series shall be convergent is that

$$\underset{n=\infty}{\text{limit}} (S_{n+p} - S_n) = 0$$

for every value of $p$. But this is the same as the left-hand member of (*C*); therefore from the right-hand member the condition may also be written

$$(D) \qquad \underset{n=\infty}{\text{limit}} (u_{n+1} + u_{n+2} + \cdots + u_{n+p}) = 0.$$

Since (*D*) is true for every value of $p$, then, letting $p = 1$, a *necessary* condition for convergence is that

$$\underset{n=\infty}{\text{limit}} (u_{n+1}) = 0;$$

or, what amounts to the same thing,

$$(E) \qquad \underset{n=\infty}{\text{limit}} (u_n) = 0.$$

Hence, if the general (or $n$th) term of a series does not approach zero as $n$ approaches infinity, we know at once that the series is non-convergent and we need proceed no further. However, (*E*) is not a *sufficient* condition; that is, even if the $n$th term does approach zero, we cannot state positively that the series is convergent; for, consider the harmonic series

$$1, \ \frac{1}{2}, \ \frac{1}{3}, \ \frac{1}{4}, \ \cdots, \ \frac{1}{n}.$$

Here $$\underset{n=\infty}{\text{limit}} (u_n) = \underset{n=\infty}{\text{limit}} \left(\frac{1}{n}\right) = 0;$$

that is, condition (*E*) is fulfilled. Yet we may show that the harmonic series is not convergent by the following comparison:

$$(F) \qquad 1 + \tfrac{1}{2} + [\tfrac{1}{3} + \tfrac{1}{4}] + [\tfrac{1}{5} + \tfrac{1}{6} + \tfrac{1}{7} + \tfrac{1}{8}] + [\tfrac{1}{9} + \cdots \tfrac{1}{16}] + \cdots.$$

$$(G) \qquad \tfrac{1}{2} + \tfrac{1}{2} + [\tfrac{1}{4} + \tfrac{1}{4}] + [\tfrac{1}{8} + \tfrac{1}{8} + \tfrac{1}{8} + \tfrac{1}{8}] + [\tfrac{1}{16} + \cdots + \tfrac{1}{16}] + \cdots.$$

We notice that every term of $(G)$ is equal to or less than the corresponding term of $(F)$, so that the sum of any number of the first terms of $(F)$ will be greater than the sum of the corresponding terms of $(G)$. But since the sum of the terms grouped in each bracket in $(G)$ equals $\frac{1}{2}$, the sum of $(G)$ may be made as large as we please by taking terms enough. The sum $(G)$ increases indefinitely as the number of terms increases without limit; hence $(G)$, and therefore also $(F)$, is divergent.

We shall now proceed to deduce special tests which, as a rule, are easier to apply than the above theorems.

**138. Comparison test for convergence.** In many cases, an example of which was given in the last section, it is easy to determine whether or not a given series is convergent by comparing it term by term with another series whose character is known. *Let*

$$(A) \qquad u_1 + u_2 + u_3 + \cdots$$

*be a series of positive terms which it is desired to test for convergence. If a series of positive terms already known to be convergent, namely,*

$$(B) \qquad a_1 + a_2 + a_3 + \cdots,$$

*can be found whose terms are never less than the corresponding terms in the series $(A)$ to be tested, then $(A)$ is a convergent series and its sum does not exceed that of $(B)$.*

**Proof.** Let $$s_n = u_1 + u_2 + u_3 + \cdots + u_n,$$
and $$S_n = a_1 + a_2 + a_3 + \cdots + a_n;$$
and suppose that $$\lim_{n=\infty} S_n = A.$$

Then, since $$S_n < A \text{ and } s_n \leqq S_n,$$
it follows that $s_n < A$. Hence, by Theorem I, p. 215, $s_n$ approaches a limit; therefore the series $(A)$ is convergent and the limit of its sum is not greater than $A$.

Illustrative Example 1. Test the series

$$(C) \qquad 1 + \frac{1}{2^2} + \frac{1}{3^3} + \frac{1}{4^4} + \frac{1}{5^5} + \cdots.$$

**Solution.** Each term after the first is less than the corresponding term of the geometric series

$$(D) \qquad 1 + \frac{1}{2} + \frac{1}{2^2} + \frac{1}{2^3} + \frac{1}{2^4} + \cdots,$$

which is known to be convergent; hence $(C)$ is also convergent.

Following a line of reasoning similar to that applied to $(A)$ and $(B)$, it is evident that, if

$$(E) \qquad u_1 + u_2 + u_3 + \cdots$$

*is a series of positive terms to be tested, which are never less than the corresponding terms of the series of positive terms, namely,*

$$(F) \qquad b_1 + b_2 + b_3 + \cdots,$$

*known to be divergent, then $(E)$ is a divergent series.*

ILLUSTRATIVE EXAMPLE 2. Test the series

$$1 + \frac{1}{\sqrt{2}} + \frac{1}{\sqrt{3}} + \frac{1}{\sqrt{4}} + \cdots.$$

**Solution.** This series is divergent, since its terms are greater than the corresponding terms of the harmonic series

$$1 + \frac{1}{2} + \frac{1}{3} + \frac{1}{4} \cdots,$$

which is known (pp. 216, 217) to be divergent.

ILLUSTRATIVE EXAMPLE 3. Test the following series (called the $p$ series) for different values of $p$:

$$(G) \qquad 1 + \frac{1}{2^p} + \frac{1}{3^p} + \frac{1}{4^p} + \cdots.$$

**Solution.** Grouping the terms, we have, when $p > 1$,

$$\frac{1}{2^p} + \frac{1}{3^p} < \frac{1}{2^p} + \frac{1}{2^p} = \frac{2}{2^p} = \frac{1}{2^{p-1}},$$

$$\frac{1}{4^p} + \frac{1}{5^p} + \frac{1}{6^p} + \frac{1}{7^p} < \frac{1}{4^p} + \frac{1}{4^p} + \frac{1}{4^p} + \frac{1}{4^p} = \frac{4}{4^p} = \left(\frac{1}{2^{p-1}}\right)^2,$$

$$\frac{1}{8^p} + \cdots + \frac{1}{15^p} < \frac{1}{8^p} + \frac{1}{8^p} + \frac{1}{8^p} + \frac{1}{8^p} + \frac{1}{8^p} + \frac{1}{8^p} + \frac{1}{8^p} + \frac{1}{8^p} = \frac{8}{8^p} = \left(\frac{1}{2^{p-1}}\right)^3,$$

and so on. Construct the series

$$(H) \qquad 1 + \frac{1}{2^{p-1}} + \left(\frac{1}{2^{p-1}}\right)^2 + \left(\frac{1}{2^{p-1}}\right)^3 + \cdots.$$

*When* $p > 1$, series $(H)$ is a geometric series with the common ratio less than unity, and is therefore convergent. But the sum of $(G)$ is less than the sum of $(H)$, as shown by the above inequalities; therefore $(G)$ is also convergent.

*When* $p = 1$, series $(G)$ becomes the harmonic series which we saw was divergent, and neither of the above tests apply.

*When* $p < 1$, the terms of series $(G)$ will, after the first, be greater than the corresponding terms of the harmonic series; hence $(G)$ is divergent.

**139. Cauchy's ratio test for convergence.** Let

$$(A) \qquad u_1 + u_2 + u_3 + \cdots$$

be a series of positive terms to be tested.

Divide any general term by the one that immediately precedes it; i.e. form the test ratio $\frac{u_{n+1}}{u_n}$.

As $n$ increases without limit, let $\underset{n=\infty}{\text{limit}} \frac{u_{n+1}}{u_n} = \rho$.

I. When $\rho < 1$. By the definition of a limit (§ 13, p. 11) we can choose $n$ so large, say $n = m$, that when $n \geqq m$ the ratio $\frac{u_{n+1}}{u_n}$ shall differ from $\rho$ by as little as we please, and therefore be less than a proper fraction $r$. Hence

$$u_{m+1} < u_m r;\ u_{m+2} < u_{m+1} r < u_m r^2;\ u_{m+3} < u_m r^3;$$

and so on. Therefore, after the term $u_m$, each term of the series $(A)$ is less than the corresponding term of the geometrical series

$$(B) \qquad u_m r + u_m r^2 + u_m r^3 + \cdots.$$

But since $r < 1$, the series $(B)$, and therefore also the series $(A)$, is convergent.*

II. When $\rho > 1$ (*or* $\rho = \infty$). Following the same line of reasoning as in I, the series $(A)$ may be shown to be divergent.

III. When $\rho = 1$, the series may be either convergent or divergent; that is, there is no test. For, consider the $p$ series, namely,

$$1 + \frac{1}{2^p} + \frac{1}{3^p} + \frac{1}{4^p} + \cdots + \frac{1}{n^p} + \frac{1}{(n+1)^p} + \cdots.$$

The test ratio is $\frac{u_{n+1}}{u_n} = \left(\frac{n}{n+1}\right)^p = \left(1 - \frac{1}{n+1}\right)^p$;

and $$\underset{n=\infty}{\text{limit}}\left(\frac{u_{n+1}}{u_n}\right) = \underset{n=\infty}{\text{limit}}\left(1 - \frac{1}{n+1}\right)^p = (1)^p = 1\ (= \rho).$$

Hence $\rho = 1$, no matter what value $p$ may have. But on p. 218 we showed that

when $p > 1$, the series converges, and

when $p \leqq 1$, the series diverges.

Thus it appears that $\rho$ can equal unity both for convergent and for divergent series, *and the ratio test for convergence fails*. There are other tests to apply in cases like this, but the scope of our book does not admit of their consideration.

* When examining a series for convergence we are at liberty to disregard any finite number of terms; the rejection of such terms would affect the *value* but not the *existence* of the limit.

Our results may then be stated as follows:

Given the series of positive terms

$$u_1 + u_2 + u_3 + \cdots + u_n + u_{n+1} + \cdots;$$

find the limit

$$\lim_{n=\infty}\left(\frac{u_{n+1}}{u_n}\right) = \rho.$$

I. When $\rho < 1$,* the series is convergent.

II. When $\rho > 1$, the series is divergent.

III. When $\rho = 1$, there is no test.

**140. Alternating series.** This is the name given to a series whose terms are alternately positive and negative. Such series occur frequently in practice and are of considerable importance.

*If*

$$u_1 - u_2 + u_3 - u_4 + \cdots$$

*is an alternating series whose terms never increase in numerical value, and if*

$$\lim_{n=\infty} u_n = 0,$$

*then the series is convergent.*

**Proof.** The sum of $2n$ (an even number) terms may be written in the two forms

$$(A)\quad S_{2n} = (u_1 - u_2) + (u_3 - u_4) + (u_5 - u_6) + \cdots + (u_{2n-1} - u_{2n}), \quad \text{or}$$

$$(B)\quad S_{2n} = u_1 - (u_2 - u_3) - (u_4 - u_5) - \cdots - u_{2n}.$$

Since each difference is positive (if it is not zero, and the assumption $\lim_{n=\infty} u_n = 0$ excludes equality of the terms of the series), series $(A)$ shows that $S_{2n}$ is positive and increases with $n$, while series $(B)$ shows that $S_{2n}$ is always less than $u_1$; therefore, by Theorem I, p. 215, $S_{2n}$ must approach a limit less than $u_1$ when $n$ increases, and the series is convergent.

ILLUSTRATIVE EXAMPLE 4. Test the alternating series $1 - \frac{1}{2} + \frac{1}{3} - \frac{1}{4} + \cdots$.

**Solution.** Since each term is less in numerical value than the preceding one, and

$$\lim_{n=\infty}(u_n) = \lim_{n=\infty}\left(\frac{1}{n}\right) = 0,$$

the series is convergent.

**141. Absolute convergence.** A series is said to be *absolutely* † or *unconditionally* convergent when the series formed from it by making all its terms positive is convergent. Other convergent series are said

* It is not enough that $u_{n+1}/u_n$ becomes and remains less than unity for all values of $n$, but this test requires that the *limit* of $u_{n+1}/u_n$ shall be less than unity. For instance, in the case of the harmonic series this ratio is always less than unity and yet the series diverges as we have seen. The *limit*, however, is not less than unity but equals unity.

† The terms of the new series are the numerical (absolute) values of the terms of the given series.

to be *not absolutely convergent* or *conditionally convergent.* To this latter class belong some convergent alternating series. For example, the series

$$1-\frac{1}{2^2}+\frac{1}{3^3}-\frac{1}{4^4}+\frac{1}{5^5}-\cdots$$

is *absolutely convergent,* since the series $(C)$, p. 217, namely,

$$1+\frac{1}{2^2}+\frac{1}{3^3}+\frac{1}{4^4}+\frac{1}{5^5}+\cdots$$

is convergent. The series

$$1-\frac{1}{2}+\frac{1}{3}-\frac{1}{4}+\frac{1}{5}-\cdots$$

is *conditionally* convergent, since the harmonic series

$$1+\frac{1}{2}+\frac{1}{3}+\frac{1}{4}+\frac{1}{5}+\cdots$$

is divergent.

*A series with terms of different signs is convergent if the series deduced from it by making all the signs positive is convergent.*

The proof of this theorem is omitted.

Assuming that the ratio test on p. 219 holds without placing any restriction on the signs of the terms of a series, we may summarize our results in the following

**General directions for testing the series**

$$u_1+u_2+u_3+u_4+\cdots+u_n+u_{n+1}+\cdots.$$

*When it is an alternating series whose terms never increase in numerical value, and if*

$$\lim_{n=\infty} u_n=0,$$

*then the series is convergent.*

*In any series in which the above conditions are not satisfied, we determine the form of* $u_n$ *and* $u_{n+1}$ *and calculate the limit*

$$\lim_{n=\infty}\left(\frac{u_{n+1}}{u_n}\right)=\rho.$$

I. *When* $|\rho|<1$, *the series is absolutely convergent.*

II. *When* $|\rho|>1$, *the series is divergent.*

III. *When* $|\rho|=1$, *there is no test, and we should compare the series with some series which we know to be convergent, as*

$$a+ar+ar^2+ar^3+\cdots;\ r<1, \qquad \textit{(geometric series)}$$

$$1+\frac{1}{2^p}+\frac{1}{3^p}+\frac{1}{4^p}+\cdots;\ p>1, \qquad (p \text{ series})$$

*or compare the given series with some series which is known to be divergent, as*

$$1+\frac{1}{2}+\frac{1}{3}+\frac{1}{4}+\cdots; \qquad \textit{(harmonic series)}$$

$$1+\frac{1}{2^p}+\frac{1}{3^p}+\frac{1}{4^p}+\cdots;\ p<1. \qquad (p \text{ series})$$

Illustrative Example 1. Test the series

$$1+\frac{1}{\lfloor\underline{1}}+\frac{1}{\lfloor\underline{2}}+\frac{1}{\lfloor\underline{3}}+\frac{1}{\lfloor\underline{4}}+\cdots.$$

**Solution.** Here $\qquad u_n=\dfrac{1}{\lfloor\underline{n-1}}, \qquad u_{n+1}=\dfrac{1}{\lfloor\underline{n}}.$

$$\therefore \lim_{n=\infty}\left(\frac{u_{n+1}}{u_n}\right)=\lim_{n=\infty}\left(\frac{\frac{1}{\lfloor\underline{n}}}{\frac{1}{\lfloor\underline{n-1}}}\right)=\lim_{n=\infty}\left(\frac{\lfloor\underline{n-1}}{\lfloor\underline{n}}\right)=\lim_{n=\infty}\left(\frac{1}{n}\right)=0\,(=\rho),$$

and by I, p. 221, the series is convergent.

Illustrative Example 2. Test the series $\dfrac{\lfloor\underline{1}}{10}+\dfrac{\lfloor\underline{2}}{10^2}+\dfrac{\lfloor\underline{3}}{10^3}+\cdots.$

**Solution.** Here $\qquad u_n=\dfrac{\lfloor\underline{n}}{10^n}, \qquad u_{n+1}=\dfrac{\lfloor\underline{n+1}}{10^{n+1}}.$

$$\therefore \lim_{n=\infty}\left(\frac{u_{n+1}}{u_n}\right)=\lim_{n=\infty}\left(\frac{\lfloor\underline{n+1}}{10^{n+1}}\times\frac{10^n}{\lfloor\underline{n}}\right)=\lim_{n=\infty}\left(\frac{n+1}{10}\right)=\infty\,(=\rho),$$

and by II, p. 221, the series is divergent.

Illustrative Example 3. Test the series

$$(C) \qquad \frac{1}{1\cdot 2}+\frac{1}{3\cdot 4}+\frac{1}{5\cdot 6}+\cdots.$$

**Solution.** Here $u_n=\dfrac{1}{(2n-1)\,2n}, \qquad u_{n+1}=\dfrac{1}{(2n+1)(2n+2)}.$

$$\therefore \lim_{n=\infty}\left(\frac{u_{n+1}}{u_n}\right)=\lim_{n=\infty}\left[\frac{(2n-1)\,2n}{(2n+1)(2n+2)}\right]=\frac{\infty}{\infty}.$$

This being an indeterminate form, we evaluate it, using the rule on p. 174.

Differentiating, $\qquad \lim_{n=\infty}\left(\dfrac{8n-2}{8n+6}\right)=\dfrac{\infty}{\infty}.$

Differentiating again, $\qquad \lim_{n=\infty}\left(\dfrac{8}{8}\right)=1\,(=\rho).$

This gives no test (III, p. 221). But if we compare series $(C)$ with $(G)$, p. 218, making $p=2$, namely,

$$(D) \qquad 1+\frac{1}{2^2}+\frac{1}{3^2}+\frac{1}{4^2}+\cdots,$$

we see that $(C)$ must be convergent, since its terms are less than the corresponding terms of $(D)$, which was proved convergent.

**EXAMPLES**

**Show that the following ten series are convergent:**

1. $\frac{1}{1^2}+\frac{1}{2^2}+\frac{1}{3^2}+\cdots.$

2. $\frac{1}{2}+\frac{2}{2^2}+\frac{3}{2^3}+\frac{4}{2^4}+\cdots.$

3. $\frac{1}{1\cdot 2}+\frac{1}{3\cdot 4}+\frac{1}{5\cdot 6}+\cdots.$

4. $\frac{1}{3}+\frac{1\cdot 3}{3\cdot 6}+\frac{1\cdot 3\cdot 5}{3\cdot 6\cdot 9}+\cdots.$

5. $\frac{1}{\underline{|3}}+\frac{1}{\underline{|5}}+\frac{1}{\underline{|7}}+\cdots.$

6. $1+\frac{1}{2\sqrt{2}}+\frac{1}{3\sqrt{3}}+\frac{1}{4\sqrt{4}}+\cdots.$

7. $1-\frac{1}{3^2}+\frac{1}{5^2}-\frac{1}{7^2}+\frac{1}{9^2}-\cdots.$

8. $\frac{1}{2}-\frac{1}{2}\cdot\frac{1}{2^2}+\frac{1}{3}\cdot\frac{1}{2^3}-\frac{1}{4}\cdot\frac{1}{2^4}+\cdots.$

9. $\frac{1}{\log 2}-\frac{1}{\log 3}+\frac{1}{\log 4}-\cdots.$

10. $\frac{1}{2^2}+\frac{1}{3^2}+\frac{1}{4^2}+\cdots.$

**Show that the following four series are divergent:**

11. $\frac{1}{2}+\frac{1}{4}+\frac{1}{6}+\cdots.$

12. $1+\frac{1+2}{1+2^2}+\frac{1+3}{1+3^2}+\frac{1+4}{1+4^2}+\cdots.$

13. $\frac{\underline{|2}}{10}+\frac{\underline{|3}}{10^2}+\frac{\underline{|4}}{10^3}+\cdots.$

14. $1+\frac{1}{3}+\frac{1}{5}+\frac{1}{7}+\cdots.$

**142. Power series.** A series of ascending integral powers of a variable, say $x$, of the form

$$(A)\qquad a_0+a_1x+a_2x^2+a_3x^3+\cdots,$$

where the coefficients $a_0$, $a_1$, $a_2$, $\cdots$ are independent of $x$, is called a *power series in x.* Such series are of prime importance in the further study of the Calculus.

In special cases a power series in $x$ may converge for all values of $x$, but in general it will converge for some values of $x$ and be divergent for other values of $x$. We shall examine $(A)$ only for the case when the coefficients are such that

$$\underset{n=\infty}{\text{limit}}\left(\frac{a_{n+1}}{a_n}\right)=L,$$

where $L$ is a definite number. In $(A)$

$$\underset{n=\infty}{\text{limit}}\left(\frac{u_{n+1}}{u_n}\right)=\underset{n=\infty}{\text{limit}}\left(\frac{a_{n+1}x^{n+1}}{a_nx^n}\right)=\underset{n=\infty}{\text{limit}}\left(\frac{a_{n+1}}{a_n}\right)\cdot x=Lx.$$

Referring to tests I, II, III, on p. 221, we have in this case $\rho=Lx$, and hence the series $(A)$ is

I. Absolutely convergent when $|Lx|<1$, or $|x|<\left|\frac{1}{L}\right|$;

II. Divergent when $|Lx|>1$, or $|x|>\left|\frac{1}{L}\right|$;

III. No test when $|Lx|=1$, or $|x|=\left|\frac{1}{L}\right|$.

We may then write down the following

**General directions for finding the interval of convergence of the power series,**

$$(A) \qquad a_0 + a_1 x + a_2 x^2 + a_3 x^3 + \cdots.$$

FIRST STEP. *Write down the series formed by coefficients, namely,*

$$a_0 + a_1 + a_2 + a_3 + \cdots + a_n + a_{n+1} + \cdots.$$

SECOND STEP. *Calculate the limit*

$$\underset{n=\infty}{\text{limit}} \left( \frac{a_{n+1}}{a_n} \right) = L.$$

THIRD STEP. *Then the power series (A) is*

I. *Absolutely convergent for all values of x lying between*

$$-\left| \frac{1}{L} \right| \text{ and } + \left| \frac{1}{L} \right|.$$

II. *Divergent for all values of x less than* $-\left|\frac{1}{L}\right|$ *or greater than* $+\left|\frac{1}{L}\right|$.

III. *No test when* $x = \pm \left|\frac{1}{L}\right|$; *but then we substitute these two values of x in the power series (A) and apply to them the general directions on p. 221.*

NOTE. *When* $L = 0$, $\pm \left|\frac{1}{L}\right| = \pm\infty$ *and the power series is absolutely convergent for all values of x.*

ILLUSTRATIVE EXAMPLE 1. Find the interval of convergence for the series

$$(B) \qquad x - \frac{x^2}{2^2} + \frac{x^3}{3^2} - \frac{x^4}{4^2} + \cdots.$$

**Solution.** *First step.* The series formed by the coefficients is

$$(C) \qquad 1 - \frac{1}{2^2} + \frac{1}{3^2} - \frac{1}{4^2} + \cdots.$$

*Second step.* $\underset{n=\infty}{\text{limit}} \left( \frac{a_{n+1}}{a_n} \right) = \underset{n=\infty}{\text{limit}} \left[ -\frac{n^2}{(n+1)^2} \right] = \frac{\infty}{\infty}.$

Differentiating, $\underset{n=\infty}{\text{limit}} \left( -\frac{2n}{2(n+1)} \right) = \frac{\infty}{\infty}.$

Differentiating again, $\underset{n=\infty}{\text{limit}} \left( -\frac{2}{2} \right) = -1 \ (= L).$

*Third step.* $\left| \frac{1}{L} \right| = \left| \frac{1}{-1} \right| = 1.$

By I the series is absolutely convergent when $x$ lies between $-1$ and $+1$.
By II the series is divergent when $x$ is less than $-1$ or greater than $+1$.
By III there is no test when $x = \pm 1$.

Substituting $x = 1$ in $(B)$, we get

$$1 - \frac{1}{2^2} + \frac{1}{3^2} - \frac{1}{4^2} + \cdots$$

which is an alternating series that converges.

Substituting $x = -1$ in $(B)$, we get

$$-1 - \frac{1}{2^2} - \frac{1}{3^2} - \frac{1}{4^2} - \cdots,$$

which is convergent by comparison with the *p series* $(p > 1)$.

The series in the above example is said to have $[-1, 1]$ as the *interval of convergence*. This may be written $-1 \leqq x \leqq 1$, or indicated graphically as follows:

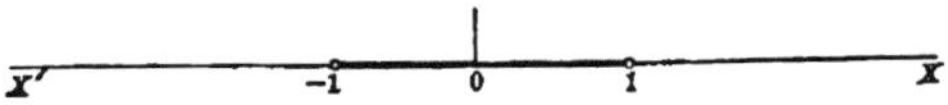


## EXAMPLES

For what values of the variable are the following series convergent?

Graphical representations of intervals of convergence *

15. $1 + x + x^2 + x^3 + \cdots$. *Ans.* $-1 < x < 1$.

16. $x - \frac{x^2}{2} + \frac{x^3}{3} - \frac{x^4}{4} + \cdots$. *Ans.* $-1 < x \leqq 1$.

17. $x + x^4 + x^9 + x^{16} + \cdots$. *Ans.* $-1 < x < 1$.

18. $x + \frac{x^2}{\sqrt{2}} + \frac{x^3}{\sqrt{3}} + \cdots$. *Ans.* $-1 \leqq x < 1$.

19. $1 + x + \frac{x^2}{\underline{|2}} + \frac{x^3}{\underline{|3}} + \cdots$. *Ans.* All values of $x$.

20. $1 - \frac{\theta^2}{\underline{|2}} + \frac{\theta^4}{\underline{|4}} - \frac{\theta^6}{\underline{|6}} + \cdots$. *Ans.* All values of $\theta$.

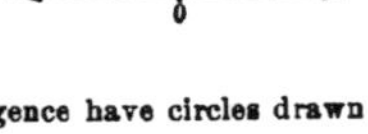


21. $\phi - \frac{\phi^3}{\underline{|3}} + \frac{\phi^5}{\underline{|5}} - \frac{\phi^7}{\underline{|7}} + \cdots$. *Ans.* All values of $\phi$.




22. $\frac{\sin \alpha}{1^2} - \frac{\sin 3\alpha}{3^2} + \frac{\sin 5\alpha}{5^2} - \cdots$.




*Ans.* All values of $\alpha$.

* End points that are not included in the interval of convergence have circles drawn about them.

Graphical representations of intervals of convergence *

**23.** $\frac{\cos x}{e^x} + \frac{\cos 2x}{e^{2x}} + \frac{\cos 3x}{e^{3x}} + \cdots.$ *Ans.* $x > 0$.

HINT. Neither the sine nor cosine can exceed 1 numerically.

**24.** $1 + x \log a + \frac{x^2 \log^2 a}{\underline{|2}} + \frac{x^3 \log^3 a}{\underline{|3}} + \cdots.$

*Ans.* All values of $x$.

**25.** $\frac{1}{1+x} + \frac{1}{1+x^2} + \frac{1}{1+x^3} + \cdots.$ *Ans.* $x > 1$.

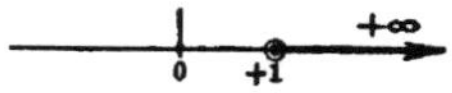


**26.** $1 + x + 2x^2 + 3x^3 + \cdots.$

**27.** $x - \frac{x^3}{3} + \frac{x^5}{5} - \frac{x^7}{7} + \cdots.$

**28.** $10x + 100x^2 + 1000x^3 + \cdots.$

**29.** $1 + x + \underline{|2}\, x^2 + \underline{|3}\, x^3 + \cdots.$

* End points that are not included in the interval of convergence have circles drawn about them.

## CHAPTER XVIII

### EXPANSION OF FUNCTIONS

**143. Introduction.** The student is already familiar with some methods of expanding certain functions into series. Thus, by the Binomial Theorem,

$$(A) \qquad (a+x)^4 = a^4 + 4a^3x + 6a^2x^2 + 4ax^3 + x^4,$$

giving a finite power series from which the exact value of $(a+x)^4$ for any value of $x$ may be calculated. Also by actual division,

$$(B) \qquad \frac{1}{1-x} = 1 + x + x^2 + x^3 + \cdots + x^{n-1} + \left(\frac{1}{1-x}\right)x^n,$$

we get an equivalent series, all of whose coefficients except that of $x^n$ are constants, $n$ being a positive integer.

Suppose we wish to calculate the value of this function when $x = .5$, not by substituting directly in

$$\frac{1}{1-x},$$

but by substituting $x = .5$ in the equivalent series

$$(C) \qquad (1 + x + x^2 + x^3 + \cdots + x^{n-1}) + \left(\frac{1}{1-x}\right)x^n.$$

Assuming $n = 8$, $(C)$ gives for $x = .5$

$$(D) \qquad \frac{1}{1-x} = 1.9921875 + .0078125.$$

If we then assume the value of the function to be the sum of the first eight terms of series $(C)$, the error we make is .0078125. However, in case we need the value of the function correct to two decimal places only, the number 1.99 is as close an approximation to the true value as we care for, since the error is less than .01. It is evident that if a greater degree of accuracy is desired, all we need to do is to use more terms of the power series

$$(E) \qquad 1 + x + x^2 + x^3 + \cdots.$$

Since, however, we see at once that

$$\left[\frac{1}{1-x}\right]_{x=.5}=2,$$

there is no necessity for the above discussion, except for purposes of illustration. As a matter of fact the process of computing the value of a function from an equivalent series into which it has been expanded is of the greatest practical importance, the values of the elementary transcendental functions such as the sine, cosine, logarithm, etc., being computed most simply in this way.

So far we have learned how to expand only a few special forms into series; we shall now consider a method of expansion applicable to an extensive and important class of functions and called *Taylor's Theorem.*

**144. Taylor's Theorem**[*] **and Taylor's Series.** Replacing $b$ by $x$ in $(E)$, p. 167, the *extended theorem of the mean* takes on the form

$$(61)\quad f(x)=f(a)+\frac{(x-a)}{\underline{|1}}f'(a)+\frac{(x-a)^2}{\underline{|2}}f''(a)+\frac{(x-a)^3}{\underline{|3}}f'''(a)+\cdots$$
$$+\frac{(x-a)^{n-1}}{\underline{|n-1}}f^{(n-1)}(a)+\frac{(x-a)^n}{\underline{|n}}f^{(n)}(x_1),$$

where $x_1$ lies between $a$ and $x$. (61), which is one of the most far-reaching theorems in the Calculus, is called *Taylor's Theorem.* We see that it expresses $f(x)$ as the sum of a finite series in $(x-a)$.

The last term in (61), namely $\frac{(x-a)^n}{\underline{|n}}f^{(n)}(x_1)$, is sometimes called the *remainder in Taylor's Theorem after n terms.* If this remainder converges toward zero as the number of terms increases without limit, then the right-hand side of (61) becomes an infinite power series called *Taylor's Series.*[†] In that case we may write (61) in the form

$$(62)\quad f(x)=f(a)+\frac{(x-a)}{\underline{|1}}f'(a)+\frac{(x-a)^2}{\underline{|2}}f''(a)+\frac{(x-a)^3}{\underline{|3}}f'''(a)+\cdots,$$

and we say that *the function has been expanded into a Taylor's Series.* For all values of $x$ for which the remainder approaches zero as $n$ increases without limit, this series converges and its sum gives the exact value of $f(x)$, because the difference (= the remainder) between the function and the sum of $n$ terms of the series approaches the limit zero (§ 15, p. 13).

[*] Also known as *Taylor's Formula.*

[†] Published by Dr. Brook Taylor (1685–1731) in his *Methodus Incrementorum,* London, 1715.

If the series converges for values of $x$ for which the remainder does not approach zero as $n$ increases without limit, then the limit of the sum of the series is not equal to the function $f(x)$.

*The infinite series* (**62**) *represents the function for those values of $x$, and those only, for which the remainder approaches zero as the number of terms increases without limit.*

It is usually easier to determine the interval of convergence of the series than that for which the remainder approaches zero; *but in simple cases the two intervals are identical.*

When the values of a function and its successive derivatives are known for some value of the variable, as $x = a$, then (**62**) is used for finding the value of the function for values of $x$ near $a$, and (**62**) is also called *the expansion of $f(x)$ in the vicinity of $x = a$.*

ILLUSTRATIVE EXAMPLE 1. Expand $\log x$ in powers of $(x - 1)$.

**Solution.**

$$f(x) = \log x, \quad f(1) = 0;$$
$$f'(x) = \frac{1}{x}, \quad f'(1) = 1;$$
$$f''(x) = -\frac{1}{x^2}, \quad f''(1) = -1;$$
$$f'''(x) = \frac{2}{x^3}, \quad f'''(1) = 2.$$
$$\cdots\cdots\cdots$$

Substituting in (62), $\log x = x - 1 - \frac{1}{2}(x-1)^2 + \frac{1}{3}(x-1)^3 - \cdots$. *Ans.*

This converges for values of $x$ between 0 and 2 and is the *expansion of $\log x$ in the vicinity of $x = 1$*, the remainder converging to zero.

When a function of the sum of two numbers $a$ and $x$ is given, say $f(a+x)$, it is frequently desirable to expand the function into a power series in one of them, say $x$. For this purpose we use another form of Taylor's Series, got by replacing $x$ by $a + x$ in (**62**), namely,

$$\textbf{(63)} \quad f(a+x) = f(a) + \frac{x}{\underline{|1}} f'(a) + \frac{x^2}{\underline{|2}} f''(a) + \frac{x^3}{\underline{|3}} f'''(a) + \cdots.$$

ILLUSTRATIVE EXAMPLE 1. Expand $\sin(a + x)$ in powers of $x$.

**Solution.** Here $f(a + x) = \sin(a + x)$.

Hence, placing

$$x = 0,$$
$$f(a) = \sin a,$$
$$f'(a) = \cos a,$$
$$f''(a) = -\sin a,$$
$$f'''(a) = -\cos a,$$
$$\cdots\cdots$$

Substituting in (61),

$$\sin(a + x) = \sin a + \frac{x}{1}\cos a - \frac{x^2}{\underline{|2}}\sin a - \frac{x^3}{\underline{|3}}\cos a + \cdots. \quad \textit{Ans.}$$

**EXAMPLES***

1. Expand $e^x$ in powers of $x-2$. *Ans.* $e^x = e^2 + e^2(x-2) + \frac{e^2}{\underline{|2}}(x-2)^2 + \cdots$.

2. Expand $x^3 - 2x^2 + 5x - 7$ in powers of $x-1$.
*Ans.* $-3 + 4(x-1) + (x-1)^2 + (x-1)^3$.

3. Expand $3y^2 - 14y + 7$ in powers of $y-3$. *Ans.* $-8 + 4(y-3) + 3(y-3)^2$.

4. Expand $5z^2 + 7z + 3$ in powers of $z-2$. *Ans.* $37 + 27(z-2) + 5(z-2)^2$.

5. Expand $4x^3 - 17x^2 + 11x + 2$ in powers of $x-4$.

6. Expand $5y^4 + 6y^3 - 17y^2 + 18y - 20$ in powers of $y+4$.

7. Expand $e^x$ in powers of $x+1$.

8. Expand $\sin x$ in powers of $x-\alpha$.

9. Expand $\cos x$ in powers of $x-\alpha$.

10. Expand $\cos(a+x)$ in powers of $x$.
*Ans.* $\cos(a+x) = \cos a - x\sin a - \frac{x^2}{\underline{|2}}\cos a + \frac{x^3}{\underline{|3}}\sin a + \cdots$.

11. Expand $\log(x+h)$ in powers of $x$.
*Ans.* $\log(x+h) = \log h + \frac{x}{h} - \frac{x^2}{2h^2} + \frac{x^3}{3h^3} + \cdots$.

12. Expand $\tan(x+h)$ in powers of $h$.
*Ans.* $\tan(x+h) = \tan x + h\sec^2 x + h^2\sec^2 x\tan x + \cdots$.

13. Expand the following in powers of $h$.

(a) $(x+h)^n = x^n + nx^{n-1}h + \frac{n(n-1)}{\underline{|2}}x^{n-2}h^2 + \frac{n(n-1)(n-2)}{\underline{|3}}x^{n-3}h^3 + \cdots$.

(b) $e^{x+h} = e^x\left(1 + h + \frac{h^2}{\underline{|2}} + \frac{h^3}{\underline{|3}} + \cdots\right)$.

**145. Maclaurin's Theorem and Maclaurin's Series.** A particular case of Taylor's Theorem is found by placing $a = 0$ in (61), p. 228, giving

$$(64)\qquad f(x) = f(0) + \frac{x}{\underline{|1}}f'(0) + \frac{x^2}{\underline{|2}}f''(0) + \frac{x^3}{\underline{|3}}f'''(0) + \cdots + \frac{x^{n-1}}{\underline{|n-1}}f^{(n-1)}(0) + \frac{x^n}{\underline{|n}}f^{(n)}(x_1),$$

where $x_1$ lies between 0 and $x$. (64) is called *Maclaurin's Theorem.* The right-hand member is evidently a series in $x$ in the same sense that (61), p. 228, is a series in $x - a$.

Placing $a = 0$ in (62), p. 228, we get *Maclaurin's Series*,†

$$(65)\qquad f(x) = f(0) + \frac{x}{\underline{|1}}f'(0) + \frac{x^2}{\underline{|2}}f''(0) + \frac{x^3}{\underline{|3}}f'''(0) + \cdots,$$

* In these examples we assume that the functions can be developed into a power series.

† Named after Colin Maclaurin (1698–1746), being first published in his *Treatise of Fluxions*, Edinburgh, 1742. The series is really due to Stirling (1692–1770).

a special case of Taylor's Series that is very useful. The statements made concerning the remainder and the convergence of Taylor's Series apply with equal force to Maclaurin's Series, the latter being merely a special case of the former.

The student should not fail to note the importance of such an expansion as (65). In all practical computations results correct to a certain number of decimal places are sought, and since the process in question replaces a function perhaps difficult to calculate by *an ordinary polynomial with constant coefficients*, it is very useful in simplifying such computations. Of course we must use terms enough to give the desired degree of accuracy.

In the case of an alternating series (§ 139, p. 218) the error made by stopping at any term is numerically less than that term, since the sum of the series after that term is numerically less than that term.

ILLUSTRATIVE EXAMPLE 1. Expand $\cos x$ into an infinite power series and determine for what values of $x$ it converges.

**Solution.** Differentiating first and then placing $x = 0$, we get

$$\begin{array}{ll} f(x) = \cos x, & f(0) = 1, \\ f'(x) = -\sin x, & f'(0) = 0, \\ f''(x) = -\cos x, & f''(0) = -1, \\ f'''(x) = \sin x, & f'''(0) = 0, \\ f^{\text{iv}}(x) = \cos x, & f^{\text{iv}}(0) = 1, \\ f^{\text{v}}(x) = -\sin x, & f^{\text{v}}(0) = 0, \\ f^{\text{vi}}(x) = -\cos x, & f^{\text{vi}}(0) = -1, \\ \text{etc.,} & \text{etc.} \end{array}$$

Substituting in (65),

$$(A) \qquad \cos x = 1 - \frac{x^2}{\underline{|2}} + \frac{x^4}{\underline{|4}} - \frac{x^6}{\underline{|6}} + \cdots.$$

Comparing with Ex. 20, p. 225, we see that the series converges for all values of $x$. In the same way for $\sin x$.

$$(B) \qquad \sin x = x - \frac{x^3}{\underline{|3}} + \frac{x^5}{\underline{|5}} - \frac{x^7}{\underline{|7}} + \cdots,$$

which converges for all values of $x$ (Ex. 21, p. 225).*

* Since here $f^{(n)}(x) = \sin\left(x + \frac{n\pi}{2}\right)$ and $f^{(n)}(x_1) = \sin\left(x_1 + \frac{n\pi}{2}\right)$, we have, by substituting in the last term of (64), p. 231,

$$\text{remainder} = \frac{x^n}{\underline{|n}} \sin\left(x_1 + \frac{n\pi}{2}\right). \qquad 0 < x_1 < x$$

But $\sin\left(x_1 + \frac{n\pi}{2}\right)$ can never exceed unity, and from Ex. 19, p. 225, $\underset{n=\infty}{\text{limit}} \frac{x^n}{\underline{|n}} = 0$ for all values of $x$. Hence

$$\underset{n=\infty}{\text{limit}} \frac{x^n}{\underline{|n}} \sin\left(x_1 + \frac{n\pi}{2}\right) = 0$$

for all values of $x$; that is, in this case *the limit of the remainder is 0 for all values of $x$ for which the series converges*. This is also the case for all the functions considered in this book.

ILLUSTRATIVE EXAMPLE 2. Using the series (*B*) found in the last example, calculate sin 1 correct to four decimal places.

**Solution.** Here $x = 1$ radian; that is, the angle is expressed in circular measure. Therefore, substituting $x = 1$ in (*B*) of the last example,

$$\sin 1 = 1 - \frac{1}{\underline{|3}} + \frac{1}{\underline{|5}} - \frac{1}{\underline{|7}} + \cdots.$$

Summing up the positive and negative terms separately,

$$\begin{array}{ll} 1 = 1.00000\cdots & \frac{1}{\underline{|3}} = 0.16667\cdots \\ \frac{1}{\underline{|5}} = 0.00833\cdots & \frac{1}{\underline{|7}} = 0.00019\cdots \\ \hline 1.00833\cdots & 0.16686\cdots \end{array}$$

Hence $\qquad \sin 1 = 1.00833 - 0.16686 = 0.84147\cdots,$

which is correct to five decimal places, since the error made must be less than $\frac{1}{\underline{|9}}$; i.e. less than .000003. Obviously the value of sin 1 may be calculated to any desired degree of accuracy by simply including a sufficient number of additional terms.

## EXAMPLES

Verify the following expansions of functions into power series by Maclaurin's Series and determine for what values of the variable they are convergent:

1. $e^x = 1 + x + \frac{x^2}{\underline{|2}} + \frac{x^3}{\underline{|3}} + \frac{x^4}{\underline{|4}} + \cdots.$ Convergent for all values of $x$.

2. $\cos x = 1 - \frac{x^2}{\underline{|2}} + \frac{x^4}{\underline{|4}} - \frac{x^6}{\underline{|6}} + \frac{x^8}{\underline{|8}} - \cdots.$ Convergent for all values of $x$.

3. $a^x = 1 + x\log a + \frac{x^2\log^2 a}{\underline{|2}} + \frac{x^3\log^3 a}{\underline{|3}} + \cdots.$ Convergent for all values of $x$.

4. $\sin kx = kx - \frac{k^3x^3}{\underline{|3}} + \frac{k^5x^5}{\underline{|5}} - \frac{k^7x^7}{\underline{|7}} + \cdots.$ Convergent for all values of $x$, $k$ being any constant.

5. $e^{-kx} = 1 - kx + \frac{k^2x^2}{\underline{|2}} - \frac{k^3x^3}{\underline{|3}} + \frac{k^4x^4}{\underline{|4}} - \cdots.$ Convergent for all values of $x$, $k$ being any constant.

6. $\log(1+x) = x - \frac{x^2}{2} + \frac{x^3}{3} - \frac{x^4}{4} + \frac{x^5}{5} - \cdots.$ Convergent if $-1 < x \leqq 1$.

7. $\log(1-x) = -x - \frac{x^2}{2} - \frac{x^3}{3} - \frac{x^4}{4} - \frac{x^5}{5} - \cdots.$ Convergent if $-1 \leqq x < 1$.

8. $\text{arc}\sin x = x + \frac{1\cdot x^3}{2\cdot 3} + \frac{1\cdot 3x^5}{2\cdot 4\cdot 5} + \cdots.$ Convergent if $-1 \leqq x \leqq 1$.

9. $\text{arc}\tan x = x - \frac{x^3}{3} + \frac{x^5}{5} - \frac{x^7}{7} + \frac{x^9}{9} - \cdots.$ Convergent if $-1 \leqq x \leqq 1$.

10. $\sin^2 x = x^2 - \frac{2x^4}{\underline{|3}} + \frac{32x^6}{\underline{|6}} + \cdots.$ Convergent for all values of $x$.

11. $e^{\sin\phi} = 1 + \phi + \frac{\phi^2}{2} - \frac{\phi^4}{8} + \cdots.$ Convergent for all values of $\phi$.

12. $e^\theta\sin\theta = \theta + \theta^2 + \frac{\theta^3}{3} - \frac{4\theta^5}{\underline{|5}} - \frac{8\theta^6}{\underline{|6}} - \cdots.$ Convergent for all values of $\theta$.

**13.** Find three terms of the expansion in each of the following functions:

(a) $\tan x$. (b) $\sec x$. (c) $e^{\cos x}$. (d) $\cos 2x$. (e) $\text{arc} \cos x$. (f) $a^{-x}$.

**14.** Show that $\log x$ cannot be expanded by Maclaurin's Theorem.

Compute the values of the following functions by substituting directly in the equivalent power series, taking terms enough until the results agree with those given below.

**15.** $e = 2.7182 \cdots$.

**Solution.** Let $x = 1$ in series of Ex. 1; then

$$e = 1 + 1 + \frac{1}{\underline{|2}} + \frac{1}{\underline{|3}} + \frac{1}{\underline{|4}} + \frac{1}{\underline{|5}} + \cdots.$$

| | | |
|---|---|---|
| First term | $= 1.00000$ | |
| Second term | $= 1.00000$ | |
| Third term | $= 0.50000$ | |
| Fourth term | $= 0.16667 \cdots$ | (Dividing third term by 3.) |
| Fifth term | $= 0.04167 \cdots$ | (Dividing fourth term by 4.) |
| Sixth term | $= 0.00833 \cdots$ | (Dividing fifth term by 5.) |
| Seventh term | $= 0.00139 \cdots$ | (Dividing sixth term by 6.) |
| Eighth term | $= 0.00019 \cdots$, etc. | (Dividing seventh term by 7.) |
| Adding, | $e = 2.71825 \cdots$ *Ans.* | |

**16.** $\text{arc} \tan (\frac{1}{2}) = 0.1973 \cdots$; use series in Ex. 9.

**17.** $\cos 1 = 0.5403 \cdots$; use series in Ex. 2.

**18.** $\cos 10^\circ = 0.9848 \cdots$; use series in Ex. 2.

**19.** $\sin .1 = .0998 \cdots$; use series $x - \frac{x^3}{\underline{|3}} + \frac{x^5}{\underline{|5}} - \frac{x^7}{\underline{|7}} + \cdots$.

**20.** $\text{arc} \sin 1 = 1.5708 \cdots$; use series in Ex. 8.

**21.** $\sin \frac{\pi}{4} = 0.7071 \cdots$; use series (*B*), p. 231.

**22.** $\sin .5 = 0.4794 \cdots$; use series (*B*), p. 231.

**23.** $e^2 = 1 + 2 + \frac{2^2}{\underline{|2}} + \frac{2^3}{\underline{|3}} + \cdots = 7.3891.$

**24.** $\sqrt{e} = 1 + \frac{1}{2} + \frac{1}{2^2\underline{|2}} + \frac{1}{2^3\underline{|3}} + \cdots = 1.6487.$

In more advanced treatises it is shown that, for values of $x$ within the interval of convergence, the sum of a power series is differentiable and that its derivative is obtained by differentiating the series term by term as in an ordinary sum. Thus from (*B*), p. 231,

$$\sin x = x - \frac{x^3}{\underline{|3}} + \frac{x^5}{\underline{|5}} - \frac{x^7}{\underline{|7}} + \cdots.$$

Differentiating both sides, we get

$$\cos x = 1 - \frac{x^2}{\underline{|2}} + \frac{x^4}{\underline{|4}} - \frac{x^6}{\underline{|6}} + \cdots,$$

which is the series of Ex. 2, p. 232. This illustrates how *we may obtain a new power series from a given power series by differentiation.*

Differentiating the power series of Ex. 6, p. 232, we obtain

$$\frac{1}{1+x}=1-x+x^2-x^3+x^4-\cdots.$$

In the same way from Ex. 8, p. 232,

$$\frac{1}{\sqrt{1-x^2}}=1+\frac{1}{2}x^2+\frac{1\cdot 3}{2\cdot 4}x^4+\frac{1\cdot 3\cdot 5}{2\cdot 4\cdot 6}x^6+\cdots.$$

**146. Computation by series.** I. *Alternating series.* Exs. 15–24 of the last exercise illustrate to what use series may be put for purposes of computation. Obviously it is very important to know the percentage of error in a result, since the computation must necessarily stop at some term in the series, the sum of the subsequent terms being thereby neglected. The *absolute error* made is of course equal to the limit of the sum of all the neglected terms. In some series this error is difficult to find, but in the case of alternating series it has been shown in § 140, p. 220, that the sum is less than the first of these terms. Hence the absolute error made is *less* than the first term neglected. Fortunately a large proportion of the series used for computation purposes are alternating series, and therefore this easy method for finding the upper limit of the absolute error and the percentage of error is available. Let us illustrate by means of an example.

ILLUSTRATIVE EXAMPLE 1. Determine the greatest possible error and percentage of error made in computing the numerical value of the sine of one radian from the sine series,

$$\sin x = x-\frac{x^3}{\underline{|3}}+\frac{x^5}{\underline{|5}}-\frac{x^7}{\underline{|7}}+\cdots;$$

(a) when all terms beyond the second are neglected;
(b) when all terms beyond the third are neglected.

**Solution.** Let $x=1$ in series; then

$$\sin 1 = 1-\frac{1}{\underline{|3}}+\frac{1}{\underline{|5}}-\frac{1}{\underline{|7}}+\cdots.$$

(a) Using only the first two terms,

$$\sin 1 = 1-\tfrac{1}{6}=\tfrac{5}{6}=.8333,$$

the absolute error is less than $\frac{1}{\underline{|5}}$; i.e. $<\frac{1}{120}$ $(=.0083)$, and the percentage of error is less than 1 per cent.*

* Since $.0083 \div .8333 = .01$.

(b) Using only the first three terms,

$$\sin 1 = 1 - \tfrac{1}{6} + \tfrac{1}{120} = .841666,$$

the absolute error is less than $\frac{1}{\lfloor 7}$; i.e. $< \frac{1}{5040}$ (= .000198), and the percentage of error is less than $\frac{1}{40}$ of 1 per cent.*

Moreover, the *exact value* of sin 1 lies between .8333 and .841666, since for an alternating series $S_n$ is alternately greater and less than $\lim\limits_{n=\infty} S_n$.

**EXAMPLES**

Determine the greatest possible error and percentage of error made in computing the numerical value of each of the following functions from its corresponding series

(a) when all terms beyond the second are neglected;

(b) when all terms beyond the third are neglected.

| | | |
|---|---|---|
| 1. cos 1. | 4. arc tan 1. | 7. $e^{-\frac{1}{2}}$. |
| 2. sin 2. | 5. $e^{-2}$. | 8. arc tan 2. |
| 3. cos $\frac{1}{2}$. | 6. $\sin\frac{\pi}{3}$. | 9. sin 15°. |

II. *The computation of π by series.*

From Ex. 8, p. 232, we have

$$\text{arc}\sin x = x + \frac{1\cdot x^3}{2\cdot 3} + \frac{1\cdot 3\, x^5}{2\cdot 4\cdot 5} + \frac{1\cdot 3\cdot 5\, x^7}{2\cdot 4\cdot 6\cdot 7} + \cdots.$$

Since this series converges for values of $x$ between $-1$ and $+1$, we may let $x = \frac{1}{2}$, giving

$$\frac{\pi}{6} = \frac{1}{2} + \frac{1}{2}\cdot\frac{1}{3}\left(\frac{1}{2}\right)^3 + \frac{1\cdot 3}{2\cdot 4}\cdot\frac{1}{5}\left(\frac{1}{2}\right)^5 + \cdots,$$

or $$\pi = 3.1415\cdots.$$

Evidently we might have used the series of Ex. 9, p. 232, instead. Both of these series converge rather slowly, but there are other series, found by more elaborate methods, by means of which the correct value of π to a large number of decimal places may be easily calculated.

III. *The computation of logarithms by series.*

Series play a very important rôle in making the necessary calculations for the construction of logarithmic tables.

From Ex. 6, p. 232, we have

$$(A) \qquad \log(1+x) = x - \frac{x^2}{2} + \frac{x^3}{3} - \frac{x^4}{4} + \frac{x^5}{5} - \cdots.$$

* Since .000198 ÷ .841666 = .00023.

This series converges for $x=1$, and we can find log 2 by placing $x=1$ in $(A)$, giving

$$\log 2 = 1 - \tfrac{1}{2} + \tfrac{1}{3} - \tfrac{1}{4} + \tfrac{1}{5} - \tfrac{1}{6} + \cdots.$$

But this series is not well adapted to numerical computation, because it converges so slowly that it would be necessary to take 2000 terms in order to get the value of log 2 correct to three decimal places. A rapidly converging series for computing logarithms will now be deduced.

By the theory of logarithms,

$$(B) \qquad \log\frac{1+x}{1-x} = \log(1+x) - \log(1-x). \qquad \text{By 8, p. 2}$$

Substituting in $(B)$ the equivalent series for $\log(1+x)$ and $\log(1-x)$ found in Exs. 6 and 7 on p. 232, we get*

$$(C) \qquad \log\frac{1+x}{1-x} = 2\left[x + \frac{x^3}{3} + \frac{x^5}{5} + \frac{x^7}{7} + \cdots\right],$$

which is convergent when $x$ is numerically less than unity. Let

$$(D) \qquad \frac{1+x}{1-x} = \frac{M}{N}, \quad \text{whence} \quad x = \frac{M-N}{M+N},$$

and we see that $x$ will always be numerically less than unity for all positive values of $M$ and $N$. Substituting from $(D)$ into $(C)$, we get

$$(E) \qquad \log\frac{M}{N} = \log M - \log N$$
$$= 2\left[\frac{M-N}{M+N} + \frac{1}{3}\left(\frac{M-N}{M+N}\right)^3 + \frac{1}{5}\left(\frac{M-N}{M+N}\right)^5 + \cdots\right],$$

a series which is convergent for all positive values of $M$ and $N$; and it is always possible to choose $M$ and $N$ so as to make it converge rapidly.

Placing $M=2$ and $N=1$ in $(E)$, we get

$$\log 2 = 2\left[\frac{1}{3} + \frac{1}{3}\cdot\frac{1}{3^3} + \frac{1}{5}\cdot\frac{1}{3^5} + \frac{1}{7}\cdot\frac{1}{3^7} + \cdots\right] = 0.69314718\cdots.$$

$$\left[\text{Since } \log N = \log 1 = 0, \text{ and } \frac{M-N}{M+N} = \frac{1}{3}.\right]$$

* The student should notice that we have treated the series as if they were ordinary sums, but they are not; they are *limits* of sums. To justify this step is beyond the scope of this book.

Placing $M=3$ and $N=2$ in $(E)$, we get

$$\log 3 = \log 2 + 2\left[\frac{1}{5} + \frac{1}{3}\cdot\frac{1}{5^3} + \frac{1}{5}\cdot\frac{1}{5^5} + \cdots\right] = 1.09861229\cdots.$$

It is only necessary to compute the logarithms of prime numbers in this way, the logarithms of composite numbers being then found by using theorems 7–10, p. 1. Thus

$$\log 8 = \log 2^3 = 3\log 2 = 2.07944154\cdots,$$
$$\log 6 = \log 3 + \log 2 = 1.79175947\cdots.$$

All the above are *Napierian* or *natural logarithms*, i.e. the base is $e=2.7182818$. If we wish to find *Briggs's* or *common logarithms*, where the base 10 is employed, all we need to do is to change the base by means of the formula

$$\log_{10} n = \frac{\log_e n}{\log_e 10}.$$

Thus $$\log_{10} 2 = \frac{\log_e 2}{\log_e 10} = \frac{0.693\cdots}{2.302\cdots} = 0.301\cdots.$$

In the actual computation of a table of logarithms only a few of the tabulated values are calculated from series, all the rest being found by employing theorems in the theory of logarithms and various ingenious devices designed for the purpose of saving work.

### EXAMPLES

Calculate by the methods of this article the following logarithms:

1. $\log_e 5 = 1.6094\cdots$.
2. $\log_e 10 = 2.3026\cdots$.
3. $\log_e 24 = 3.1781\cdots$.
4. $\log_{10} 5 = 0.6990\cdots$.

**147. Approximate formulas derived from series. Interpolation.** In the two preceding sections we evaluated a function from its equivalent power series by substituting the given value of $x$ in a certain number of the first terms of that series, the number of terms taken depending on the degree of accuracy required. It is of great practical importance to note that this really means that *we are considering the function as approximately equal to an ordinary polynomial with constant coefficients.* For example, consider the series

$$(A) \qquad \sin x = x - \frac{x^3}{\lfloor 3} + \frac{x^5}{\lfloor 5} - \frac{x^7}{\lfloor 7} + \cdots.$$

This is an alternating series for both positive and negative values of $x$. Hence the error made if we assume $\sin x$ to be approximately equal to the sum of the first $n$ terms is numerically less than the $(n+1)$th term (§ 139, p. 218). For example, assume

$$(B) \qquad \sin x = x,$$

and let us find for what values of $x$ this is correct to three places of decimals. To do this, set

$$(C) \qquad \left|\frac{x^3}{\underline{|3}}\right| < .001.$$

This gives $x$ numerically less than $\sqrt[3]{.006}\,(=.1817)$; i.e. $(B)$ is correct to three decimal places when $x$ lies between $+10.4°$ and $-10.4°$.

The error made in neglecting all terms in $(A)$ after the one in $x^{n-1}$ is given by the remainder (see (**64**), p. 230)

$$(D) \qquad R = \frac{x^n}{\underline{|n}} f^{(n)}(x_1);$$

hence we can find for what values of $x$ a polynomial represents the functions to any desired degree of accuracy by writing the inequality

$$(E) \qquad |R| < \textit{limit of error},$$

and solving for $x$, provided we know the maximum value of $f^{(n)}(x_1)$. Thus if we wish to find for what values of $x$ the formula

$$(F) \qquad \sin x = x - \frac{x^3}{6}$$

is correct to two decimal places (i.e. error $< .01$), knowing that $|f^{(v)}(x_1)| \leqq 1$, we have, from $(D)$ and $(E)$,

$$\frac{|x^5|}{120} < .01; \text{ i.e. } |x| < \sqrt[5]{1.2}; \text{ or } |x| \leqq 1.$$

Therefore $x - \frac{x^3}{6}$ gives the correct value of $\sin x$ to two decimal places if $|x| \leqq 1$; i.e. if $x$ lies between $+57°$ and $-57°$. This agrees with the discussion of $(A)$ as an alternating series.

Since in a great many practical problems accuracy to two or three decimal places only is required, the usefulness of such approximate formulas as $(B)$ and $(F)$ is apparent.

Again, if we expand $\sin x$ by Taylor's Series, (**62**), p. 228, in powers of $x - a$, we get

$$\sin x = \sin a + \cos a\,(x-a) - \frac{\sin a}{\underline{|2}}(x-a)^2 + \cdots,$$

**Hence for all values of $x$ *in the neighborhood of some fixed value* $a$ we have the approximate formula**

$$(G) \qquad \sin x = \sin a + \cos a\,(x - a).$$

Transposing $\sin a$ and dividing by $x - a$, we get

$$\frac{\sin x - \sin a}{x - a} = \cos a.$$

Since $\cos a$ is constant, this means that

*The change in the value of the sine is proportional to the change in the angle for values of the angle near* $a$.

For example, let $a = 30° = .5236$ radians, and suppose it is required to calculate the sines of 31° and 32° by the approximate formula $(G)$. Then

$$\begin{aligned}\sin 31° &= \sin 30° + \cos 30° (.01745)^{*}\\ &= .5000 + .8660 \times .01745\\ &= .5000 + .0151\\ &= .5151.\end{aligned}$$

Similarly, $\sin 32° = \sin 30° + \cos 30° (.03490) = .5302$.

In general, then, by Taylor's Series, we have the approximate formula

$$(H) \qquad f(x) = f(a) + f'(a)(x - a).$$

If the constant $f'(a) \neq 0$, this formula asserts that *the ratio of the increments of function and variable for all values of the latter differing little from the fixed value* $a$ *is constant.*

Care must, however, be observed in applying $(H)$; for while the absolute error made in using it in a given case may be small, the percentage of error may be so large that the results are worthless.

Using one more term in Taylor's Series we get the approximate formula

$$(I) \qquad f(x) = f(a) + f'(a)(x - a) + \frac{f''(a)}{\underline{|2}}(x - a)^2.$$

* $x - a = 1° = .01745$ radian.

The values of sin 31° and sin 32° calculated on p. 239 from $(G)$ are correct to only three decimal places. If greater accuracy than this is desired, we may use $(I)$, which gives, for $f(x) = \sin x$,

$$(J) \qquad \sin x = \sin a + \cos a (x - a) - \frac{\sin a}{\lfloor 2}(x-a)^2.$$

Let $\qquad a = 30° = .5236$ radian.

Then
$$\sin 31° = \sin 30° + \cos 30°(.01745) - \frac{\sin 30°}{2}(.01745)^2$$
$$= .50000 + .01511 - .00008$$
$$= .51503.$$
$$\sin 32° = \sin 30° + \cos 30°(.03490) - \frac{\sin 30°}{2}(.03490)^2$$
$$= .50000 + .03022 - .00030$$
$$= .52992.$$

These results are correct to four decimal places.

**EXAMPLES**

1. Using formula $(H)$ for interpolation by first differences, calculate the following functions:

(a) cos 61°, taking $a = 60°$. (c) sin 85.1°, taking $a = 85°$.
(b) tan 46°, taking $a = 45°$. (d) cot 70.3°, taking $a = 70°$.

2. Using formula $(I)$ for interpolation by second differences, calculate the following functions:

(a) sin 11°, taking $a = 10°$. (c) cot 15.2°, taking $a = 15°$.
(b) cos 86°, taking $a = 85°$. (d) tan 69°, taking $a = 70°$.

3. Draw the graphs of the functions $x$, $x - \frac{x^3}{\lfloor 3}$, $x - \frac{x^3}{\lfloor 3} + \frac{x^5}{\lfloor 5}$ respectively, and compare them with the graph of $\sin x$.

**148. Taylor's Theorem for functions of two or more variables.** The scope of this book will allow only an elementary treatment of the expansion of functions involving more than one variable by Taylor's Theorem. The expressions for the remainder are complicated and will not be written down.

Having given the function

$$(A) \qquad f(x, y),$$

it is required to expand the function

$$(B) \qquad f(x+h, y+k)$$

in powers of $h$ and $k$.

Consider the function

$$(C) \qquad f(x+ht, y+kt).$$

Evidently $(B)$ is the value of $(C)$ when $t=1$. Considering $(C)$ as a function of $t$, we may write

$$(D) \qquad f(x+ht,\ y+kt)=F(t),$$

which may then be expanded in powers of $t$ by Maclaurin's Theorem, (**64**), p. 230, giving

$$(E) \qquad F(t)=F(0)+tF'(0)+\frac{t^2}{\lfloor 2}F''(0)+\frac{t^3}{\lfloor 3}F'''(0)+\cdots.$$

Let us now express the successive derivatives of $F(t)$ with respect to $t$ in terms of the partial derivatives of $F(t)$ with respect to $x$ and $y$. Let

$$(F) \qquad \alpha=x+ht, \quad \beta=y+kt;$$

then by (**51**), p. 195,

$$(G) \qquad F'(t)=\frac{\partial F}{\partial \alpha}\frac{d\alpha}{dt}+\frac{\partial F}{\partial \beta}\frac{d\beta}{dt}.$$

But from $(F)$,

$$(H) \qquad \frac{d\alpha}{dt}=h \quad \text{and} \quad \frac{d\beta}{dt}=k;$$

and since $F(t)$ is a function of $x$ and $y$ through $\alpha$ and $\beta$,

$$\frac{\partial F}{\partial x}=\frac{\partial F}{\partial \alpha}\frac{\partial \alpha}{\partial x} \quad \text{and} \quad \frac{\partial F}{\partial y}=\frac{\partial F}{\partial \beta}\frac{\partial \beta}{\partial y};$$

or, since from $(F)$, $\dfrac{\partial \alpha}{\partial x}=1$ and $\dfrac{\partial \beta}{\partial y}=1$,

$$(I) \qquad \frac{\partial F}{\partial x}=\frac{\partial F}{\partial \alpha} \quad \text{and} \quad \frac{\partial F}{\partial y}=\frac{\partial F}{\partial \beta}.$$

Substituting in $(G)$ from $(I)$ and $(H)$,

$$(J) \qquad F'(t)=h\frac{\partial F}{\partial x}+k\frac{\partial F}{\partial y}.$$

Replacing $F(t)$ by $F'(t)$ in $(J)$, we get

$$F''(t)=h\frac{\partial F'}{\partial x}+k\frac{\partial F'}{\partial y}=h\left\{h\frac{\partial^2 F}{\partial x^2}+k\frac{\partial^2 F}{\partial x \partial y}\right\}+k\left\{h\frac{\partial^2 F}{\partial x \partial y}+k\frac{\partial^2 F}{\partial y^2}\right\}.$$

$$(K) \qquad \therefore F''(t)=h^2\frac{\partial^2 F}{\partial x^2}+2hk\frac{\partial^2 F}{\partial x \partial y}+k^2\frac{\partial^2 F}{\partial y^2}.$$

In the same way the third derivative is

$$(L) \qquad F'''(t)=h^3\frac{\partial^3 F}{\partial x^3}+3h^2k\frac{\partial^3 F}{\partial x^2 \partial y}+3hk^2\frac{\partial^3 F}{\partial x \partial y^2}+k^3\frac{\partial^3 F}{\partial y^3},$$

and so on for higher derivatives.

When $t=0$, we have from $(D)$, $(G)$, $(J)$, $(K)$, $(L)$,

$$F(0)=f(x, y), \text{ i.e. } F(t) \text{ is replaced by } f(x, y),$$

$$F'(0)=h\frac{\partial f}{\partial x}+k\frac{\partial f}{\partial y},$$

$$F''(0)=h^2\frac{\partial^2 f}{\partial x^2}+2\,hk\frac{\partial^2 f}{\partial x\partial y}+k^2\frac{\partial^2 f}{\partial y^2},$$

$$F'''(0)=h^3\frac{\partial^3 f}{\partial x^3}+3\,h^2k\frac{\partial^3 f}{\partial x^2\partial y}+3\,hk^2\frac{\partial^3 f}{\partial x\partial y^2}+k^3\frac{\partial^3 f}{\partial y^3},$$

and so on.

Substituting these results in $(E)$, we get

$$(66)\qquad f(x+ht, y+kt)=f(x, y)+t\left(h\frac{\partial f}{\partial x}+k\frac{\partial f}{\partial y}\right)$$
$$+\frac{t^2}{\underline{|2}}\left(h^2\frac{\partial^2 f}{\partial x^2}+2\,hk\frac{\partial^2 f}{\partial x\partial y}+k^2\frac{\partial^2 f}{\partial y^2}\right)+\cdots.$$

To get $f(x+h, y+k)$, replace $t$ by 1 in (66), giving *Taylor's Theorem for a function of two independent variables,*

$$(67)\qquad f(x+h, y+k)=f(x, y)+h\frac{\partial f}{\partial x}+k\frac{\partial f}{\partial y}$$
$$+\frac{1}{\underline{|2}}\left(h^2\frac{\partial^2 f}{\partial x^2}+2\,hk\frac{\partial^2 f}{\partial x\partial y}+k^2\frac{\partial^2 f}{\partial y^2}\right)+\cdots,$$

which is the required expansion in powers of $h$ and $k$. Evidently (67) is also adapted to the expansion of $f(x+h, y+k)$ in powers of $x$ and $y$ by simply interchanging $x$ with $h$ and $y$ with $k$. Thus

$$(67\,a)\qquad f(x+h, y+k)=f(h, k)+x\frac{\partial f}{\partial h}+y\frac{\partial f}{\partial k}$$
$$+\frac{1}{\underline{|2}}\left(x^2\frac{\partial^2 f}{\partial h^2}+2\,xy\frac{\partial^2 f}{\partial h\partial k}+y^2\frac{\partial^2 f}{\partial k^2}\right)+\cdots.$$

Similarly, for three variables we shall find

$$(68)\quad f(x+h, y+k, z+l)=f(x, y, z)+h\frac{\partial f}{\partial x}+k\frac{\partial f}{\partial y}+l\frac{\partial f}{\partial z}$$
$$+\frac{1}{\underline{|2}}\left(h^2\frac{\partial^2 f}{\partial x^2}+k^2\frac{\partial^2 f}{\partial y^2}+l^2\frac{\partial^2 f}{\partial z^2}+2\,hk\frac{\partial^2 f}{\partial x\partial y}\right.$$
$$\left.+2\,lh\frac{\partial^2 f}{\partial z\partial x}+2\,kl\frac{\partial^2 f}{\partial y\partial z}\right)+\cdots,$$

and so on for any number of variables.

**EXAMPLES**

1. Given $f(x, y) \equiv Ax^2 + Bxy + Cy^2$, expand $f(x+h, y+k)$ in powers of $h$ and $k$.

**Solution.**

$$\frac{\partial f}{\partial x} = 2Ax + By, \quad \frac{\partial f}{\partial y} = Bx + 2Cy;$$

$$\frac{\partial^2 f}{\partial x^2} = 2A, \quad \frac{\partial^2 f}{\partial x \partial y} = B, \quad \frac{\partial^2 f}{\partial y^2} = 2C.$$

The third and higher partial derivatives are all zero. Substituting in (67),

$$f(x+h, y+k) \equiv Ax^2 + Bxy + Cy^2 + (2Ax + By)h + (Bx + 2Cy)k$$
$$+ Ah^2 + Bhk + Ck^2. \quad \textit{Ans.}$$

2. Given $f(x, y, z) \equiv Ax^2 + By^2 + Cz^2$, expand $f(x+l, y+m, z+n)$ in powers of $l, m, n$.

**Solution.**

$$\frac{\partial f}{\partial x} = 2Ax, \quad \frac{\partial f}{\partial y} = 2By, \quad \frac{\partial f}{\partial z} = 2Cz;$$

$$\frac{\partial^2 f}{\partial x^2} = 2A, \quad \frac{\partial^2 f}{\partial y^2} = 2B, \quad \frac{\partial^2 f}{\partial z^2} = 2C, \quad \frac{\partial^2 f}{\partial x \partial y} = \frac{\partial^2 f}{\partial y \partial z} = \frac{\partial^2 f}{\partial z \partial x} = 0.$$

The third and higher partial derivatives are all zero. Substituting in (68),

$$f(x+l, y+m, z+n) \equiv Ax^2 + By^2 + Cz^2 + 2Axl + 2Bym + 2Czn$$
$$+ Al^2 + Bm^2 + Cn^2. \quad \textit{Ans.}$$

3. Given $f(x, y) \equiv \sqrt{x} \tan y$, expand $f(x+h, y+k)$ in powers of $h$ and $k$.

4. Given $f(x, y, z) \equiv Ax^2 + By^2 + Cz^2 + Dxy + Eyz + Fzx$, expand $f(x+h, y+k, z+l)$ in powers of $h, k, l$.

**149. Maxima and minima of functions of two independent variables.** The function $f(x, y)$ is said to be a *maximum* at $x = a$, $y = b$ when $f(a, b)$ is greater than $f(x, y)$ for all values of $x$ and $y$ in the neighborhood of $a$ and $b$. Similarly, $f(a, b)$ is said to be a *minimum* at $x = a$, $y = b$ when $f(a, b)$ is less than $f(x, y)$ for all values of $x$ and $y$ in the neighborhood of $a$ and $b$.

These definitions may be stated in analytical form as follows:

If, for all values of $h$ and $k$ numerically less than some small positive quantity,

(*A*) $f(a+h, b+k) - f(a, b) =$ *a negative number*, then $f(a, b)$ is a *maximum* value of $f(x, y)$. If

(*B*) $f(a+h, b+k) - f(a, b) =$ *a positive number*, then $f(a, b)$ is a *minimum* value of $f(x, y)$.

These statements may be interpreted geometrically as follows: a point $P$ on the surface

$$z = f(x, y)$$

is a maximum point when it is "higher" than *all* other points on the surface in its neighborhood, the coördinate plane $XOY$ being assumed horizontal. Similarly, $P'$ is a minimum point on the surface when it is "lower" than *all* other points on the surface in its neighborhood. It is therefore evident that all vertical planes through $P$ cut the surface in curves (as $APB$ or $DPE$ in the figure), each of which has a maximum ordinate $z (= MP)$ at $P$. In the same manner all vertical planes through $P'$ cut the surface in curves (as $BP'C$ or $FP'G$), each of which has a minimum ordinate $z (= NP')$ at $P'$. Also,

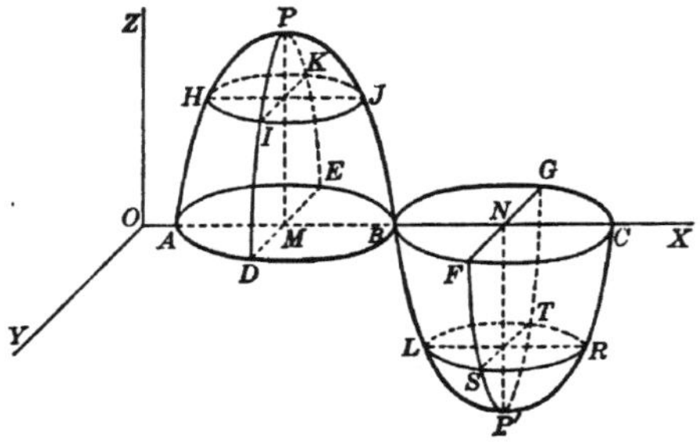


any contour (as $HIJK$) cut out of the surface by a horizontal plane in the immediate neighborhood of $P$ must be a small closed curve. Similarly, we have the contour $LSRT$ near the minimum point $P'$.

It was shown in §§ 81, 82, pp. 108, 109, that a *necessary* condition that a function of one variable should have a maximum or a minimum for a given value of the variable was that its first derivative should be zero for the given value of the variable. Similarly, for a function $f(x, y)$ of two independent variables, a *necessary* condition that $f(a, b)$ should be a maximum or a minimum (i.e. a turning value) is that for $x = a$, $y = b$,

$$(C) \qquad \frac{\partial f}{\partial x} = 0, \qquad \frac{\partial f}{\partial y} = 0.$$

**Proof.** Evidently $(A)$ and $(B)$ must hold when $k = 0$; that is,

$$f(a + h, b) - f(a, b)$$

is always negative or always positive for all values of $h$ sufficiently small numerically. By §§ 81, 82, a necessary condition for this is that $\frac{d}{dx} f(x, b)$ shall vanish for $x = a$, or, what amounts to the same thing, $\frac{\partial}{\partial x} f(x, y)$ shall vanish for $x = a$, $y = b$. Similarly, $(A)$ and $(B)$ must hold when $h = 0$, giving as a second necessary condition that $\frac{\partial}{\partial y} f(x, y)$ shall vanish for $x = a$, $y = b$.

In order to determine *sufficient* conditions that $f(a, b)$ shall be a maximum or a minimum, it is necessary to proceed to higher derivatives. To derive sufficient conditions for all cases is beyond the scope of this book.* The following discussion, however, will suffice for all the problems given here.

Expanding $f(a+h, b+k)$ by Taylor's Theorem, (67), p. 242, replacing $x$ by $a$ and $y$ by $b$, we get

$$(D)\quad f(a+h, b+k)=f(a, b)+h\frac{\partial f}{\partial x}+k\frac{\partial f}{\partial y}$$
$$+\frac{1}{\underline{|2}}\left(h^2\frac{\partial^2 f}{\partial x^2}+2\,hk\frac{\partial^2 f}{\partial x\partial y}+k^2\frac{\partial^2 f}{\partial y^2}\right)+R,$$

where the partial derivatives are evaluated for $x=a$, $y=b$, and $R$ denotes the sum of all the terms not written down. All such terms are of a degree higher than the second in $h$ and $k$.

Since $\frac{\partial f}{\partial x}=0$ and $\frac{\partial f}{\partial y}=0$, from $(C)$, p. 244, we get, after transposing $f(a, b)$,

$$(E)\quad f(a+h, b+k)-f(a, b)=\frac{1}{2}\left(h^2\frac{\partial^2 f}{\partial x^2}+2\,hk\frac{\partial^2 f}{\partial x\partial y}+k^2\frac{\partial^2 f}{\partial y^2}\right)+R.$$

If $f(a, b)$ is a turning value, the expression on the left-hand side of $(E)$ must retain the same sign for all values of $h$ and $k$ sufficiently small in numerical value,—the negative sign for a maximum value (see $(A)$, p. 243) and the positive sign for a minimum value (see $(B)$, p. 243); i.e. $f(a, b)$ will be a maximum or a minimum according as the right-hand side of $(E)$ is negative or positive. Now $R$ is of a degree higher than the second in $h$ and $k$. Hence as $h$ and $k$ diminish in numerical value, it seems plausible to conclude that *the numerical value of $R$ will eventually become and remain less than the numerical value of the sum of the three terms of the second degree written down on the right-hand side of $(E)$*.† Then the sign of the right-hand side (and therefore also of the left-hand side) will be the same as the sign of the expression

$$(F)\qquad h^2\frac{\partial^2 f}{\partial x^2}+2\,hk\frac{\partial^2 f}{\partial x\partial y}+k^2\frac{\partial^2 f}{\partial y^2}.$$

But from Algebra we know that the quadratic expression

$$h^2A+2\,hkC+k^2B$$

always has the same sign as $A$ (or $B$) when $AB-C^2>0$.

* See *Cours d'Analyse*, Vol. I, by C. Jordan.

† Peano has shown that this conclusion does not always hold. See the article on "Maxima and Minima of Functions of Several Variables," by Professor James Pierpont in the *Bulletin of the American Mathematical Society*, Vol. IV.

Applying this to $(F)$, $A=\frac{\partial^2 f}{\partial x^2}$, $B=\frac{\partial^2 f}{\partial y^2}$, $C=\frac{\partial^2 f}{\partial x \partial y}$, and we see that $(F)$, and therefore also the left-hand member of $(E)$, has the same sign as $\frac{\partial^2 f}{\partial x^2}\left(\text{or } \frac{\partial^2 f}{\partial y^2}\right)$ when

$$\frac{\partial^2 f}{\partial x^2}\frac{\partial^2 f}{\partial y^2}-\left(\frac{\partial^2 f}{\partial x \partial y}\right)^2>0.$$

Hence the following **rule for finding maximum and minimum values of a function $f(x, y)$.**

FIRST STEP. *Solve the simultaneous equations*

$$\frac{\partial f}{\partial x}=0, \qquad \frac{\partial f}{\partial y}=0.$$

SECOND STEP. *Calculate for these values of $x$ and $y$ the value of*

$$\Delta=\frac{\partial^2 f}{\partial x^2}\frac{\partial^2 f}{\partial y^2}-\left(\frac{\partial^2 f}{\partial x \partial y}\right)^2.$$

THIRD STEP. *The function will have a*

**maximum** *if* $\Delta>0$ and $\frac{\partial^2 f}{\partial x^2}\left(\text{or } \frac{\partial^2 f}{\partial y^2}\right)<0$;

**minimum** *if* $\Delta>0$ and $\frac{\partial^2 f}{\partial x^2}\left(\text{or } \frac{\partial^2 f}{\partial y^2}\right)>0$;

**neither a maximum nor a minimum** *if* $\Delta<0$.

**The question is undecided** *if* $\Delta=0$.*

The student should notice that this *rule* does not necessarily give *all* maximum and minimum values. For a pair of values of $x$ and $y$ determined by the First Step may cause $\Delta$ to vanish, and may lead to a maximum or a minimum or neither. Further investigation is therefore necessary for such values. The rule is, however, sufficient for solving many important examples.

The question of maxima and minima of functions of three or more independent variables must be left to more advanced treatises.

ILLUSTRATIVE EXAMPLE 1. Examine the function $3axy-x^3-y^3$ for maximum and minimum values.

**Solution.** $$f(x, y)=3axy-x^3-y^3.$$

***First step.*** $$\frac{\partial f}{\partial x}=3ay-3x^2=0, \qquad \frac{\partial f}{\partial y}=3ax-3y^2=0.$$

Solving these two simultaneous equations, we get

$$x=0, \quad y=0; \qquad x=a, \quad y=a.$$

* The discussion of the text merely renders the given rule plausible. The student should observe that the case $\Delta=0$ is omitted in the discussion.

*Second step.* $\frac{\partial^2 f}{\partial x^2} = -6x,\quad \frac{\partial^2 f}{\partial x \partial y} = 3a,\quad \frac{\partial^2 f}{\partial y^2} = -6y;$

$$\Delta = \frac{\partial^2 f}{\partial x^2}\frac{\partial^2 f}{\partial y^2} - \left(\frac{\partial^2 f}{\partial x \partial y}\right)^2 = 36xy - 9a^2.$$

***Third step.*** When $x = 0$ and $y = 0$, $\Delta = -9a^2$, and there can be neither a maximum nor a minimum at (0, 0).

When $x = a$ and $y = a$, $\Delta = +27a^2$; and since $\frac{\partial^2 f}{\partial x^2} = -6a$, we have the conditions for a maximum value of the function fulfilled at $(a, a)$. Substituting $x = a$, $y = a$ in the given function, we get its maximum value equal to $a^3$.

ILLUSTRATIVE EXAMPLE 2. Divide $a$ into three parts such that their product shall be a maximum.

**Solution.** Let $x$ = first part, $y$ = second part; then $a - (x + y) = a - x - y$ = third part, and the function to be examined is

$$f(x, y) = xy(a - x - y).$$

*First step.* $\frac{\partial f}{\partial x} = ay - 2xy - y^2 = 0,\quad \frac{\partial f}{\partial y} = ax - 2xy - x^2 = 0.$

Solving simultaneously, we get as one pair of values $x = \frac{a}{3}$, $y = \frac{a}{3}$.*

*Second step.* $\frac{\partial^2 f}{\partial x^2} = -2y,\quad \frac{\partial^2 f}{\partial x \partial y} = a - 2x - 2y,\quad \frac{\partial^2 f}{\partial y^2} = -2x;$

$$\Delta = 4xy - (a - 2x - 2y)^2.$$

*Third step.* When $x = \frac{a}{3}$ and $y = \frac{a}{3}$, $\Delta = \frac{a^2}{3}$; and since $\frac{\partial^2 f}{\partial x^2} = -\frac{2a}{3}$, it is seen that our product is a maximum when $x = \frac{a}{3}$, $y = \frac{a}{3}$. Therefore the third part is also $\frac{a}{3}$, and the maximum value of the product is $\frac{a^3}{27}$.

## EXAMPLES

1. Find the minimum value of $x^2 + xy + y^2 - ax - by$. *Ans.* $\frac{1}{3}(ab - a^2 - b^2)$.

2. Show that $\sin x + \sin y + \cos(x + y)$ is a minimum when $x = y = \frac{3\pi}{2}$, and a maximum when $x = y = \frac{\pi}{6}$.

3. Show that $xe^{y + x\sin y}$ has neither a maximum nor a minimum.

4. Show that the maximum value of $\frac{(ax + by + c)^2}{x^2 + y^2 + 1}$ is $a^2 + b^2 + c^2$.

5. Find the greatest rectangular parallelepiped that can be inscribed in an ellipsoid. That is, find the maximum value of $8xyz$ (= volume) subject to the condition

$$\frac{x^2}{a^2} + \frac{y^2}{b^2} + \frac{z^2}{c^2} = 1. \qquad \textit{Ans. } \frac{8abc}{3\sqrt{3}}.$$

HINT. Let $u = xyz$, and substitute the value of $z$ from the equation of the ellipsoid. This gives

$$u^2 = x^2y^2c^2\left(1 - \frac{x^2}{a^2} - \frac{y^2}{b^2}\right),$$

where $u$ is a function of only two variables.

* $x = 0$, $y = 0$ are not considered, since from the nature of the problem we would then have a minimum.

6. Show that the surface of a rectangular parallelepiped of given volume is least when the solid is a cube.

7. Examine $x^4 + y^4 - x^2 + xy - y^2$ for maximum and minimum values.

*Ans.* Maximum when $x = 0$, $y = 0$;
minimum when $x = y = \pm \frac{1}{2}$, and when $x = - y = \pm \frac{1}{2}\sqrt{3}$.

8. Show that the most economical dimensions for a rectangular tank to hold a given volume are a square base and a depth equal to one half the side of the base.

9. The electric time constant of a cylindrical coil of wire is

$$u = \frac{mxyz}{ax + by + cz},$$

where $x$ is the mean radius, $y$ is the difference between the internal and external radii, $z$ is the axial length, and $m$, $a$, $b$, $c$ are known constants. The volume of the coil is $nxyz = g$. Find the values of $x$, $y$, $z$ which make $u$ a maximum if the volume of the coil is fixed.

*Ans.* $ax = by = cz = \sqrt[3]{\frac{abcg}{n}}$.

# CHAPTER XIX

## ASYMPTOTES. SINGULAR POINTS

**150. Rectilinear asymptotes.** An *asymptote* to a curve is the limiting position* of a tangent whose point of contact moves off to an infinite distance from the origin.†

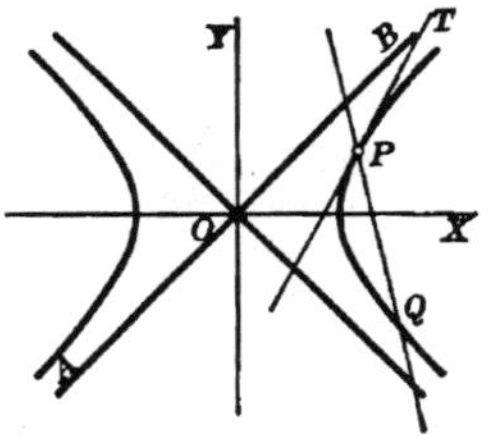


Thus, in the hyperbola, the asymptote $AB$ is the limiting position of the tangent $PT$ as the point of contact $P$ moves off to the right to an infinite distance. In the case of algebraic curves the following definition is useful: an asymptote is the limiting position of a secant as two points of intersection of the secant with a branch of the curve move off in the same direction along that branch to an infinite distance. For example, the asymptote $AB$ is the limiting position of the secant $PQ$ as $P$ and $Q$ move upwards to an infinite distance.

**151. Asymptotes found by method of limiting intercepts.** The equation of the tangent to a curve at $(x_1, y_1)$ is, by (1), p. 76,

$$y - y_1 = \frac{dy_1}{dx_1}(x - x_1).$$

First placing $y = 0$ and solving for $x$, and then placing $x = 0$ and solving for $y$, and denoting the intercepts by $x_i$ and $y_i$ respectively, we get

$$x_i = x_1 - y_1\frac{dx_1}{dy_1} = \textit{intercept on } OX;$$

$$y_i = y_1 - x_1\frac{dy_1}{dx_1} = \textit{intercept on } OY.$$

Since an asymptote must pass within a finite distance of the origin, one or both of these intercepts must approach finite values as limits when the point of contact $(x_1, y_1)$ moves off to an infinite distance. If

$$\text{limit}\,(x_i) = a \quad \text{and} \quad \text{limit}\,(y_i) = b,$$

* A line that approaches a fixed straight line as a limiting position cannot be wholly at infinity; hence it follows that an asymptote must pass within a finite distance of the origin. It is evident that a curve which has no infinite branch can have no real asymptote.

† Or, less precisely, an asymptote to a curve is sometimes defined as a tangent whose point of contact is at an infinite distance.

then the equation of the asymptote is found by substituting the limiting values $a$ and $b$ in the equation

$$\frac{x}{a}+\frac{y}{b}=1.$$

If only one of these limits exists, but

$$\text{limit}\left(\frac{dy_1}{dx_1}\right)=m,$$

then we have one intercept and the slope given, so that the equation of the asymptote is

$$y=mx+b, \quad \text{or} \quad x=\frac{y}{m}+a.$$

ILLUSTRATIVE EXAMPLE 1. Find the asymptotes to the hyperbola $\frac{x^2}{a^2}-\frac{y^2}{b^2}=1.$

**Solution.** $\frac{dy}{dx}=\frac{b^2x}{a^2y}=\pm\frac{b}{a}\frac{1}{\sqrt{1-\frac{a^2}{x^2}}}$, and $m=\underset{x=\infty}{\text{limit}}\left(\frac{dy}{dx}\right)=\pm\frac{b}{a}.$

Also $x_i=\frac{a^2}{x}$ and $y_i=-\frac{b^2}{y}$; hence these intercepts are zero when $x=y=\infty$. Therefore the asymptotes pass through the origin (see figure on p. 249) and their equations are

$$y-0=\pm\frac{b}{a}(x-0x), \quad \text{or} \quad ay=\pm bx. \quad \textit{Ans.}$$

This method is frequently too complicated to be of practical use. The most convenient method of determining the asymptotes to algebraic curves is given in the next section.

**152. Method of determining asymptotes to algebraic curves.** Given the algebraic equation in two variables,

$$(A) \qquad f(x, y)=0.$$

If this equation when cleared of fractions and radicals is of degree $n$, then it may be arranged according to descending powers of one of the variables, say $y$, in the form

$$(B) \qquad ay^n+(bx+c)y^{n-1}+(dx^2+ex+f)y^{n-2}+\cdots=0.^*$$

For a given value of $x$ this equation determines in general $n$ values of $y$.

* For use in this section the attention of the student is called to the following theorem from Algebra: Given an algebraic equation of degree $n$,

$$Ay^n+By^{n-1}+Cy^{n-2}+Dy^{n-3}+\cdots=0.$$

When $A$ approaches zero, one root (value of $y$) approaches $\infty$.
When $A$ and $B$ approach zero, two roots approach $\infty$.
When $A$, $B$, and $C$ approach zero, three roots approach $\infty$, etc.

CASE I. **To determine the asymptotes to the curve (B) which are parallel to the coördinate axes.** Let us first investigate for asymptotes parallel to $OY$. The equation of any such asymptote is of the form

$$(C) \qquad x = k,$$

and it must have two points of intersection with $(B)$ having infinite ordinates.

*First.* Suppose $a$ is not zero in $(B)$, that is, the term in $y^n$ is present. Then for any finite value of $x$, $(B)$ gives $n$ values of $y$, *all finite.* Hence all such lines as $(C)$ will intersect $(B)$ in points having finite ordinates, and *there are no asymptotes parallel to* $OY$.

*Second.* Next suppose $a = 0$, but $b$ and $c$ are not zero. Then we know from Algebra that one root $(= y)$ of $(B)$ is infinite for every finite value of $x$; that is, any arbitrary line $(C)$ intersects $(B)$ at only one point having an infinite ordinate. If now, in addition,

$$bx + c = 0, \text{ or}$$

$$(D) \qquad x = -\frac{c}{b},$$

then the first two terms in $(B)$ will drop out, and hence two of its roots are infinite. That is, $(D)$ and $(B)$ intersect in two points having infinite ordinates, and therefore $(D)$ *is the equation of an asymptote to* $(B)$ *which is parallel to* $OY$.

*Third.* If $a = b = c = 0$, there are two values of $x$ that make $y$ in $(B)$ infinite, namely, those satisfying the equation

$$(E) \qquad dx^2 + ex + f = 0.$$

*Solving* $(E)$ *for* $x$, *we get two asymptotes parallel to* $OY$, and so on in general.

In the same way, by arranging $f(x, y)$ according to descending powers of $x$, we may find the asymptotes parallel to $OX$. Hence the following **rule for finding the asymptotes parallel to the coördinate axis:**

FIRST STEP. *Equate to zero the coefficient of the highest power of* $x$ *in the equation. This gives all asymptotes parallel to* $OX$.

SECOND STEP. *Equate to zero the coefficient of the highest power of* $y$ *in the equation. This gives all asymptotes parallel to* $OY$.

NOTE. *Of course if one or both of these coefficients do not involve* $x$ *(or* $y$*), they cannot be zero, and there will be no corresponding asymptote.*

ILLUSTRATIVE EXAMPLE 1. Find the asymptotes of the curve $a^2x = y(x-a)^2$.

**Solution.** Arranging the terms according to powers of $x$,

$$yx^2 - (2ay + a^2)x + a^2y = 0.$$

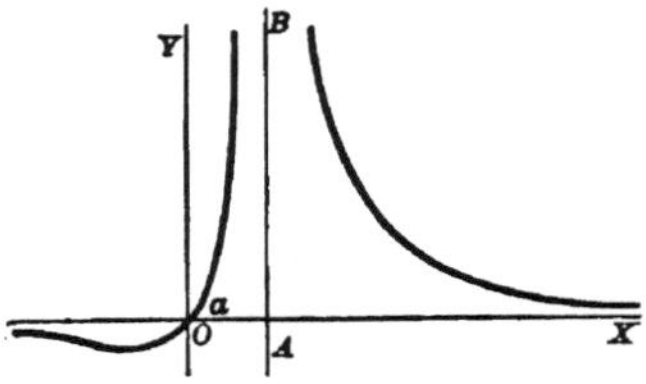


Equating to zero the coefficient of the highest power of $x$, we get $y = 0$ as the asymptote parallel to $OX$. In fact, the asymptote coincides with the axis of $x$. Arranging the terms according to the powers of $y$,

$$(x-a)^2y - a^2x = 0.$$

Placing the coefficient of $y$ equal to zero, we get $x = a$ twice, showing that $AB$ is a double asymptote parallel to $OY$. If this curve is examined for asymptotes oblique to the axes by the method explained below, it will be seen that there are none. Hence $y = 0$ and $x = a$ are the only asymptotes of the given curve.

CASE II. **To determine asymptotes oblique to the coördinate axes.** Given the algebraic equation

$$(F) \qquad f(x, y) = 0.$$

Consider the straight line

$$(G) \qquad y = mx + k.$$

It is required to determine $m$ and $k$ so that the line $(G)$ shall be an asymptote to the curve $(F)$.

Since an asymptote is the limiting position of a secant as two points of intersection on the same branch of the curve move off to an infinite distance, if we eliminate $y$ between $(F)$ and $(G)$, the resulting equation in $x$, namely,

$$(H) \qquad f(x, mx + k) = 0,$$

must have two infinite roots. But this requires that the coefficients of the two highest powers of $x$ shall vanish. Equating these coefficients to zero, we get two equations from which the required values of $m$ and $k$ may be determined. Substituting these values in $(G)$ gives the equation of an asymptote. Hence the following **rule for finding asymptotes oblique to the coördinate axes:**

FIRST STEP. *Replace $y$ by $mx + k$ in the given equation and expand.*

SECOND STEP. *Arrange the terms according to descending powers of $x$.*

THIRD STEP. *Equate to zero the coefficients of the two highest powers* of $x$, and solve for $m$ and $k$.*

* If the term involving $x^{n-1}$ is missing, or if the value of $m$ obtained by placing the first coefficient equal to zero causes the second coefficient to vanish, then by placing the coefficients of $x^n$ and $x^{n-2}$ equal to zero we obtain two equations from which the values of $m$ and $k$ may be found. In this case we shall, in general, obtain two $k$'s for each $m$, that is, pairs of parallel oblique asymptotes. Similarly, if the term in $x^{n-2}$ is also missing, each value of $m$ furnishes three parallel oblique asymptotes, and so on.

FOURTH STEP. *Substitute these values of m and k in*

$$y = mx + k.$$

*This gives the required asymptotes.*

ILLUSTRATIVE EXAMPLE 2. Examine $y^3 = 2ax^2 - x^3$ for asymptotes.

**Solution.** Since none of the terms involve both $x$ and $y$, it is evident that there are no asymptotes parallel to the coördinate axes. To find the oblique asymptotes, eliminate $y$ between the given equation and $y = mx + k$. This gives

$$(mx + k)^3 = 2ax^2 - x^3;$$

and arranging the terms in powers of $x$,

$$(1 + m^3)x^3 + (3m^2k - 2a)x^2 + 3k^2mx + k^3 = 0.$$

Placing the first two coefficients equal to zero,

$$1 + m^3 = 0 \quad \text{and} \quad 3m^2k - 2a = 0.$$

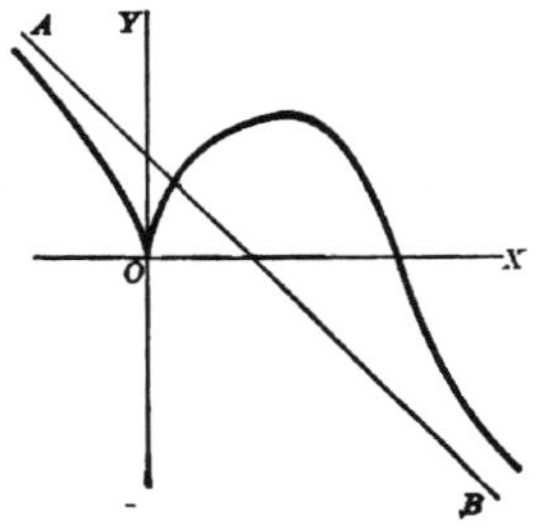


Solving, we get $m = -1$, $k = \frac{2a}{3}$. Substituting in $y = mx + k$, we have $y = -x + \frac{2a}{3}$, the equation of asymptote $AB$.

## EXAMPLES

Examine the first eight curves for asymptotes by the method of § 151, and the remaining ones by the method of § 152:

1. $y = e^x$. *Ans.* $y = 0$.
2. $y = e^{-x^2}$. *Ans.* $y = 0$.
3. $y = \log x$. *Ans.* $x = 0$.
4. $y = \left(1 + \frac{1}{x}\right)^x$. $y = e$, $x = -1$.
5. $y = \tan x$. $n$ being any odd integer, $x = \frac{n\pi}{2}$.
6. $y = e^{\frac{1}{x}} - 1$. $x = 0$, $y = 0$.
7. $y^3 = 6x^2 + x^3$. $y = x + 2$.
8. Show that the parabola has no asymptotes.
9. $y^3 = a^3 - x^3$. $y + x = 0$.
10. The cissoid $y^2 = \frac{x^3}{2r - x}$. $x = 2r$.
11. $y^2a = y^2x + x^3$. $x = a$.
12. $y^2(x^2 + 1) = x^2(x^2 - 1)$. $y = \pm x$.
13. $y^2(x - 2a) = x^3 - a^3$. $x = 2a$, $y = \pm(x + a)$.
14. $x^2y^2 = a^2(x^2 + y^2)$. $x = \pm a$, $y = \pm a$.
15. $y(x^2 - 3bx + 2b^2) = x^3 - 3ax^2 + a^3$. $x = b$, $x = 2b$ $y + 3a = x + 3b$.
16. $y = c + \frac{a^3}{(x - b)^2}$. $y = c$, $x = b$.
17. The folium $x^3 + y^3 - 3axy = 0$. $y + x + a = 0$.
18. The witch $x^2y = 4a^2(2a - y)$. $y = 0$.
19. $xy^2 + x^2y = a^3$. $x = 0$, $y = 0$, $x + y = 0$.
20. $x^3 + 2x^2y - xy^2 - 2y^3 + 4y^2 + 2xy + y = 1$. $x + 2y = 0$, $x + y = 1$, $x - y = -1$.

**153. Asymptotes in polar coördinates.** Let $f(\rho, \theta) = 0$ be the equation of the curve $PQ$ having the asymptote $CD$. As the asymptote must pass within a finite distance (as $OE$) of the origin, and the point of contact is at an infinite distance, it is evident that the radius vector $OF$ drawn to the point of contact is parallel to the asymptote, and the subtangent $OE$ is perpendicular to it. Or, more precisely, the distance of the asymptote from the origin is the limiting value of the polar subtangent as the point of contact moves off an infinite distance.

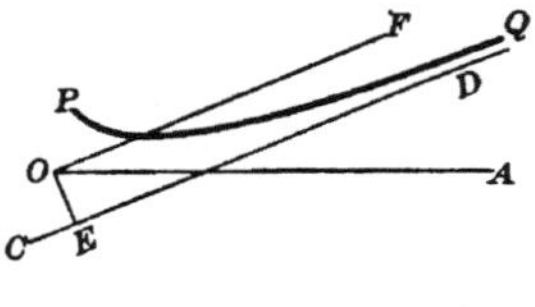


**To determine the asymptotes to a polar curve,** proceed as follows:

FIRST STEP. *Find from the equation of the curve the values of $\theta$ which make $\rho = \infty$.** *These values of $\theta$ give the directions of the asymptotes.*

SECOND STEP. *Find the limit of the polar subtangent*

$$\rho^2 \frac{d\theta}{d\rho}, \qquad \text{by (7), p. 86}$$

*as $\theta$ approaches each such value, remembering that $\rho$ approaches $\infty$ at the same time.*

THIRD STEP. *If the limiting value of the polar subtangent is finite, there is a corresponding asymptote at that distance from the origin and parallel to the radius vector drawn to the point of contact. When this limit is positive the asymptote is to the right of the origin, and when negative, to the left, looking in the direction of the infinite radius vector.*

**EXAMPLES**

1. Examine the hyperbolic spiral $\rho = \frac{a}{\theta}$ for asymptotes.

**Solution.** When $\theta = 0$, $\rho = \infty$. Also $\frac{d\rho}{d\theta} = -\frac{a}{\theta^2}$; hence

$$\text{subtangent} = \rho^2 \frac{d\theta}{d\rho} = \frac{a^2}{\theta^2} \cdot -\frac{\theta^2}{a} = -a.$$

$$\therefore \underset{\theta = 0}{\text{limit}} \left[\rho^2 \frac{d\theta}{d\rho}\right] = -a, \text{ which is finite.}$$

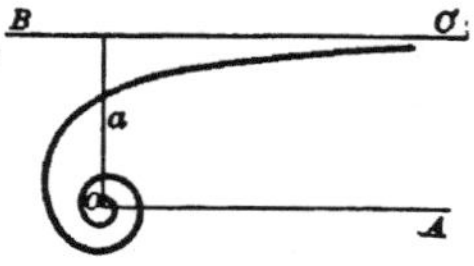


It happens in this case that the subtangent is the same for all values of $\theta$. The curve has therefore an asymptote $BC$ parallel to the initial line $OA$ and at a distance $a$ above it.

* If the equation can be written as a polynomial in $\rho$, these values of $\theta$ may be found by equating to zero the coefficient of the highest power of $\rho$.

**Examine the following curves for asymptotes:**

**2.** $\rho \cos\theta = a \cos 2\theta$.

*Ans.* There is an asymptote perpendicular to the initial line at a distance $a$ to the left of the origin.

**3.** $\rho = a \tan\theta$.

*Ans.* There are two asymptotes perpendicular to the initial line and at a distance $a$ from the origin, on either side of it.

**4.** The lituus $\rho\theta^{\frac{1}{2}} = a$. *Ans.* The initial line.

**5.** $\rho = a \sec 2\theta$.

*Ans.* There are four asymptotes at the same distance $\frac{a}{2}$ from the origin, and inclined 45° to the initial line.

**6.** $(\rho - a)\sin\theta = b$.

*Ans.* There is an asymptote parallel to the initial line at the distance $b$ above it.

**7.** $\rho = a(\sec 2\theta + \tan 2\theta)$.

*Ans.* Two asymptotes parallel to $\theta = \frac{\pi}{4}$, at distance $a$ on each side of origin.

**8.** Show that the initial line is an asymptote to two branches of the curve $\rho^2 \sin\theta = a^2 \cos 2\theta$.

**9.** Parabola $\rho = \dfrac{a}{1 - \cos\theta}$. *Ans.* There is no asymptote.

**154. Singular points.** Given a curve whose equation is

$$f(x, y) = 0.$$

Any point on the curve for which

$$\frac{\partial f}{\partial x} = 0 \quad \text{and} \quad \frac{\partial f}{\partial y} = 0$$

is called a *singular point* of the curve. All other points are called *ordinary points* of the curve. Since by (**57** *a*), p. 199, we have

$$\frac{dy}{dx} = -\frac{\dfrac{\partial f}{\partial x}}{\dfrac{\partial f}{\partial y}},$$

it is evident that at a singular point the direction of the curve (or tangent) is indeterminate, for the slope takes the form $\frac{0}{0}$. In the next section it will be shown how tangents at such points may be found.

**155. Determination of the tangent to an algebraic curve at a given point by inspection.** If we transform the given equation to a new set of parallel coördinate axes having as origin the point in question on the curve, we know that the new equation will have no constant term. Hence it may be written in the form

$$(A) \qquad f(x, y) = ax + by + (cx^2 + dxy + ey^2) + (fx^3 + gx^2y + hxy^2 + iy^3) + \cdots = 0.$$

The equation of a tangent to the curve at the given point (now the origin) will be

$$(B) \qquad y = \left(\frac{dy}{dx}\right)x. \qquad \text{By (1), p. 76}$$

Let $y = mx$ be the equation of a line through the origin $O$ and a second point $P$ on the locus of $(A)$. If then $P$ approaches $O$ along the curve, we have, from $(B)$,

$$(C) \qquad \text{limit } m = \frac{dy}{dx}.$$

Let $O$ be an ordinary point. Then, by § 155, $a$ and $b$ do not both vanish, since at (0, 0), from $(A)$, p. 255,

$$\frac{\partial f}{\partial x} = a, \qquad \frac{\partial f}{\partial y} = b.$$

Replace $y$ in $(A)$ by $mx$, divide out the factor $x$, and let $x$ approach zero as a limit. Then $(A)$ will become *

$$a + bm = 0.$$

Hence we have, from $(B)$ and $(C)$,

$$ax + by = 0,$$

the equation of the tangent. The left-hand member is seen to consist of the terms of the first degree in $(A)$.

When $O$ is not an ordinary point we have $a = b = 0$. Assume that $c, d, e$ do not all vanish. Then, proceeding as before (except that we divide out the factor $x^2$), we find, after letting $x$ approach the limit zero, that $(A)$ becomes

$$c + dm + em^2 = 0,$$

or, from $(C)$,

$$(D) \qquad c + d\left(\frac{dy}{dx}\right) + e\left(\frac{dy}{dx}\right)^2 = 0.$$

Substituting from $(B)$, we see that

$$(E) \qquad cx^2 + dxy + ey^2 = 0$$

is the equation of the pair of tangents at the origin. The left-hand member is seen to consist of the terms of the second degree in $(A)$. Such a singular point of the curve is called a *double point* from the fact that there are two tangents to the curve at that point.

* After dividing by $x$ an algebraic equation in $m$ remains whose coefficients are functions of $x$. If now $x$ approaches zero as a limit, the theorem holds that one root of this equation in $m$ will approach the limit $-a \div b$.

Since at (0, 0), from ($A$),

$$\frac{\partial^2 f}{\partial x^2} = 2c, \quad \frac{\partial^2 f}{\partial x \partial y} = d, \quad \frac{\partial^2 f}{\partial y^2} = 2e,$$

it is evident that ($D$) may be written in the form

$$(F) \qquad \frac{\partial^2 f}{\partial x^2} + 2\frac{\partial^2 f}{\partial x \partial y}\left(\frac{dy}{dx}\right) + \frac{\partial^2 f}{\partial y^2}\left(\frac{dy}{dx}\right)^2 = 0.$$

In the same manner, if

$$a = b = c = d = e = 0,$$

there is a *triple point* at the origin, the equation of the three tangents being

$$fx^3 + gx^2y + hxy^2 + iy^3 = 0,$$

and so on in general.

If we wish to investigate the appearance of a curve at a given point, it is of fundamental importance to solve the tangent problem for that point. The above results indicate that this can be done *by simple inspection* after we have transformed the origin to that point.

Hence we have the following **rule for finding the tangents at a given point.**

FIRST STEP. *Transform the origin to the point in question.*

SECOND STEP. *Arrange the terms of the resulting equation according to ascending powers of $x$ and $y$.*

THIRD STEP. *Set the group of terms of lowest degree equal to zero. This gives the equation of the tangents at the point (origin).*

ILLUSTRATIVE EXAMPLE 1. Find the equation of the tangent to the ellipse

$$5x^2 + 5y^2 + 2xy - 12x - 12y = 0$$

at the origin.

**Solution.** Placing the terms of lowest (first) degree equal to zero, we get

$$-12x - 12y = 0,$$

or

$$x + y = 0,$$

which is then the equation of the tangent $PT$ at the origin.

ILLUSTRATIVE EXAMPLE 2. Examine the curve $3x^2 - xy - 2y^2 + x^3 - 8y^3 = 0$ for tangents at the origin.

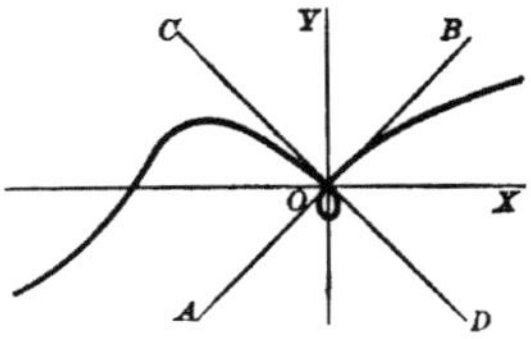


**Solution.** Placing the terms of lowest (second) degree equal to zero,

$$3x^2 - xy - 2y^2 = 0,$$

or

$$(x - y)(3x + 2y) = 0,$$

$x - y = 0$ being the equation of the tangent $AB$, and $3x + 2y = 0$ the equation of the tangent $CD$. The origin is, then, a double point of the curve.

Since the roots of the quadratic equation ($F$), p. 257, namely,

$$\frac{\partial^2 f}{\partial y^2}\left(\frac{dy}{dx}\right)^2 + 2\frac{\partial^2 f}{\partial x \partial y}\left(\frac{dy}{dx}\right) + \frac{\partial^2 f}{\partial x^2} = 0,$$

may be real and unequal, real and equal, or imaginary, there are three cases of double points to be considered, according as

$$(G) \qquad \left(\frac{\partial^2 f}{\partial x \partial y}\right)^2 - \frac{\partial^2 f}{\partial x^2}\frac{\partial^2 f}{\partial y^2}$$

is positive, zero, or negative (see 3, p. 1).

**156. Nodes.** $\left(\frac{\partial^2 f}{\partial x \partial y}\right)^2 - \frac{\partial^2 f}{\partial x^2}\frac{\partial^2 f}{\partial y^2} > 0.$

In this case there are two real and unequal values of the slope $\left(=\frac{dy}{dx}\right)$ found from ($F$), so that we have two distinct real tangents to the curve at the singular point in question. This means that the curve passes through the point in two different directions, or, in other words, two branches of the curve cross at this point. Such a singular point we call a *real double point* of the curve, or a *node*. **Hence the conditions to be satisfied at a node are**

$$f(x, y) = 0, \quad \frac{\partial f}{\partial x} = 0, \quad \frac{\partial f}{\partial y} = 0, \quad \left(\frac{\partial^2 f}{\partial x \partial y}\right)^2 - \frac{\partial^2 f}{\partial x^2}\frac{\partial^2 f}{\partial y^2} > 0.$$

ILLUSTRATIVE EXAMPLE 1. Examine the lemniscate $y^2 = x^2 - x^4$ for singular points.

**Solution.** Here $f(x, y) = y^2 - x^2 + x^4 = 0.$

Also $\frac{\partial f}{\partial x} = -2x + 4x^3 = 0, \quad \frac{\partial f}{\partial y} = 2y = 0.$

The point (0, 0) is a singular point, since its coördinates satisfy the above three equations. We have at (0, 0)

$$\frac{\partial^2 f}{\partial x^2} = -2, \quad \frac{\partial^2 f}{\partial x \partial y} = 0, \quad \frac{\partial^2 f}{\partial y^2} = 2.$$

$$\therefore \left(\frac{\partial^2 f}{\partial x \partial y}\right)^2 - \frac{\partial^2 f}{\partial x^2}\frac{\partial^2 f}{\partial y^2} = 4,$$

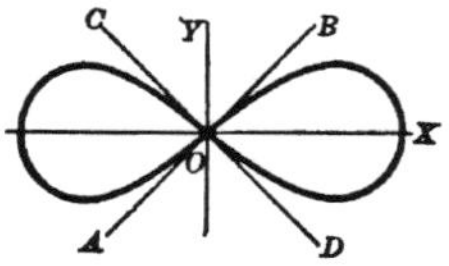


and the origin is a double point (node) through which two branches of the curve pass in different directions. By placing the terms of the lowest (second) degree equal to zero we get

$$y^2 - x^2 = 0, \text{ or } y = x \text{ and } y = -x,$$

the equations of the two tangents $AB$ and $CD$ at the singular point or node (0, 0).

**157. Cusps.** $$\left(\frac{\partial^2 f}{\partial x \partial y}\right)^2 - \frac{\partial^2 f}{\partial x^2}\frac{\partial^2 f}{\partial y^2} = 0.$$

In this case there are two real and equal values of the slope found from $(F)$; hence there are two coincident tangents. This means that the two branches of the curve which pass through the point are tangent. When the curve recedes from the tangent in both directions from the point of tangency, the singular point is called a *point of osculation;* if it recedes from the point of tangency in one direction only, it is called a *cusp*. There are two kinds of cusps.

***First kind.*** When the two branches lie on opposite sides of the common tangent.

***Second kind.*** When the two branches lie on the same side of the common tangent.*

The following examples illustrate how we may determine the nature of singular points coming under this head.

ILLUSTRATIVE EXAMPLE 1. Examine $a^4y^2 = a^2x^4 - x^6$ for singular points.

**Solution.** Here $$f(x, y) = a^4y^2 - a^2x^4 + x^6 = 0,$$

$$\frac{\partial f}{\partial x} = -4a^2x^3 + 6x^5 = 0, \quad \frac{\partial f}{\partial y} = 2a^4y = 0,$$

and (0, 0) is a singular point, since it satisfies the above three equations. Also, at (0, 0) we have

$$\frac{\partial^2 f}{\partial x^2} = 0, \quad \frac{\partial^2 f}{\partial x \partial y} = 0, \quad \frac{\partial^2 f}{\partial y^2} = 2a^4.$$

$$\therefore \left(\frac{\partial^2 f}{\partial x \partial y}\right)^2 - \frac{\partial^2 f}{\partial x^2}\frac{\partial^2 f}{\partial y^2} = 0;$$

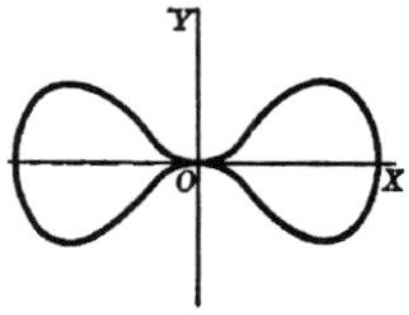


and since the curve is symmetrical with respect to $OY$, the origin is a point of osculation. Placing the terms of lowest (second) degree equal to zero, we get $y^2 = 0$, showing that the two common tangents coincide with $OX$.

ILLUSTRATIVE EXAMPLE 2. Examine $y^2 = x^3$ for singular points.

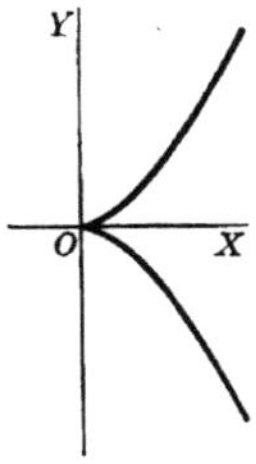


**Solution.** Here $$f(x, y) = y^2 - x^3 = 0,$$

$$\frac{\partial f}{\partial x} = -3x^2 = 0, \quad \frac{\partial f}{\partial y} = 2y = 0,$$

showing that (0, 0) is a singular point. Also, at (0, 0) we have

$$\frac{\partial^2 f}{\partial x^2} = 0, \quad \frac{\partial^2 f}{\partial x \partial y} = 0, \quad \frac{\partial^2 f}{\partial y^2} = 2. \quad \therefore \left(\frac{\partial^2 f}{\partial x \partial y}\right)^2 - \frac{\partial^2 f}{\partial x^2}\frac{\partial^2 f}{\partial y^2} = 0.$$

This is not a point of osculation, however, for if we solve the given equation for $y$ we get $$y = \pm\sqrt{x^3},$$

* Meaning in the neighborhood of the singular point.

which shows that the curve extends to the right only of $OY$, for negative values of $x$ make $y$ imaginary. The origin is therefore a cusp, and since the branches lie on opposite sides of the common tangent, it is a cusp of the first kind. Placing the terms of lowest (second) degree equal to zero, we get $y^2 = 0$, showing that the two common tangents coincide with $OX$.

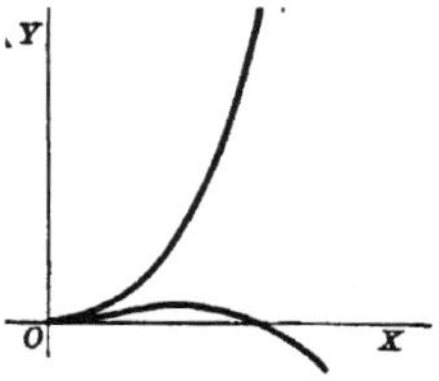


ILLUSTRATIVE EXAMPLE 3. Examine $(y - x^2)^2 = x^5$ for singular points.

**Solution.** Proceeding as in the last example, we find a cusp at (0, 0), the common tangents to the two branches coinciding with $OX$. Solving for $y$,

$$y = x^2 \pm x^{\frac{5}{2}}.$$

If we let $x$ take on any value between 0 and 1, $y$ takes on two different positive values, showing that in the vicinity of the origin both branches lie above the common tangent. Hence the singular point (0, 0) is a cusp of the second kind.

**158. Conjugate or isolated points.** $\left(\dfrac{\partial^2 f}{\partial x \partial y}\right)^2 - \dfrac{\partial^2 f}{\partial x^2}\dfrac{\partial^2 f}{\partial y^2} < 0.$

In this case the values of the slope found are imaginary. Hence there are no real tangents; the singular point is the real intersection of imaginary branches of the curve, and the coördinates of no other real point in the immediate vicinity satisfy the equation of the curve. Such an isolated point is called a *conjugate point.*

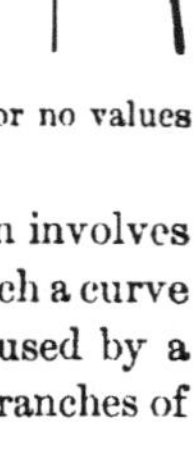


ILLUSTRATIVE EXAMPLE 1. Examine the curve $y^2 = x^3 - x^2$ for singular points.

**Solution.** Here (0, 0) is found to be a singular point of the curve at which $\dfrac{dy}{dx} = \pm\sqrt{-1}$. Hence the origin is a conjugate point. Solving the equation for $y$,

$$y = \pm x\sqrt{x-1}.$$

This shows clearly that the origin is an isolated point of the curve, for no values of $x$ between 0 and 1 give real values of $y$.

**159. Transcendental singularities.** A curve whose equation involves transcendental functions is called a transcendental curve. Such a curve may have an *end point* at which it terminates abruptly, caused by a discontinuity in the function; or a *salient point* at which two branches of the curve terminate without having a common tangent, caused by a discontinuity in the derivative.

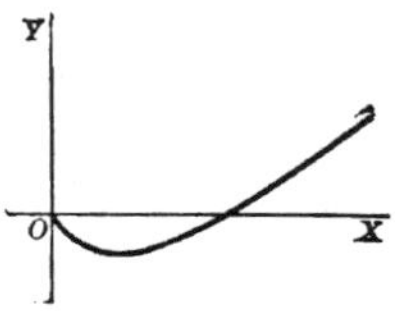


ILLUSTRATIVE EXAMPLE 1. Show that $y = x \log x$ has an end point at the origin.

**Solution.** $x$ cannot be negative, since negative numbers have no logarithms; hence the curve extends only to the right of $OY$. When $x = 0$, $y = 0$. There being only one value of $y$ for each positive value of $x$, the curve consists of a single branch terminating at the origin, which is therefore an end point.

ILLUSTRATIVE EXAMPLE 2. Show that $y = \frac{x}{1 + e^{\frac{1}{x}}}$ has a salient point at the origin.

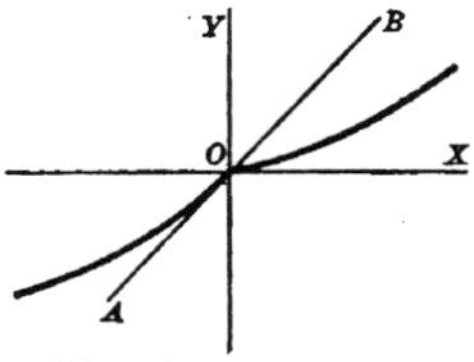


**Solution.** Here $\frac{dy}{dx} = \frac{1}{1 + e^{\frac{1}{x}}} + \frac{e^{\frac{1}{x}}}{x(1 + e^{\frac{1}{x}})^2}$.

If $x$ is positive and approaches zero as a limit, we have ultimately

$$y = 0 \quad \text{and} \quad \frac{dy}{dx} = 0.$$

If $x$ is negative and approaches zero as a limit, we get ultimately

$$y = 0 \quad \text{and} \quad \frac{dy}{dx} = 1.$$

Hence at the origin two branches meet, one having $OX$ as its tangent and the other, $AB$, making an angle of 45° with $OX$.

## EXAMPLES

**1.** Show that $y^2 = 2x^2 + x^3$ has a node at the origin, the slopes of the tangents being $\pm\sqrt{2}$.

**2.** Show that the origin is a node of $y^2(a^2 + x^2) = x^2(a^2 - x^2)$, and that the tangents bisect the angles between the axes.

**3.** Prove that $(a, 0)$ is a node of $y^2 = x(x - a)^2$, and that the slopes of the tangents are $\pm\sqrt{a}$.

**4.** Prove that $a^3y^2 - 2abx^2y - x^5 = 0$ has a point of osculation at the origin.

**5.** Show that the curve $y^2 = x^5 + x^4$ has a point of osculation at the origin.

**6.** Show that the cissoid $y^2 = \frac{x^3}{2a - x}$ has a cusp of the first kind at the origin.

**7.** Show that $y^3 = 2ax^2 - x^3$ has a cusp of the first kind at the origin.

**8.** In the curve $(y - x^2)^2 = x^n$ show that the origin is a cusp of the first or second kind according as $n$ is $< $ or $> 4$.

**9.** Prove that the curve $x^4 - 2ax^2y - axy^2 + a^2y^2 = 0$ has a cusp of the second kind at the origin.

**10.** Show that the origin is a conjugate point on the curve $y^2(x^2 - a^2) = x^4$

**11.** Show that the curve $y^2 = x(a + x)^2$ has a conjugate point at $(-a, 0)$.

**12.** Show that the origin is a conjugate point on the curve $ay^2 - x^3 + bx^2 = 0$ when $a$ and $b$ have the same sign, and a node when they have opposite signs.

**13.** Show that the curve $x^4 + 2ax^2y - ay^3 = 0$ has a triple point at the origin, and that the slopes of the tangents are 0, $+\sqrt{2}$, and $-\sqrt{2}$.

**14.** Show that the points of intersection of the curve $\left(\frac{x}{a}\right)^{\frac{2}{3}} + \left(\frac{y}{b}\right)^{\frac{2}{3}} = 1$ with the axes are cusps of the first kind.

**15.** Show that no curve of the second or third degree in $x$ and $y$ can have a cusp of the second kind.

**16.** Show that $y = e^{-\frac{1}{x}}$ has an end point at the origin.

**17.** Show that $y = x \arctan\frac{1}{x}$ has a salient point at the origin, the slopes of the tangents being $\pm\frac{\pi}{2}$.

# CHAPTER XX

## APPLICATIONS TO GEOMETRY OF SPACE

**160. Tangent line and normal plane to a skew curve whose equations are given in parametric form.** The student is already familiar with the parametric representation of a plane curve. In order to extend this notion to curves in space, let the coördinates of any point $P(x, y, z)$ on a skew curve be given as functions of some fourth variable which we shall denote by $t$, thus,

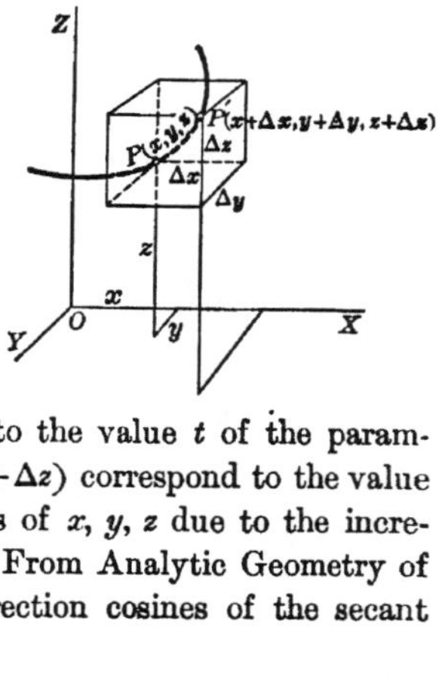


$$(A) \quad x = \phi(t), \quad y = \psi(t), \quad z = \chi(t).$$

The elimination of the parameter $t$ between these equations two by two will give us the equations of the projecting cylinders of the curve on the coördinate planes.

Let the point $P(x, y, z)$ correspond to the value $t$ of the parameter, and the point $P'(x+\Delta x, y+\Delta y, z+\Delta z)$ correspond to the value $t+\Delta t$; $\Delta x$, $\Delta y$, $\Delta z$ being the increments of $x$, $y$, $z$ due to the increment $\Delta t$ as found from equations $(A)$. From Analytic Geometry of three dimensions we know that the direction cosines of the secant (diagonal) $PP'$ are proportional to

$$\Delta x, \quad \Delta y, \quad \Delta z;$$

or, dividing through by $\Delta t$ and denoting the direction angles of the secant by $\alpha'$, $\beta'$, $\gamma'$,

$$\cos\alpha' : \cos\beta' : \cos\gamma' :: \frac{\Delta x}{\Delta t} : \frac{\Delta y}{\Delta t} : \frac{\Delta z}{\Delta t}.$$

Now let $P'$ approach $P$ along the curve. Then $\Delta t$, and therefore also $\Delta x$, $\Delta y$, $\Delta z$, will approach zero as a limit, the secant $PP'$ will approach the tangent line to the curve at $P$ as a limiting position, and we shall have

$$\cos\alpha : \cos\beta : \cos\gamma :: \frac{dx}{dt} : \frac{dy}{dt} : \frac{dz}{dt},$$

where $\alpha$, $\beta$, $\gamma$ are the direction angles of the tangent (or curve) at $P$. Hence *the equations of the tangent line to the curve*

$$x=\phi(t), \quad y=\psi(t), \quad z=\chi(t)$$

*at the point* $(x, y, z)$ *are given by*

$$\text{(69)} \qquad \frac{X-x}{\frac{dx}{dt}}=\frac{Y-y}{\frac{dy}{dt}}=\frac{Z-z}{\frac{dz}{dt}};$$

*and the equation of the normal plane, i.e. the plane passing through* $(x, y, z)$ *perpendicular to the tangent, is*

$$\text{(70)} \qquad \frac{dx}{dt}(X-x)+\frac{dy}{dt}(Y-y)+\frac{dz}{dt}(Z-z)=0,$$

$X$, $Y$, $Z$ *being the variable coördinates.*

Illustrative Example 1. **Find the equations of the tangent and the equation of the normal plane to the helix* ($\theta$ being the parameter)**

$$\begin{cases} x = a\cos\theta, \\ y = a\sin\theta, \\ z = b\theta, \end{cases}$$

**(a) at any point; (b) when $\theta = 2\pi$.**

**Solution.** $\qquad \dfrac{dx}{d\theta}=-a\sin\theta=-y, \quad \dfrac{dy}{d\theta}=a\cos\theta=x, \quad \dfrac{dz}{d\theta}=b.$

Substituting in (69) and (70), we get at $(x, y, z)$

$$\frac{X-x}{-y}=\frac{Y-y}{x}=\frac{Z-z}{b}, \text{ tangent line;}$$

and $-y(X-x)+x(Y-y)+b(Z-z)=0$, normal plane.

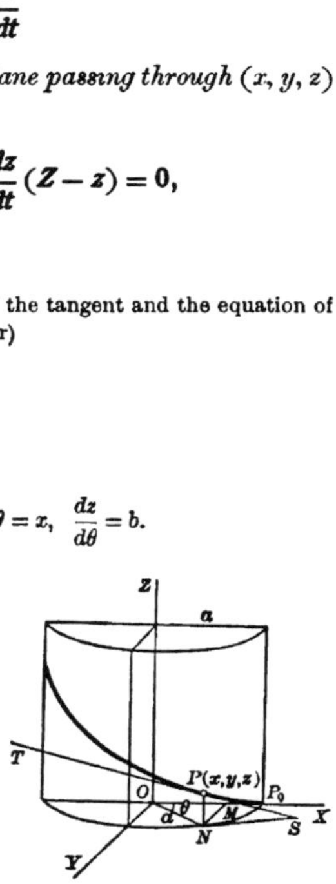


When $\theta = 2\pi$, the point of contact is $(a, 0, 2b\pi)$, giving

$$\frac{X-a}{0}=\frac{Y-0}{a}=\frac{Z-2b\pi}{b},$$

or, $\qquad X=a, \quad bY=aZ-2ab\pi,$

the equations of the tangent line; and

$$aY+bZ-2b^2\pi=0,$$

the equation of the normal plane.

* The helix may be defined as a curve traced on a right circular cylinder so as to cut all the elements at the same angle.

Take $OZ$ as the axis of the cylinder, and the point of starting in $OX$ at $P_0$. Let $a$ = radius of base of cylinder and $\theta$ = angle of rotation. By definition,

$$\frac{PN}{SN}=\frac{PN}{\text{arc } P_0N}=\frac{z}{a\theta}=k\text{ (const.)}, \text{ or } z=ak\theta.$$

Let $ak=b$; then $z=b\theta$. Also $y=MN=a\sin\theta$, $x=OM=a\cos\theta$.

**EXAMPLES**

Find the equations of the tangent line and the equation of the normal plane to each of the following skew curves at the point indicated:

1. $x = 2t,\ y = t^2,\ z = 4t^4;\ t = 1.$ *Ans.* $\frac{x-2}{2} = \frac{y-1}{2} = \frac{z-4}{16}$; $x + y + 8z - 35 = 0.$

2. $x = t^2 - 1,\ y = t + 1,\ z = t^3;\ t = 2.$ *Ans.* $\frac{x-3}{4} = \frac{y-3}{1} = \frac{z-8}{12}$; $4x + y + 12z - 111 = 0.$

3. $x = t^3 - 1,\ y = t + t^2,\ z = 4t^3 - 3t + 1;\ t = 1.$ *Ans.* $\frac{x}{3} = \frac{y-2}{3} = \frac{z-2}{9}$; $x + y + 3z - 8 = 0.$

4. $x = t,\ y = \sin t,\ z = \cos t;\ t = \frac{\pi}{4}.$ *Ans.* $\frac{4x-\pi}{4} = \frac{\sqrt{2}y-1}{1} = \frac{\sqrt{2}z-1}{-1}$; $4x + 2\sqrt{2}y - 2\sqrt{2}z - \pi = 0.$

5. $x = at,\ y = bt^2,\ z = ct^3;\ t = 1.$

6. $x = t,\ y = 1 - t^2,\ z = 3t^2 + 4t;\ t = -2.$

7. $x = t,\ y = e^t,\ z = e^{-t};\ t = 0.$

8. $x = a \sin t,\ y = b \cos t,\ z = t;\ t = \frac{\pi}{6}.$

9. Find the direction cosines of the tangent to the curve $x = t^2,\ y = t^3,\ z = t^4$ at point $x = 1$.

**161. Tangent plane to a surface.** A straight line is said to be *tangent to a surface* at a point $P$ if it is the limiting position of a secant through $P$ and a neighboring point $P'$ on the surface, when $P'$ is made to approach $P$ along the surface. We now proceed to establish a theorem of fundamental importance.

**Theorem.** *All tangent lines to a surface at a given point * lie in general in a plane called the tangent plane at that point.*

**Proof.** Let

$$(A) \qquad F(x, y, z) = 0$$

be the equation of the given surface, and let $P(x, y, z)$ be the given point on the surface. If now $P'$ be made to approach $P$ along a curve $C$ lying on the surface and passing through $P$ and $P'$, then evidently the secant $PP'$ approaches the position of a tangent to the curve $C$ at $P$. Now let the equations of the curve $C$ be

$$(B) \qquad x = \phi(t), \quad y = \psi(t), \quad z = \chi(t).$$

* The point in question is assumed to be an ordinary (nonsingular) point of the surface, i.e. $\frac{\partial F}{\partial x}, \frac{\partial F}{\partial y}, \frac{\partial F}{\partial z}$ are not all zero at the point.

Then the equation $(A)$ must be satisfied identically by these values, and since the total differential of $(A)$ when $x$, $y$, $z$ are defined by $(B)$ must vanish, we have

$$(C) \qquad \frac{\partial F}{\partial x}\frac{dx}{dt}+\frac{\partial F}{\partial y}\frac{dy}{dt}+\frac{\partial F}{\partial z}\frac{dz}{dt}=0. \qquad \text{By (52), p. 196}$$

This equation shows that the tangent line to $C$, whose direction cosines are proportional to

$$\frac{dx}{dt}, \quad \frac{dy}{dt}, \quad \frac{dz}{dt},$$

is perpendicular * to a line whose direction cosines are determined by the ratios

$$\frac{\partial F}{\partial x}:\frac{\partial F}{\partial y}:\frac{\partial F}{\partial z};$$

and since $C$ is any curve on the surface through $P$, it follows at once, if we replace the point $P(x, y, z)$ by $P_1(x_1, y_1, z_1)$, that all tangent lines to the surface at $P_1$ lie in the plane †

$$(71) \qquad \frac{\partial F_1}{\partial x_1}(x-x_1)+\frac{\partial F_1}{\partial y_1}(y-y_1)+\frac{\partial F_1}{\partial z_1}(z-z_1)=0, \text{‡}$$

which is then *the formula for finding the equation of a plane tangent at* $(x_1, y_1, z_1)$ *to a surface whose equation is given in the form*

$$F(x, y, z)=0.$$

In case the equation of the surface is given in the form $z=f(x, y)$, let

$$(D) \qquad F(x, y, z)=f(x, y)-z=0.$$

Then $$\frac{\partial F}{\partial x}=\frac{\partial f}{\partial x}=\frac{\partial z}{\partial x}, \quad \frac{\partial F}{\partial y}=\frac{\partial f}{\partial y}=\frac{\partial z}{\partial y}, \quad \frac{\partial F}{\partial z}=-1.$$

* From Solid Analytic Geometry we know that if two lines having the direction cosines $\cos\alpha_1$, $\cos\beta_1$, $\cos\gamma_1$ and $\cos\alpha_2$, $\cos\beta_2$, $\cos\gamma_2$ are perpendicular, then

$$\cos\alpha_1\cos\alpha_2+\cos\beta_1\cos\beta_2+\cos\gamma_1\cos\gamma_2=0.$$

† The direction cosines of the normal to the plane (71) are proportional to $\frac{\partial F_1}{\partial x_1}, \frac{\partial F_1}{\partial y_1}, \frac{\partial F_1}{\partial z_1}$. Hence from Analytic Geometry we see that $(C)$ is the condition that the tangents whose direction cosines are $\cos\alpha$, $\cos\beta$, $\cos\gamma$ are perpendicular to the normal; i.e. the tangents must lie in the plane.

‡ In agreement with our former practice,

$$\frac{\partial F_1}{\partial x_1}, \quad \frac{\partial F_1}{\partial y_1}, \quad \frac{\partial F_1}{\partial z_1}, \quad \frac{\partial z_1}{\partial x_1}, \quad \frac{\partial z_1}{\partial y_1}$$

denote the values of the partial derivatives at the point $(x_1, y_1, z_1)$.

If we evaluate these at $(x_1, y_1, z_1)$ and substitute in **(71)**, we get

$$\textbf{(72)} \qquad \frac{\partial z_1}{\partial x_1}(x - x_1) + \frac{\partial z_1}{\partial y_1}(y - y_1) - (z - z_1) = 0,$$

which is then *the formula for finding the equation of a plane tangent at* $(x_1, y_1, z_1)$ *to a surface whose equation is given in the form* $z = f(x, y)$.

In § 126, p. 197, we found (55) the total differential of a function $u$ (or $z$) of $x$ and $y$, namely,

$$(E) \qquad dz = \frac{\partial z}{\partial x}dx + \frac{\partial z}{\partial y}dy.$$

We have now a means of interpreting this result geometrically. For the tangent plane to the surface $z = f(x, y)$ at $(x, y, z)$ is, from (72),

$$(F) \qquad Z - z = \frac{\partial z}{\partial x}(X - x) + \frac{\partial z}{\partial y}(Y - y),$$

$X$, $Y$, $Z$ denoting the variable coördinates at any point on the plane. If we substitute

$$X = x + dx \quad \text{and} \quad Y = y + dy$$

in $(F)$, there results

$$(G) \quad Z - z = \frac{\partial z}{\partial x}dx + \frac{\partial z}{\partial y}dy.$$

Comparing $(E)$ and $(G)$, we get

$(H)$ $\quad dz = Z - z$. Hence

**Theorem.** *The total differential of a function* $f(x, y)$ *corresponding to the increments* $dx$ *and* $dy$ *equals the corresponding increment of the* $z$*-coördinate of the tangent plane to the surface* $z = f(x, y)$.

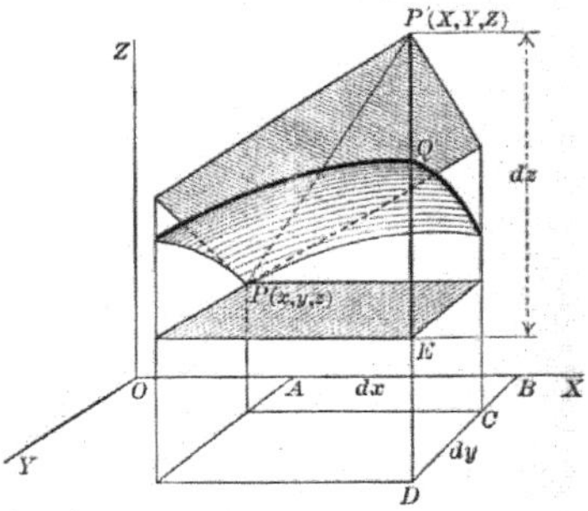


Thus, in the figure, $PP'$ is the plane tangent to surface $PQ$ at $P(x, y, z)$.

Let $\qquad AB = dx \quad$ and $\quad CD = dy$;

then $\qquad dz = Z - z = DP' - DE = EP'$.

**162. Normal line to a surface.** The normal line to a surface at a given point is the line passing through the point perpendicular to the tangent plane to the surface at that point.

The direction cosines of any line perpendicular to the tangent plane **(71)** are proportional to

$$\frac{\partial F_1}{\partial x_1}, \quad \frac{\partial F_1}{\partial y_1}, \quad \frac{\partial F_1}{\partial z_1}.$$

$$\textbf{(73)} \qquad \therefore \frac{x - x_1}{\dfrac{\partial F_1}{\partial x_1}} = \frac{y - y_1}{\dfrac{\partial F_1}{\partial y_1}} = \frac{z - z_1}{\dfrac{\partial F_1}{\partial z_1}}.$$

*are the equations of the normal line* to the surface* $F(x, y, z) = 0$ at $(x_1, y_1, z_1)$.

Similarly, from (72),

$$\frac{x - x_1}{\dfrac{\partial z_1}{\partial x_1}} = \frac{y - y_1}{\dfrac{\partial z_1}{\partial y_1}} = \frac{z - z_1}{-1} \tag{74}$$

*are the equations of the normal line* to the surface* $z = f(x, y)$ *at* $(x_1, y_1, z_1)$.

## EXAMPLES

**1.** Find the equation of the tangent plane and the equations of the normal line to the sphere $x^2 + y^2 + z^2 = 14$ at the point (1, 2, 3).

**Solution.** Let $F(x, y, z) = x^2 + y^2 + z^2 - 14$;

then
$$\frac{\partial F}{\partial x} = 2x,\ \frac{\partial F}{\partial y} = 2y,\ \frac{\partial F}{\partial z} = 2z;\ x_1 = 1,\ y_1 = 2,\ z_1 = 3.$$

$$\therefore \frac{\partial F_1}{\partial x_1} = 2,\ \frac{\partial F_1}{\partial y_1} = 4,\ \frac{\partial F_1}{\partial z_1} = 6.$$

Substituting in (71), $2(x - 1) + 4(y - 2) + 6(z - 3) = 0$, $x + 2y + 3z = 14$, the tangent plane.

Substituting in (73),
$$\frac{x-1}{2} = \frac{y-2}{4} = \frac{z-3}{6},$$

giving $z = 3x$ and $2z = 3y$, equations of the normal line.

**2.** Find the equation of the tangent plane and the equations of the normal line to the ellipsoid $4x^2 + 9y^2 + 36z^2 = 36$ at point of contact where $x = 2$, $y = 1$, and $z$ is positive. *Ans.* Tangent plane, $8(x - 2) + 9(y - 1) + 6\sqrt{11}(z - \frac{1}{6}\sqrt{11}) = 0$;

normal line, $\dfrac{x-2}{8} = \dfrac{y-1}{9} = \dfrac{z - \frac{1}{6}\sqrt{11}}{6\sqrt{11}}$.

**3.** Find the equation of the tangent plane to the elliptic parabola $z = 2x^2 + 4y^2$ at the point (2, 1, 12). *Ans.* $8x + 8y - z = 12$.

**4.** Find the equations of the normal line to the hyperboloid of one sheet $x^2 - 4y^2 + 2z^2 = 6$ at (2, 2, 3). *Ans.* $y + 4x = 10$, $3x - z = 3$.

**5.** Find the equation of the tangent plane to the hyperboloid of two sheets $\dfrac{x^2}{a^2} - \dfrac{y^2}{b^2} - \dfrac{z^2}{c^2} = 1$ at $(x_1, y_1, z_1)$. *Ans.* $\dfrac{x_1x}{a^2} - \dfrac{y_1y}{b^2} - \dfrac{z_1z}{c^2} = 1$.

**6.** Find the equation of the tangent plane at the point $(x_1, y_1, z_1)$ on the surface $ax^2 + by^2 + cz^2 + d = 0$. *Ans.* $ax_1x + by_1y + cz_1z + d = 0$.

**7.** Show that the equation of the plane tangent to the sphere

$$x^2 + y^2 + z^2 + 2Lx + 2My + 2Nz + D = 0$$

at the point $(x_1, y_1, z_1)$ is

$$x_1x + y_1y + z_1z + L(x + x_1) + M(y + y_1) + N(z + z_1) + D = 0.$$

* See second footnote, p. 265.

**8.** Find the equation of the tangent plane at any point of the surface

$$x^{\frac{2}{3}} + y^{\frac{2}{3}} + z^{\frac{2}{3}} = a^{\frac{2}{3}},$$

and show that the sum of the squares of the intercepts on the axes made by the tangent plane is constant.

**9.** Prove that the tetrahedron formed by the coördinate planes and any tangent plane to the surface $xyz = a^3$ is of constant volume.

**10.** Find the equation of the tangent plane and the equations of the normal line to the following surfaces at the points indicated:

(a) $2x^2 + 4y^2 - z = 0$; $(2, 1, 12)$.
(b) $x^2 + 4y^2 - z^2 = 16$; $(1, 2, -1)$.
(c) $x^2 + y^2 + z^2 = 11$; $(3, 1, 1)$.
(d) $3x^2 + y^2 - 2z = 0$; $x = 1, y = 1$.
(e) $x^2y^2 + 2x + z^3 = 16$; $x = 2, y = 1$.
(f) $x^2 + 3y^2 + 2z^2 = 9$; $y = 1, z = 1$.

**163. Another form of the equations of the tangent line to a skew curve.** If the curve in question be the curve of intersection $AB$ of the two surfaces $F(x, y, z) = 0$ and $G(x, y, z) = 0$, the tangent line $PT$ at $P(x_1, y_1, z_1)$ is the intersection of the tangent planes $CD$ and $CE$ at that point, for it is also tangent to both surfaces and hence must lie in both tangent planes. The equations of the two tangent planes at $P$ are, from **(71)**,

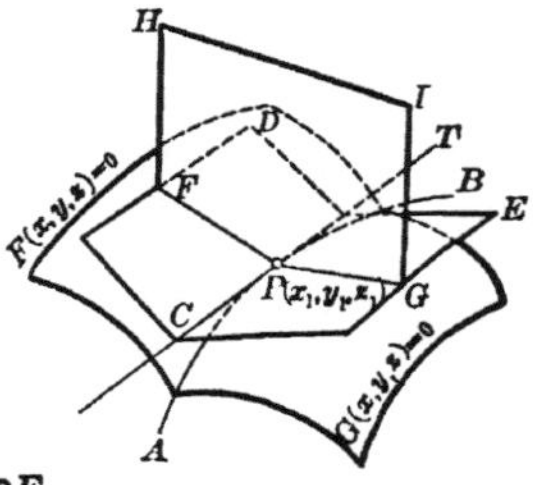


**(75)**
$$\begin{cases} \dfrac{\partial F_1}{\partial x_1}(x - x_1) + \dfrac{\partial F_1}{\partial y_1}(y - y_1) + \dfrac{\partial F_1}{\partial z_1}(z - z_1) = 0, \\[2ex] \dfrac{\partial G_1}{\partial x_1}(x - x_1) + \dfrac{\partial G_1}{\partial y_1}(y - y_1) + \dfrac{\partial G_1}{\partial z_1}(z - z_1) = 0. \end{cases}$$

Taken simultaneously, the equations **(75)** are *the equations of the tangent line $PT$ to the skew curve $AB$.* Equations **(75)** in more compact form are

**(76)**
$$\frac{x - x_1}{\dfrac{\partial F_1}{\partial y_1}\dfrac{\partial G_1}{\partial z_1} - \dfrac{\partial F_1}{\partial z_1}\dfrac{\partial G_1}{\partial y_1}} = \frac{y - y_1}{\dfrac{\partial F_1}{\partial z_1}\dfrac{\partial G_1}{\partial x_1} - \dfrac{\partial F_1}{\partial x_1}\dfrac{\partial G_1}{\partial z_1}} = \frac{z - z_1}{\dfrac{\partial F_1}{\partial x_1}\dfrac{\partial G_1}{\partial y_1} - \dfrac{\partial F_1}{\partial y_1}\dfrac{\partial G_1}{\partial x_1}},$$

or,

**(77)**
$$\frac{x - x_1}{\begin{vmatrix} \dfrac{\partial F_1}{\partial y_1} & \dfrac{\partial F_1}{\partial z_1} \\[2ex] \dfrac{\partial G_1}{\partial y_1} & \dfrac{\partial G_1}{\partial z_1} \end{vmatrix}} = \frac{y - y_1}{\begin{vmatrix} \dfrac{\partial F_1}{\partial z_1} & \dfrac{\partial F_1}{\partial x_1} \\[2ex] \dfrac{\partial G_1}{\partial z_1} & \dfrac{\partial G_1}{\partial x_1} \end{vmatrix}} = \frac{z - z_1}{\begin{vmatrix} \dfrac{\partial F_1}{\partial x_1} & \dfrac{\partial F_1}{\partial y_1} \\[2ex] \dfrac{\partial G_1}{\partial x_1} & \dfrac{\partial G_1}{\partial y_1} \end{vmatrix}},$$

using the notation of determinants.

**164. Another form of the equation of the normal plane to a skew curve.** The *normal plane* to a skew curve at a given point has already been defined as the plane passing through that point perpendicular to the tangent line to the curve at that point. Thus, in the above figure, $PHI$ is the normal plane to the curve $AB$ at $P$. Since this plane is perpendicular to (77), we have at once

$$(78)\qquad \begin{vmatrix} \frac{\partial F_1}{\partial y_1} & \frac{\partial F_1}{\partial z_1} \\ \frac{\partial G_1}{\partial y_1} & \frac{\partial G_1}{\partial z_1} \end{vmatrix}(x-x_1)+\begin{vmatrix} \frac{\partial F_1}{\partial z_1} & \frac{\partial F_1}{\partial x_1} \\ \frac{\partial G_1}{\partial z_1} & \frac{\partial G_1}{\partial x_1} \end{vmatrix}(y-y_1)+\begin{vmatrix} \frac{\partial F_1}{\partial x_1} & \frac{\partial F_1}{\partial y_1} \\ \frac{\partial G_1}{\partial x_1} & \frac{\partial G_1}{\partial y_1} \end{vmatrix}(z-z_1)=0,$$

*the equation of the normal plane to a skew curve.*

## EXAMPLES

1. Find the equations of the tangent line and the equation of the normal plane at $(r, r, r\sqrt{2})$ to the curve of intersection of the sphere and cylinder whose equations are respectively $x^2+y^2+z^2=4r^2$, $x^2+y^2=2rx$.

**Solution.** Let $F = x^2+y^2+z^2-4r^2$ and $G = x^2+y^2-2rx$.

$$\frac{\partial F_1}{\partial x_1}=2r,\quad \frac{\partial F_1}{\partial y_1}=2r,\quad \frac{\partial F_1}{\partial z_1}=2\sqrt{2}r;$$

$$\frac{\partial G_1}{\partial x_1}=0,\quad \frac{\partial G_1}{\partial y_1}=2r,\quad \frac{\partial G_1}{\partial z_1}=0.$$

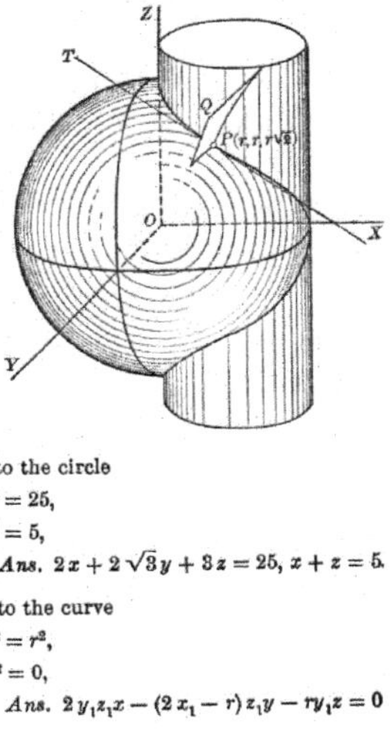


Substituting in (77),

$$\frac{x-r}{-\sqrt{2}}=\frac{y-r}{0}=\frac{z-r\sqrt{2}}{1};$$

or, $\quad y=r,\ x+\sqrt{2}z=3r,$

the equations of the tangent $PT$ at $P$ to the curve of intersection.

Substituting in (78), we get the equation of the normal plane,

$$-\sqrt{2}(x-r)+0(y-r)+(z-r\sqrt{2})=0,$$

or, $\quad \sqrt{2}x-z=0.$

2. Find the equations of the tangent line to the circle

$$x^2+y^2+z^2=25,$$
$$x+z=5,$$

at the point $(2, 2\sqrt{3}, 3)$. *Ans.* $2x+2\sqrt{3}y+3z=25,\ x+z=5.$

3. Find the equation of the normal plane to the curve

$$x^2+y^2+z^2=r^2,$$
$$x^2-rx+y^2=0,$$

at $(x_1, y_1, z_1)$. *Ans.* $2y_1z_1x-(2x_1-r)z_1y-ry_1z=0$

**4.** Find the equations of the tangent line and the normal plane to the curve

$$2x^2 + 3y^2 + z^2 = 9, \quad z^2 = 3x^2 + y^2$$

at $(1, -1, 2)$.

**5.** Find the direction of the curve

$$xyz = 1, \quad y^2 = x$$

at the point $(1, 1, 1)$.

**6.** What is the direction of the tangent to the curve

$$y = x^2, \quad z^2 = 1 - y$$

at $(0, 0, 1)$?

**7.** The equations of a helix (spiral) are

$$x^2 + y^2 = r^2,$$
$$y = x \tan \frac{z}{c}.$$

Show that at the point $(x_1, y_1, z_1)$ the equations of the tangent line are

$$c(x - x_1) + y_1(z - z_1) = 0,$$
$$c(y - y_1) - x_1(z - z_1) = 0;$$

and the equation of the normal plane is

$$y_1 x - x_1 y - c(z - z_1) = 0.$$

**8.** A skew curve is formed by the intersection of the cone $\frac{x^2}{a^2} + \frac{y^2}{b^2} - \frac{z^2}{c^2} = 0$ and the sphere $x^2 + y^2 + z^2 = r^2$. Show that at the point $(x_1, y_1, z_1)$ the equations of the tangent line to the curve are

$$c^2(a^2 - b^2)x_1(x - x_1) = -a^2(b^2 + c^2)z_1(z - z_1),$$
$$c^2(a^2 - b^2)y_1(y - y_1) = +b^2(c^2 + a^2)z_1(z - z_1);$$

and the equation of the normal plane is

$$a^2(b^2 + c^2)y_1 z_1 x - b^2(c^2 + a^2)z_1 x_1 y - c^2(a^2 - b^2)x_1 y_1 z = 0.$$

# CHAPTER XXI

## CURVES FOR REFERENCE

For the convenience of the student a number of the more common curves employed in the text are collected here.

CUBICAL PARABOLA

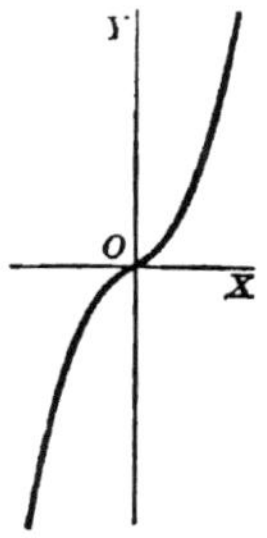


$y = ax^3.$

SEMICUBICAL PARABOLA

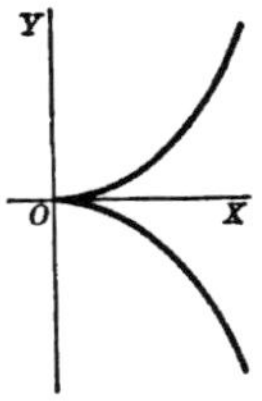


$y^2 = ax^3.$

THE WITCH OF AGNESI

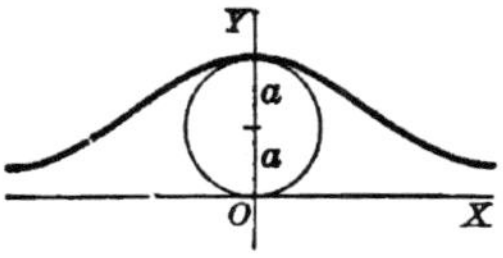


$x^2y = 4\,a^2(2\,a - y).$

THE CISSOID OF DIOCLES

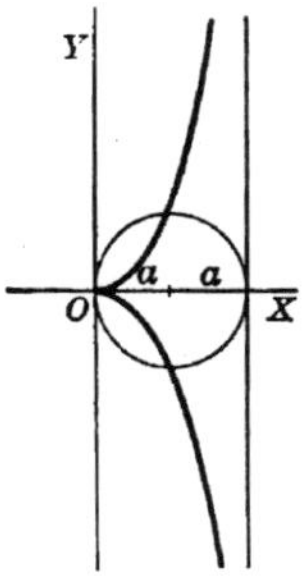


$y^2(2\,a - x) = x^3.$

THE LEMNISCATE OF BERNOULLI

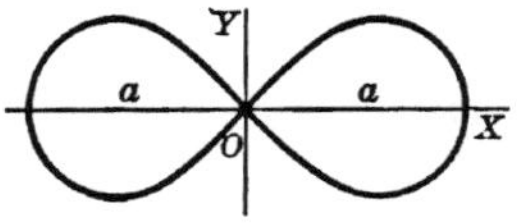


$$(x^2+y^2)^2=a^2(x^2-y^2).$$

$$\rho^2=a^2\cos 2\theta.$$

THE CONCHOID OF NICOMEDES

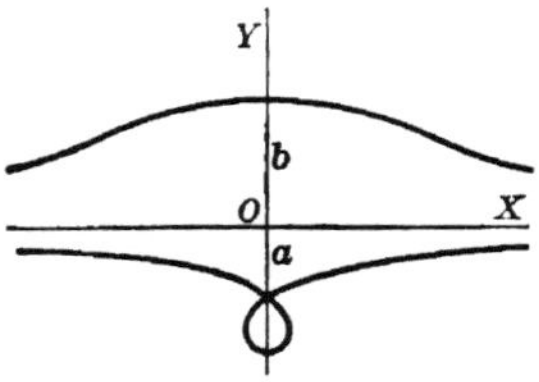


$$x^2y^2=(y+a)^2(b^2-y^2).$$

$$\rho=a\csc\theta+b.$$

CYCLOID, ORDINARY CASE

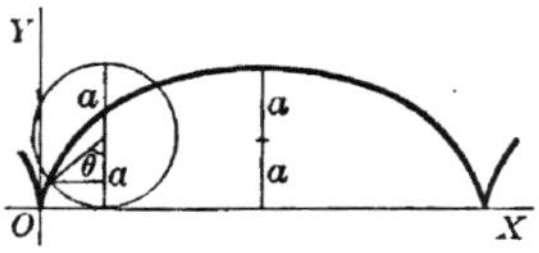


$$x=a\text{ arc vers }\frac{y}{a}-\sqrt{2ay-y^2}.$$

$$\begin{cases}x=a(\theta-\sin\theta),\\ y=a(1-\cos\theta).\end{cases}$$

CYCLOID, VERTEX AT ORIGIN

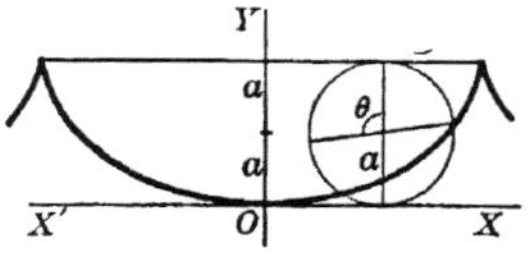


$$x=a\text{ arc vers }\frac{y}{a}+\sqrt{2ay-y^2}.$$

$$\begin{cases}x=a(\theta+\sin\theta),\\ y=a(1-\cos\theta).\end{cases}$$

CATENARY

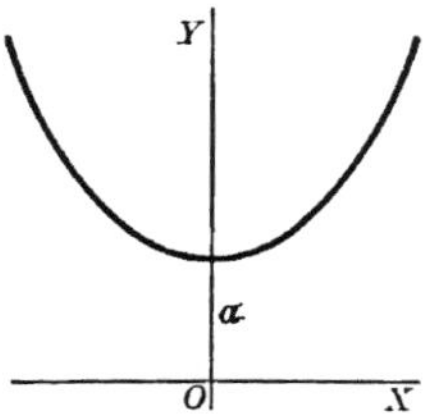


$$y=\frac{a}{2}\left(e^{\frac{x}{a}}+e^{-\frac{x}{a}}\right).$$

PARABOLA

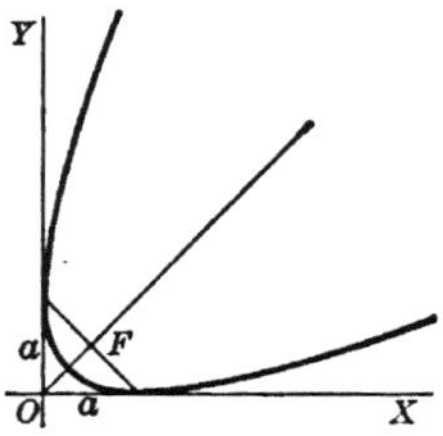


$$x^{\frac{1}{2}}+y^{\frac{1}{2}}=a^{\frac{1}{2}}.$$

**Hypocycloid of Four Cusps**

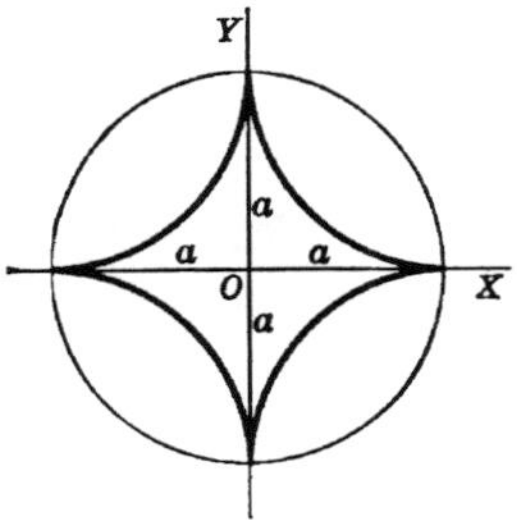


$$x^{\frac{2}{3}} + y^{\frac{2}{3}} = a^{\frac{2}{3}}.$$

$$\begin{cases} x = a\cos^3\theta, \\ y = a\sin^3\theta. \end{cases}$$

**Evolute of Ellipse**

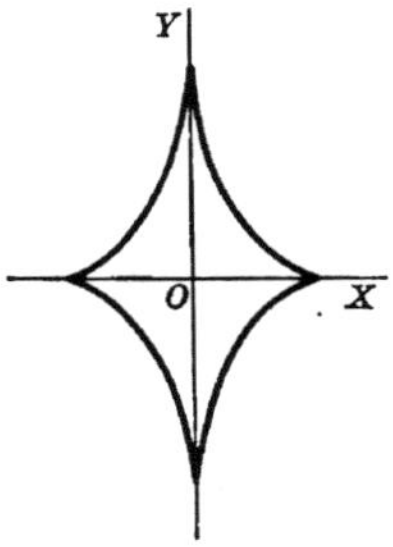


$$(ax)^{\frac{2}{3}} + (by)^{\frac{2}{3}} = (a^2 - b^2)^{\frac{2}{3}}.$$

**Cardioid**

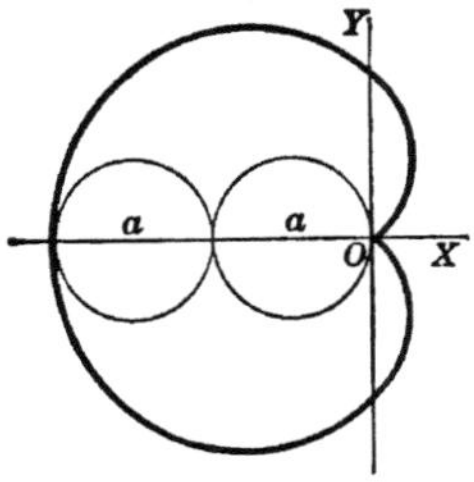


$$x^2 + y^2 + ax = a\sqrt{x^2 + y^2},$$

$$\rho = a(1 - \cos\theta).$$

**Folium of Descartes**

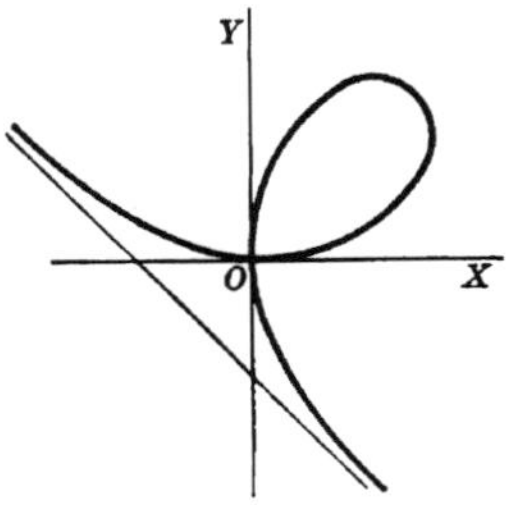


$$x^3 + y^3 - 3\,axy = 0.$$

**Sine Curve**

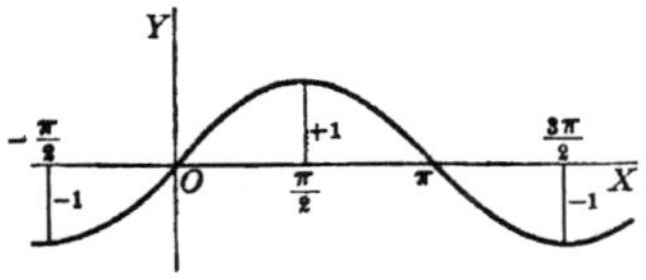


$$y = \sin x.$$

**Cosine Curve**

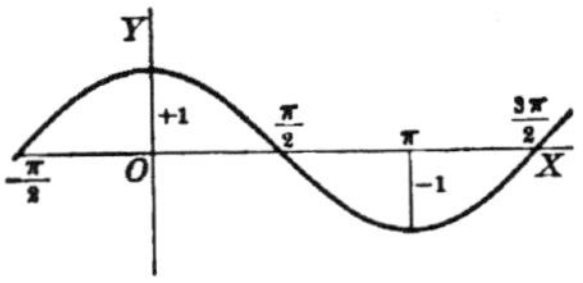


$$y = \cos x.$$

LIMAÇON

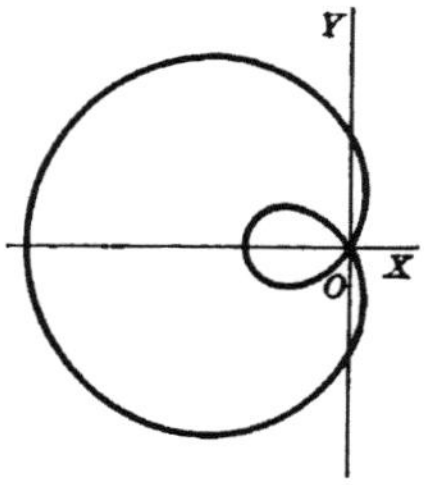


$\rho = b - a\cos\theta.$

STROPHOID

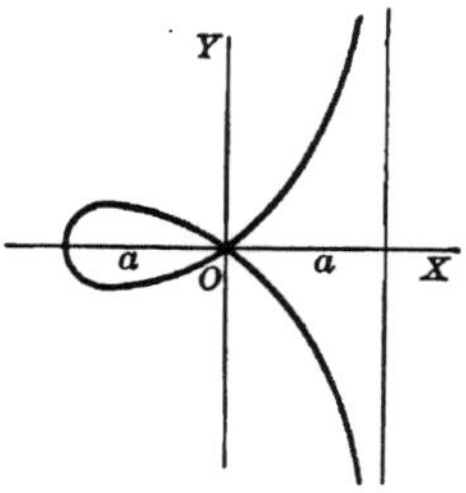


$y^2 = x^2 \frac{a + x}{a - x}.$

SPIRAL OF ARCHIMEDES

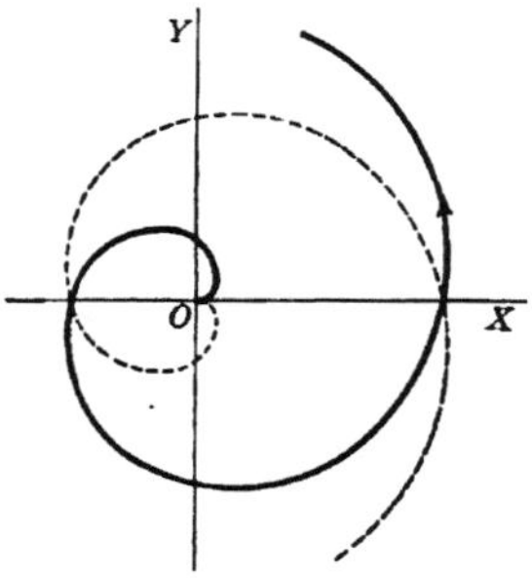


$\rho = a\theta.$

LOGARITHMIC OR EQUIANGULAR SPIRAL

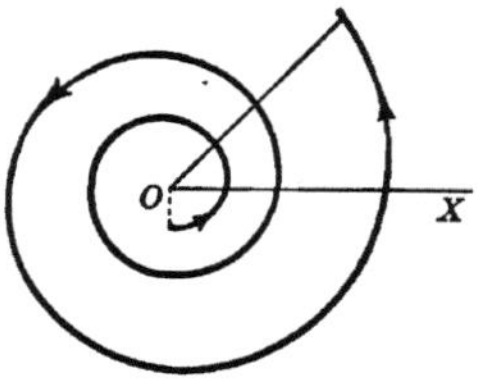


$\rho = e^{a\theta}$, or
$\log \rho = a\theta.$

HYPERBOLIC OR RECIPROCAL SPIRAL

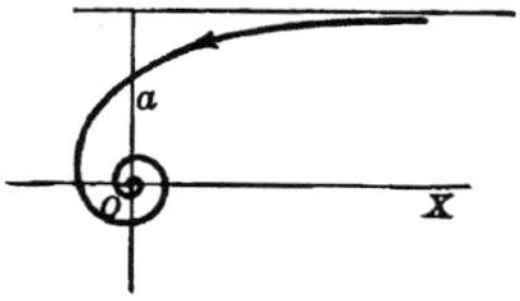


$\rho\theta = a.$

LITUUS

$\rho^2\theta = a^2.$

PARABOLIC SPIRAL

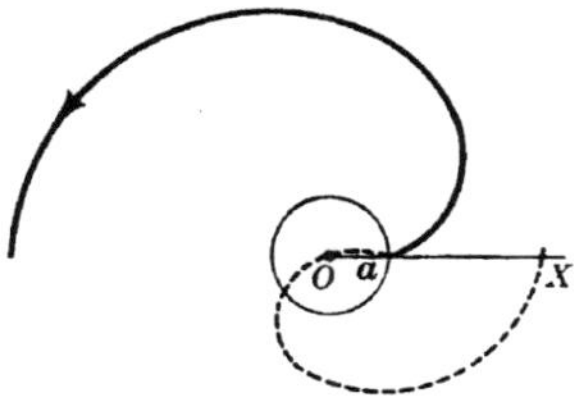


$(\rho - a)^2 = 4\,ac\theta.$

LOGARITHMIC CURVE

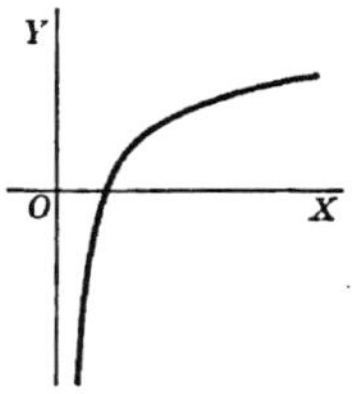


$y = \log x.$

EXPONENTIAL CURVE

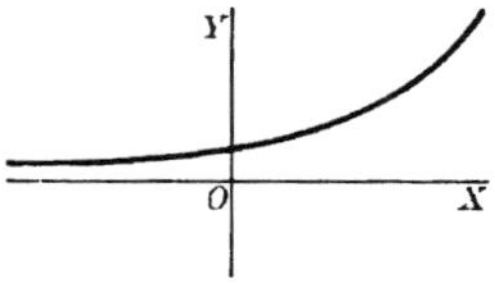


$y = e^x.$

PROBABILITY CURVE

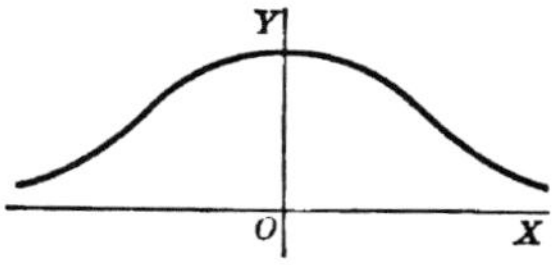


$y = e^{-x^2}.$

SECANT CURVE

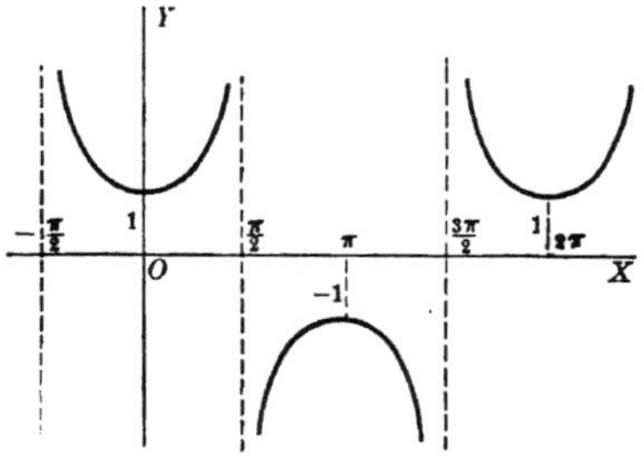


$y = \sec x.$

TANGENT CURVE

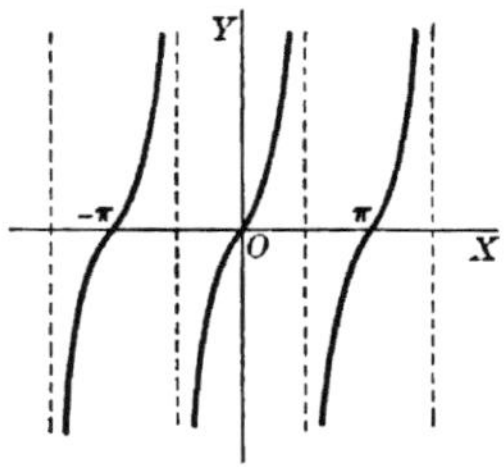


$y = \tan x.$

THREE-LEAVED ROSE

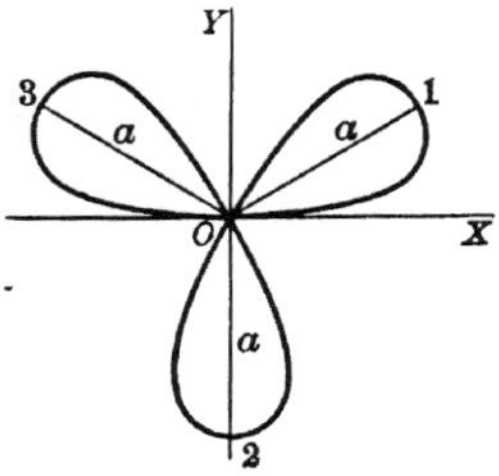


$\rho = a \sin 3\,\theta.$

THREE-LEAVED ROSE

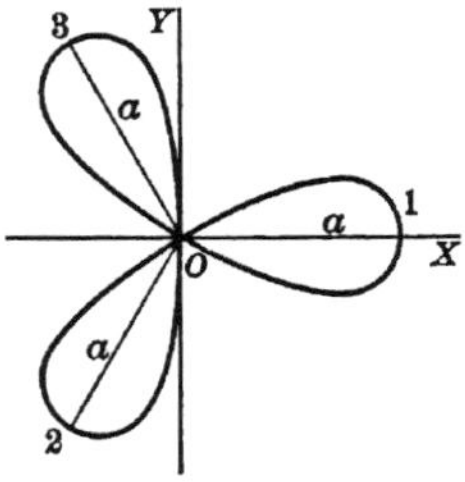


$\rho = a \cos 3\,\theta.$

FOUR-LEAVED ROSE

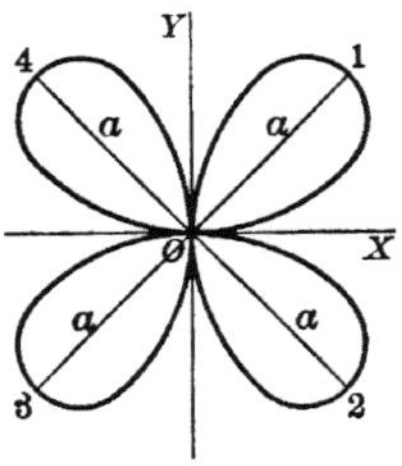


$\rho = a \sin 2\,\theta.$

FOUR-LEAVED ROSE

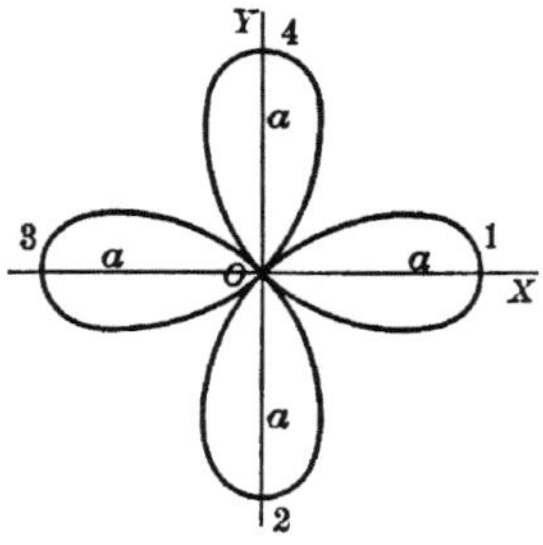


$\rho = a \cos 2\,\theta.$

TWO-LEAVED ROSE LEMNISCATE

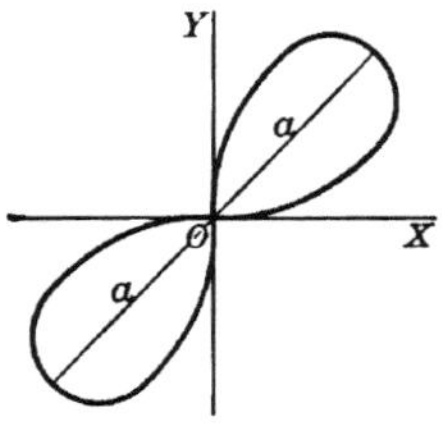


$\rho^2 = a^2 \sin 2\,\theta.$

EIGHT-LEAVED ROSE

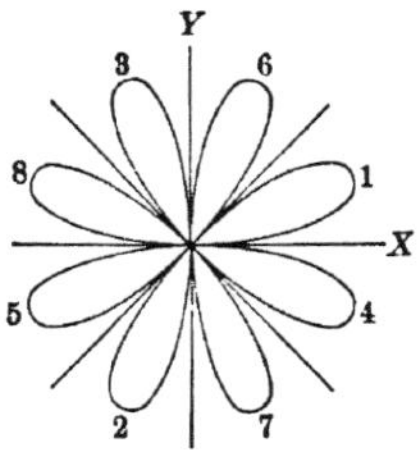


$\rho = a \sin 4\,\theta.$

**Curve with End Point at Origin**

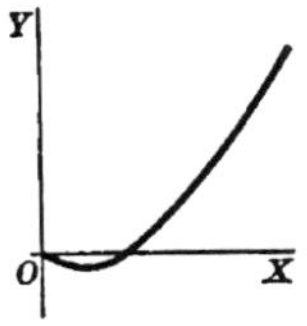


$y = x \log x.$

**Curve with Salient Point at Origin**

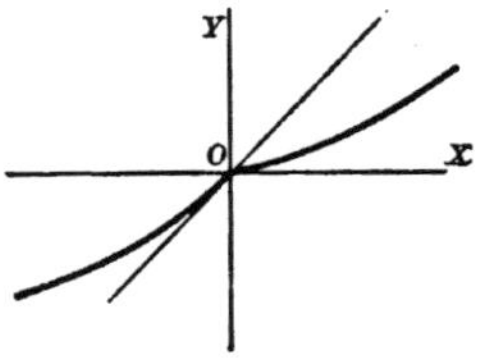


$y(1 + e^{\frac{1}{x}}) = x.$

**Curve with Conjugate (Isolated) Point at the Origin**

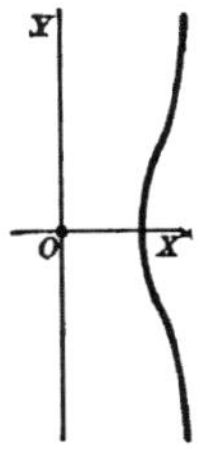


$y^2 = x^3 - x^2.$

**Curve with Cusp of Second Kind at Origin**

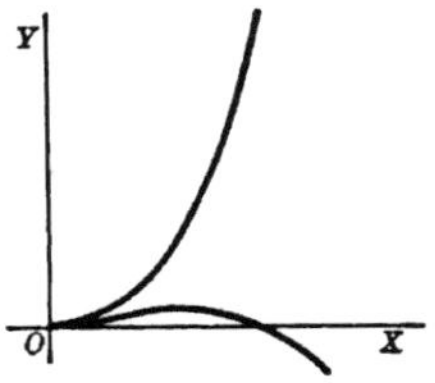


$(y - x^2)^2 = x^5.$

**Parabola**

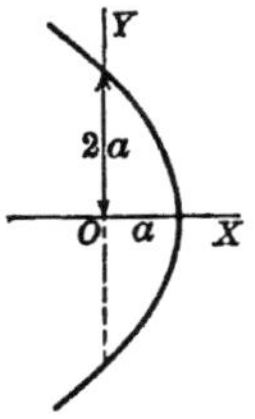


$\rho = a \sec^2 \frac{\theta}{2}.$

**Equilateral Hyperbola**

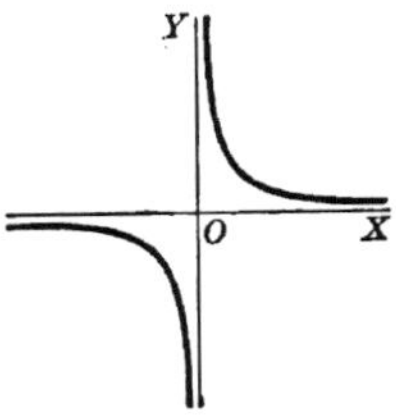


$xy = a.$

# INTEGRAL CALCULUS

## CHAPTER XXII

### INTEGRATION. RULES FOR INTEGRATING STANDARD ELEMENTARY FORMS

**165. Integration.** The student is already familiar with the mutually inverse operations of addition and subtraction, multiplication and division, involution and evolution. In the examples which follow, the second members of one column are respectively the inverse of the second members of the other column:

$$y = x^2 + 1, \qquad x = \pm\sqrt{y-1};$$

$$y = a^x, \qquad x = \log_a y;$$

$$y = \sin x, \qquad x = \text{arc} \sin y.$$

From the Differential Calculus we have learned how to calculate the derivative $f'(x)$ of a given function $f(x)$, an operation indicated by

$$\frac{d}{dx} f(x) = f'(x),$$

or, if we are using differentials, by

$$df(x) = f'(x)\, dx.$$

The problems of the Integral Calculus depend on the *inverse operation*, namely:

*To find a function $f(x)$ whose derivative*

$$(A) \qquad f'(x) = \phi(x)$$

*is given.*

Or, since it is customary to use differentials in the Integral Calculus, we may write

$$(B) \qquad df(x) = f'(x)\, dx = \phi(x)\, dx,$$

and state the problem as follows:

*Having given the differential of a function, to find the function itself.*

The function $f(x)$ thus found is called an *integral* * of the given differential expression, the process of finding it is called *integration*, and the operation is indicated by writing the *integral sign* † $\int$ in front of the given differential expression; thus

$$(C) \qquad \int f'(x)\,dx\ddagger = f(x),$$

read *an integral of* $f'(x)\,dx$ *equals* $f(x)$. The differential $dx$ indicates that $x$ is *the variable of integration.* For example,

(a) If $f(x) = x^3$, then $f'(x)\,dx = 3\,x^2dx$, and

$$\int 3\,x^2dx = x^3.$$

(b) If $f(x) = \sin x$, then $f'(x)\,dx = \cos x dx$, and

$$\int \cos x dx = \sin x.$$

(c) If $f(x) = \text{arc}\tan x$, then $f'(x)\,dx = \dfrac{dx}{1+x^2}$, and

$$\int \frac{dx}{1+x^2} = \text{arc}\tan x.$$

Let us now emphasize what is apparent from the preceding explanations, namely, that

*Differentiation and integration are inverse operations.*

Differentiating $(C)$ gives

$$(D) \qquad d\int f'(x)\,dx = f'(x)\,dx.$$

Substituting the value of $f'(x)\,dx\,[= df(x)]$ from $(B)$ in $(C)$, we get

$$(E) \qquad \int df(x) = f(x).$$

Therefore, considered as symbols of operation, $\dfrac{d}{dx}$ and $\int \cdots dx$ are *inverse to each other*; or, if we are using differentials, $d$ and $\int$ are inverse to each other.

* Called *anti-differential* by some writers.

† Historically this sign is a distorted $S$, the initial letter of the word *sum*. Instead of defining integration as the inverse of differentiation, we may define it as a process of summation, a very important notion which we will consider in Chapter XXVIII.

‡ Some authors write this $D_x^{-1}f'(x)$ when they wish to emphasize the fact that it is an inverse operation.

When $d$ is followed by $\int$ they annul each other, as in ($D$), but when $\int$ is followed by $d$, as in ($E$), that will not in general be the case unless we ignore the *constant of integration.* The reason for this will appear at once from the definition of the constant of integration given in the next section.

**166. Constant of integration. Indefinite integral.** From the preceding section it follows that

$$\text{since } d(x^3) = 3x^2dx, \text{ we have } \int 3x^2dx = x^3;$$

$$\text{since } d(x^3+2) = 3x^2dx, \text{ we have } \int 3x^2dx = x^3+2;$$

$$\text{since } d(x^3-7) = 3x^2dx, \text{ we have } \int 3x^2dx = x^3-7.$$

In fact, since $$d(x^3+C) = 3x^2dx,$$

where $C$ is any arbitrary constant, we have

$$\int 3x^2dx = x^3 + C.$$

A constant $C$ arising in this way is called a *constant of integration.** Since we can give $C$ as many values as we please, it follows that if a given differential expression has one integral, it has infinitely many differing only by constants. Hence

$$\int f'(x)\,dx = f(x) + C;$$

and since $C$ is unknown and *indefinite,* the expression

$$f(x) + C$$

is called the *indefinite integral of* $f'(x)\,dx$.

It is evident that if $\phi(x)$ is a function the derivative of which is $f(x)$, then $\phi(x) + C$, where $C$ is any constant whatever, is likewise a function the derivative of which is $f(x)$. Hence the

**Theorem.** *If two functions differ by a constant, they have the same derivative.*

It is, however, not obvious that if $\phi(x)$ is a function the derivative of which is $f(x)$, then *all* functions having the same derivative $f(x)$ are of the form $$\phi(x) + C,$$

where $C$ is any constant. In other words, there remains to be proved the

* Constant here means that it is independent of the *variable of integration.*

**Converse theorem.** *If two functions have the same derivative, their difference is a constant.*

**Proof.** Let $\phi(x)$ and $\psi(x)$ be two functions having the common derivative $f(x)$. Place

$$F(x) = \phi(x) - \psi(x); \text{ then}$$

$$(A) \quad F'(x) = \frac{d}{dx}[\phi(x) - \psi(x)] = f(x) - f(x) = 0. \text{ By hypothesis}$$

But from the Theorem of Mean Value (**46**), p. 166, we have

$$F(x + \Delta x) - F(x) = \Delta x F'(x + \theta \cdot \Delta x). \qquad 0 < \theta < 1$$

$$\therefore F(x + \Delta x) - F(x) = 0,$$

[Since by (*A*) the derivative of $F(x)$ is zero for all values of $x$.]

and
$$F(x + \Delta x) = F(x).$$

This means that the function

$$F(x) = \phi(x) - \psi(x)$$

does not change in value at all when $x$ takes on the increment $\Delta x$, i.e. $\phi(x)$ and $\psi(x)$ differ only by a constant.

In any given case the value of $C$ can be found when we know the value of the integral for some value of the variable, and this will be illustrated by numerous examples in the next chapter. For the present we shall content ourselves with first learning how to find the indefinite integrals of given differential expressions. In what follows we shall assume that *every continuous function has an indefinite integral*, a statement the rigorous proof of which is beyond the scope of this book. For all elementary functions, however, the truth of the statement will appear in the chapters which follow.

In all cases of indefinite integration the test to be applied in verifying the results is that *the differential of the integral must be equal to the given differential expression.*

**167. Rules for integrating standard elementary forms.** The Differential Calculus furnished us with a *General Rule* for differentiation (p. 29). The Integral Calculus gives us no corresponding general rule that can be readily applied in practice for performing the inverse operation of integration.* Each case requires special treatment and we arrive at the integral of a given differential expression through

* Even though the integral of a given differential expression may be known to exist, yet it may not be possible for us actually to find it in terms of known functions, because there are functions other than the elementary functions whose derivatives are elementary functions.

our previous knowledge of the known results of differentiation. That is, we must be able to answer the question, *What function, when differentiated, will yield the given differential expression?*

Integration then is essentially a tentative process, and to expedite the work, tables of known integrals are formed called *standard forms.* To effect any integration we compare the given differential expression with these forms, and if it is found to be identical with one of them, the integral is known. If it is not identical with one of them, we strive to reduce it to one of the standard forms by various methods, many of which employ artifices which can be suggested by practice only. Accordingly a large portion of our treatise on the Integral Calculus will be devoted to the explanation of methods for integrating those functions which frequently appear in the process of solving practical problems.

From any result of differentiation may always be derived a formula for integration.

The following two rules are useful in reducing differential expressions to standard forms:

(a) *The integral of any algebraic sum of differential expressions equals the same algebraic sum of the integrals of these expressions taken separately.*

**Proof.** Differentiating the expression

$$\int du + \int dv - \int dw,$$

$u$, $v$, $w$ being functions of a single variable, we get

$$du + dv - dw. \qquad \text{By III, p. 34}$$

(1) $$\therefore \int (du + dv - dw) = \int du + \int dv - \int dw.$$

(b) *A constant factor may be written either before or after the integral sign.*

**Proof.** Differentiating the expression

$$a\int dv$$

gives

$$adv. \qquad \text{By IV, p. 34}$$

(2) $$\therefore \int adv = a\int dv.$$

On account of their importance we shall write the above two rules as formulas at the head of the following list of

## Standard Elementary Forms

(1) $$\int (du + dv - dw) = \int du + \int dv - \int dw.$$

(2) $$\int a\,dv = a \int dv.$$

(3) $$\int dx = x + C.$$

(4) $$\int v^n dv = \frac{v^{n+1}}{n+1} + C. \qquad n \neq -1$$

(5) $$\int \frac{dv}{v} = \log v + C$$
$$= \log v + \log c = \log cv.$$

[Placing $C = \log c$.]

(6) $$\int a^v dv = \frac{a^v}{\log a} + C.$$

(7) $$\int e^v dv = e^v + C.$$

(8) $$\int \sin v\, dv = -\cos v + C.$$

(9) $$\int \cos v\, dv = \sin v + C.$$

(10) $$\int \sec^2 v\, dv = \tan v + C.$$

(11) $$\int \csc^2 v\, dv = -\cot v + C.$$

(12) $$\int \sec v \tan v\, dv = \sec v + C.$$

(13) $$\int \csc v \cot v\, dv = -\csc v + C.$$

(14) $$\int \tan v\, dv = \log \sec v + C.$$

(15) $$\int \cot v\, dv = \log \sin v + C.$$

(16) $$\int \sec v\, dv = \log (\sec v + \tan v) + C.$$

(17) $$\int \csc v\, dv = \log (\csc v - \cot v) + C.$$

(18) $$\int \frac{dv}{v^2+a^2} = \frac{1}{a}\text{arc tan}\frac{v}{a} + C.$$

(19) $$\int \frac{dv}{v^2-a^2} = \frac{1}{2a}\log\frac{v-a}{v+a} + C.$$

(20) $$\int \frac{dv}{\sqrt{a^2-v^2}} = \text{arc sin}\frac{v}{a} + C.$$

(21) $$\int \frac{dv}{\sqrt{v^2 \pm a^2}} = \log(v + \sqrt{v^2 \pm a^2}) + C.$$

(22) $$\int \frac{dv}{\sqrt{2av-v^2}} = \text{arc vers}\frac{v}{a} + C.$$

(23) $$\int \frac{dv}{v\sqrt{v^2-a^2}} = \frac{1}{a}\text{arc sec}\frac{v}{a} + C.$$

**Proof of (3).** Since $d(x+C) = dx,$ II, p. 34

we get $$\int dx = x + C.$$

**Proof of (4).** Since $d\left(\frac{v^{n+1}}{n+1} + C\right) = v^n dv,$ VI, p. 34

we get $$\int v^n dv = \frac{v^{n+1}}{n+1} + C.$$

This holds true for all values of $n$ except $n = -1$. For, when $n = -1$, (4) gives

$$\int v^{-1}dv = \frac{v^{-1+1}}{-1+1} + C = \frac{1}{0} + C = \infty + C,$$

which has no meaning.

The case when $n = -1$ comes under (5).

**Proof of (5).** Since $d(\log v + C) = \frac{dv}{v},$ VIII *a*, p. 35

we get $$\int \frac{dv}{v} = \log v + C.$$

The results we get from (5) may be put in more compact form if we denote the constant of integration by $\log c$. Thus

$$\int \frac{dv}{v} = \log v + \log c = \log cv.$$

Formula (5) states that *if the expression under the integral sign is a fraction whose numerator is the differential of the denominator, then the integral is the natural logarithm of the denominator.*

## EXAMPLES *

For formulas (1)–(5).

Verify the following integrations:

1. $\int x^6 dx = \frac{x^{6+1}}{6+1} + C = \frac{x^7}{7} + C$, by (4), where $v = x$ and $n = 6$.

2. $\int \sqrt{x}\, dx = \int x^{\frac{1}{2}}\, dx = \frac{x^{\frac{3}{2}}}{\frac{3}{2}} + C = \frac{2}{3} x^{\frac{3}{2}} + C,$ by (4)

   where $v = x$ and $n = \frac{1}{2}$.

3. $\int \frac{dx}{x^3} = \int x^{-3}\, dx = \frac{x^{-2}}{-2} + C = -\frac{1}{2x^2} + C,$ by (4)

   where $v = x$ and $n = -3$.

4. $\int ax^5\, dx = a \int x^5\, dx = \frac{ax^6}{6} + C.$ By (2) and (4)

5. $\int x^2\, dx = \frac{x^3}{3} + C.$

6. $\int x^{\frac{2}{3}}\, dx = \frac{3x^{\frac{5}{3}}}{5} + C.$

7. $\int at^{\frac{5}{2}}\, dt = \frac{2at^{\frac{7}{2}}}{7} + C.$

8. $\int \frac{dx}{3x^2} = -\frac{1}{3x} + C.$

9. $\int \frac{2\, dx}{ax^{\frac{1}{5}}} = \frac{5x^{\frac{4}{5}}}{2a} + C.$

10. $\int 5y dy = \frac{5y^2}{2} + C.$

11. $\int \sqrt{2px}\, dx = \frac{2}{3} x \sqrt{2px} + C.$

12. $\int \sqrt[3]{x}\, dx = \frac{3x^{\frac{4}{3}}}{4} + C.$

13. $\int s^{-\frac{1}{2}}\, ds = 2\sqrt{s} + C.$

14. $\int 3a\theta^2 d\theta = a\theta^3 + C.$

15. $\int 5m^2 z^6 dz = \frac{5m^2 z^7}{7} + C.$

16. $\int \frac{bd\phi}{\sqrt[3]{\phi}} = \frac{3b\phi^{\frac{2}{3}}}{2} + C.$

17. $\int (nx)^{\frac{1-n}{n}}\, dx = (nx)^{\frac{1}{n}} + C.$

18. $\int y^{-m-1} dy = -\frac{1}{my^m} + C.$

19.
$$\int (2x^3 - 5x^2 - 3x + 4)\, dx = \int 2x^3 dx - \int 5x^2 dx - \int 3x dx + \int 4\, dx \qquad \text{by (1)}$$
$$= 2\int x^3 dx - 5\int x^2 dx - 3\int x dx + 4\int dx \qquad \text{by (2)}$$
$$= \frac{x^4}{2} - \frac{5x^3}{3} - \frac{3x^2}{2} + 4x + C.$$

NOTE. Although each separate integration requires an arbitrary constant, we write down only a single constant denoting their algebraic sum.

20.
$$\int \left(\frac{2a}{\sqrt{x}} - \frac{b}{x^2} + 3c\sqrt[3]{x^2}\right) dx = \int 2ax^{-\frac{1}{2}}\, dx - \int bx^{-2}\, dx + \int 3cx^{\frac{2}{3}}\, dx \qquad \text{by (1)}$$
$$= 2a \int x^{-\frac{1}{2}}\, dx - b\int x^{-2}\, dx + 3c \int x^{\frac{2}{3}}\, dx \qquad \text{by (2)}$$
$$= 2a \cdot \frac{x^{\frac{1}{2}}}{\frac{1}{2}} - b \cdot \frac{x^{-1}}{-1} + 3c \cdot \frac{x^{\frac{5}{3}}}{\frac{5}{3}} + C \qquad \text{by (4)}$$
$$= 4a\sqrt{x} + \frac{b}{x} + \frac{9}{5} cx^{\frac{5}{3}} + C.$$

* When learning to integrate, the student should have oral drill in integrating simple functions.

21. $\int (2x^9 - 3x^6 + 12x^3 - 3)\,dx = \frac{x^{10}}{5} - \frac{3x^7}{7} + 3x^4 - 3x + C.$

22. $\int \left(\sqrt[3]{x^2} - \frac{1}{\sqrt[3]{x^2}} + \frac{2}{x^5}\right) dx = \frac{3x^{\frac{5}{3}}}{5} - 3x^{\frac{1}{3}} - \frac{1}{2x^4} + C.$

23. $\int (a^{\frac{2}{3}} - x^{\frac{2}{3}})^3\,dx = a^2x + \frac{9}{7}a^{\frac{2}{3}}x^{\frac{7}{3}} - \frac{9}{5}a^{\frac{4}{3}}x^{\frac{5}{3}} - \frac{x^3}{3} + C.$

HINT. First expand.

24. $\int (a^2 - y^2)^3\sqrt{y}\,dy = 2y^{\frac{3}{2}}\left(\frac{a^6}{3} - \frac{3a^4y^2}{7} + \frac{3a^2y^4}{11} - \frac{y^6}{15}\right) + C.$

25. $\int (\sqrt{a} - \sqrt{t})^3\,dt = a^{\frac{3}{2}}t - 2at^{\frac{3}{2}} + \frac{3a^{\frac{1}{2}}t^2}{2} - \frac{2t^{\frac{5}{2}}}{5} + C.$

26. $\int (x^2 - 2)^3x^3dx = \frac{x^{10}}{10} - \frac{3x^8}{4} + 2x^6 - 2x^4 + C.$

27. $\int (a^2 + b^2x^2)^{\frac{1}{2}}x\,dx = \frac{(a^2 + b^2x^2)^{\frac{3}{2}}}{3b^2} + C.$

HINT. This may be brought to form (4). For let $v = a^2 + b^2x^2$ and $n = \frac{1}{2}$; then $dv = 2b^2x\,dx$. If we now insert the constant factor $2b^2$ before $x\,dx$, and its reciprocal $\frac{1}{2b^2}$ before the integral sign (so as not to change the value of the expression), the expression may be integrated, using (4), namely,

$$\int v^n dv = \frac{v^{n+1}}{n+1} + C.$$

Thus,
$$\int (a^2 + b^2x^2)^{\frac{1}{2}}x\,dx = \frac{1}{2b^2}\int (a^2 + b^2x^2)^{\frac{1}{2}}2b^2x\,dx = \frac{1}{2b^2}\int (a^2 + b^2x^2)^{\frac{1}{2}}d(a^2 + b^2x^2)$$
$$= \frac{1}{2b^2}\cdot\frac{(a^2 + b^2x^2)^{\frac{3}{2}}}{\frac{3}{2}} + C = \frac{(a^2 + b^2x^2)^{\frac{3}{2}}}{3b^2} + C.$$

NOTE. The student is warned against transferring any function of the variable from one side of the integral sign to the other, since that would change the value of the integral.

28. $\int \sqrt{a^2 - x^2}\,x\,dx = \int (a^2 - x^2)^{\frac{1}{2}}x\,dx = -\frac{1}{3}(a^2 - x^2)^{\frac{3}{2}} + C.$

29. $\int (3ax^2 + 4bx^3)^{\frac{4}{3}}(2ax + 4bx^2)\,dx = \frac{1}{7}(3ax^2 + 4bx^3)^{\frac{7}{3}} + C.$

HINT. Use (4), making $v = 3ax^2 + 4bx^3$, $dv = (6ax + 12bx^2)\,dx$ and $n = \frac{4}{3}$.

30. $\int b(6ax^2 + 8bx^3)^{\frac{5}{3}}(2ax + 4bx^2)\,dx = \frac{b}{16}(6ax^2 + 8bx^3)^{\frac{8}{3}} + C.$

31. $\int \frac{x^2dx}{(a^2 + x^3)^{\frac{1}{2}}} = \frac{2}{3}(a^2 + x^3)^{\frac{1}{2}} + C.$

HINT. Write this $\int (a^2 + x^3)^{-\frac{1}{2}}x^2dx$ and apply (4).

32. $\int \frac{dx}{\sqrt{1 - x}} = -2\sqrt{1 - x} + C.$

33. $\int 2\pi y\left(\frac{y^2}{p^2} + 1\right)^{\frac{1}{2}}dy = \frac{2\pi}{3p}(y^2 + p^2)^{\frac{3}{2}} + C.$

34. $\int (1 + e^x)^{\frac{1}{2}}e^xdx = \frac{2}{3}(1 + e^x)^{\frac{3}{2}} + C.$

35. $\int \sin^2 x\cos x\,dx = \int (\sin x)^2\cos x\,dx = \frac{(\sin x)^3}{3} + C = \frac{\sin^3 x}{3} + C.$

HINT. Use (4), making $v = \sin x$, $dv = \cos x\,dx$, and $n = 2$.

36. $\int \cos^5 x \sin x dx = -\frac{\cos^6 x}{6} + C.$

37. $\int \sin^3 ax \cos ax dx = \frac{1}{4a} \sin^4 ax + C.$

38. $\int \cos^4 3x \sin 3x dx = -\frac{1}{15} \cos^5 3x + C.$

39. $\int \frac{x dx}{\sqrt{a^2 - x^2}} = -\sqrt{a^2 - x^2} + C.$

40. $\int \frac{5 a dt}{(b - t)^6} = \frac{a}{(b - t)^5} + C.$

41. $\int \sqrt[3]{1 + x^2}\, x dx = \frac{3}{8}(1 + x^2)^{\frac{4}{3}} + C.$

42. $\int \frac{s ds}{\sqrt[3]{1 - s^2}} = -\frac{3}{4}(1 - s^2)^{\frac{2}{3}} + C.$

43. $\int \frac{u^{n-1} du}{(a + bu^n)^m} = \frac{(a + bu^n)^{1-m}}{bn(1 - m)} + C.$

44. $\int \frac{2 as ds}{(b^2 - c^2 s^2)^2} = \frac{a}{c^2(b^2 - c^2 s^2)} + C.$

45. $\int \frac{3 ax dx}{b^2 + e^2 x^2} = \frac{3a}{2e^2} \log(b^2 + e^2 x^2) + C.$

**Solution.** $\int \frac{3 ax dx}{b^2 + e^2 x^2} = 3a \int \frac{x dx}{b^2 + e^2 x^2}.$ By (2)

This resembles (5). For let $v = b^2 + e^2 x^2$; then $dv = 2 e^2 x dx$. If we introduce the factor $2e^2$ after the integral sign, and $\frac{1}{2e^2}$ before it, we have not changed the value of the expression, but the numerator is now seen to be the differential of the denominator. Therefore

$$3a \int \frac{x dx}{b^2 + e^2 x^2} = \frac{3a}{2e^2} \int \frac{2 e^2 x dx}{b^2 + e^2 x^2} = \frac{3a}{2e^2} \int \frac{d(b^2 + e^2 x^2)}{b^2 + e^2 x^2} = \frac{3a}{2e^2} \log(b^2 + e^2 x^2) + C. \quad \text{By (5).}$$

46. $\int \frac{x dx}{x^2 - 1} = \frac{1}{2} \log(x^2 - 1) + C.$

47. $\int \frac{(x^2 - a^2) dx}{x^3 - 3a^2 x} = \log(x^3 - 3a^2 x)^{\frac{1}{3}} + C.$

48. $\int \frac{5x^2 dx}{10x^3 + 15} = \log(10x^3 + 15)^{\frac{1}{6}} + C.$

49. $\int \frac{5 bx dx}{8a - 6bx^2} = -\frac{5}{12} \log(8a - 6bx^2) + C.$

50. $\int \frac{x^3 dx}{x + 1} = x - \frac{x^2}{2} + \frac{x^3}{3} - \log(x + 1) + C.$

HINT. First divide the numerator by the denominator.

51. $\int \frac{2x - 1}{2x + 3} dx = x - \log(2x + 3)^2 + C.$

52. $\int \frac{x^{n-1}-1}{x^n - nx}dx = \frac{1}{n}\log(x^n - nx) + C.$

53. $\int \frac{(y^2-2)^3 dy}{y^5} = \frac{2}{y^4} - \frac{6}{y^2} + \frac{y^2}{2} - \log y^6 + C.$

54. $\int \frac{t^{n-1}dt}{a + bt^n} = \frac{1}{nb}\log(a + bt^n) + C.$

55. $\int (\log \alpha)^3 \frac{d\alpha}{\alpha} = \frac{1}{4}(\log \alpha)^4 + C.$

56. $\int \frac{r^2+1}{r-1}dr = \frac{r^2}{2} + r + 2\log(r-1) + C.$

57. $\int \frac{2e^x dx}{e^x + 1} = 2\log(e^x + 1) + C.$

58. $\int \frac{\sin x dx}{a + b\cos x} = -\frac{1}{b}\log(a + b\cos x) + C.$

59. $\int \frac{\sec^2\theta d\theta}{1 + 3\tan\theta} = \frac{1}{3}\log(1 + 3\tan\theta) + C.$

60. $\int \frac{e^{3s}ds}{e^s - 1} = \frac{e^{2s}}{2} + e^s + \log(e^s - 1) + C.$

61. $\int \frac{e^r - 1}{e^r + 1}dr = \log(e^r + 1)^2 - r + C.$

**62.** **Integrate the following and verify your results by differentiation:**

(a) $\int\left(4x^2 - \frac{2}{x}\right)dx.$

**Solution.** $\int\left(4x^2 - \frac{2}{x}\right)dx = 4\int x^2 dx - 2\int\frac{dx}{x} = \frac{4x^3}{3} - 2\log x + C.$

*Verification.* $d\left(\frac{4x^3}{3} - 2\log x + C\right) = \left(\frac{4}{3}\cdot 3x^2 - 2\cdot\frac{1}{x}\right)dx = \left(4x^2 - \frac{2}{x}\right)dx.$

(b) $\int x^4 dx.$ (h) $\int s^{m+n}ds.$ (n) $\int \frac{ax^{n-1}dx}{(b - cx^n)^m}.$ (t) $\int \sin^3\frac{2x}{3}\cos\frac{2x}{3}dx.$

(c) $\int 5\sqrt{x}dx.$ (i) $\int a\phi^{\frac{1}{3}}d\phi.$ (o) $\int \frac{aydy}{b - cy^2}.$ (u) $\int \frac{9s^3 ds}{s - 3}.$

(d) $\int x^{\frac{p}{q}}dx.$ (j) $\int \frac{a d\theta}{b\theta}.$ (p) $\int \frac{(a^{\frac{1}{3}} - z^{\frac{1}{3}})^2 dz}{\sqrt{z}}.$ (v) $\int \sqrt{a - bx}\,dx.$

(e) $\int y^{\frac{3}{5}}dy.$ (k) $\int t^{-2}dt.$ (q) $\int \frac{(x-1)dx}{x^2 - 2x + 5}.$ (w) $\int \frac{\csc^2\phi d\phi}{b - a\cot\phi}.$

(f) $\int 7\theta^{\frac{3}{4}}d\theta.$ (l) $\int b\sqrt[3]{x^2}dx.$ (r) $\int \frac{(x^3+1)dx}{x+2}.$ (x) $\int (e^{\frac{x}{a}} + 1)^{\frac{1}{2}}e^{\frac{x}{a}}dx.$

(g) $\int \frac{4dy}{y^3}.$ (m) $\int \frac{8dz}{\sqrt[5]{z^2}}.$ (s) $\int \frac{atdt}{\sqrt{b^2 - c^2t^2}}.$ (y) $\int (\log t)^3\frac{dt}{t}.$

**Proofs of (6) and (7).** These follow at once from the corresponding formulas for differentiation, **IX** and **IX *a***, p. 35.

**EXAMPLES**

For formulas (6) and (7).

Verify the following integrations:

1. $\int ba^{2x}dx = \frac{ba^{2x}}{2\log a} + C.$

**Solution.** $\int ba^{2x}dx = b\int a^{2x}dx.$ By (2)

This resembles (6). Let $v = 2x$; then $dv = 2\,dx$. If we then insert the factor 2 before $dx$ and the factor $\frac{1}{2}$ before the integral sign, we have

$$b\int a^{2x}dx = \frac{b}{2}\int a^{2x}2\,dx = \frac{b}{2}\int a^{2x}d(2x) = \frac{b}{2}\cdot\frac{a^{2x}}{\log a} + C. \qquad \text{By (6)}$$

2. $\int 3e^{x}dx = 3e^{x} + C.$

3. $\int e^{\frac{x}{n}}dx = ne^{\frac{x}{n}} + C.$

4. $\int e^{\sin x}\cos x\,dx = e^{\sin x} + C.$

5. $\int e^{2\cos x}\sin x\,dx = -\frac{e^{2\cos x}}{2} + C.$

6. $\int 3^{2y-1}dy = \frac{3^{2y-1}}{2\log 3} + C.$

7. $\int e^{-x}dx = -e^{-x} + C.$

8. $\int e^{ax}dx = \frac{e^{ax}}{a} + C.$

9. $\int a^{2x}dx = \frac{a^{2x}}{2\log a} + C.$

10. $\int 5^{x}dx = \frac{5^{x}}{\log 5} + C.$

11. $\int a^{x}e^{x}dx = \frac{a^{x}e^{x}}{1+\log a} + C.$

12. $\int (e^{5x} + a^{5x})\,dx = \frac{1}{5}\left(e^{5x} + \frac{a^{5x}}{\log a}\right) + C.$

13. $\int e^{x^2+4x+3}(x+2)\,dx = \frac{1}{2}e^{x^2+4x+3} + C.$

14. $\int (a^{nx} - b^{mx})\,dx = \frac{a^{nx}}{n\log a} - \frac{b^{mx}}{m\log b} + C.$

15. $\int \left(e^{\frac{x}{a}} + e^{-\frac{x}{a}}\right)dx = a\left(e^{\frac{x}{a}} - e^{-\frac{x}{a}}\right) + C.$

16. $\int (e^{y} + e^{-y})^2\,dy = \frac{1}{2}(e^{2y} - e^{-2y}) + 2y + C.$

17. $\int \frac{(a^{x} - b^{x})^2}{a^{x}b^{x}}dx = \frac{a^{x}b^{-x} - a^{-x}b^{x}}{\log a - \log b} - 2x + C.$

18. $\int (e^{4x} + a^{5x} + 3b^{-2x})\,dx = \frac{e^{4x}}{4} + \frac{a^{5x}}{5\log a} - \frac{3b^{-2x}}{2\log b} + C.$

19. $\int (e^{at} + e^{-at})^3dt = \frac{1}{a}\left[\frac{e^{3at}}{3} + 3e^{at} - 3e^{-at} - \frac{e^{-3at}}{3}\right] + C.$

20. Integrate the following and verify your results by differentiation:

(a) $\int e^{2s}ds.$ (e) $\int e^{-3x}dx.$ (i) $\int 5e^{ax}dx.$ (m) $\int a^{x^2}x\,dx.$

(b) $\int b^{-4x}dx.$ (f) $\int 2^{t^2}t\,dt.$ (j) $\int e^{\frac{2x}{a}}dx.$ (n) $\int e^{\frac{a\theta}{b}}d\theta.$

(c) $\int c^{ax}dx.$ (g) $\int 3^{x}e^{x}dx.$ (k) $\int ae^{-mx}dx.$ (o) $\int (e^{2x})^2dx.$

(d) $\int \frac{3\,dx}{e^{x}}.$ (h) $\int \frac{dy}{a^{2y}}.$ (l) $\int \frac{3\,dt}{\sqrt{e^{t}}}.$ (p) $\int \frac{a\,d\theta}{b^{3\theta}}.$

(q) $\int a^{2\sin\phi}\cos\phi d\phi$. (s) $\int e^{a\cos\theta}\sin\theta d\theta$. (u) $\int e^{\tan t}\sec^2 t dt$.

(r) $\int (e^{\frac{x}{a}} + e^{-\frac{x}{a}})^2 dx$. (t) $\int e^{x^2-4} x dx$. (v) $\int a^{\log x}\frac{dx}{x}$.

**Proofs of (8)–(13).** These follow at once from the corresponding formulas for differentiation, **XI,** etc., p. 35.

**Proof of (14).**

$$\begin{aligned}\int \tan v dv &= \int \frac{\sin v dv}{\cos v}\\ &= -\int \frac{-\sin v dv}{\cos v}\\ &= -\int \frac{d(\cos v)}{\cos v}\\ &= -\log\cos v + C \qquad \text{by (5)}\\ &= \log\sec v + C.\end{aligned}$$

$$\left[\text{Since } -\log\cos v = -\log\frac{1}{\sec v} = -\log 1 + \log\sec v = \log\sec v.\right]$$

**Proof of (15).**

$$\begin{aligned}\int \cot v dv &= \int \frac{\cos v dv}{\sin v} = \int \frac{d(\sin v)}{\sin v}\\ &= \log\sin v + C. \qquad \text{By (5)}\end{aligned}$$

**Proof of (16).** Since

$$\begin{aligned}\sec v &= \sec v \frac{\sec v + \tan v}{\sec v + \tan v}\\ &= \frac{\sec v\tan v + \sec^2 v}{\sec v + \tan v},\end{aligned}$$

$$\begin{aligned}\int \sec v\, dv &= \int \frac{\sec v\tan v + \sec^2 v}{\sec v + \tan v} dv\\ &= \int \frac{d(\sec v + \tan v)}{\sec v + \tan v}\\ &= \log(\sec v + \tan v) + C. \qquad \text{By (5)}\end{aligned}$$

**Proof of (17).** Since

$$\begin{aligned}\csc v &= \csc v \frac{\csc v - \cot v}{\csc v - \cot v}\\ &= \frac{-\csc v\cot v + \csc^2 v}{\csc v - \cot v},\end{aligned}$$

$$\begin{aligned}\int \csc v dv &= \int \frac{-\csc v\cot v + \csc^2 v}{\csc v - \cot v} dv\\ &= \int \frac{d(\csc v - \cot v)}{\csc v - \cot v}\\ &= \log(\csc v - \cot v) + C. \qquad \text{By (5)}\end{aligned}$$

**EXAMPLES**

For formulas (8)–(17).

Verify the following integrations:

1. $\int \sin 2\,ax dx = -\frac{\cos 2\,ax}{2\,a} + C.$

**Solution.** This resembles (8). For let $v = 2\,ax$; then $dv = 2\,adx$. If we now insert the factor $2\,a$ before $dx$ and the factor $\frac{1}{2\,a}$ before the integral sign, we get

$$\int \sin 2\,ax dx = \frac{1}{2\,a}\int \sin 2\,ax \cdot 2\,adx$$

$$= \frac{1}{2\,a}\int \sin 2\,ax \cdot d(2\,ax) = \frac{1}{2\,a} \cdot -\cos 2\,ax + C. \qquad \text{By (8)}$$

$$= -\frac{\cos 2\,ax}{2\,a} + C.$$

2. $\int \cos mx dx = \frac{1}{m}\sin mx + C.$

3. $\int \tan bx dx = \frac{1}{b}\log \sec bx + C.$

4. $\int \sec ax dx = \frac{1}{a}\log(\sec ax + \tan ax) + C.$

5. $\int \csc\frac{x}{a}dx = a\log\left(\csc\frac{x}{a} - \cot\frac{x}{a}\right) + C.$

6. $\int \sec 3\,t \tan 3\,t dt = \frac{1}{3}\sec 3\,t + C.$

7. $\int \csc ay \cot ay dy = -\frac{1}{a}\csc ay + C.$

8. $\int \csc^2 3\,x dx = -\frac{1}{3}\cot 3\,x + C.$

9. $\int \cot\frac{x}{2}dx = 2\log\sin\frac{x}{2} + C.$

10. $\int \sec^2 x^3 \cdot x^2 dx = \frac{1}{3}\tan x^3 + C.$

11. $\int \frac{dx}{\sin^2 x} = -\cot x + C.$

12. $\int \frac{ds}{\cos^2 s} = \tan s + C.$

13. $\int (\tan\theta + \cot\theta)^2 d\theta = \tan\theta - \cot\theta + C.$

14. $\int (\sec\alpha - \tan\alpha)^2 d\alpha = 2(\tan\alpha - \sec\alpha) - \alpha + C.$

15. $\int (\tan 2\,s - 1)^2 ds = \frac{1}{2}\tan 2\,s + \log\cos 2\,s + C.$

16. $\int \left(\cos\frac{\theta}{3} - \sin 3\,\theta\right)d\theta = 3\sin\frac{\theta}{3} + \frac{1}{3}\cos 3\,\theta + C.$

17. $\int \left(\sin ax + \sin\frac{x}{a}\right)dx = -\frac{1}{a}\cos ax - a\cos\frac{x}{a} + C.$

18. $\int k\cos(a + by)\,dy = \frac{k}{b}\sin(a + by) + C.$

19. $\int \text{cosec}^2 x^3 \cdot x^2 dx = -\frac{1}{3}\cot x^3 + C.$

20. $\int \cos(\log x)\frac{dx}{x} = \sin(\log x) + C.$

21. $\int \frac{dx}{1 + \cos x} = -\cot x + \csc x + C = \tan\frac{x}{2} + C.$

HINT. Multiply both numerator and denominator by $1 - \cos x$ and reduce before integrating.

22. $\int \frac{dx}{1 + \sin x} = \tan x - \sec x + C.$

**23.** Integrate the following and verify the results by differentiation:

(a) $\int \sin\frac{2x}{3}dx.$

(b) $\int \cot e^x \cdot e^x dx.$

(c) $\int \sec\frac{\theta}{2}\tan\frac{\theta}{2}d\theta.$

(d) $\int \csc\frac{a\phi}{b}\cot\frac{a\phi}{b}d\phi.$

(e) $\int \cos(b+ax)\,dx.$

(f) $\int \sec^2 2\,ax dx.$

(g) $\int \frac{dx}{\cos^2 3x}.$

(h) $\int \frac{dt}{\tan 5t}.$

(i) $\int \tan\frac{x}{3}dx.$

(j) $\int \csc^2(a-bx)\,dx.$

(k) $\int \frac{d\theta}{\sin^2 4\theta}.$

(l) $\int \frac{dy}{\cot 7y}.$

(m) $\int \left(\sec 2\theta - \csc\frac{\theta}{2}\right)d\theta.$

(n) $\int (\tan\phi + \sec\phi)^2 d\phi.$

(o) $\int \left(\tan 4s - \cot\frac{s}{4}\right)ds.$

(p) $\int (\cot x - 1)^2 dx.$

(q) $\int (\sec t - 1)^2 dt.$

(r) $\int (1-\csc y)^2 dy.$

(s) $\int \frac{dx}{1-\cos x}.$

(t) $\int \frac{dx}{1-\sin x}.$

(u) $\int \frac{2\,adt}{\sin bt}.$

(v) $\int \frac{5\,bd\theta}{\cos 8\theta}.$

**Proof of (18).** Since

$$d\left(\frac{1}{a}\text{arc tan}\frac{v}{a}+C\right)=\frac{1}{a}\frac{d\left(\frac{v}{a}\right)}{1+\left(\frac{v}{a}\right)^2}=\frac{dv}{v^2+a^2},\qquad \text{by XI, p. 36}$$

we get

$$\int\frac{dv}{v^2+a^2}=\frac{1}{a}\text{arc tan}\frac{v}{a}+C.^*$$

**Proof of (19).** Since $\frac{1}{v^2-a^2}=\frac{1}{2a}\left(\frac{1}{v-a}-\frac{1}{v+a}\right),$†

$$\int\frac{dv}{v^2-a^2}=\frac{1}{2a}\int\left(\frac{1}{v-a}-\frac{1}{v+a}\right)dv$$

$$=\frac{1}{2a}\{\log(v-a)-\log(v+a)\}+C\qquad \text{by (5)}$$

$$=\frac{1}{2a}\log\frac{v-a}{v+a}+C.$$

* Also $d\left(\frac{1}{a}\text{arc cot}\frac{v}{a}+C\right)=-\frac{dv}{v^2+a^2}$ and $\int\frac{dv}{v^2+a^2}=-\frac{1}{a}\text{arc cot}\frac{v}{a}+C'$. Hence

$$\int\frac{dv}{v^2+a^2}=\frac{1}{a}\text{arc tan}\frac{v}{a}+C=-\frac{1}{a}\text{arc cot}\frac{v}{a}+C'.$$

Since $\text{arc tan}\frac{v}{a}+\text{arc cot}\frac{v}{a}=\frac{\pi}{2}$, we see that one result may be easily transformed into the other. The same kind of discussion may be given for (20) involving $\text{arc sin}\frac{v}{a}$ and $\text{arc cos}\frac{v}{a}$, and for (23) involving $\text{arc sec}\frac{v}{a}$ and $\text{arc csc}\frac{v}{a}$.

† By breaking the fraction up into partial fractions (see Case I, p. 325).

**Proof of (20).** Since

$$d\left(\text{arc}\sin\frac{v}{a}+C\right)=\frac{d\left(\frac{v}{a}\right)}{\sqrt{1-\left(\frac{v}{a}\right)^2}}=\frac{dv}{\sqrt{a^2-v^2}},\qquad \text{by } \mathbf{XVIII},\text{ p. 35}$$

we get

$$\int\frac{dv}{\sqrt{a^2-v^2}}=\text{arc}\sin\frac{v}{a}+C.$$

**Proof of (21).** Assume $v=a\tan z$, where $z$ is a new variable; differentiating, $dv=a\sec^2 z dz$. Hence, by substitution,

$$\int\frac{dv}{\sqrt{v^2+a^2}}=\int\frac{a\sec^2 z dz}{\sqrt{a^2\tan^2 z+a^2}}=\int\frac{\sec^2 z dz}{\sqrt{\tan^2 z+1}}$$

$$=\int\sec z dz=\log(\sec z+\tan z)+C \qquad \text{by } \mathbf{(16)}$$

$$=\log\left(\tan z+\sqrt{\tan^2 z+1}\right)+c. \qquad \text{By 28, p. 2}$$

But $\tan z=\frac{v}{a}$; hence,

$$\int\frac{dv}{\sqrt{v^2+a^2}}=\log\left(\frac{v}{a}+\sqrt{\frac{v^2}{a^2}+1}\right)+c$$

$$=\log\frac{v+\sqrt{v^2+a^2}}{a}+c$$

$$=\log\left(v+\sqrt{v^2+a^2}\right)-\log a+c.$$

Placing $C=-\log a+c$, we get

$$\int\frac{dv}{\sqrt{v^2+a^2}}=\log\left(v+\sqrt{v^2+a^2}\right)+C.$$

In the same manner, by assuming $v=a\sec z$, $dv=a\sec z\tan z dz$ we get

$$\int\frac{dv}{\sqrt{v^2-a^2}}=\int\frac{a\sec z\tan z dz}{\sqrt{a^2\sec^2 z-a^2}}=\int\sec z dz$$

$$=\log(\sec z+\tan z)+c \qquad \text{by } \mathbf{(16)}$$

$$=\log\left(\sec z+\sqrt{\sec^2 z-1}\right)+c \qquad \text{by 28, p. 2}$$

$$=\log\left(\frac{v}{a}+\sqrt{\frac{v^2}{a^2}-1}\right)+c=\log\left(v+\sqrt{v^2-a^2}\right)+C.$$

**Proofs of (22) and (23).** These follow at once from the corresponding formulas for differentiation, **XXII** and **XXIV**, p. 36.

A large number of the fractional forms to be integrated have a single term in the numerator, while the denominator is a quadratic expression with or without a square root sign over it. The following outline will assist the student in choosing the right formula.

| | NUMERATOR OF FIRST DEGREE | NUMERATOR OF ZERO DEGREE |
|---|---|---|
| No radical in denominator | $\int \frac{dv}{v} = \log v + C$ | $\int \frac{dv}{v^2 + a^2} = \frac{1}{a} \text{arc tan} \frac{v}{a} + C$, or, $\int \frac{dv}{v^2 - a^2} = \frac{1}{2a} \log \frac{v-a}{v+a} + C$ |
| Radical in denominator | $\int v^n dv = \frac{v^{n+1}}{n+1} + C$ $(n = -\frac{1}{2})$ | $\int \frac{dv}{\sqrt{a^2 - v^2}} = \text{arc sin} \frac{v}{a} + C$, or, $\int \frac{dv}{\sqrt{v^2 \pm a^2}} = \log(v + \sqrt{v^2 \pm a^2}) + C$ |

Students should be drilled in integrating the simple forms orally and to tell by inspection what formulas may be applied in integrating examples chosen at random.

### EXAMPLES

For formulas (18)–(23).

Verify the following integrations:

1. $\int \frac{dx}{4x^2 + 9} = \frac{1}{6} \text{arc tan} \frac{2x}{3} + C.$

**Solution.** This resembles (18). For, let $v^2 = 4x^2$ and $a^2 = 9$; then $v = 2x$, $dv = 2\,dx$, and $a = 3$. Hence if we multiply the numerator by 2 and divide in front of the integral sign by 2, we get

$$\int \frac{dx}{4x^2+9} = \frac{1}{2}\int \frac{2\,dx}{(2x)^2 + (3)^2} = \frac{1}{2}\int \frac{d(2x)}{(2x)^2 + (3)^2}$$

$$= \frac{1}{6} \text{arc tan} \frac{2x}{3} + C. \qquad \text{By (18)}$$

2. $\int \frac{dx}{9x^2 - 4} = \frac{1}{12} \log \frac{3x-2}{3x+2} + C.$

3. $\int \frac{dx}{\sqrt{16 - 9x^2}} = \frac{1}{3} \text{arc sin} \frac{3x}{4} + C.$

4. $\int \frac{dx}{\sqrt{9 - x^2}} = \text{arc sin} \frac{x}{3} + C.$

5. $\int \frac{dx}{\sqrt{x^2 - 9}} = \log(x + \sqrt{x^2 - 9}) + C.$

6. $\int \frac{dx}{\sqrt{x^2 + 9}} = \log(x + \sqrt{x^2 + 9}) + C.$

7. $\int \frac{5\,dx}{x^2 + 9} = \frac{5}{3} \text{arc tan} \frac{x}{3} + C.$

8. $\int \frac{b\,dx}{a^2x^2 - c^2} = \frac{b}{2ac} \log \frac{ax - c}{ax + c} + C.$

9. $\int \frac{7x^2 dx}{5 - x^6} = \frac{7}{6\sqrt{5}} \log \frac{x^3 + \sqrt{5}}{x^3 - \sqrt{5}} + C.$

10. $\int \frac{5x\,dx}{\sqrt{1-x^4}} = \frac{5}{2}\text{arc}\sin x^2 + C.$

11. $\int \frac{dx}{x\sqrt{4x^2-9}} = \frac{1}{3}\text{arc}\sec\frac{2x}{3} + C.$

12. $\int \frac{ax\,dx}{x^4+e^4} = \frac{a}{2e^2}\text{arc}\tan\frac{x^2}{e^2} + C.$

13. $\int \frac{dx}{\sqrt{6x-x^2}} = \text{arc vers}\,\frac{x}{3} + C.$

14. $\int \frac{e\,dt}{a^2-b^2t^2} = \frac{e}{2ab}\log\frac{bt+a}{bt-a} + C$

15. $\int \frac{e^t dt}{\sqrt{1-e^{2t}}} = \text{arc}\sin e^t + C.$

16. $\int \frac{7\,ds}{\sqrt{3-5s^2}} = \frac{7}{\sqrt{5}}\text{arc}\sin\sqrt{\frac{5}{3}}\,s + C.$

17. $\int \frac{dv}{\sqrt{av^2-b}} = \frac{1}{\sqrt{a}}\log\left(\sqrt{a}v + \sqrt{av^2-b}\right) + C.$

18. $\int \frac{\cos\alpha\,d\alpha}{a^2+\sin^2\alpha} = \frac{1}{a}\text{arc}\tan\left(\frac{\sin\alpha}{a}\right) + C.$

19. $\int \frac{dx}{x\sqrt{1-\log^2 x}} = \text{arc}\sin(\log x) + C.$

20. $\int \frac{dx}{\sqrt{b^2+e^2x^2}} = \frac{1}{e}\log\left(ex + \sqrt{b^2+e^2x^2}\right) + C.$

21. $\int \frac{dy}{\sqrt{b^2y^2-a^2}} = \frac{1}{b}\log\left(by + \sqrt{b^2y^2-a^2}\right) + C.$

22. $\int \frac{du}{\sqrt{a^2-(u+b)^2}} = \text{arc}\sin\frac{u+b}{a} + C.$

23. $\int \frac{a\,dz}{(z-e)^2+b^2} = \frac{a}{b}\text{arc}\tan\frac{z-e}{b} + C.$

24. $\int \frac{dx}{x^2+2x+5} = \frac{1}{2}\text{arc}\tan\frac{x+1}{2} + C.$

HINT. By completing the square in the denominator, this expression may be brought to a form similar to that of Ex. 17. Thus,

$$\int \frac{dx}{x^2+2x+5} = \int \frac{dx}{(x^2+2x+1)+4} = \int \frac{dx}{(x+1)^2+4} = \frac{1}{2}\text{arc}\tan\frac{x+1}{2} + C. \qquad \text{By (18)}$$

Here $v = x+1$ and $a = 2$.

25. $\int \frac{dx}{\sqrt{2+x-x^2}} = \text{arc}\sin\frac{2x-1}{3} + C.$

HINT. Bring this to the form of Ex. 16 by completing the square. Thus,

$$\int \frac{dx}{\sqrt{2+x-x^2}} = \int \frac{dx}{\sqrt{2-(x^2-x)}} = \int \frac{dx}{\sqrt{2-(x^2-x+\frac{1}{4})+\frac{1}{4}}} = \int \frac{dx}{\sqrt{\frac{9}{4}-(x-\frac{1}{2})^2}} = \text{arc}\sin\frac{2x-1}{3} + C. \quad \text{By (20)}$$

Here $v = x - \frac{1}{2}$ and $a = \frac{3}{2}$.

26. $\int \frac{dx}{1+x+x^2} = \frac{2}{\sqrt{3}}\text{arc}\tan\frac{2x+1}{\sqrt{3}} + C.$

27. $\int \frac{dx}{3x^2-2x+4} = \frac{1}{3}\int \frac{dx}{x^2-\frac{2}{3}x+\frac{4}{3}} = \frac{1}{\sqrt{11}}\text{arc}\tan\frac{3x-1}{\sqrt{11}} + C.$

28. $\int \frac{dx}{\sqrt{2 - 3x - 4x^2}} = \frac{1}{2}\int \frac{dx}{\sqrt{\frac{1}{2} - \frac{3}{4}x - x^2}} = \frac{1}{2}\int \frac{dx}{\sqrt{\frac{1}{2} - (x^2 + \frac{3}{4}x + \frac{9}{64}) + \frac{9}{64}}}$

$$= \frac{1}{2}\text{arc sin}\frac{8x + 3}{\sqrt{41}} + C.$$

29. $\int \frac{dx}{\sqrt{8x - x^2 - 2}} = \text{arc sin}\,(2x - 3) + C.$

30. $\int \frac{dv}{v^2 - 6v + 5} = \frac{1}{4}\log\frac{v - 5}{v - 1} + C.$

31. $\int \frac{dy}{y^2 + 3y + 1} = \frac{1}{\sqrt{5}}\log\frac{2y + 3 - \sqrt{5}}{2y + 3 + \sqrt{5}} + C.$

32. $\int \frac{dt}{\sqrt{1 + t + t^2}} = \log\left(t + \frac{1}{2} + \sqrt{t^2 + t + 1}\right) + C.$

33. $\int \frac{dz}{2z^2 - 2z + 1} = \text{arc tan}\,(2z - 1) + C.$

34. $\int \frac{ds}{\sqrt{2as + s^2}} = \log\left(s + a + \sqrt{2as + s^2}\right) + C.$

35. $\int \frac{dx}{x\sqrt{c^2x^2 - a^2b^2}} = \frac{1}{ab}\text{arc sec}\frac{cx}{ab} + C.$

36. $\int \frac{3x^2dx}{\sqrt{x^3 - 9x^6}} = \frac{1}{3}\text{arc vers}\,18x^3 + C.$

37. $\int \frac{(b + ex)dx}{a^2 + x^2} = \frac{b}{a}\text{arc tan}\frac{x}{a} + \frac{e}{2}\log(a^2 + x^2) + C.$

HINT. A fraction with more than one term in the numerator may be broken up into the sum of two or more fractions having the several terms of the original numerator as numerators, all the denominators being the same as the denominator of the original fraction. Thus, the last example may be written

$$\int \frac{(b + ex)\,dx}{a^2 + x^2} = \int \frac{bdx}{a^2 + x^2} + \int \frac{exdx}{a^2 + x^2} = b\int \frac{dx}{a^2 + x^2} + e\int \frac{xdx}{a^2 + x^2},$$

each term being integrated separately.

38. $\int \frac{(3x - 1)dx}{x^2 + 9} = \frac{3}{2}\log(x^2 + 9) - \frac{1}{3}\text{arc tan}\frac{x}{3} + C.$

39. $\int \frac{2x - 5}{3x^2 - 2}dx = \frac{1}{3}\log(3x^2 - 2) - \frac{5}{2\sqrt{6}}\log\frac{x\sqrt{3} - \sqrt{2}}{x\sqrt{3} + \sqrt{2}} + C.$

40. $\int \frac{3s - 2}{\sqrt{9 - s^2}}ds = -3\sqrt{9 - s^2} - 2\,\text{arc sin}\frac{s}{3} + C.$

41. $\int \frac{x + 3}{\sqrt{x^2 + 4}}dx = \sqrt{x^2 + 4} + 3\log\left(x + \sqrt{x^2 + 4}\right) + C.$

42. $\int \frac{(5t - 1)\,dt}{\sqrt{3t^2 - 9}} = \frac{5}{3}\sqrt{3t^2 - 9} - \frac{1}{\sqrt{3}}\log\left(t\sqrt{3} + \sqrt{3t^2 - 9}\right) + C.$

**43.** Integrate the following expressions and verify your results by differentiation :

(a) $\int \frac{dx}{\sqrt{4-25x^2}}$.

(b) $\int \frac{adx}{3-12x^2}$.

(c) $\int \frac{2\,dt}{3t^2-5t+2}$.

(d) $\int \frac{dx}{x\sqrt{9x^2-4}}$.

(e) $\int \frac{\sin\theta\,d\theta}{\sqrt{9-4\cos^2\theta}}$.

(f) $\int \frac{(2x-3)\,dx}{x^2+4}$.

(g) $\int \frac{y+3}{\sqrt{9y^2+1}}dy$.

(h) $\int \frac{dx}{x^2+6x+13}$.

(i) $\int \frac{2\,dx}{\sqrt{25x^2-4}}$.

(j) $\int \frac{bdy}{12y^2+3}$.

(k) $\int \frac{d\phi}{\sqrt{3\phi^2-2}}$.

(l) $\int \frac{dx}{x\sqrt{9x^2-16}}$.

(m) $\int \frac{dz}{z\sqrt{4-(\log z)^2}}$.

(n) $\int \frac{(t+2)\,dt}{4t^2-3}$.

(o) $\int \frac{(ax-b)\,dx}{\sqrt{1+9x^2}}$.

(p) $\int \frac{dx}{\sqrt{8+4x-4x^2}}$.

(q) $\int \frac{3\,dx}{\sqrt{5x^2+1}}$.

(r) $\int \frac{dw}{12w^2-3}$.

(s) $\int \frac{3\,d\theta}{\sqrt{2-3\theta^2}}$.

(t) $\int \frac{dt}{\sqrt{7t-4t^2+5}}$.

(u) $\int \frac{e^{ax}dx}{e^{2ax}+1}$.

(v) $\int \frac{(3s-5)\,ds}{\sqrt{1-9s^2}}$.

(w) $\int \frac{(2x+3)\,dx}{\sqrt{a^2x^2-b^2}}$.

(x) $\int \frac{dt}{\sqrt{t^2-4t+2}}$.

**168. Trigonometric differentials.** We shall now consider some trigonometric differentials of frequent occurrence which may be readily integrated by being transformed into standard forms by means of simple trigonometric reductions.

**Example I.** *To find* $\int \sin^m x \cos^n x dx$.

When either $m$ or $n$ is a positive odd integer, no matter what the other may be, this integration may be performed by means of formula (**4**),

$$\int v^n dv = \frac{v^{n+1}}{n+1}.$$

For the integral is reducible to the form

$$\int (\textit{terms involving only } \cos x) \sin x dx,$$

when $\sin x$ has the odd exponent, and to the form

$$\int (\textit{terms involving only } \sin x) \cos x dx,$$

when $\cos x$ has the odd exponent. We shall illustrate this by means of examples.

ILLUSTRATIVE EXAMPLE 1. Find $\int \sin^2 x \cos^5 x dx$.

**Solution.**

$$\begin{aligned}\int \sin^2 x \cos^5 x dx &= \int \sin^2 x \cos^4 x \cos x dx \\ &= \int \sin^2 x (1 - \sin^2 x)^2 \cos x dx && \text{by 28, p. 2} \\ &= \int (\sin^2 x - 2\sin^4 x + \sin^6 x)\cos x dx \\ &= \int (\sin x)^2 \cos x dx - 2\int (\sin x)^4 \cos x dx + \int (\sin x)^6 \cos x dx \\ &= \frac{\sin^3 x}{3} - \frac{2\sin^5 x}{5} + \frac{\sin^7 x}{7} + C. && \text{By (4)}\end{aligned}$$

Here $v = \sin x$, $dv = \cos x dx$, and $n = 2$, 4, and 6 respectively.

ILLUSTRATIVE EXAMPLE 2. Find $\int \cos^3 x dx$.

**Solution.**

$$\begin{aligned}\int \cos^3 x dx &= \int \cos^2 x \cos x dx = \int (1 - \sin^2 x)\cos x dx \\ &= \int \cos x dx - \int \sin^2 x \cos x dx \\ &= \sin x - \frac{\sin^3 x}{3} + C.\end{aligned}$$

## EXAMPLES

1. $\int \sin^3 x dx = \frac{1}{3}\cos^3 x - \cos x + C.$

2. $\int \sin^2 x \cos x dx = \frac{\sin^3 x}{3} + C.$

3. $\int \sin x \cos x dx = \frac{\sin^2 x}{2} + C.$*

4. $\int \cos^2 \alpha \sin \alpha d\alpha = -\frac{\cos^3 \alpha}{3} + C.$

5. $\int \sin^3 6\theta \cos 6\theta d\theta = \frac{\sin^4 6\theta}{24} + C.$

6. $\int \cos^3 2\theta \sin 2\theta d\theta = -\frac{\cos^4 2\theta}{8} + C.$

7. $\int \frac{\cos^3 x dx}{\sin^4 x} = \csc x - \frac{1}{3}\csc^3 x + C$

8. $\int \frac{\sin^3 \alpha d\alpha}{\cos^2 \alpha} = \sec \alpha + \cos \alpha + C.$

9. $\int \cos^4 x \sin^3 x dx = -\frac{1}{5}\cos^5 x + \frac{1}{7}\cos^7 x + C.$

10. $\int \sin^5 x dx = -\cos x + \frac{2}{3}\cos^3 x - \frac{\cos^5 x}{5} + C.$

11. $\int \cos^5 x dx = \sin x - \frac{2}{3}\sin^3 x + \frac{\sin^5 x}{5} + C.$

12. $\int \sin^{\frac{3}{7}} \phi \cos^3 \phi d\phi = \frac{7}{10}\sin^{\frac{10}{7}} \phi - \frac{7}{24}\sin^{\frac{24}{7}} \phi + C.$

* This was integrated by the power formula taking $n = 1$, $v = \sin x$, $dv = \cos x dx$. To illustrate how an answer may take on different forms when more than one method of integration is possible, let us take $n = 1$, $v = \cos x$, $dv = -\sin x dx$, and again integrate by the power formula. Then

$$\int \sin x \cos x dx = -\int (\cos x)(-\sin x dx) = -\frac{\cos^2 x}{2} + C',$$

a result which differs from the first one in the arbitrary constant only. For,

$$-\frac{\cos^2 x}{2} + C' = -\frac{1 - \sin^2 x}{2} + C' = -\frac{1}{2} + \frac{\sin^2 x}{2} + C' = \frac{\sin^2 x}{2} - \frac{1}{2} + C'.$$

Hence, comparing the two answers, $C = -\frac{1}{2} + C'$.

13. $\int \sin^{\frac{2}{3}}\theta \cos^5\theta d\theta = \frac{3}{5}\sin^{\frac{5}{3}}\theta - \frac{6}{11}\sin^{\frac{11}{3}}\theta + \frac{3}{17}\sin^{\frac{17}{3}}\theta + C.$

14. $\int \frac{\sin^5 y}{\sqrt{\cos y}} dy = -2\sqrt{\cos y}\left(1 - \frac{2}{5}\cos^2 y + \frac{1}{9}\cos^4 y\right) + C.$

15. $\int \frac{\cos^5 t dt}{\sqrt[3]{\sin t}} = \frac{3}{2}\sin^{\frac{2}{3}} t\left(1 - \frac{1}{2}\sin^2 t + \frac{1}{7}\sin^4 t\right) + C.$

16. Integrate the following expressions and prove your results by differentiation:

(a) $\int \sin^3 2\theta d\theta.$ (f) $\int \cos^5 ax \sin axdx.$ (k) $\int \sin^3 mt \cos^2 mtdt.$

(b) $\int \cos^3 \frac{\theta}{2} d\theta.$ (g) $\int \sin^2 \frac{2x}{3} \cos \frac{2x}{3} dx.$ (l) $\int \sin^5 ntdt.$

(c) $\int \sin 2x \cos 2xdx.$ (h) $\int \cos^2 3x \sin 3xdx.$ (m) $\int \sin^4 x \cos xdx.$

(d) $\int \sin^3 t \cos^3 tdt.$ (i) $\int \sin^5 bs \cos bsds.$ (n) $\int \cos^4 y \sin ydy.$

(e) $\int \cos\frac{x}{a} \sin\frac{x}{a} dx.$ (j) $\int \cos^3 \frac{\phi}{2} \sin^2 \frac{\phi}{2} d\phi.$ (o) $\int \cos^3 (a + bt) dt$

**Example II.** *To find* $\int \tan^n xdx$, *or* $\int \cot^n xdx$.

These forms can be readily integrated, when $n$ is an integer, on somewhat the same plan as the previous examples.

Illustrative Example 1. Find $\int \tan^4 xdx$.

**Solution.** $\int \tan^4 xdx = \int \tan^2 x (\sec^2 x - 1) dx$ by 28, p. 2

$= \int \tan^2 x \sec^2 xdx - \int \tan^2 xdx$

$= \int (\tan x)^2 d(\tan x) - \int (\sec^2 x - 1) dx$

$= \frac{\tan^3 x}{3} - \tan x + x + C.$

**Example III.** *To find* $\int \sec^n xdx$, *or* $\int \csc^n xdx$.

These can be easily integrated when $n$ is a positive even integer, as follows:

Illustrative Example 2. Find $\int \sec^6 xdx$.

**Solution.** $\int \sec^6 xdx = \int (\tan^2 x + 1)^2 \sec^2 xdx$ by 28, p. 2

$= \int (\tan x)^4 \sec^2 xdx + 2\int (\tan x)^2 \sec^2 xdx + \int \sec^2 xdx$

$= \frac{\tan^5 x}{5} + 2\frac{\tan^3 x}{3} + \tan x + C.$

When $n$ is an odd positive integer greater than unity, the best plan is to reduce to sine or cosine and then use reduction formulas on p. 303.

**Example IV.** *To find* $\int \tan^m x \sec^n x dx$, *or* $\int \cot^m x \csc^n x dx$.

When $n$ is a positive even integer we proceed as in Example III.

Illustrative Example 3. Find $\int \tan^6 x \sec^4 x dx$.

**Solution.**

$$\int \tan^6 x \sec^4 x dx = \int \tan^6 x (\tan^2 x + 1) \sec^2 x dx \quad \text{by 28, p. 2}$$

$$= \int (\tan x)^8 \sec^2 x dx + \int \tan^6 x \sec^2 x dx$$

$$= \frac{\tan^9 x}{9} + \frac{\tan^7 x}{7} + C. \quad \text{By (4)}$$

Here $v = \tan x$, $dv = \sec^2 x dx$, etc.

When $m$ is odd we may proceed as in the following example.

Illustrative Example 4. Find $\int \tan^5 x \sec^3 x dx$.

**Solution.**

$$\int \tan^5 x \sec^3 x dx = \int \tan^4 x \sec^2 x \sec x \tan x dx$$

$$= \int (\sec^2 x - 1)^2 \sec^2 x \sec x \tan x dx \quad \text{by 28, p. 2}$$

$$= \int (\sec^6 x - 2 \sec^4 x + \sec^2 x) \sec x \tan x dx$$

$$= \frac{\sec^7 x}{7} - \frac{2 \sec^5 x}{5} + \frac{\sec^3 x}{3} + C. \quad \text{By (4)}$$

Here $v = \sec x$, $dv = \sec x \tan x dx$, etc.

**EXAMPLES**

1. $\int \tan^3 x dx = \frac{\tan^2 x}{2} + \log \cos x + C.$

2. $\int \tan^2 2x dx = \frac{\tan 2x}{2} - x + C.$

3. $\int \cot^3 x dx = -\frac{\cot^2 x}{2} - \log \sin x + C.$

4. $\int \cot^2 x dx = - \cot x - x + C.$

5. $\int \cot^4 \frac{x}{3} dx = - \cot^3 \frac{x}{3} + 3 \cot \frac{x}{3} + x + C.$

6. $\int \cot^5 \alpha d\alpha = - \frac{1}{4} \cot^4 \alpha + \frac{1}{4} \cot^2 \alpha + \log \sin \alpha + C.$

7. $\int \tan^5 \frac{y}{4} dy = \tan^4 \frac{y}{4} - 2 \tan^2 \frac{y}{4} + 4 \log \sec \frac{y}{4} + C.$

8. $\int \sec^8 x dx = \frac{\tan^7 x}{7} + \frac{3 \tan^5 x}{5} + \tan^3 x + \tan x + C.$

9. $\int \csc^6 x dx = - \cot x - \frac{2}{3} \cot^3 x - \frac{1}{5} \cot^5 x + C.$

10. $\int \tan^4 \phi \sec^4 \phi d\phi = \frac{\tan^7 \phi}{7} + \frac{\tan^5 \phi}{5} + C.$

11. $\int \tan^3 \theta \sec^5 \theta d\theta = \frac{1}{7} \sec^7 \theta - \frac{1}{5} \sec^5 \theta + C.$

12. $\int \cot^5 x \csc^4 x dx = - \frac{\cot^6 x}{6} - \frac{\cot^8 x}{8} + C.$

13. $\int \tan^{\frac{3}{2}} x \sec^4 x dx = \frac{2 \tan^{\frac{5}{2}} x}{5} + \frac{2 \tan^{\frac{9}{2}} x}{9} + C.$

14. $\int \tan^5 y \sec^{\frac{3}{2}} y dy = 2 \sec^{\frac{3}{2}} y \left(\frac{\sec^4 y}{11} - \frac{2 \sec^2 y}{7} + \frac{1}{3}\right) + C.$

15. $\int \frac{\sec^6 \alpha d\alpha}{\tan^4 \alpha} = \tan \alpha - 2 \cot \alpha - \frac{\cot^3 \alpha}{3} + C.$

16. $\int (\tan^2 z + \tan^4 z) dz = \frac{1}{3} \tan^3 z + C.$

17. $\int (\tan t + \cot t)^3 dt = \frac{1}{2} (\tan^2 t - \cot^2 t) + \log \tan^2 t + C.$

18. Integrate the following expressions and prove your results by differentiation:

(a) $\int \tan^2 2\,t dt.$

(b) $\int \cot^2 \frac{t}{2} dt.$

(c) $\int \tan^3 ax dx.$

(d) $\int \cot^3 \frac{x}{a} dx.$

(e) $\int \frac{2\,dt}{\tan^4 t}.$

(f) $\int \frac{3\,d\theta}{\cot^2 4\theta}.$

(g) $\int \sec^2 \theta \tan^2 \theta d\theta.$

(h) $\int \csc^2 \phi \cot^2 \phi\, d\phi.$

(i) $\int \frac{a dx}{\tan^2 4\,x}.$

(j) $\int \tan^3 t \sec^2 t dt.$

(k) $\int \cot^5 y \csc^2 y dy.$

(l) $\int \frac{b d\theta}{\cot^3 \theta}.$

(m) $\int \frac{\sqrt{\tan x}}{\cos^2 x} dx.$

(n) $\int \frac{\cot^{\frac{2}{3}} x}{\sin^2 x} dx.$

(o) $\int \sec^4 x dx.$

(p) $\int \csc^4 x dx.$

(q) $\int \tan x \sec^2 x dx.$

(r) $\int \cot x \csc^2 x dx.$

**Example V.** *To find* $\int \sin^m x \cos^n x dx$ *by means of multiple angles.*

When either $m$ or $n$ is a positive odd integer, the shortest method is that shown in Example I, p. 298. When $m$ and $n$ are both positive even integers, the given differential expression may be transformed by suitable trigonometric substitutions into an expression involving sines and cosines of multiple angles, and then integrated. For this purpose we employ the following formulas:

$$\sin u \cos u = \tfrac{1}{2} \sin 2\,u, \qquad 36, \text{p. } 2$$

$$\sin^2 u = \tfrac{1}{2} - \tfrac{1}{2} \cos 2\,u, \qquad 38, \text{p. } 2$$

$$\cos^2 u = \tfrac{1}{2} + \tfrac{1}{2} \cos 2\,u. \qquad 39, \text{p. } 2$$

ILLUSTRATIVE EXAMPLE 1. Find $\int \cos^2 x dx.$

**Solution.** $\int \cos^2 x dx = \int (\frac{1}{2} + \frac{1}{2} \cos 2\,x)\, dx$ 38, p. 2

$$= \frac{1}{2} \int dx + \frac{1}{2} \int \cos 2\,x dx = \frac{x}{2} + \frac{1}{4} \sin 2\,x + C.$$

ILLUSTRATIVE EXAMPLE 2. Find $\int \sin^2 x \cos^2 x dx.$

**Solution.** $\int \sin^2 x \cos^2 x dx = \frac{1}{4} \int \sin^2 2\,x dx$ 36, p. 2

$= \frac{1}{4} \int (\frac{1}{2} - \frac{1}{2} \cos 4\,x)\, dx$ 38, p. 2

$$= \frac{x}{8} - \frac{1}{32} \sin 4\,x + C.$$

ILLUSTRATIVE EXAMPLE 3. Find $\int \sin^4 x \cos^2 x dx$.

**Solution.** $\int \sin^4 x \cos^2 x dx = \int (\sin x \cos x)^2 \sin^2 x dx$

$$= \int \tfrac{1}{4} \sin^2 2x \left(\tfrac{1}{2} - \tfrac{1}{2} \cos 2x\right) dx \qquad 36, \text{p. } 2 ; \ 38, \text{p. } 2$$

$$= \tfrac{1}{8} \int \sin^2 2x dx - \tfrac{1}{8} \int \sin^2 2x \cos 2x dx$$

$$= \tfrac{1}{8} \int \left(\tfrac{1}{2} - \tfrac{1}{2} \cos 4x\right) dx - \tfrac{1}{8} \int \sin^2 2x \cos 2x dx$$

$$= \frac{x}{16} - \frac{\sin 4x}{64} - \frac{\sin^3 2x}{48} + C.$$

**Example VI.** *To find* $\int \sin mx \cos nx dx$, $\int \sin mx \sin nx dx$, *or* $\int \cos mx \cos nx dx$, *when* $m \neq n$.

By 41, p. 2, $\sin mx \cos nx = \frac{1}{2} \sin (m + n) x + \frac{1}{2} \sin (m - n) x$.

$$\therefore \int \sin mx \cos nx dx = \tfrac{1}{2} \int \sin (m+n) x dx + \tfrac{1}{2} \int \sin (m - n) x dx$$

$$= -\frac{\cos (m+n) x}{2(m+n)} - \frac{\cos (m-n) x}{2(m-n)} + C.$$

Similarly, we find

$$\int \sin mx \sin nx dx = -\frac{\sin (m+n) x}{2(m+n)} + \frac{\sin (m-n) x}{2(m-n)} + C,$$

$$\int \cos mx \cos nx dx = \frac{\sin (m+n) x}{2(m+n)} + \frac{\sin (m-n) x}{2(m-n)} + C.$$

**EXAMPLES**

1. $\int \cos^2 x dx = \frac{x}{2} + \frac{1}{4} \sin 2x + C.$

2. $\int \sin^4 x dx = \frac{3x}{8} - \frac{\sin 2x}{4} + \frac{\sin 4x}{32} + C.$

3. $\int \cos^4 x dx = \frac{3x}{8} + \frac{\sin 2x}{4} + \frac{\sin 4x}{32} + C.$

4. $\int \sin^6 x dx = \frac{1}{16}\left(5x - 4 \sin 2x + \frac{\sin^3 2x}{3} + \frac{3}{4} \sin 4x\right) + C.$

5. $\int \cos^6 x dx = \frac{1}{16}\left(5x + 4 \sin 2x - \frac{\sin^3 2x}{3} + \frac{3}{4} \sin 4x\right) + C.$

6. $\int \sin^4 \alpha \cos^2 \alpha d\alpha = -\frac{\sin^3 2\alpha}{48} + \frac{\alpha}{16} - \frac{\sin 4\alpha}{64} + C.$

7. $\int \sin^4 t \cos^4 t dt = \frac{1}{128}\left(3t - \sin 4t + \frac{\sin 8t}{8}\right) + C.$

8. $\int \cos^6 x \sin^2 x dx = \frac{1}{128}\left(5x + \frac{8}{3} \sin^3 2x - \sin 4x - \frac{\sin 8x}{8}\right) + C.$

9. $\int \cos 3y \sin 5y dy = -\frac{\cos 8y}{16} - \frac{\cos 2y}{4} + C.$

10. $\int \sin 5z \sin 6z dz = -\frac{\sin 11z}{22} + \frac{\sin z}{2} + C.$

11. $\int \cos 4s \cos 7s ds = \frac{\sin 11s}{22} + \frac{\sin 3s}{6} + C.$

**169. Integration of expressions containing $\sqrt{a^2 - x^2}$ or $\sqrt{x^2 \pm a^2}$ by a trigonometric substitution.** In many cases the shortest method of integrating such expressions is to change the variable as follows:

*When $\sqrt{a^2 - x^2}$ occurs, let $x = a \sin z$.*
*When $\sqrt{a^2 + x^2}$ occurs, let $x = a \tan z$.*
*When $\sqrt{x^2 - a^2}$ occurs, let $x = a \sec z$.**

ILLUSTRATIVE EXAMPLE 1. Find $\int \frac{dx}{(a^2 - x^2)^{\frac{3}{2}}}$.

**Solution.** Let $x = a \sin z$; then $dx = a \cos z dz$, and

$$\int \frac{dx}{(a^2 - x^2)^{\frac{3}{2}}} = \int \frac{a \cos z dz}{(a^2 - a^2 \sin^2 z)^{\frac{3}{2}}} = \int \frac{a \cos z dz}{a^3 \cos^3 z}$$
$$= \frac{1}{a^2}\int \frac{dz}{\cos^2 z} = \frac{1}{a^2}\int \sec^2 z dz = \frac{\tan z}{a^2} + C$$
$$= \frac{x}{a^2 \sqrt{a^2 - x^2}} + C.$$

[Since $\sin z = \frac{x}{a}$, draw a right triangle with $x$ as the opposite leg to the acute angle $z$, and $a$ as the hypotenuse. Then the adjacent leg will be $\sqrt{a^2 - x^2}$ and $\tan z = \frac{x}{\sqrt{a^2 - x^2}}$.

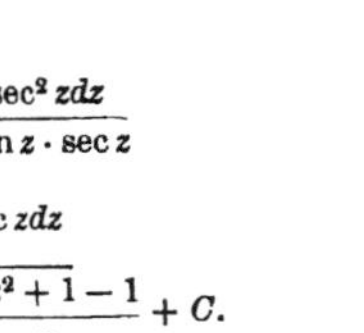

]

ILLUSTRATIVE EXAMPLE 2. Find $\int \frac{dx}{x\sqrt{x^2 + 1}}$.

**Solution.** Let $x = \tan z$ †; then $dx = \sec^2 z dz$, and

$$\int \frac{dx}{x\sqrt{x^2 + 1}} = \int \frac{\sec^2 z dz}{\tan z \sqrt{\tan^2 z + 1}} = \int \frac{\sec^2 z dz}{\tan z \cdot \sec z}$$
$$= \int \frac{\sec z}{\tan z} dz = \int \frac{dz}{\sin z} = \int \csc z dz$$
$$= \log(\csc z - \cot z) = \log \frac{\sqrt{x^2 + 1} - 1}{x} + C.$$

[Since $\tan z = x$, $\cot z = \frac{1}{x}$, and $\csc z = \frac{\sqrt{x^2 + 1}}{x}$.

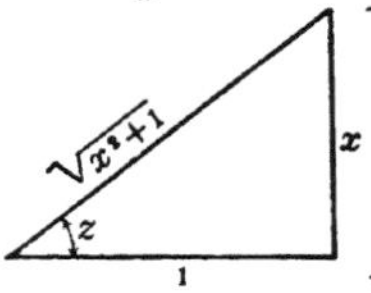

]

* We may also use the substitutions $x = a \cos z$, $x = a \cot z$, and $x = a \csc z$ respectively.
† In this example $a = 1$.

**EXAMPLES**

1. $\int \frac{\sqrt{x^2-a^2}}{x}dx = \sqrt{x^2-a^2} - a \operatorname{arc sec}\frac{x}{a} + C.$

2. $\int \sqrt{a^2-x^2}\,dx = \frac{x}{2}\sqrt{a^2-x^2} + \frac{a^2}{2}\operatorname{arc sin}\frac{x}{a} + C.$

3. $\int \frac{\sqrt{x^2+a^2}}{x^2}dx = \log\left(x+\sqrt{x^2+a^2}\right) - \frac{\sqrt{x^2+a^2}}{x} + C.$

4. $\int \frac{x^2dx}{\sqrt{1-x^2}} = \frac{1}{2}\operatorname{arc sin} x - \frac{x\sqrt{1-x^2}}{2} + C.$

5. $\int \frac{dx}{x^4\sqrt{x^2+1}} = \frac{(2x^2-1)\sqrt{x^2+1}}{3x^3} + C.$

6. $\int \frac{dx}{x^2\sqrt{x^2+a^2}} = -\frac{\sqrt{x^2+a^2}}{a^2x} + C.$

7. $\int \frac{\sqrt{x^2-a^2}}{x^4}dx = \frac{(x^2-a^2)^{\frac{3}{2}}}{3a^2x^3} + C.$

**MISCELLANEOUS EXAMPLES**

1. $\int \frac{dx}{\sin^4 x}.$

2. $\int \frac{(x^2+1)\,dx}{x+2}.$

3. $\int \frac{(ax+b)\,dx}{\sqrt{x^2-a^2}}.$

4. $\int \tan^3\frac{\theta}{3}\,d\theta.$

5. $\int \frac{(4x-1)\,dx}{\sqrt{1-5x^2}}.$

6. $\int \frac{dx}{\sqrt{2+2x-x^2}}.$

7. $\int \frac{5-3t}{\sqrt{1-a^2t^2}}dt.$

8. $\int \frac{d\phi}{2\phi^2-\phi+1}.$

9. $\int \frac{4xdx}{1-4x^4}.$

10. $\int (\tan 3x - 1)^2 dx.$

11. $\int \tan^3\theta \sec^3\theta d\theta.$

12. $\int \sin^4\frac{x}{2}\,dx.$

13. $\int \frac{d\theta}{\cos^4\theta}.$

14. $\int \frac{dt}{t^2+6t+5}.$

15. $\int \frac{3\cos\theta d\theta}{5-7\sin\theta}.$

16. $\int \frac{(a^{2x}+1)^2}{\sqrt{a^x}}dx.$

17. $\int \frac{ds}{\sqrt{1+3s-s^2}}.$

18. $\int \cos^5\frac{x}{5}\,dx.$

19. $\int \frac{dx}{x^2+2x+1}.$

20. $\int \frac{x^3dx}{x-3}.$

21. $\int \frac{d\theta}{\sin 2\theta}.$

22. $\int \frac{dt}{\cos 3t}.$

23. $\int \frac{5\,dx}{\sqrt{x-3}}.$

24. $\int \frac{dy}{\sqrt{y^2-6y+10}}.$

25. $\int \frac{axdx}{b-cx^2}.$

26. $\int \frac{xdx}{(1+x^2)^3}.$

27. $\int \frac{dx}{(a+bx)^n}.$

28. $\int \frac{1+\sec^2\theta}{1+\tan\theta}d\theta.$

29. $\int \frac{dx}{x^4\sqrt{x^2-1}}.$

30. $\int (a-3x^2)^m\,2\,xdx.$

31. $\int \frac{2x^2dx}{(a^3-x^3)^{\frac{3}{2}}}.$

32. $\int \frac{(a+x)^3}{\sqrt{x}}dx.$

33. $\int \frac{\log^3 xdx}{x}.$

34. $\int e^{-ax^3}x^2dx.$

35. $\int \frac{ax-b}{x^2+4m^2}dx.$

36. $\int \frac{1-2x}{9x^2-n^2}dx.$

37. $\int \cos^3 ax \sin axdx.$

38. $\int \cot^4 3\,aydy.$

39. $\int \sin^2 6\,xdx.$

**40.** The following functions have been obtained by differentiating certain functions. Find the functions and verify your results by differentiation.

(a) $5x^3 + \sin 2x$.

**Solution.** In this example $(5x^3 + \sin 2x)\,dx$ is the differential expression to be integrated. Thus $\int (5x^3 + \sin 2x)\,dx = \frac{5x^4}{4} - \frac{1}{2}\cos 2x + C$. *Ans.*

*Verification.* $\frac{d}{dx}\left(\frac{5x^4}{4} - \frac{1}{2}\cos 2x + C\right) = 5x^3 + \sin 2x$.

(b) $5x^3 - 6x$.

(c) $2x^2 - 3x - 4$.

(d) $\cos^2 ax + \sin\frac{x}{a}$.

(e) $\sqrt{a + bx}$.

(f) $\frac{ax + b}{bx + a}$.

(g) $\frac{x}{5 + 2x}$.

(h) $\frac{3 + 2x}{x^2 + 1}$.

(i) $\frac{1 - 3x}{4x^2 - 7}$.

(j) $\frac{mx + n}{\sqrt{g^2 - h^2x^2}}$.

(k) $\frac{ax - m}{\sqrt{3 + 4x^2}}$.

(l) $\frac{bt + c}{\sqrt{a^2t^2 - b^2}}$.

(m) $\frac{5 - 6s}{9 - 4s^2}$.

(n) $\frac{z^3}{5 - 2z}$.

(o) $\sin mx \cos mx$.

(p) $\cos^2 4px$.

(q) $\tan^3\frac{x}{a}$.

(r) $\frac{(1 - 2y)^3}{\sqrt[3]{y}}$.

(s) $\frac{1}{x^2 + 4x - 1}$.

(t) $\sec^4\frac{ax}{b}$.

(u) $\frac{1}{\sqrt{4 - x^2 + 2x}}$.

(v) $\left(e^{\frac{x}{a}} - e^{-\frac{x}{a}}\right)^2$.

(w) $x^3(1 + x^2)^{\frac{1}{2}}$.

(x) $\frac{1}{x^3\sqrt{x^2 - 1}}$.

(y) $x^8\sqrt{1 + x^2}$.

# CHAPTER XXIII

## CONSTANT OF INTEGRATION

**170. Determination of the constant of integration by means of initial conditions.** As was pointed out on p. 281, the constant of integration may be found in any given case when we know the value of the integral for some value of the variable. In fact, it is necessary, in order to be able to determine the constant of integration, to have some data given in addition to the differential expression to be integrated. Let us illustrate this by means of an example.

ILLUSTRATIVE EXAMPLE 1. Find a function whose first derivative is $3x^2 - 2x + 5$, and which shall have the value 12 when $x = 1$.

**Solution.** $(3x^2 - 2x + 5)\,dx$ is the differential expression to be integrated. Thus

$$\int (3x^2 - 2x + 5)\,dx = x^3 - x^2 + 5x + C,$$

where $C$ is the constant of integration. From the conditions of our problem this result must equal 12 when $x = 1$; that is,

$$12 = 1 - 1 + 5 + C, \text{ or } C = 7.$$

Hence $x^3 - x^2 + 5x + 7$ is the required function.

**171. Geometrical signification of the constant of integration.** We shall illustrate this by means of examples.

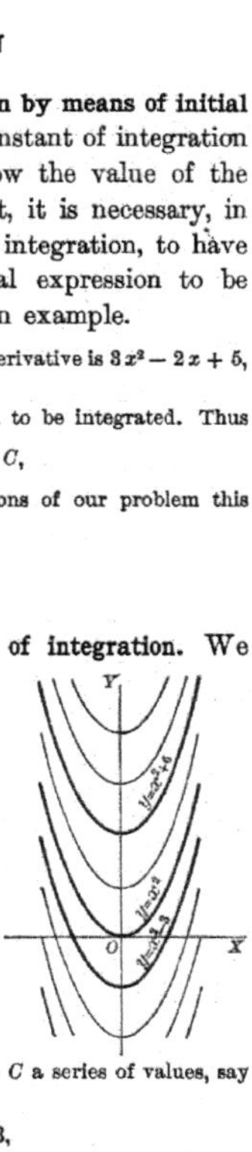


ILLUSTRATIVE EXAMPLE 1. Determine the equation of the curve at every point of which the tangent has the slope $2x$.

**Solution.** Since the slope of the tangent to a curve at any point is $\frac{dy}{dx}$, we have, by hypothesis,

$$\frac{dy}{dx} = 2x,$$

or, $$dy = 2x\,dx.$$

Integrating, $$y = 2\int x\,dx, \text{ or,}$$

(A) $$y = x^2 + C,$$

where $C$ is the constant of integration. Now if we give to $C$ a series of values, say $6, 0, -3$, (A) yields the equations

$$y = x^2 + 6, \quad y = x^2, \quad y = x^2 - 3,$$

whose loci are parabolas with axes coinciding with the axis of $y$ and having $6, 0, -3$ respectively as intercepts on the axis of $Y$.

All of the parabolas ($A$) (there are an infinite number of them) have the same value of $\frac{dy}{dx}$; that is, they have the same direction (or slope) for the same value of $x$. It will also be noticed that the difference in the lengths of their ordinates remains the same for all values of $x$. Hence all the parabolas can be obtained by moving any one of them vertically up or down, the value of $C$ in this case not affecting the slope of the curve.

If in the above example we impose the additional condition that the curve shall pass through the point (1, 4), then the coördinates of this point must satisfy ($A$), giving

$$4 = 1 + C, \text{ or } C = 3.$$

Hence the particular curve required is the parabola $y = x^2 + 3$.

ILLUSTRATIVE EXAMPLE 2. Determine the equation of a curve such that the slope of the tangent to the curve at any point is the negative ratio of the abscissa to the ordinate.

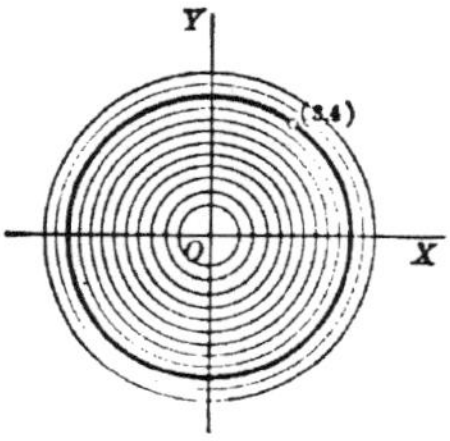


**Solution.** The condition of the problem is expressed by the equation

$$\frac{dy}{dx} = -\frac{x}{y},$$

or, separating the variables,

$$ydy = -xdx.$$

Integrating, $$\frac{y^2}{2} = -\frac{x^2}{2} + C,$$

or, $$x^2 + y^2 = 2C.$$

This we see represents a series of concentric circles with their centers at the origin.

If, in addition, we impose the condition that the curve must pass through the point (3, 4), then

$$9 + 16 = 2C.$$

Hence the particular curve required is the circle $x^2 + y^2 = 25$.

The orthogonal trajectories of a system of curves are another system of curves each of which cuts all the curves of the first system at right angles. Hence the slope of the tangent to a curve of the new system at a point will be the negative reciprocal of the slope of the tangent to that curve of the given system which passes through that point. Let us illustrate by an example.

ILLUSTRATIVE EXAMPLE 3. Find the equation of the orthogonal trajectories of the system of circles in Illustrative Example 2.

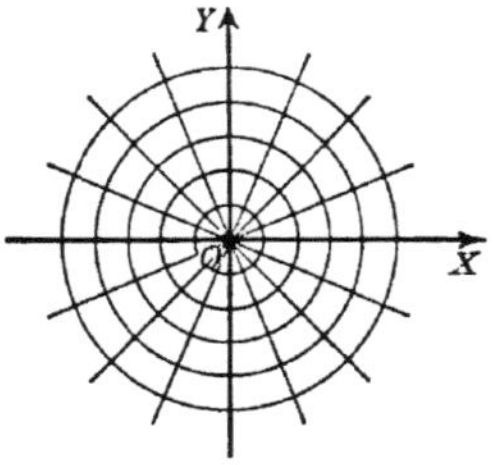


**Solution.** For the orthogonal system we will then have

$$\frac{dy}{dx} = \frac{y}{x},$$

or, separating the variables,

$$\frac{dy}{y} = \frac{dx}{x}.$$

Integrating, $$\log y = \log x + \log c = \log cx,$$

or, $$y = cx.$$

Hence the orthogonal trajectories of the system of circles $x^2 + y^2 = C$ is the system of straight lines which pass through the origin, as shown in the figure.

**172. Physical signification of the constant of integration.** The following examples will illustrate what is meant.

ILLUSTRATIVE EXAMPLE 1. Find the laws governing the motion of a point which moves in a straight line with constant acceleration.

**Solution.** Since the acceleration $\left[=\frac{dv}{dt} \text{ from (14), p. 92}\right]$ is constant, say $f$, we have

$$\frac{dv}{dt}=f,$$

or, $dv = fdt$. Integrating,

(A) $v = ft + C.$

To determine $C$, suppose that the *initial* velocity be $v_0$; that is, let

$$v = v_0 \quad \text{when} \quad t = 0.$$

These values substituted in (A) give

$$v_0 = 0 + C, \quad \text{or,} \quad C = v_0.$$

Hence (A) becomes

(B) $v = ft + v_0.$

Since $v = \frac{ds}{dt}$ [(9), p. 90], we get from (B)

$$\frac{ds}{dt} = ft + v_0,$$

or, $ds = ftdt + v_0 dt$. Integrating,

(C) $s = \frac{1}{2}ft^2 + v_0 t + C.$

To determine $C$, suppose that the *initial* space (= distance) be $s_0$; that is, let

$$s = s_0 \quad \text{when} \quad t = 0.$$

These values substituted in (C) give

$$s_0 = 0 + 0 + C, \quad \text{or,} \quad C = s_0.$$

Hence (C) becomes

(D) $s = \frac{1}{2}ft^2 + v_0 t + s_0.$

By substituting the values $f = g$, $v_0 = 0$, $s_0 = 0$, $s = h$ in (B) and (D), we get the laws of motion of a body falling from rest in a vacuum, namely,

(B a) $v = gt$, and

(D a) $h = \frac{1}{2}gt^2.$

Eliminating $t$ between (B a) and (D a) gives

$$v = \sqrt{2gh}.$$

ILLUSTRATIVE EXAMPLE 2. Discuss the motion of a projectile having an initial velocity $v_0$ inclined at an angle $\alpha$ with the horizontal, the resistance of the air being neglected.

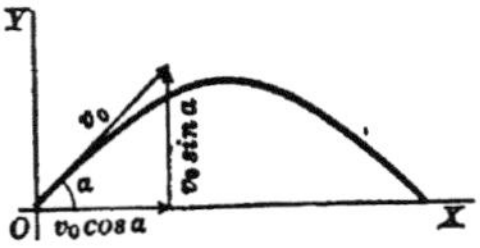


**Solution.** Assume the $XY$-plane as the plane of motion, $OX$ as horizontal, and $OY$ as vertical, and let the projectile be thrown from the origin.

Suppose the projectile to be acted upon by gravity alone. Then the acceleration in the horizontal direction will be zero and in the vertical direction $-g$. Hence from (15), p. 93,

$$\frac{dv_x}{dt} = 0, \quad \text{and} \quad \frac{dv_y}{dt} = -g.$$

Integrating, $v_x = C_1$, and $v_y = -gt + C_2$.

But $v_0 \cos \alpha$ = initial velocity in the horizontal direction,

and $v_0 \sin \alpha$ = initial velocity in the vertical direction.

Hence $C_1 = v_0 \cos \alpha$, and $C_2 = v_0 \sin \alpha$, giving

(*E*) $v_x = v_0 \cos \alpha$, and $v_y = -gt + v_0 \sin \alpha$.

But from (10) and (11), p. 92, $v_x = \frac{dx}{dt}$, and $v_y = \frac{dy}{dt}$; therefore (*E*) gives

$$\frac{dx}{dt} = v_0 \cos \alpha, \text{ and } \frac{dy}{dt} = -gt + v_0 \sin \alpha,$$

or,

$$dx = v_0 \cos \alpha dt, \text{ and } dy = -gtdt + v_0 \sin \alpha dt.$$

Integrating, we get

(*F*) $x = v_0 \cos \alpha \cdot t + C_3$, and $y = -\frac{1}{2} gt^2 + v_0 \sin \alpha \cdot t + C_4$.

To determine $C_3$ and $C_4$, we observe that when

$$t = 0, \quad x = 0 \quad \text{and} \quad y = 0.$$

Substituting these values in (*F*) gives

$$C_3 = 0, \quad \text{and} \quad C_4 = 0.$$

Hence

(*G*) $x = v_0 \cos \alpha \cdot t$, and

(*H*) $y = -\frac{1}{2} gt^2 + v_0 \sin \alpha \cdot t$.

Eliminating $t$ between (*G*) and (*H*), we obtain

(*I*) $$y = x \tan \alpha - \frac{gx^2}{2 v_0^2 \cos^2 \alpha},$$

which is the equation of the *trajectory*, and shows that the projectile will move in a parabola.

**EXAMPLES**

1. The following expressions have been obtained by differentiating certain functions. Find the function in each case for the given values of the variable and the function:

| Derivative of function | Value of variable | Corresponding value of function | Answers |
|---|---|---|---|
| (a) $x - 3$. | 2. | 9. | $\frac{x^2}{2} - 3x + 13$. |
| (b) $3 + x - 5x^2$. | 6. | $-20$. | $304 + 3x + \frac{x^2}{2} - \frac{5x^3}{3}$. |
| (c) $y^3 - b^2 y$. | 2. | 0. | $\frac{y^4}{4} - \frac{b^2y^2}{2} + 2b^2 - 4$. |
| (d) $\sin \alpha + \cos \alpha$. | $\frac{\pi}{2}$. | 2. | $\sin \alpha - \cos \alpha + 1$. |
| (e) $\frac{1}{t} - \frac{1}{2-t}$. | 1. | 0. | $\log (2t - t^2)$. |
| (f) $\sec^2 \theta + \tan \theta$. | 0. | 5. | $\tan \theta + \log \sec \theta + 5$. |
| (g) $\frac{1}{x^2 + a^2}$. | $a$. | $\frac{\pi}{2a}$. | $\frac{1}{a} \text{arc tan} \frac{x}{a} + \frac{\pi}{4a}$. |
| (h) $bx^3 + ax + 4$. | $b$. | 10. | |
| (i) $\sqrt{t} + \frac{1}{\sqrt{t}}$. | 4. | 0. | |
| (j) $\cot \phi - \csc^2 \phi$. | $\frac{\pi}{2}$. | 8. | |
| (k) $3 e^{2t^2} t$. | 0. | $\frac{7}{4}$. | |

**2.** Find the equation of the system of curves such that the slope of the tangent at any point is:

(a) $x$. *Ans.* Parabolas, $y = \frac{x^2}{2} + C$.

(b) $2x - 2$. Parabolas, $y = x^2 - 2x + C$.

(c) $\frac{1}{y}$. Parabolas, $\frac{y^2}{2} = x + C$.

(d) $\frac{x^2}{y}$. Semicubical parabolas, $\frac{y^2}{2} = \frac{x^3}{3} + C$.

(e) $\frac{x}{y^2}$. Semicubical parabolas, $\frac{y^3}{3} = \frac{x^2}{2} + C$.

(f) $3x^2$. Cubical parabolas, $y = x^3 + C$.

(g) $x^2 + 5x$. Cubical parabolas, $y = \frac{x^3}{3} + \frac{5}{2}x^2 + C$.

(h) $\frac{1}{y^2}$. Cubical parabolas, $\frac{y^3}{3} = x + C$.

(i) $\frac{x}{y}$. Equilateral hyperbolas, $y^2 - x^2 = C$.

(j) $-\frac{y}{x}$. Equilateral hyperbolas, $xy = C$.

(k) $\frac{b^2x}{a^2y}$. Hyperbolas, $a^2y^2 - b^2x^2 = C$.

(l) $-\frac{a^2x}{b^2y}$. Ellipses, $b^2y^2 + a^2x^2 = C$.

(m) $xy$. $\log y = \frac{x^2}{2} + C$, or $y = ce^{\frac{x^2}{2}}$.

(n) $y$. $\log y = x + C$, or $y = ce^x$.

(o) $m$. Straight lines, $y = mx + C$.

(p) $\frac{1+x}{1-y}$. Circles, $x^2 + y^2 + 2x - 2y + C = 0$.

**3.** Find the equations of those curves of the systems found in Ex. 2 (a), (c), (d), (i), (j), (m), which pass through the point $(2, -1)$.

*Ans.* (a) $x^2 - 2y - 6 = 0$; (m) $y = -e^{\frac{x^2-4}{2}}$; etc.

**4.** Find the equations of those curves of the systems found in Ex. 2 (b), (e), (g), (h), (o), (p), which pass through the origin. *Ans.* (b) $y = x^2 - 2x$; (o) $y = mx$; etc.

**5.** Find the equations of the orthogonal trajectories of the following systems of curves found in Ex. 2:

(a) $y = \frac{x^2}{2} + C$, Ex. 2 (a). *Ans.* $y = -\log x + C$.

(b) $\frac{y^2}{2} = x + C$, Ex. 2 (c). $\log y = -x + C$.

(c) $\frac{y^2}{2} = \frac{x^3}{3} + C$, Ex. 2 (d). $\log y = \frac{1}{x} + C$.

(d) $y^2 - x^2 = C$, Ex. 2 (i). $xy = C$.

(e) $xy = C$, Ex. 2 (j). $y^2 - x^2 = C$.

(f) $y = ce^x$, Ex. 2 (n). $\frac{y^2}{2} = -x + C$.

(g) $y = mx + C$, Ex. 2 (o). $my + x = C$.

(h) $x^2 + y^2 + 2x - 2y + C = 0$, Ex. 2 (p). $y - 1 = c(x + 1)$.

**6.** Find the equation of the curve whose subnormal is constant and equal to $2a$.

*Ans.* $y^2 = 4ax + C$, a parabola.

HINT. From (4), p. 77, subnormal $= y\frac{dy}{dx}$.

**7.** Find the curve whose subtangent is constant and equal to $a$ (see (3), p. 77).

*Ans.* $a \log y = x + C$.

**8.** Find the curve whose subnormal equals the abscissa of the point of contact.

*Ans.* $y^2 - x^2 = 2C$, an equilateral hyperbola.

**9.** Find the curve whose normal is constant ($= R$), assuming that $y = R$ when $x = 0$.

*Ans.* $x^2 + y^2 = R^2$, a circle.

HINT. From (6), p. 77, length of normal $= y\sqrt{1+\left(\frac{dy}{dx}\right)^2}$, or $dx = \pm(R^2 - y^2)^{-\frac{1}{2}}y\,dy$.

**10.** Find the curve whose subtangent equals three times the abscissa of the point of contact.

*Ans.* $x = cy^3$.

**11.** Show that the curve whose polar subtangent (see (7), p. 86) is constant is the reciprocal spiral.

**12.** Show that the curve whose polar subnormal (see (8), p. 86) is constant is the spiral of Archimedes.

**13.** Find the curve in which the polar subnormal is proportional to the length of the radius vector.

*Ans.* $\rho = ce^{a\theta}$.

**14.** Find the curve in which the polar subnormal is proportional to the sine of the vectorial angle.

*Ans.* $\rho = c - a\cos\theta$.

**15.** Find the curve in which the polar subtangent is proportional to the length of the radius vector.

*Ans.* $\rho = ce^{a\theta}$.

**16.** Determine the curve in which the polar subtangent and the polar subnormal are in a constant ratio.

*Ans.* $\rho = ce^{a\theta}$.

**17.** Find the equation of the curve in which the angle between the radius vector and the tangent is one half the vectorial angle.

*Ans.* $\rho = c(1 - \cos\theta)$.

**18.** Determine the curves in which the subtangent is $n$ times the subnormal; and find the particular curve which passes through (2, 3).

*Ans.* $\sqrt{n}y = x + C$; $\sqrt{n}(y - 3) = x - 2$.

**19.** Determine the curves in which the length of the subnormal is proportional to the square of the ordinate.

*Ans.* $y = ce^{kx}$.

**20.** Find the curves in which the angle between the radius vector and the tangent at any point is $n$ times the vectorial angle.

*Ans.* $\rho^n = c\sin n\theta$.

Assuming that $v = v_0$ when $t = 0$, find the relation between $v$ and $t$, knowing that the acceleration is:

**21.** Zero. *Ans.* $v = v_0$.

**22.** Constant $= k$. $v = v_0 + kt$.

**23.** $a + bt$. $v = v_0 + at + \frac{bt^2}{2}$.

Assuming that $s = 0$ when $t = 0$, find the relation between $s$ and $t$, knowing that the velocity is:

**24.** Constant ($= v_0$). *Ans.* $s = v_0t$.

**25.** $m + nt$. $s = mt + \frac{nt^2}{2}$.

**26.** $3 + 2t - 3t^2$. $s = 3t + t^2 - t^3$.

**27.** The velocity of a body starting from rest is $5t^2$ feet per second after $t$ seconds. (a) How far will it be from the point of starting in 3 seconds? (b) In what time will it pass over a distance of 360 feet measured from the starting point?

*Ans.* (a) 45 ft.; (b) 6 seconds.

**28.** Assuming that $s = 2$ when $t = 1$, find the relation between $s$ and $t$, knowing that the velocity is:

(a) 3. *Ans.* $s = 3t - 1$.

(b) $2t - 3$. $s = t^2 - 3t + 4$.

(c) $t^2 + 2t - 1$. $s = \frac{t^3}{3} + t^2 - t + \frac{5}{3}$.

(d) $\frac{1}{t}$. $s = \log t + 2$.

(e) $4t^3 - 4$. $s = t^4 - 4t + 5$.

(f) $\frac{k}{t^2}$. $s = -\frac{k}{t} + k + 2$.

**29.** Assuming that $v = 3$ when $t = 2$, find the relation between $v$ and $t$, knowing that the acceleration is:

(a) 2. *Ans.* $v = 2t - 1$.

(b) $3t^2 + 1$. $v = t^3 + t - 7$.

(c) $t^3 - 2t$. $v = \frac{t^4}{4} - t^2 + 3$.

(d) $\frac{1}{t} + t$. $v = \log\frac{t}{2} + \frac{t^2}{2} + 1$.

**30.** A train starting from a station has, after $t$ hours, a speed of $t^3 - 21t^2 + 80t$ miles per hour. Find (a) its distance from the station; (b) during what interval the train was moving backwards; (c) when the train repassed the station; (d) the distance the train had traveled when it passed the station the last time.

*Ans.* (a) $\frac{1}{4}t^4 - 7t^3 + 40t^2$ miles; (b) from 5th to 16th hour; (c) in 8 and 20 hours; (d) $4658\frac{1}{4}$ miles.

**31.** A body starts from the origin and in $t$ seconds its velocity in the $X$ direction is $12t$ and in the $Y$ direction $4t^2 - 9$. Find (a) the distances traversed parallel to each axis; (b) the equation of the path.

*Ans.* (a) $x = 6t^2$, $y = \frac{4}{3}t^3 - 9t$; (b) $y = \left(\frac{2}{9}x - 9\right)\sqrt{\frac{x}{6}}$.

**32.** The equation giving the strength of the current $i$ for the time $t$ after the source of E.M.F. is removed is ($R$ and $L$ being constants)

$$Ri = -L\frac{di}{dt}.$$

Find $i$, assuming that $I$ = current when $t = 0$. *Ans.* $i = Ie^{-\frac{Rt}{L}}$.

**33.** Find the current of discharge $i$ from a condenser of capacity $C$ in a circuit of resistance $R$, assuming the initial current to be $I_0$, having given the relation ($C$ and $R$ being constants)

$$\frac{di}{i} = \frac{dt}{CR}.$$

*Ans.* $i = I_0 e^{\frac{t}{CR}}$.

**34.** If a particle moves so that its velocities parallel to the axes of $X$ and $Y$ are $ky$ and $kx$ respectively, prove that its path is an equilateral hyperbola.

**35.** A body starts from the origin of coördinates, and in $t$ seconds its velocity parallel to the axis of $X$ is $6t$, and its velocity parallel to the axis of $Y$ is $3t^2 - 3$. Find (a) the distance traversed parallel to each axis in $t$ seconds; (b) the equation of the path.

*Ans.* (a) $x = 3t^2$, $y = t^3 - 3t$; (b) $27y^2 = x(x - 9)^2$.

# CHAPTER XXIV

## THE DEFINITE INTEGRAL

**173. Differential of an area.** Consider the continuous function $\phi(x)$, and let

$$y = \phi(x)$$

be the equation of the curve $AB$. Let $CD$ be a fixed and $MP$ a variable ordinate, and let $u$ be the measure of the area $CMPD$.* When $x$ takes on a sufficiently small increment $\Delta x$, $u$ takes on an increment $\Delta u$ (= area $MNQP$). Completing the rectangles $MNRP$ and $MNQS$, we see that

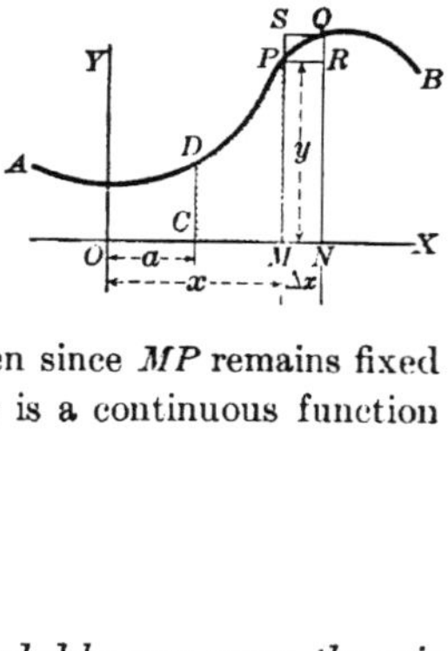


area $MNRP <$ area $MNQP <$ area $MNQS$,

or, $\quad MP \cdot \Delta x < \Delta u < NQ \cdot \Delta x$;

and, dividing by $\Delta x$,

$$MP < \frac{\Delta u}{\Delta x} < NQ.\dagger$$

Now let $\Delta x$ approach zero as a limit; then since $MP$ remains fixed and $NQ$ approaches $MP$ as a limit (since $y$ is a continuous function of $x$), we get

$$\frac{du}{dx} = y\,(= MP),$$

or, using differentials,

$$du = ydx.$$

**Theorem.** *The differential of the area bounded by any curve, the axis of X, and two ordinates is equal to the product of the ordinate terminating the area and the differential of the corresponding abscissa.*

**174. The definite integral.** It follows from the theorem in the last section that if $AB$ is the locus of

$$y = \phi(x),$$

then $\qquad du = ydx,$ or

(A) $\qquad du = \phi(x)\,dx,$

* We may suppose this area to be generated by a variable ordinate starting out from $CD$ and moving to the right; hence $u$ will be a function of $x$ which vanishes when $x = a$.

† In this figure $MP$ is less than $NQ$; if $MP$ happens to be greater than $NQ$, simply reverse the inequality signs.

where $du$ is the differential of the area between the curve, the axis of $x$, and any two ordinates. Integrating ($A$), we get

$$u=\int\phi(x)\,dx.$$

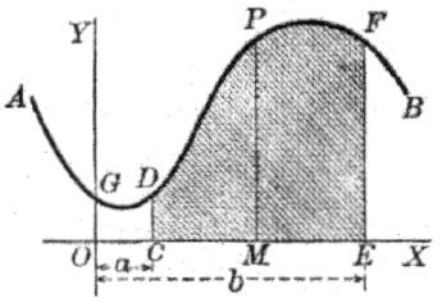


Since $\int\phi(x)\,dx$ exists (it is here represented geometrically as an area), denote it by $f(x)+C$.

($B$) $\therefore u=f(x)+C.$

We may determine $C$, as in Chapter XXIII, if we know the value of $u$ for some value of $x$. If we agree to reckon the area from the axis of $y$, i.e. when

($C$) $x=a,\quad u=$ area $OCDG$,

and when $x=b,\quad u=$ area $OEFG$, etc.,

it follows that if

($D$) $x=0,\quad$ then $u=0$.

Substituting ($D$) in ($B$), we get

$$0=f(0)+C,\quad\text{or,}\quad C=-f(0).$$

Hence from ($B$) we obtain

($E$) $u=f(x)-f(0)$,

giving the area from the axis of $y$ to any ordinate (as $MP$).

To find the area between the ordinates $CD$ and $EF$, substitute the values ($C$) in ($E$), giving

($F$) area $OCDG=f(a)-f(0)$,

($G$) area $OEFG=f(b)-f(0)$.

Subtracting ($F$) from ($G$),

($H$) area $CEFD=f(b)-f(a)$.*

**Theorem.** *The difference of the values of* $\int y\,dx$ *for* $x=a$ *and* $x=b$ *gives the area bounded by the curve whose ordinate is* $y$, *the axis of* $X$, *and the ordinates corresponding to* $x=a$ *and* $x=b$.

This difference is represented by the symbol †

($I$) $$\int_a^b y\,dx,\quad\text{or,}\quad\int_a^b\phi(x)\,dx,$$

* The student should observe that under the present hypothesis $f(x)$ will be a *single-valued* function which changes *continuously* from $f(a)$ to $f(b)$ as $x$ changes from $a$ to $b$.

† This notation is due to Joseph Fourier (1768–1830).

and is read "the integral from $a$ to $b$ of $ydx$." The operation is called *integration between limits*, $a$ being the *lower* and $b$ the *upper* limit.*

Since $(I)$ always has a *definite* value, it is called a *definite integral.* For, if

$$\int \phi(x)\,dx = f(x) + C,$$

then

$$\int_a^b \phi(x)\,dx = \Big[f(x) + C\Big]_a^b$$

$$= [f(b) + C] - [f(a) + C],$$

or

$$\int_a^b \phi(x)\,dx = f(b) - f(a),$$

the *constant of integration* having disappeared.

*We may accordingly define the symbol*

$$\int_a^b \phi(x)\,dx \quad \text{or} \quad \int_a^b y\,dx$$

*as the numerical measure of the area bounded by the curve* $y = \phi(x)$,† *the axis of X, and the ordinates of the curve at* $x = a$, $x = b$. *This definition presupposes that these lines bound an area, i.e. the curve does not rise or fall to infinity, and both a and b are finite.*

We have shown that the numerical value of the definite integral is always $f(b) - f(a)$, but we shall see in Illustrative Example 2, p. 324, that $f(b) - f(a)$ may be a number when the definite integral has no meaning.

**175. Calculation of a definite integral.** The process may be summarized as follows:

First Step. *Find the indefinite integral of the given differential expression.*

Second Step. *Substitute in this indefinite integral first the upper limit and then the lower limit for the variable, and subtract the last result from the first.*

It is not necessary to bring in the constant of integration, since it always disappears in subtracting.

* The word *limit* in this connection means merely the value of the variable at one end of its range (end value), and should not be confused with the meaning of the word in the Theory of Limits.

† $\phi(x)$ is continuous and single-valued throughout the interval $[a, b]$.

ILLUSTRATIVE EXAMPLE 1. Find $\int_1^4 x^2dx$.

**Solution.** $\int_1^4 x^2dx = \left[\frac{x^3}{3}\right]_1^4 = \frac{64}{3} - \frac{1}{3} = 21.$ *Ans.*

ILLUSTRATIVE EXAMPLE 2. Find $\int_0^\pi \sin x dx$.

**Solution.** $\int_0^\pi \sin x dx = \Big[-\cos x\Big]_0^\pi = \Big[-(-1)\Big] - \Big[-1\Big] = 2.$ *Ans.*

ILLUSTRATIVE EXAMPLE 3. Find $\int_0^a \frac{dx}{a^2+x^2}$.

**Solution.**

$$\int_0^a \frac{dx}{a^2+x^2} = \left[\frac{1}{a}\operatorname{arc}\tan\frac{x}{a}\right]_0^a = \frac{1}{a}\operatorname{arc}\tan 1 - \frac{1}{a}\operatorname{arc}\tan 0$$

$$= \frac{\pi}{4a} - 0 = \frac{\pi}{4a}. \quad \textit{Ans.}$$

## EXAMPLES

1. $\int_2^3 6x^2dx = 38.$

2. $\int_0^a (a^2x - x^3)\,dx = \frac{a^4}{4}.$

3. $\int_1^4 \frac{dx}{x^{\frac{3}{2}}} = 1.$

4. $\int_1^e \frac{dx}{x} = 1.$

5. $\int_0^1 (x^2 - 2x + 2)(x-1)\,dx = -\frac{3}{4}.$

6. $\int_0^1 \frac{dx}{\sqrt{3-2x}} = \sqrt{3} - 1.$

7. $\int_0^2 \frac{x^3dx}{x+1} = \frac{8}{3} - \log 3.$

8. $\int_0^{\frac{1}{\sqrt{3}}} \frac{dx}{\sqrt{2-3x^2}} = \frac{\pi}{4\sqrt{3}}.$

9. $\int_2^3 \frac{3x dx}{2\sqrt[4]{x^2-4}} = \sqrt[4]{125}.$

10. $\int_0^1 \frac{dy}{y^2-y+1} = \frac{2\pi}{3\sqrt{3}}.$

11. $\int_2^3 \frac{tdt}{1+t^2} = \frac{\log 2}{2}.$

12. $\int_0^{\frac{\pi}{2}} \sin\phi d\phi = 1.$

13. $\int_0^{\frac{\pi}{4}} \sec^4\theta d\theta = \frac{4}{3}.$

14. $\int_0^{2r} \frac{\sqrt{2r}}{\sqrt{x}}dx = 4r.$

15. $\int_0^5 \left(\frac{3}{5}\sqrt{t} - \frac{3}{25}t^2\right)dt = 2\sqrt{5} - 5$

16. $\int_0^r \frac{rdx}{\sqrt{r^2-x^2}} = \frac{\pi r}{2}.$

17. $\int_0^{2r} \frac{2\sqrt{2r}dy}{\sqrt{2r-y}} = 8r.$

18. $\int_{-b}^b \frac{\pi}{a^4}(y^2-b^2)^4dy = \frac{256\pi b^9}{315a^4}.$

19. $2a\int_0^\pi (2+2\cos\theta)^{\frac{1}{2}}d\theta = 8a.$

20. $\int_0^{\frac{\pi}{2}} \sin^3\alpha\cos^3\alpha d\alpha = \frac{1}{12}.$

21. $\int_{-\frac{\pi}{4}}^{\frac{\pi}{4}} \tan\alpha d\alpha = 0.$

22. $\int_{\frac{\pi}{6}}^{\frac{\pi}{4}} \sec\theta d\theta = \log\left(\frac{1+\sqrt{2}}{\sqrt{3}}\right).$

23. $\int_0^{\frac{\pi}{2}} \frac{\cos\theta d\theta}{1+\sin^2\theta} = \frac{\pi}{4}.$

**176. Calculation of areas.** On p. 316 it was shown that the area between a curve, the axis of $X$, and the ordinates $x = a$ and $x = b$ is given by the formula

$$\textbf{Area} = \int_a^b y\,dx,$$

where the value of $y$ in terms of $x$ is substituted from the equation of the given curve.

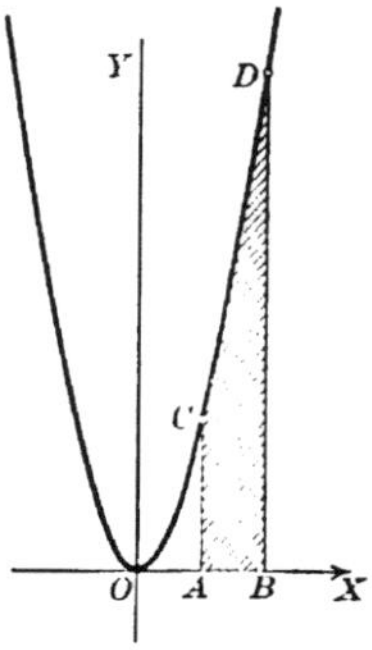


ILLUSTRATIVE EXAMPLE 1. Find the area bounded by the parabola $y = x^2$, the axis of $X$, and the ordinates $x = 2$ and $x = 4$.

**Solution.** Substituting in the formula

$$\text{Area } ABDC = \int_2^4 x^2\,dx = \left[\frac{x^3}{3}\right]_2^4$$

$$= \frac{64}{3} - \frac{8}{3} = \frac{56}{3} = 18\tfrac{2}{3}. \quad \textit{Ans.}$$

## EXAMPLES

1. Find the area bounded by the parabola $y = x^2$, the axis of $X$, and the ordinate $x = 3$. *Ans.* 9.

2. Find the area above the axis of $X$, under the parabola $y^2 = 4x$, and included between the ordinates $x = 4$ and $x = 9$. *Ans.* $25\frac{1}{3}$.

3. Find the area bounded by the equilateral hyperbola $xy = a^2$, the axis of $X$, and the ordinates $x = a$ and $x = 2a$. *Ans.* $a^2 \log 2$.

4. Find the area between the parabola $y = 4 - x^2$ and the axis of $X$. *Ans.* $10\frac{2}{3}$.

5. Find the area intercepted between the coördinate axes and the parabola $x^{\frac{1}{2}} + y^{\frac{1}{2}} = a^{\frac{1}{2}}$. *Ans.* $\frac{a^2}{6}$.

6. Find the area by integration of the triangle bounded by the line $y = 5x$, the axis of $X$, and the ordinate $x = 2$. Verify your result by finding the area as one half the product of the base and altitude.

7. Find the area by integration of the triangle bounded by the line $y = 2x + 6$, the axis of $X$, and the ordinate $x = 4$. Verify your result as in the last example.

8. Find the area by integration of the trapezoid bounded by the line $x - y + 4 = 0$, the axis of $X$, and the ordinates $x = -2$ and $x = 4$. Verify your result by finding the area as one half the product of the sum of the parallel sides and the altitude.

9. Find the area by integration of the trapezoid bounded by the line $x + 2y - 6 = 0$, the axis of $X$, and the ordinates $x = 0$ and $x = 3$. Verify your result as in the last example.

10. Find the area by integration of the rectangle bounded by the line $y = 5$, the axis of $X$, and the ordinates $x = 2$ and $x = 6$. Verify your result geometrically.

11. Find by integration the area bounded by the lines $x = 0$, $x = 9$, $y = 0$, $y = 7$. Verify your result geometrically.

12. Find the area bounded by the semicubical parabola $y^3 = x^2$, the axis of $X$, and the line $x = 4$. *Ans.* $\frac{3}{5}\sqrt[3]{1024}$.

**13.** Find the area bounded by the cubical parabola $y = x^3$, the axis of $X$, and the ordinate $x = 4$. *Ans.* 64.

**14.** Find in each of the following cases the area bounded by the given curve, the axis of $X$, and the given ordinates:

| | | |
|---|---|---|
| (a) $y = 9 - x^2$. | $x = -3,\ x = 3$. | *Ans.* 36. |
| (b) $y = \dfrac{x}{1+x^2}$. | $x = 0,\ x = 8$. | $\log\sqrt{65}$. |
| (c) $y = \sin x$. | $x = 0,\ x = \dfrac{\pi}{2}$. | 1. |
| (d) $y = x^3 + 3x^2 + 2x$. | $x = 0,\ x = 3$. | $56\frac{1}{4}$. |
| (e) $y = x^2 + x + 1$. | $x = 2,\ x = 3$. | $9\frac{5}{6}$. |
| (f) $y = x^4 + 4x^3 + 2x^2 + 3$. | $x = 1,\ x = 2$. | $28\frac{13}{15}$. |
| (g) $y^2 = -4x$. | $x = -1,\ x = 0$. | $\frac{4}{3}$. |
| (h) $xy = k^2$. | $x = a,\ x = b$. | $k^2 \log\dfrac{b}{a}$. |
| (i) $y = 2x + 3$. | $x = 0,\ x = 4$. | |
| (j) $y^2 = 4x + 16$. | $x = -2,\ x = 0$. | |
| (k) $y = x^2 + 4x$. | $x = -4,\ x = -2$. | |
| (l) $y = \cos x$. | $x = 0,\ x = \dfrac{\pi}{4}$. | |
| (m) $xy = 12$. | $x = 1,\ x = 4$. | |

**15.** Find the area included between the parabolas $y^2 = 4x$ and $x^2 = 4y$. *Ans.* $5\frac{1}{3}$.

**16.** Find the total area included between the cubical parabola $y = x^3$ and the line $y = 2x$. *Ans.* 2.

**17.** Prove that the area bounded by a parabola and one of its double ordinates equals two thirds of the circumscribing rectangle having the double ordinate as one side.

**18.** Find the area included between the parabolas $y^2 = 4 + x$ and $y^2 = 4 - x$.

**19.** Find the area between the curve $y = \dfrac{x}{1+x^2}$ and the line $y = \dfrac{x}{4}$.

*Ans.* $\log 4 - \frac{3}{4}$.

**20.** Find by integration the area of the triangle bounded by the lines

$$x + 3y - 8 = 0, \qquad 5x - y - 15 = 0, \qquad x - y + 1 = 0. \qquad \textit{Ans. } 8.$$

**177. Geometrical representation of an integral.** In the last section we represented the definite integral as an area. This does not necessarily mean that every integral *is* an area, for the physical interpretation of the result depends on the nature of the quantities represented by the abscissa and the ordinate. Thus, if $x$ and $y$ are considered as simply the coördinates of a point and nothing more, then the integral is indeed an area. But suppose the ordinate represents the speed of a moving point, and the corresponding abscissa the time at which the point has that speed; then the graph is the speed curve of the motion, and the area under it and between any two ordinates will *represent* the distance passed through in the corresponding interval of time. That is, the *number* which denotes the area equals the *number* which denotes the distance (or value of the integral).

Similarly, a definite integral standing for volume, surface, mass, force, etc., may be represented geometrically by an area. On p. 366 the algebraic sign of an area is interpreted.

**178. Mean value of $\phi(x)$.** This is defined as follows:

$$\left.\begin{array}{l}\textbf{Mean value of } \phi(x)\\ \textbf{from } x=a \textbf{ to } x=b\end{array}\right\} = \frac{\int_a^b \phi(x)\,dx}{b-a}.$$

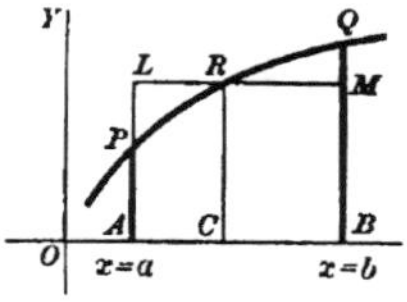


Since from the figure

$$\int_a^b \phi(x)\,dx = \text{area } APQB,$$

this definition means that if we construct on the base $AB (= b - a)$ a rectangle (as $ALMB$) whose area equals the area of $APQB$, then

$$\text{mean value} = \frac{\text{area } ALMB}{b-a} = \frac{AB \cdot CR}{AB} = \text{altitude } CR.$$

**179. Interchange of limits.**

Since $$\int_a^b \phi(x)\,dx = f(b) - f(a),$$

and $$\int_b^a \phi(x)\,dx = f(a) - f(b) = -[f(b) - f(a)],$$

we have $$\int_a^b \phi(x)\,dx = -\int_b^a \phi(x)\,dx.$$

**Theorem.** *Interchanging the limits is equivalent to changing the sign of the definite integral.*

**180. Decomposition of the interval of integration of the definite integral.**

Since $$\int_a^{x_1} \phi(x)\,dx = f(x_1) - f(a),$$

and $$\int_{x_1}^b \phi(x)\,dx = f(b) - f(x_1),$$

we get, by addition,

$$\int_a^{x_1} \phi(x)\,dx + \int_{x_1}^b \phi(x)\,dx = f(b) - f(a).$$

But $$\int_a^b \phi(x)\,dx = f(b) - f(a);$$

therefore, by comparing the last two expressions, we obtain

$$\int_a^b \phi(x)\,dx = \int_a^{x_1} \phi(x)\,dx + \int_{x_1}^b \phi(x)\,dx.$$

Interpreting this theorem geometrically, as in § 174, p. 315, we see that the integral on the left-hand side represents the whole area *CEFD*, the first integral on the right-hand side the area *CMPD*, and the second integral on the right-hand side the area *MEFP*. The truth of the theorem is therefore obvious.

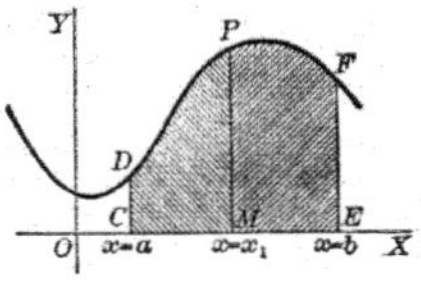


Even if $x_1$ does not lie in the interval between $a$ and $b$, the truth of the theorem is apparent when the sign as well as the magnitude of the areas is taken into account. Evidently the definite integral may be decomposed into any number of separate definite integrals in this way.

**181. The definite integral a function of its limits.**

From
$$\int_a^b \phi(x)\,dx = f(b) - f(a)$$

we see that the definite integral is a function of its limits. Thus $\int_a^b \phi(z)\,dz$ has precisely the same value as $\int_a^b \phi(x)\,dx$.

**Theorem.** *A definite integral is a function of its limits.*

**182. Infinite limits.** So far the limits of the integral have been assumed as finite. Even in elementary work, however, it is sometimes desirable to remove this restriction and to consider integrals with infinite limits. This is possible in certain cases by making use of the following *definitions*.

When the upper limit is infinite,

$$\int_a^{+\infty} \phi(x)\,dx = \underset{b=+\infty}{\text{limit}} \int_a^b \phi(x)\,dx,$$

and when the lower limit is infinite,

$$\int_{-\infty}^b \phi(x)\,dx = \underset{a=-\infty}{\text{limit}} \int_a^b \phi(x)\,dx,$$

provided the limits exist.

Illustrative Example 1. Find $\int_1^{+\infty} \frac{dx}{x^2}$.

**Solution.** $\int_1^{\infty} \frac{dx}{x^2} = \underset{b=+\infty}{\text{limit}} \int_1^b \frac{dx}{x^2} = \underset{b=+\infty}{\text{limit}} \left[-\frac{1}{x}\right]_1^b$

$= \underset{b=+\infty}{\text{limit}} \left[-\frac{1}{b}+1\right] = 1.$ *Ans.*

ILLUSTRATIVE EXAMPLE 2. Find $\int_0^{+\infty}\frac{8a^3dx}{x^2+4a^2}$.

**Solution.** $\int_0^{+\infty}\frac{8a^3dx}{x^2+4a^2}=\lim_{b=+\infty}\int_0^b\frac{8a^3dx}{x^2+4a^2}=\lim_{b=+\infty}\left[4a^2\arctan\frac{x}{2a}\right]_0^b$

$$=\lim_{b=+\infty}\left[4a^2\arctan\frac{b}{2a}\right]=4a^2\cdot\frac{\pi}{2}=2\pi a^2.\quad \textit{Ans.}$$

Let us interpret this result geometrically. The graph of our function is the witch, the locus of

$$y=\frac{8a^3}{x^2+4a^2}.$$

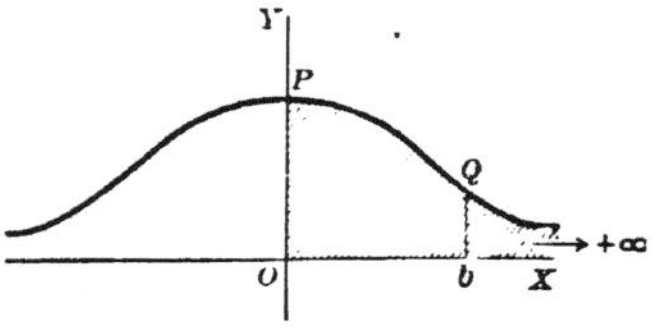


$$\text{area } OPQb=\int_0^b\frac{8a^3dx}{x^2+4a^2}=4a^2\arctan\frac{b}{2a}.$$

Now as the ordinate $Qb$ moves indefinitely to the right,

$$4a^2\arctan\frac{b}{2a}$$

is always finite, and

$$\lim_{b=+\infty}\left[4a^2\arctan\frac{b}{2a}\right]=2\pi a^2,$$

which is also finite. In such cases *we call the result the area bounded by the curve, the ordinate OP, and OX*, although strictly speaking this area is not completely bounded.

ILLUSTRATIVE EXAMPLE 3. Find $\int_1^{+\infty}\frac{dx}{x}$.

**Solution.** $$\int_1^{+\infty}\frac{dx}{x}=\lim_{b=+\infty}\int_1^b\frac{dx}{x}=\lim_{b=+\infty}(\log b).$$

The limit of $\log b$ as $b$ increases without limit does not exist; hence the integral has in this case no meaning.

**183. When $y=\phi(x)$ is discontinuous.** Let us now consider cases when the function to be integrated is discontinuous for isolated values of the variable lying within the limits of integration.

Consider first the case where the function to be integrated is continuous for all values of $x$ between the limits $a$ and $b$ except $x=a$.

If $a<b$ and $\epsilon$ is positive, we use the *definition*

$$(A)\qquad \int_a^b\phi(x)\,dx=\lim_{\epsilon=0}\int_{a+\epsilon}^b\phi(x)\,dx,$$

and when $\phi(x)$ is continuous except at $x=b$, we use the *definition*

$$(B)\qquad \int_a^b\phi(x)\,dx=\lim_{\epsilon=0}\int_a^{b-\epsilon}\phi(x)\,dx,$$

provided the limits are definite quantities.

ILLUSTRATIVE EXAMPLE 1. Find $\int_0^a \frac{dx}{\sqrt{a^2 - x^2}}$.

**Solution.** Here $\frac{1}{\sqrt{a^2 - x^2}}$ becomes infinite for $x = a$. Therefore, by (*B*),

$$\int_0^a \frac{dx}{\sqrt{a^2 - x^2}} = \underset{\epsilon = 0}{\text{limit}} \int_0^{a-\epsilon} \frac{dx}{\sqrt{a^2 - x^2}} = \underset{\epsilon = 0}{\text{limit}} \left[\text{arc sin} \frac{x}{a}\right]_0^{a-\epsilon}$$

$$= \underset{\epsilon = 0}{\text{limit}} \left[\text{arc sin}\left(1 - \frac{\epsilon}{a}\right)\right] = \text{arc sin } 1 = \frac{\pi}{2}. \quad \textit{Ans.}$$

ILLUSTRATIVE EXAMPLE 2. Find $\int_0^1 \frac{dx}{x^2}$.

**Solution.** Here $\frac{1}{x^2}$ becomes infinite for $x = 0$. Therefore, by (*A*),

$$\int_0^1 \frac{dx}{x^2} = \underset{\epsilon = 0}{\text{limit}} \int_\epsilon^1 \frac{dx}{x^2} = \underset{\epsilon = 0}{\text{limit}} \left(\frac{1}{\epsilon} - 1\right).$$

In this case there is no limit and therefore the integral does not exist.

If $c$ lies between $a$ and $b$, and $\phi(x)$ is continuous except at $x = c$, then, $\epsilon$ and $\epsilon'$ being positive numbers, the integral between $a$ and $b$ is *defined by*

$$(C) \qquad \int_a^b \phi(x)\,dx = \underset{\epsilon = 0}{\text{limit}} \int_a^{c-\epsilon} \phi(x)\,dx + \underset{\epsilon' = 0}{\text{limit}} \int_{c+\epsilon'}^b \phi(x)\,dx,$$

provided each separate limit is a definite quantity.

ILLUSTRATIVE EXAMPLE 1. Find $\int_0^{3a} \frac{2x\,dx}{(x^2 - a^2)^{\frac{2}{3}}}$.

**Solution.** Here the function to be integrated becomes infinite for $x = a$, i.e. for a value of $x$ between the limits of integration 0 and $3a$. Hence the above definition (*C*) must be employed. Thus

$$\int_0^{3a} \frac{2x\,dx}{(x^2 - a^2)^{\frac{2}{3}}} = \underset{\epsilon = 0}{\text{limit}} \int_0^{a-\epsilon} \frac{2x\,dx}{(x^2 - a^2)^{\frac{2}{3}}} + \underset{\epsilon' = 0}{\text{limit}} \int_{a+\epsilon'}^{3a} \frac{2x\,dx}{(x^2 - a^2)^{\frac{2}{3}}}$$

$$= \underset{\epsilon = 0}{\text{limit}} \left[3(x^2 - a^2)^{\frac{1}{3}}\right]_0^{a-\epsilon} + \underset{\epsilon' = 0}{\text{limit}} \left[3(x^2 - a^2)^{\frac{1}{3}}\right]_{a+\epsilon'}^{3a}$$

$$= \underset{\epsilon = 0}{\text{limit}} \left[3\sqrt[3]{(a - \epsilon)^2 - a^2} + 3a^{\frac{2}{3}}\right] + \underset{\epsilon' = 0}{\text{limit}} \left[3\sqrt[3]{8a^2} - 3\sqrt[3]{(a + \epsilon')^2 - a^2}\right]$$

$$= 3a^{\frac{2}{3}} + 6a^{\frac{2}{3}} = 9a^{\frac{2}{3}}. \quad \textit{Ans.}$$

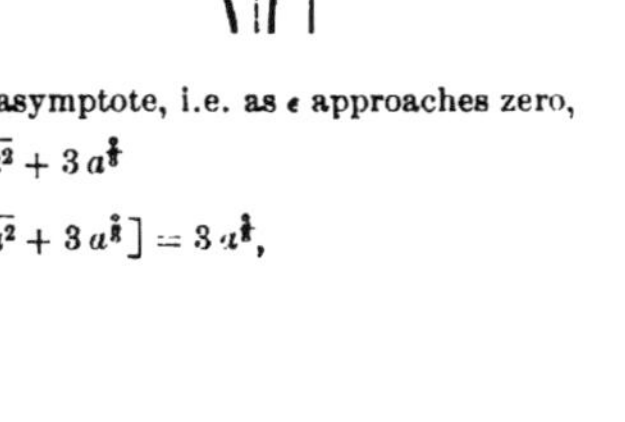


To interpret this geometrically, let us plot the graph, i.e. the locus, of

$$y = \frac{2x}{(x^2 - a^2)^{\frac{2}{3}}},$$

and note that $x = a$ is an asymptote.

$$\text{area } OPE = \int_0^{a-\epsilon} \frac{2x\,dx}{(x^2 - a^2)^{\frac{2}{3}}}$$

$$= 3\sqrt[3]{(a - \epsilon)^2 - a^2} + 3a^{\frac{2}{3}}.$$

Now as $PE$ moves to the right toward the asymptote, i.e. as $\epsilon$ approaches zero,

$$3\sqrt[3]{(a - \epsilon)^2 - a^2} + 3a^{\frac{2}{3}}$$

is always finite, and

$$\underset{\epsilon = 0}{\text{limit}} \left[3\sqrt[3]{(a - \epsilon)^2 - a^2} + 3a^{\frac{2}{3}}\right] = 3a^{\frac{2}{3}},$$

which is also finite. As in Illustrative Example 1, p. 323, $3a^{\frac{4}{3}}$ is called the area bounded by $OP$, the asymptote, and $OX$. Similarly,

$$\text{area } E'QRG = \int_{a+\epsilon'}^{3a} \frac{2x\,dx}{(x^2-a^2)^{\frac{2}{3}}} = 3\sqrt[3]{8a^2} - 3\sqrt[3]{(a+\epsilon')^2 - a^2}$$

is always finite as $QE'$ moves to the left toward the asymptote, and as $\epsilon'$ approaches zero, the result $6a^{\frac{2}{3}}$ is also finite. Hence $6a^{\frac{2}{3}}$ is called the area between $QR$, the asymptote, the ordinate $x = 3a$, and $OX$. Adding these results, we get $9a^{\frac{2}{3}}$, which is then called the area to the right of $OY$ between the curve, the ordinate $x = 3a$, and $OX$.

Illustrative Example 2. Find $\int_0^{2a} \frac{dx}{(x-a)^2}$.

**Solution.** This function also becomes infinite between the limits of integration. Hence, by (C),

$$\int_0^{2a} \frac{dx}{(x-a)^2} = \underset{\epsilon=0}{\text{limit}} \int_0^{a-\epsilon} \frac{dx}{(x-a)^2} + \underset{\epsilon'=0}{\text{limit}} \int_{a+\epsilon'}^{2a} \frac{dx}{(x-a)^2}$$

$$= \underset{\epsilon=0}{\text{limit}} \left[-\frac{1}{x-a}\right]_0^{a-\epsilon} + \underset{\epsilon'=0}{\text{limit}} \left[-\frac{1}{x-a}\right]_{a+\epsilon'}^{2a}$$

$$= \underset{\epsilon=0}{\text{limit}} \left(\frac{1}{\epsilon} - \frac{1}{a}\right) + \underset{\epsilon'=0}{\text{limit}} \left(-\frac{1}{a} + \frac{1}{\epsilon'}\right).$$

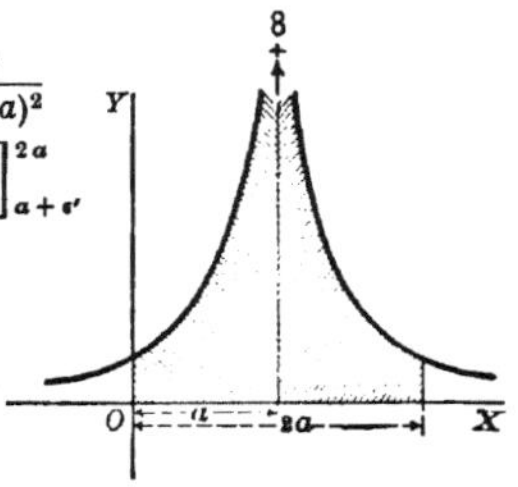


In this case the limits do not exist and the integral has no meaning.

If we plot the graph of this function and note the limits, the condition of things appears very much the same as in the last example. It turns out, however, that the shaded portion cannot be properly spoken of as an area, and the integral sign has no meaning in this case.

That it is important to note whether or not the given function becomes infinite within the limits of integration will appear at once if we apply our integration formula without any investigation. Thus

$$\int_0^{2a} \frac{dx}{(x-a)^2} = \left[-\frac{1}{x-a}\right]_0^{2a} = -\frac{2}{a},$$

a result which is absurd in view of the above discussions.

## EXAMPLES

1. $\int_0^{+\infty} \frac{dx}{a^2+x^2} = \frac{\pi}{2a}$.

2. $\int_1^{+\infty} \frac{dx}{x\sqrt{2x^2-1}} = \frac{\pi}{4}$.

3. $\int_1^{+\infty} \frac{dx}{x^4} = \frac{1}{3}$.

4. $\int_0^{\infty} \frac{dx}{a^2+b^2x^2} = \frac{\pi}{2ab}$.

5. $\int_1^{\infty} \frac{dx}{x^2} = 1$.

6. $\int_0^{+\infty} e^{-x}dx = 1$.

7. $\int_0^{+\infty} e^{-ax}dx = \frac{1}{a}$.

8. $\int_a^{+\infty} \frac{dx}{(a+x)^n} = \frac{1}{(n-1)(2a)^{n-1}}$.

9. $\int_{-\infty}^{1} e^x dx = e$.

10. $\int_{-\infty}^{+\infty} \frac{dx}{x^2+2x+2} = \pi$.

11. $\int_0^a x^2\sqrt{a^2-x^2}\,dx = \frac{\pi a^4}{16}$.

12. $\int_0^a \frac{x^3dx}{(a^2+x^2)^{\frac{3}{2}}} = \frac{3\sqrt{2}-4}{2}a$.

# CHAPTER XXV

## INTEGRATION OF RATIONAL FRACTIONS

**184. Introduction.** A rational fraction is a fraction the numerator and denominator of which are integral rational functions.* If the degree of the numerator is equal to or greater than that of the denominator, the fraction may be reduced to a mixed quantity by dividing the numerator by the denominator. For example,

$$\frac{x^4+3x^3}{x^2+2x+1}=x^2+x-3+\frac{5x+3}{x^2+2x+1}.$$

The last term is a fraction reduced to its lowest terms, having the degree of the numerator less than that of the denominator. It readily appears that the other terms are at once integrable, and hence we need consider only the fraction.

In order to integrate a differential expression involving such a fraction, it is often necessary to resolve it into simpler partial fractions, i.e. to replace it by the algebraic sum of fractions of forms such that we can complete the integration. That this is always possible when the denominator can be broken up into its real prime factors is shown in Algebra.†

**185. Case I.** *When the factors of the denominators are all of the first degree and none repeated.*

To each nonrepeated linear factor, such as $x-a$, there corresponds a partial fraction of the form

$$\frac{A}{x-a}.$$

Such a partial fraction may be integrated at once as follows:

$$\int\frac{Adx}{x-a}=A\int\frac{dx}{x-a}$$
$$=A\log(x-a)+C.$$

* That is, the variable is not affected with fractional or negative exponents.

† See Chap. XIX in Hawkes's "Advanced Algebra," Ginn and Company, Boston.

ILLUSTRATIVE EXAMPLE 1. Find $\int \frac{(2x+3)\,dx}{x^3+x^2-2x}$.

**Solution.** The factors of the denominator being $x$, $x-1$, $x+2$, we assume*

$$(A)\qquad \frac{2x+3}{x(x-1)(x+2)} = \frac{A}{x} + \frac{B}{x-1} + \frac{C}{x+2},$$

where $A$, $B$, $C$ are constants to be determined.

Clearing $(A)$ of fractions, we get

$$(B)\qquad 2x+3 = A(x-1)(x+2) + B(x+2)x + C(x-1)x,$$

$$2x+3 = (A+B+C)x^2 + (A+2B-C)x - 2A.$$

Since this equation is an identity, we equate the coefficients of the like powers of $x$ in the two members according to the method of Undetermined Coefficients, and obtain three simultaneous equations

$$(C)\qquad \begin{cases} A+B+C=0, \\ A+2B-C=2, \\ -2A=3. \end{cases}$$

Solving equations $(C)$, we get

$$A = -\tfrac{3}{2},\quad B = \tfrac{5}{3},\quad C = -\tfrac{1}{6}.$$

Substituting these values in $(A)$,

$$\frac{2x+3}{x(x-1)(x+2)} = -\frac{3}{2x} + \frac{5}{3(x-1)} - \frac{1}{6(x+2)}.$$

$$\therefore \int \frac{2x+3}{x(x-1)(x+2)}\,dx = -\frac{3}{2}\int\frac{dx}{x} + \frac{5}{3}\int\frac{dx}{x-1} - \frac{1}{6}\int\frac{dx}{x+2}$$

$$= -\tfrac{3}{2}\log x + \tfrac{5}{3}\log(x-1) - \tfrac{1}{6}\log(x+2) + \log c$$

$$= \log\frac{c(x-1)^{\frac{5}{3}}}{x^{\frac{3}{2}}(x+2)^{\frac{1}{6}}}.\quad \textit{Ans.}$$

A shorter method of finding the values of $A$, $B$, and $C$ from $(B)$ is the following

Let factor $x=0$; then $3 = -2A$, or $A = -\tfrac{3}{2}$.

Let factor $x-1=0$, or $x=1$; then $5 = 3B$, or $B = \tfrac{5}{3}$.

Let factor $x+2=0$, or $x=-2$; then $-1 = 6C$, or $C = -\tfrac{1}{6}$.

A useful exercise is to integrate without determining the constants $A$, $B$, $C$, etc. For instance, in the above example,

$$\int\frac{(2x+3)\,dx}{x(x-1)(x+2)} = \int\frac{A\,dx}{x} + \int\frac{B\,dx}{x-1} + \int\frac{C\,dx}{x+2}$$

$$= A\log x + B\log(x-1) + C\log(x+2).$$

* In the process of decomposing the fractional part of the given differential neither the integral sign nor $dx$ enters.

**EXAMPLES**

1. $\int \frac{(2x-1)\,dx}{(x-1)(x-2)} = \log \frac{(x-2)^3}{x-1} + C.$

2. $\int \frac{x\,dx}{(x+1)(x+3)(x+5)} = \frac{1}{8} \log \frac{(x+3)^6}{(x+5)^5 (x+1)} + C.$

3. $\int \frac{(x-1)\,dx}{x^2+6x+8} = \log \frac{c\,(x+4)^{\frac{5}{2}}}{(x+2)^{\frac{3}{2}}}.$

4. $\int \frac{(3x-1)\,dx}{x^2+x-6} = \log [c\,(x+3)^2 (x-2)].$

5. $\int \frac{(x^2+x-1)\,dx}{x^3+x^2-6x} = \log \sqrt[6]{x(x-2)^3 (x+3)^2} + C.$

6. $\int \frac{(2x^3+1)\,dx}{x^2+3x+2} = x^2 - 6x + \log \frac{(x+2)^{15}}{x+1} + C.$

7. $\int \frac{x^5+x^4-8}{x^3-4x} dx = \frac{x^3}{3} + \frac{x^2}{2} + 4x + \log \frac{x^2 (x-2)^5}{(x+2)^3} + C.$

8. $\int \frac{x^4\,dx}{(x^2-1)(x+2)} = \frac{x^2}{2} - 2x + \frac{1}{6} \log \frac{x-1}{(x+1)^3} + \frac{16}{3} \log (x+2) + C.$

9. $\int \frac{(a-b)\,y\,dy}{y^2-(a+b)\,y+ab} = \log \frac{(y-a)^a}{(y-b)^b} + C.$

10. $\int \frac{(t^2+pq)\,dt}{t\,(t-p)(t+q)} = \log \frac{(t-p)(t+q)}{t} + C.$

11. $\int \frac{(2z^2-5)\,dz}{z^4-5z^2+6} = \frac{1}{2\sqrt{2}} \log \frac{z-\sqrt{2}}{z+\sqrt{2}} + \frac{1}{2\sqrt{3}} \log \frac{z-\sqrt{3}}{z+\sqrt{3}} + C.$

12. $\int_1^2 \frac{dx}{x^2+4x} = \frac{1}{4} \log \frac{5}{3}.$

13. $\int_0^5 \frac{dx}{1+3x+2x^2} = \log \frac{11}{6}.$

14. $\int_3^4 \frac{(x^2+6x-8)\,dx}{x^3-4x} = \log \frac{200}{81}.$

**186. Case II.** *When the factors of the denominator are all of the first degree and some repeated.*

To every $n$-fold linear factor, such as $(x-a)^n$, there corresponds the $n$ partial fractions

$$\frac{A}{(x-a)^n} + \frac{B}{(x-a)^{n-1}} + \cdots + \frac{L}{x-a}.$$

The last one is integrated as in Case I. The rest are all integrated by means of the power formula. Thus

$$\int \frac{A\,dx}{(x-a)^n} = A \int (x-a)^{-n} dx = \frac{A}{(1-n)(x-a)^{n-1}} + C.$$

ILLUSTRATIVE EXAMPLE 1. Find $\int \frac{x^3+1}{x(x-1)^3}dx$.

**Solution.** Since $x-1$ occurs three times as a factor, we assume

$$\frac{x^3+1}{x(x-1)^3}=\frac{A}{x}+\frac{B}{(x-1)^3}+\frac{C}{(x-1)^2}+\frac{D}{x-1}.$$

Clearing of fractions,

$$x^3+1=A(x-1)^3+Bx+Cx(x-1)+Dx(x-1)^2.$$

$$x^3+1=(A+D)x^3+(-3A+C-2D)x^2+(3A+B-C+D)x-A.$$

Equating the coefficients of like powers of $x$, we get the simultaneous equations

$$A+D=1,$$
$$-3A+C-2D=0,$$
$$3A+B-C+D=0,$$
$$-A=1.$$

Solving, $A=-1$, $B=2$, $C=1$, $D=2$, and

$$\frac{x^3+1}{x(x-1)^3}=-\frac{1}{x}+\frac{2}{(x-1)^3}+\frac{1}{(x-1)^2}+\frac{2}{x-1}.$$

$$\therefore \int\frac{x^3+1}{x(x-1)^3}dx=-\log x-\frac{1}{(x-1)^2}-\frac{1}{x-1}+2\log(x-1)+C$$

$$=-\frac{x}{(x-1)^2}+\log\frac{(x-1)^2}{x}+C.$$

**EXAMPLES**

1. $\int\frac{dx}{(x-1)^2(x-2)}=\frac{1}{x-1}+\log\frac{x-2}{x-1}+C.$

2. $\int\frac{x^2dx}{(x+2)^2(x+1)}=\frac{4}{x+2}+\log(x+1)+C.$

3. $\int\frac{(x-8)\,dx}{x^3-4x^2+4x}=\frac{3}{x-2}+\log\frac{(x-2)^2}{x^2}+C.$

4. $\int\frac{x^2+1}{(x-1)^3}dx=-\frac{1}{(x-1)^2}-\frac{2}{x-1}+\log(x-1)+C.$

5. $\int\frac{(x^5-x^3+1)\,dx}{x^4-x^3}=\frac{x^2}{2}+x+\frac{1}{2x^2}+\frac{1}{x}+\log\frac{x-1}{x}+C.$

6. $\int\frac{(3x+2)\,dx}{x(x+1)^3}=\frac{4x+3}{2(x+1)^2}+\log\frac{x^2}{(x+1)^2}+C.$

7. $\int\frac{x^2dx}{(x+2)^2(x+4)^2}=-\frac{5x+12}{x^2+6x+8}+\log\left(\frac{x+4}{x+2}\right)^2+C.$

8. $\int\frac{y^2dy}{y^3+5y^2+8y+4}=\frac{4}{y+2}+\log(y+1)+C.$

9. $\int \frac{dt}{(t^2-2)^2} = -\frac{t}{4(t^2-2)} + \frac{1}{8\sqrt{2}}\log\frac{t+\sqrt{2}}{t-\sqrt{2}} + C.$

10. $\int \frac{as^2ds}{(s+a)^3} = a\log(s+a) + \frac{2a^2}{s+a} - \frac{a^3}{2(s+a)^2} + C.$

11. $\int\left(\frac{m}{z+m} - \frac{nz}{(z+n)^2}\right)dz = \log(z+m)^m(z+n)^{-n} - \frac{n^2}{z+n} + C.$

12. $\int_1^\infty \frac{dx}{x^2(1+x)} = 1 - \log 2.$

13. $\int_1^\infty \frac{dt}{t(1+t)^2} = \log 2 - \frac{1}{2}.$

14. $\int_1^4 \frac{(1+3x)\,dx}{x+2x^2+x^3} = \log\frac{8}{5} + \frac{3}{5}.$

**187. Case III.** *When the denominator contains factors of the second degree but none repeated.*

To every nonrepeated quadratic factor, such as $x^2+px+q$, there corresponds a partial fraction of the form

$$\frac{Ax+B}{x^2+px+q}.$$

This may be integrated as follows:

$$\int\frac{(Ax+B)\,dx}{x^2+px+q} = \int\frac{\left(Ax+\frac{Ap}{2}-\frac{Ap}{2}+B\right)dx}{x^2+px+q}$$

[Adding and subtracting $\frac{Ap}{2}$ in the numerator.]

$$= \int\frac{\left(Ax+\frac{Ap}{2}\right)dx}{x^2+px+q} + \int\frac{\left(-\frac{Ap}{2}+B\right)dx}{x^2+px+q}$$

$$= \frac{A}{2}\int\frac{(2x+p)\,dx}{x^2+px+q} + \left(\frac{2B-Ap}{2}\right)\int\frac{dx}{\left(x+\frac{p}{2}\right)^2+\left(q-\frac{p^2}{4}\right)}$$

[Completing the square in the denominator of the second integral.]

$$= \frac{A}{2}\log(x^2+px+q) + \frac{2B-Ap}{\sqrt{4q-p^2}}\arctan\frac{2x+p}{\sqrt{4q-p^2}} + C.$$

Since $x^2+px+q=0$ has imaginary roots, we know from 3, p. 1, that $4q-p^2>0$.

ILLUSTRATIVE EXAMPLE 1. Find $\int \frac{4\,dx}{x^3 + 4x}$.

**Solution.** Assume
$$\frac{4}{x(x^2+4)} = \frac{A}{x} + \frac{Bx + C}{x^2 + 4}.$$

Clearing of fractions, $4 = A(x^2+4) + x(Bx + C) = (A + B)x^2 + Cx + 4A$.
Equating the coefficients of like powers of $x$, we get
$$A + B = 0, \quad C = 0, \quad 4A = 4.$$

This gives $A = 1$, $B = -1$, $C = 0$, so that $\frac{4}{x(x^2+4)} = \frac{1}{x} - \frac{x}{x^2+4}$.
$$\therefore \int \frac{4\,dx}{x(x^2+4)} = \int \frac{dx}{x} - \int \frac{x\,dx}{x^2+4}$$
$$= \log x - \frac{1}{2}\log(x^2+4) + \log c = \log \frac{cx}{\sqrt{x^2+4}}. \quad \textit{Ans.}$$

**EXAMPLES**

1. $\int \frac{dx}{x(x^2+1)} = \log \frac{x}{\sqrt{x^2+1}} + C.$

2. $\int \frac{x\,dx}{(x+1)(x^2+4)} = \frac{1}{10}\log\frac{x^2+4}{(x+1)^2} + \frac{2}{5}\arctan\frac{x}{2} + C.$

3. $\int \frac{(2x^2 - 3x - 3)\,dx}{(x-1)(x^2-2x+5)} = \log\frac{(x^2-2x+5)^{\frac{3}{2}}}{x-1} + \frac{1}{2}\arctan\frac{x-1}{2} + C.$

4. $\int \frac{x^2dx}{1-x^4} = \frac{1}{4}\log\frac{1+x}{1-x} - \frac{1}{2}\arctan x + C.$

5. $\int \frac{dx}{(x^2+1)(x^2+x)} = \frac{1}{4}\log\frac{x^4}{(x+1)^2(x^2+1)} - \frac{1}{2}\arctan x + C.$

6. $\int \frac{(x^3-6)\,dx}{x^4+6x^2+8} = \log\frac{x^2+4}{\sqrt{x^2+2}} + \frac{3}{2}\arctan\frac{x}{2} - \frac{3}{\sqrt{2}}\arctan\frac{x}{\sqrt{2}} + C.$

7. $\int \frac{(5x^2-1)\,dx}{(x^2+3)(x^2-2x+5)} = \log\frac{x^2-2x+5}{x^2+3} + \frac{5}{2}\arctan\frac{x-1}{2} - \frac{2}{\sqrt{3}}\arctan\frac{x}{\sqrt{3}} + C.$

8. $\int \frac{dx}{x^3+1} = \frac{1}{6}\log\frac{(x+1)^2}{x^2-x+1} + \frac{1}{\sqrt{3}}\arctan\frac{2x-1}{\sqrt{3}} + C.$

9. $\int \frac{z^2dz}{z^4+z^2-2} = \frac{1}{6}\log\left(\frac{z-1}{z+1}\right) + \frac{\sqrt{2}}{3}\arctan\frac{z}{\sqrt{2}} + C.$

10. $\int \frac{4\,dt}{t^4+1} = \frac{1}{\sqrt{2}}\log\frac{t^2+t\sqrt{2}+1}{t^2-t\sqrt{2}+1} + \sqrt{2}\arctan\frac{t\sqrt{2}}{1-t^2} + C.$

11. $\int \frac{dy}{1-y^3} = \frac{1}{6}\log\frac{y^2+y+1}{y^2-2y+1} + \frac{1}{\sqrt{3}}\arctan\frac{2y+1}{\sqrt{3}} + C.$

12. $\int_0^3 \frac{2x\,dx}{(1+x^2)(3+x^2)} = \log\sqrt{\frac{5}{2}}.$

13. $\int_0^1 \frac{(2x^2+x+3)\,dx}{(x+1)(x^2+1)} = \log 4 + \frac{\pi}{4}.$

**188. Case IV.** *When the denominator contains factors of the second degree some of which are repeated.*

To every $n$-fold quadratic factor, such as $(x^2 + px + q)^n$, there correspond the $n$ partial fractions

$$(A)\qquad \frac{Ax+B}{(x^2+px+q)^n} + \frac{Cx+D}{(x^2+px+q)^{n-1}} + \cdots + \frac{Lx+M}{x^2+px+q}.$$

To derive a formula for integrating the first one we proceed as follows:

$$\int \frac{Ax+B}{(x^2+px+q)^n}dx = \int \frac{\left(Ax + \frac{Ap}{2} - \frac{Ap}{2} + B\right)dx}{(x^2+px+q)^n}$$

[Adding and subtracting $\frac{Ap}{2}$ in the numerator.]

$$= \int \frac{\left(Ax + \frac{Ap}{2}\right)dx}{(x^2+px+q)^n} + \int \frac{\left(-\frac{Ap}{2} + B\right)dx}{(x^2+px+q)^n}$$

$$= \frac{A}{2}\int (x^2+px+q)^{-n}(2x+p)\,dx + \left(\frac{2B - Ap}{2}\right)\int \frac{dx}{(x^2+px+q)^n}.$$

The first one of these may be integrated by (4) p. 284; hence

$$(B)\qquad \int \frac{Ax+B}{(x^2+px+q)^n}dx = \frac{A}{2(1-n)(x^2+px+q)^{n-1}}$$
$$+ \left(\frac{2B-Ap}{2}\right)\int \frac{dx}{(x^2+px+q)^n}.$$

Let us now differentiate the function $\dfrac{\left(x+\frac{p}{2}\right)}{(x^2+px+q)^{n-1}}$. Thus

$$\frac{d}{dx}\left(\frac{x+\frac{p}{2}}{(x^2+px+q)^{n-1}}\right) = \frac{1}{(x^2+px+q)^{n-1}} - \frac{2(n-1)\left(x+\frac{p}{2}\right)^2}{(x^2+px+q)^n},\text{ or}$$

$$(C)\qquad d\left(\frac{x+\frac{p}{2}}{(x^2+px+q)^{n-1}}\right) = \left(\frac{-(2n-3)}{(x^2+px+q)^{n-1}} + \frac{2(n-1)\left(q-\frac{p^2}{4}\right)}{(x^2+px+q)^n}\right)dx.$$

[Since $x^2+px+q = \left(x+\frac{p}{2}\right)^2 + \left(q-\frac{p^2}{4}\right)$, and $\left(x+\frac{p}{2}\right)^2 = (x^2+px+q) - \left(q - \frac{p^2}{4}\right)$.]

Integrating both sides of $(C)$,

$$\frac{x+\frac{p}{2}}{(x^2+px+q)^{n-1}}=-(2n-3)\int\frac{dx}{(x^2+px+q)^{n-1}}$$
$$+2(n-1)\left(q-\frac{p^2}{4}\right)\int\frac{dx}{(x^2+px+q)^n},$$

or, solving for the last integral,

$$(D)\quad \int\frac{dx}{(x^2+px+q)^n}=\frac{x+\frac{p}{2}}{2(n-1)\left(q-\frac{p^2}{4}\right)(x^2+px+q)^{n-1}}$$
$$+\frac{2n-3}{2(n-1)\left(q-\frac{p^2}{4}\right)}\int\frac{dx}{(x^2+px+q)^{n-1}}.$$

Substituting this result in the second member of $(B)$, we get *

$$(E)\quad \int\frac{(Ax+B)\,dx}{(x^2+px+q)^n}=\frac{A(p^2-4q)+(2B-Ap)(2x+p)}{2(n-1)(4q-p^2)(x^2+px+q)^{n-1}}$$
$$+\frac{(2B-Ap)(2n-3)}{(n-1)(4q-p^2)}\int\frac{dx}{(x^2+px+q)^{n-1}}.$$

It is seen that our integral has been made to depend on the integration of a rational fraction of the same type in which, however, the quadratic factor occurs only $n-1$ times. By applying the formula $(E)$ $n-1$ times successively it is evident that our integral may be made ultimately to depend on

$$\int\frac{dx}{x^2+px+q},$$

and this may be integrated by completing the square, as shown on p. 296.

In the same manner all but the last fraction of $(A)$ may be integrated. But this last fraction, namely,

$$\frac{Lx+M}{x^2+px+q},$$

may be integrated by the method already given under the previous case (p. 329).

* $4q-p^2>0$, since $x^2+px+q=0$ has imaginary roots.

ILLUSTRATIVE EXAMPLE 1. Find $\int \frac{(x^3 + x^2 + 2)\,dx}{(x^2 + 2)^2}$.

**Solution.** Since $x^2 + 2$ occurs twice as a factor, we assume

$$\frac{x^3 + x^2 + 2}{(x^2 + 2)^2} = \frac{Ax + B}{(x^2 + 2)^2} + \frac{Cx + D}{x^2 + 2}.$$

Clearing of fractions, we get

$$x^3 + x^2 + 2 = Ax + B + (Cx + D)(x^2 + 2).$$

$$x^3 + x^2 + 2 = Cx^3 + Dx^2 + (A + 2C)x + B + 2D.$$

Equating the coefficients of like powers of $x$,

$$C = 1, \quad D = 1, \quad A + 2C = 0, \quad B + 2D = 2.$$

This gives $\quad A = -2, \quad B = 0, \quad C = 1, \quad D = 1.$

Hence $\quad \frac{x^3 + x^2 + 2}{(x^2 + 2)^2} = -\frac{2x}{(x^2 + 2)^2} + \frac{x + 1}{x^2 + 2}$, and

$$\int \frac{(x^3 + x^2 + 2)\,dx}{(x^2 + 2)^2} = -\int \frac{2x\,dx}{(x^2 + 2)^2} + \int \frac{x\,dx}{x^2 + 2} + \int \frac{dx}{x^2 + 2}$$

$$= \frac{1}{x^2 + 2} + \frac{1}{\sqrt{2}} \arctan \frac{x}{\sqrt{2}} + \frac{1}{2} \log (x^2 + 2) + C.$$

ILLUSTRATIVE EXAMPLE 2. Find $\int \frac{2x^3 + x + 3}{(x^2 + 1)^2} dx$.

**Solution.** Since $x^2 + 1$ occurs twice as a factor, we assume

$$\frac{2x^3 + x + 3}{(x^2 + 1)^2} = \frac{Ax + B}{(x^2 + 1)^2} + \frac{Cx + D}{x^2 + 1}.$$

Clearing of fractions,

$$2x^3 + x + 3 = Ax + B + (Cx + D)(x^2 + 1).$$

Equating the coefficients of like powers of $x$ and solving, we get

$$A = -1, \quad B = 3, \quad C = 2, \quad D = 0.$$

Hence $\quad \int \frac{2x^3 + x + 3}{(x^2 + 1)^2} dx = \int \frac{-x + 3}{(x^2 + 1)^2} dx + \int \frac{2x\,dx}{x^2 + 1}$

$$= \log (x^2 + 1) + \int \frac{-x + 3}{(x^2 + 1)^2} dx.$$

Now apply formula (E), p. 332, to the remaining integral. Here

$$A = -1, \quad B = 3, \quad p = 0, \quad q = 1, \quad n = 2.$$

Substituting, we get

$$\int \frac{-x + 3}{(x^2 + 1)^2} dx = \frac{1 + 3x}{2(x^2 + 1)} + \frac{3}{2} \int \frac{dx}{x^2 + 1} = \frac{1 + 3x}{2(x^2 + 1)} + \frac{3}{2} \arctan x.$$

Therefore

$$\int \frac{2x^3 + x + 3}{(x^2 + 1)^2} dx = \log (x^2 + 1) + \frac{1 + 3x}{2(x^2 + 1)} + \frac{3}{2} \arctan x + C.$$

**EXAMPLES**

1. $\int \frac{dx}{(x^2+1)^2} = \frac{x}{2(x^2+1)} + \frac{1}{2}\operatorname{arc}\tan x + C.$

2. $\int \frac{x^3+x-1}{(x^2+2)^2}dx = \frac{2-x}{4(x^2+2)} + \log(x^2+2)^{\frac{1}{2}} - \frac{1}{4\sqrt{2}}\operatorname{arc}\tan\frac{x}{\sqrt{2}} + C.$

3. $\int \frac{2\,xdx}{(1+x)(1+x^2)^2} = \frac{1}{4}\log\frac{x^2+1}{(x+1)^2} + \frac{x-1}{2(x^2+1)} + C.$

4. $\int \left(\frac{x^2-a^2}{x^2+a^2}\right)^2 dx = x + \frac{2\,a^2x}{x^2+a^2} - 2\,a\operatorname{arc}\tan\frac{x}{a} + C.$

5. $\int \frac{(4\,x+3)\,dx}{(4\,x^2+3)^3} = \frac{4\,x^3+5\,x-2}{8(4\,x^2+3)^2} + \frac{1}{16\sqrt{3}}\operatorname{arc}\tan\frac{2\,x}{\sqrt{3}} + C.$

6. $\int \frac{9\,x^3\,dx}{(x^3+1)^2} = -\frac{3\,x}{x^3+1} + \frac{1}{2}\log\frac{(x+1)^2}{x^2-x+1} + \sqrt{3}\operatorname{arc}\tan\frac{2\,x-1}{\sqrt{3}} + C.$

7. $\int \frac{x^7+x^5+x^3+x}{(x^2+2)^2(x^2+3)^2}dx = \frac{5}{2(x^2+2)} + \frac{10}{x^2+3} + \frac{19}{2}\log(x^2+2) - 9\log(x^2+3) + C.$

8. $\int \frac{(4\,x^2-8\,x)\,dx}{(x-1)^2(x^2+1)^2} = \frac{3\,x^2-x}{(x-1)(x^2+1)} + \log\frac{(x-1)^2}{x^2+1} + \operatorname{arc}\tan x + C.$

9. $\int \frac{(3\,x+2)\,dx}{(x^2-3\,x+3)^2} = \frac{13\,x-24}{3(x^2-3\,x+3)} + \frac{26}{3\sqrt{3}}\operatorname{arc}\tan\frac{2\,x-3}{\sqrt{3}} + C.$

Since a rational function may always be reduced to the quotient of two integral rational functions, i.e. to a rational fraction, it follows from the preceding sections in this chapter that any rational function whose denominator can be broken up into real quadratic and linear factors may be expressed as the algebraic sum of integral rational functions and partial fractions. The terms of this sum have forms all of which we have shown how to integrate. Hence the

**Theorem.** *The integral of every rational function whose denominator can be broken up into real quadratic and linear factors may be found, and is expressible in terms of algebraic, logarithmic, and inverse-trigonometric functions; that is, in terms of the elementary functions.*

# CHAPTER XXVI

## INTEGRATION BY SUBSTITUTION OF A NEW VARIABLE. RATIONALIZATION

**189. Introduction.** In the last chapter it was shown that all rational functions whose denominators can be broken up into real quadratic and linear factors may be integrated. Of algebraic functions which are *not rational*, that is, such as contain radicals, only a small number, relatively speaking, can be integrated in terms of elementary functions. By substituting a new variable, however, these functions can in some cases be transformed into equivalent functions that are either in the list of standard forms (pp. 284, 285) or else are rational. The method of integrating a function that is not rational by substituting for the old variable such a function of a new variable that the result is a rational function is sometimes called *integration by rationalization.* This is a very important artifice in integration and we will now take up some of the more important cases coming under this head.

**190. Differentials containing fractional powers of $x$ only.**

*Such an expression can be transformed into a rational form by means of the substitution*

$$x = z^n,$$

*where $n$ is the least common denominator of the fractional exponents of $x$.*

For $x$, $dx$, and each radical can then be expressed rationally in terms of $z$.

ILLUSTRATIVE EXAMPLE 1. Find $\int \frac{x^{\frac{2}{3}} - x^{\frac{1}{4}}}{x^{\frac{1}{2}}} dx.$

**Solution.** Since 12 is the L.C.M. of the denominators of the fractional exponents, we assume

$$x = z^{12}.$$

Here $\quad dx = 12\, z^{11} dz, \quad x^{\frac{2}{3}} = z^8, \quad x^{\frac{1}{4}} = z^3, \quad x^{\frac{1}{2}} = z^6.$

$$\therefore \int \frac{x^{\frac{2}{3}} - x^{\frac{1}{4}}}{x^{\frac{1}{2}}} dx = \int \frac{z^8 - z^3}{z^6} 12\, z^{11} dz = 12 \int (z^{13} - z^8)\, dz$$

$$= \tfrac{6}{7} z^{14} - \tfrac{4}{3} z^9 + C = \tfrac{6}{7} x^{\frac{7}{6}} - \tfrac{4}{3} x^{\frac{3}{4}} + C.$$

[Substituting back the value of $z$ in terms of $x$, namely, $z = x^{\frac{1}{12}}$.]

The general form of the irrational expression here treated is then

$$R\left(x^{\frac{1}{n}}\right) dx,$$

where $R$ denotes a rational function of $x^{\frac{1}{n}}$.

**191. Differentials containing fractional powers of $a + bx$ only.**

*Such an expression can be transformed into a rational form by means of the substitution*

$$a + bx = z^n,$$

*where n is the least common denominator of the fractional exponents of the expression $a + bx$.*

For $x$, $dx$, and each radical can then be expressed rationally in terms of $z$.

ILLUSTRATIVE EXAMPLE 1. Find $\int \frac{dx}{(1+x)^{\frac{3}{2}} + (1+x)^{\frac{1}{2}}}$.

**Solution.** Assume $$1 + x = z^2;$$

then $$dx = 2\,zdz, \quad (1+x)^{\frac{3}{2}} = z^3, \quad \text{and} \quad (1+x)^{\frac{1}{2}} = z.$$

$$\therefore \int \frac{dx}{(1+x)^{\frac{3}{2}} + (1+x)^{\frac{1}{2}}} = \int \frac{2\,zdz}{z^3 + z} = 2\int \frac{dz}{z^2+1}$$

$$= 2 \text{ arc tan } z + C = 2 \text{ arc tan } (1+x)^{\frac{1}{2}} + C,$$

when we substitute back the value of $z$ in terms of $x$.

The general integral treated here has then the form

$$R\left[x, (a+bx)^{\frac{1}{n}}\right] dx,$$

where $R$ denotes a rational function.

**192. Change in limits corresponding to change in variable.** When integrating by the substitution of a new variable it is sometimes rather troublesome to translate the result back into the original variable. When integrating between limits, however, we may avoid the process of restoring the original variable by changing the limits to correspond with the new variable.* This process will now be illustrated by an example.

ILLUSTRATIVE EXAMPLE 1. Calculate $\int_0^{16} \frac{x^{\frac{1}{2}}dx}{1+x^{\frac{1}{4}}}$.

**Solution.** Assume $x = z^4$.

Then $dx = 4\,z^3dz$, $x^{\frac{1}{2}} = z^2$, $x^{\frac{1}{4}} = z$. Also to change the limits we observe that

when $$x = 0, \quad z = 0,$$

and when $$x = 16, \quad z = 2.$$

$$\therefore \int_0^{16} \frac{x^{\frac{1}{2}}dx}{1+x^{\frac{1}{4}}} = \int_0^2 \frac{z^2 \cdot 4\,z^3dz}{1+z^2} = 4\int_0^2 \left(z^3 - z + \frac{z}{1+z^2}\right)dz$$

$$= 4\int_0^2 z^2dz - 4\int_0^2 dz + 4\int_0^2 \frac{dz}{1+z^2} = \left[\frac{4\,z^3}{3} - 4\,z + 4 \text{ arc tan } z\right]_0^2$$

$$= \tfrac{8}{3} + 4 \text{ arc tan } 2. \quad \textit{Ans.}$$

* The relation between the old and the new variable should be such that to each value of one within the limits of integration there is always one, and only one, finite value of the other. When one is given as a many-valued function of the other, care must be taken to choose the right values.

**EXAMPLES**

1. $\int \frac{x^{\frac{1}{2}}dx}{x^{\frac{3}{4}}+1} = \frac{4}{3}x^{\frac{3}{4}} - \frac{4}{3}\log(x^{\frac{3}{4}}+1) + C.$

2. $\int \frac{x^{\frac{3}{2}} - x^{\frac{1}{3}}}{6x^{\frac{1}{4}}}dx = \frac{1}{3}\left(\frac{2}{9}x^{\frac{9}{4}} - \frac{6}{13}x^{\frac{13}{12}}\right) + C.$

3. $\int_0^1 \frac{x^{\frac{3}{2}}dx}{1+x} = \frac{\pi}{2} - \frac{4}{3}.$

4. $\int_0^3 \frac{dx}{(2+x)\sqrt{1+x}} = 2 \text{ arc tan } 2 - \frac{\pi}{2}.$

5. $\int \frac{x^{\frac{1}{3}}+1}{x^{\frac{7}{6}}+x^{\frac{5}{4}}}dx = -\frac{6}{x^{\frac{1}{6}}} + \frac{12}{x^{\frac{1}{12}}} + 2\log x - 24\log(x^{\frac{1}{12}}+1) + C.$

6. $\int \frac{dx}{x^{\frac{5}{8}} - x^{\frac{1}{8}}} = \frac{8}{3}x^{\frac{3}{8}} + 2\log\frac{x^{\frac{1}{8}}-1}{x^{\frac{1}{8}}+1} + 4 \text{ arc tan } x^{\frac{1}{8}} + C.$

7. $\int \frac{3\sqrt{x}\,dx}{2\sqrt{x} - \sqrt[3]{x^2}} = -18\left[\frac{x^{\frac{5}{6}}}{5} + \frac{x^{\frac{2}{3}}}{2} + \frac{4x^{\frac{1}{2}}}{3} + 4x^{\frac{1}{3}} + 16x^{\frac{1}{6}} + 32\log(x^{\frac{1}{6}} - 2)\right] + C.$

8. $\int_0^4 \frac{dx}{1+\sqrt{x}} = 4 - 2\log 3.$

9. $\int_3^{29} \frac{(x-2)^{\frac{2}{3}}dx}{(x-2)^{\frac{2}{3}}+3} = 8 + \frac{3\sqrt{3}}{2}\pi.$

10. $\int_1^4 \frac{ydy}{\sqrt{2+4y}} = \frac{3\sqrt{2}}{2}.$

11. $\int_0^{12} \frac{xdx}{(2x+3)^{\frac{4}{3}}} = \frac{3}{8}\left(11 - \frac{9}{\sqrt[3]{3}}\right).$

12. $\int \frac{y^{\frac{1}{7}} + y^{\frac{1}{2}}}{y^{\frac{8}{7}} + y^{\frac{15}{14}}}dy = 14\left[y^{\frac{1}{14}} - \frac{y^{\frac{1}{7}}}{2} + \frac{y^{\frac{3}{14}}}{3} - \frac{y^{\frac{2}{7}}}{4} + \frac{y^{\frac{5}{14}}}{5}\right] + C.$

13. $\int \frac{dx}{x(x+1)^{\frac{1}{2}}} = \log\frac{(x+1)^{\frac{1}{2}}-1}{(x+1)^{\frac{1}{2}}+1} + C.$

14. $\int \frac{xdx}{(a+bx)^{\frac{3}{2}}} = \frac{2(2a+bx)}{b^2\sqrt{a+bx}} + C.$

15. $\int \frac{x^2dx}{(4x+1)^{\frac{5}{2}}} = \frac{6x^2+6x+1}{12(4x+1)^{\frac{3}{2}}} + C.$

16. $\int y\sqrt[3]{a+y}\,dy = \frac{3}{28}(4y-3a)(a+y)^{\frac{4}{3}} + C.$

17. $\int \frac{\sqrt{x+1}+1}{\sqrt{x+1}-1}dx = x + 1 + 4\sqrt{x+1} + 4\log(\sqrt{x+1} - 1) + C.$

18. $\int \frac{dx}{1+\sqrt[3]{x+1}} = \frac{3}{2}(x+1)^{\frac{2}{3}} - 3(x+1)^{\frac{1}{3}} + 3\log(1+\sqrt[3]{x+1}) + C.$

19. $\int \frac{x+1}{x\sqrt{x-2}}dx = 2\sqrt{x-2} + \sqrt{2} \text{ arc tan}\sqrt{\frac{x-2}{2}} + C.$

20. $\int_1^{64} \frac{dt}{2t^{\frac{1}{2}} + t^{\frac{1}{3}}} = 5.31.$

21. $\int_4^9 \frac{dx}{1-\sqrt{x}} = -3.386.$

22. $\int \frac{dx}{(x+1)^{\frac{2}{3}} - (x+1)^{\frac{1}{2}}} = 3\{(x+1)^{\frac{1}{3}} + 2(x+1)^{\frac{1}{6}} + 2\log[(x+1)^{\frac{1}{6}} - 1]\} + C.$

**193. Differentials containing no radical except $\sqrt{a+bx+x^2}$.***

*Such an expression can be transformed into a rational form by means of the substitution*

$$\sqrt{a+bx+x^2}=z-x.$$

For, squaring and solving for $x$,

$$x=\frac{z^2-a}{b+2z};\ \text{then } dx=\frac{2(z^2+bz+a)\,dz}{(b+2z)^2},$$

and
$$\sqrt{a+bx+x^2}\,(=z-x)=\frac{z^2+bz+a}{b+2z}.$$

Hence $x$, $dx$, and $\sqrt{a+bx+x^2}$ are rational when expressed in terms of $z$.

ILLUSTRATIVE EXAMPLE 1. Find $\int\frac{dx}{\sqrt{1+x+x^2}}$.

**Solution.** Assume
$$\sqrt{1+x+x^2}=z-x.$$

Squaring and solving for $x$,

$$x=\frac{z^2-1}{2z+1};\ \text{then } dx=\frac{2(z^2+z+1)\,dz}{(2z+1)^2},$$

and
$$\sqrt{1+x+x^2}\,(=z-x)=\frac{z^2+z+1}{2z+1}.$$

$$\therefore\int\frac{dx}{\sqrt{1+x+x^2}}=\int\frac{\dfrac{2(z^2+z+1)\,dz}{(2z+1)^2}}{\dfrac{z^2+z+1}{2z+1}}=\int\frac{2\,dz}{2z+1}=\log\left[(2z+1)c\right]$$
$$=\log\left[\left(2x+1+2\sqrt{1+x+x^2}\right)c\right],$$

when we substitute back the value of $z$ in terms of $x$.

**194. Differentials containing no radical except $\sqrt{a+bx-x^2}$.**†

*Such an expression can be transformed into a rational form by means of the substitution*

$$\sqrt{a+bx-x^2}\left[=\sqrt{(x-\alpha)(\beta-x)}\right]=(x-\alpha)z\left[\text{or}=(\beta-x)z\right],$$

*where $x-\alpha$ and $\beta-x$ are real* ‡ *factors of $a+bx-x^2$.*

* If the radical is of the form $\sqrt{n+px+qx^2}$, $q>0$, it may be written $\sqrt{q}\sqrt{\frac{n}{q}+\frac{p}{q}x+x^2}$, and therefore comes under the above head, where $a=\frac{n}{q}$, $b=\frac{p}{q}$.

† If the radical is of the form $\sqrt{n+px-qx^2}$, $q>0$, it may be written $\sqrt{q}\sqrt{\frac{n}{q}+\frac{p}{q}x-x^2}$, and therefore comes under the above head, where $a=\frac{n}{q}$, $b=\frac{p}{q}$.

‡ If the factors of $a+bx-x^2$ are imaginary, $\sqrt{a+bx-x^2}$ is imaginary for all values of $x$. For if one of the factors is $x-m+in$, the other must be $-(x-m-in)$, and therefore

$$b+ax-x^2=-(x-m+in)(x-m-in)=-\left[(x-m)^2+n^2\right],$$

which is negative for all values of $x$. We shall consider only those cases where the factors are real.

For if $\sqrt{a+bx-x^2}=\sqrt{(x-\alpha)(\beta-x)}=(x-\alpha)z$, by squaring, cancelling out $(x-\alpha)$, and solving for $x$, we get

$$x=\frac{\alpha z^2+\beta}{z^2+1};\quad \text{then}\quad dx=\frac{2(\alpha-\beta)zdz}{(z^2+1)^2},$$

and $$\sqrt{a+bx-x^2}\,[=(x-\alpha)z]=\frac{(\beta-\alpha)z}{z^2+1}.$$

Hence $x$, $dx$, and $\sqrt{a+bx-x^2}$ are rational when expressed in terms of $z$.

Illustrative Example 1. Find $\int\frac{dx}{\sqrt{2+x-x^2}}$.

**Solution.** Since $$2+x-x^2=(x+1)(2-x),$$
we **assume** $$\sqrt{(x+1)(2-x)}=(x+1)z.$$

Squaring and solving for $x$, $x=\frac{2-z^2}{z^2+1}$.

Hence $$dx=\frac{-6zdz}{(z^2+1)^2},\quad \text{and}\quad \sqrt{2+x-x^2}\,[=(x+1)z]=\frac{3z}{z^2+1}.$$

$$\therefore \int\frac{dx}{\sqrt{2+x-x^2}}=-2\int\frac{dz}{z^2+1}=-2\arctan z+C$$
$$=-2\arctan\sqrt{\frac{2-x}{x+1}}+C,$$

when we substitute back the value of $z$ in terms of $x$.

## EXAMPLES

1. $\int\frac{dx}{x\sqrt{x^2-x+2}}=\frac{1}{\sqrt2}\log\frac{\sqrt{x^2-x+2}+x-\sqrt2}{\sqrt{x^2-x+2}+x+\sqrt2}+C.$

2. $\int\frac{dx}{x\sqrt{x^2+2x-1}}=2\arctan\left(x+\sqrt{x^2+2x-1}\right)+C.$

3. $\int\frac{dx}{x\sqrt{2+x-x^2}}=\frac{1}{\sqrt2}\log\frac{\sqrt{2+2x}-\sqrt{2-x}}{\sqrt{2+2x}+\sqrt{2-x}}+C.$

4. $\int\frac{dx}{x\sqrt{x^2+4x-4}}=\arctan\frac{x+\sqrt{x^2+4x-4}}{2}+C.$

5. $\int\frac{\sqrt{x^2+4x}}{x^2}dx=-\frac{8}{x+\sqrt{x^2+4x}}+\log\left(x+2+\sqrt{x^2+4x}\right)+C.$

6. $\int\frac{xdx}{(2+3x-2x^2)^{\frac32}}=\frac{8+6x}{25\sqrt{2+3x-2x^2}}+C.$

7. $\int\frac{dx}{(2ax-x^2)^{\frac32}}=\frac{x-a}{a^2\sqrt{2ax-x^2}}+C.$

8. $\int\frac{(2x+x^2)^{\frac12}dx}{x^2}=\log\left(x+1+\sqrt{2x+x^2}\right)-\frac{4}{x+\sqrt{2x+x^2}}+C.$

9. $\int\frac{dx}{x\sqrt{x^2+x+1}}=\log\frac{x-1+\sqrt{x^2+x+1}}{x+1+\sqrt{x^2+x+1}}+C.$

10. $\int\frac{dx}{x\sqrt{5x-6-x^2}}=-\sqrt{\frac23}\arctan\sqrt{\frac{2(3-x)}{3(x-2)}}+C.$

The general integral treated in the last two sections has then the form

$$R\left(x, \sqrt{a+bx+cx^2}\right) dx,$$

where $R$ denotes a rational function.

Combining the results of this chapter with the theorem on p. 334, we can then state the following

**Theorem.** *Every rational function of $x$ and the square root of a polynomial of degree not higher than the second can be integrated and the result expressed in terms of the elementary functions.**

**195. Binomial differentials.** A differential of the form

$$x^m(a+bx^n)^p dx,$$

where $a$ and $b$ are any constants and the exponents $m$, $n$, $p$ are rational numbers, is called a *binomial differential.*

Let $$x = z^\alpha; \quad \text{then} \quad dx = \alpha z^{\alpha-1} dz,$$

and $$x^m(a+bx^n)^p dx = \alpha z^{m\alpha+\alpha-1}(a+bz^{n\alpha})^p dz.$$

If an integer $\alpha$ be chosen such that $m\alpha$ and $n\alpha$ are also integers,† we see that the given differential is equivalent to another of the same form where $m$ and $n$ have been replaced by integers. Also

$$x^m(a+bx^n)^p dx = x^{m+np}(ax^{-n}+b)^p dx$$

transforms the given differential into another of the same form where the exponent $n$ of $x$ has been replaced by $-n$. Therefore, no matter what the algebraic sign of $n$ may be, in one of the two differentials the exponent of $x$ inside the parentheses will surely be positive.

When $p$ is an integer the binomial may be expanded and the differential integrated termwise. In what follows $p$ is regarded as a fraction; hence we replace it by $\frac{r}{s}$, where $r$ and $s$ are integers.‡

We may then make the following statement:

*Every binomial differential may be reduced to the form*

$$x^m(a+bx^n)^{\frac{r}{s}} dx,$$

*where $m$, $n$, $r$, $s$ are integers and $n$ is positive.*

* As before, however, it is assumed that in each case the denominator of the rational function can be broken up into real quadratic and linear factors.

† It is always possible to choose $\alpha$ so that $m\alpha$ and $n\alpha$ are integers, for we can take $\alpha$ as the L.C.M. of the denominators of $m$ and $n$.

‡ The case where $p$ is an integer is not excluded, but appears as a special case, namely, $r = p$, $s = 1$.

**196. Conditions of integrability of the binomial differential**

$$(A) \qquad x^m(a+bx^n)^{\frac{r}{s}}dx.$$

CASE I. Assume $a+bx^n=z^s$.

Then $$(a+bx^n)^{\frac{1}{s}}=z, \quad \text{and} \quad (a+bx^n)^{\frac{r}{s}}=z^r;$$

also $$x=\left(\frac{z^s-a}{b}\right)^{\frac{1}{n}}, \quad \text{and} \quad x^m=\left(\frac{z^s-a}{b}\right)^{\frac{m}{n}};$$

hence $$dx=\frac{s}{bn}z^{s-1}\left(\frac{z^s-a}{b}\right)^{\frac{1}{n}-1}dz.$$

Substituting in $(A)$, we get

$$x^m(a+bx^n)^{\frac{r}{s}}dx=\frac{s}{bn}z^{r+s-1}\left(\frac{z^s-a}{b}\right)^{\frac{m+1}{n}-1}dz.$$

The second member of this expression is rational when

$$\frac{m+1}{n}$$

is an integer or zero.

CASE II. Assume $a+bx^n=z^sx^n$.

Then $$x^n=\frac{a}{z^s-b}, \quad \text{and} \quad a+bx^n=z^sx^n=\frac{az^s}{z^s-b}.$$

Hence $$(a+bx^n)^{\frac{r}{s}}=a^{\frac{r}{s}}(z^s-b)^{-\frac{r}{s}}z^r;$$

also $$x=a^{\frac{1}{n}}(z^s-b)^{-\frac{1}{n}}, \quad x^m=a^{\frac{m}{n}}(z^s-b)^{-\frac{m}{n}};$$

and $$dx=-\frac{s}{n}a^{\frac{1}{n}}z^{s-1}(z^s-b)^{-\frac{1}{n}-1}dz.$$

Substituting in $(A)$, we get

$$x^m(a+bx^n)^{\frac{r}{s}}dx=-\frac{s}{n}a^{\frac{m+1}{n}+\frac{r}{s}}(z^s-b)^{-\left(\frac{m+1}{n}+\frac{r}{s}+1\right)}z^{r+s-1}dz.$$

The second member of this expression is rational when $\frac{m+1}{n}+\frac{r}{s}$ is an integer or zero.

*Hence the binomial differential*

$$x^m(a+bx^n)^{\frac{r}{s}}dx$$

*can be integrated by rationalization in the following cases:* *

* Assuming as before that the denominator of the resulting rational function can be broken up into real quadratic and linear factors.

CASE I. *When* $\dfrac{m+1}{n}$ = *an integer or zero, by assuming*

$$a + bx^n = z^s.$$

CASE II. *When* $\dfrac{m+1}{n} + \dfrac{r}{s}$ = *an integer or zero, by assuming*

$$a + bx^n = z^s x^n.$$

**EXAMPLES**

1. $\displaystyle\int \frac{x^3 dx}{(a + bx^2)^{\frac{3}{2}}} = \int x^3 (a + bx^2)^{-\frac{3}{2}} dx = \frac{1}{b^2}\frac{2a + bx^2}{\sqrt{a + bx^2}} + C.$

**Solution.** $m = 3$, $n = 2$, $r = -3$, $s = 2$; and here $\dfrac{m+1}{n} = 2$, an integer. Hence this comes under Case I and we assume

$$a + bx^2 = z^2;\ \text{whence } x = \left(\frac{z^2 - a}{b}\right)^{\frac{1}{2}},\ dx = \frac{zdz}{b^{\frac{1}{2}}(z^2 - a)^{\frac{1}{2}}},\ \text{and } (a + bx^2)^{\frac{3}{2}} = z^3.$$

$$\therefore \int \frac{x^3 dx}{(a + bx^2)^{\frac{3}{2}}} = \int \left(\frac{z^2 - a}{b}\right)^{\frac{3}{2}} \cdot \frac{zdz}{b^{\frac{1}{2}}(z^2 - a)^{\frac{1}{2}}} \cdot \frac{1}{z^3}$$

$$= \frac{1}{b^2}\int (1 - az^{-2})\, dz = \frac{1}{b^2}(z + az^{-1}) + C$$

$$= \frac{1}{b^2}\frac{2a + bx^2}{\sqrt{a + bx^2}} + C.$$

2. $\displaystyle\int \frac{dx}{x^4\sqrt{1 + x^2}} = \frac{(2x^2 - 1)(1 + x^2)^{\frac{1}{2}}}{3x^3} + C.$

**Solution.** $m = -4$, $n = 2$, $\dfrac{r}{s} = -\dfrac{1}{2}$; and here $\dfrac{m+1}{n} + \dfrac{r}{s} = -2$, an integer. Hence this comes under Case II and we assume

$$1 + x^2 = z^2 x^2, \quad z = \frac{(1 + x^2)^{\frac{1}{2}}}{x};$$

whence $\quad x^2 = \dfrac{1}{z^2 - 1}, \quad 1 + x^2 = \dfrac{z^2}{z^2 - 1}, \quad \sqrt{1 + x^2} = \dfrac{z}{(z^2 - 1)^{\frac{1}{2}}};$

also $\quad x = \dfrac{1}{(z^2 - 1)^{\frac{1}{2}}}, \quad x^4 = \dfrac{1}{(z^2 - 1)^2};$ and $dx = -\dfrac{zdz}{(z^2 - 1)^{\frac{3}{2}}}.$

$$\therefore \int \frac{dx}{x^4\sqrt{1 + x^2}} = -\int \frac{\dfrac{zdz}{(z^2 - 1)^{\frac{3}{2}}}}{\dfrac{1}{(z^2 - 1)^2} \cdot \dfrac{z}{(z^2 - 1)^{\frac{1}{2}}}} = -\int (z^2 - 1)\, dz$$

$$= z - \frac{z^3}{3} + C = \frac{(2x^2 - 1)(1 + x^2)^{\frac{1}{2}}}{3x^3} + C.$$

3. $\displaystyle\int x^3 (1 + x^2)^{\frac{1}{2}} dx = \frac{(3x^2 - 2)(1 + x^2)^{\frac{3}{2}}}{15} + C.$

4. $\displaystyle\int \frac{dx}{(1 + x^2)^{\frac{3}{2}}} = \frac{x}{\sqrt{1 + x^2}} + C.$

5. $\int \frac{x^3 dx}{\sqrt{1+x^2}} = (1+x^2)^{\frac{1}{2}} \frac{(x^2-2)}{3} + C.$

6. $\int \frac{dx}{x\sqrt{a^2-x^2}} = \frac{1}{2a} \log \frac{\sqrt{a^2-x^2}-a}{\sqrt{a^2-x^2}+a} + C.$

7. $\int \frac{a dx}{x^2(1+x^2)^{\frac{3}{2}}} = -a(1+x^2)^{-\frac{1}{2}}\left(2x+\frac{1}{x}\right) + C.$

8. $\int_0^a \frac{x^5 dx}{\sqrt{a^2-x^2}} = \frac{8}{15}a^5.$

9. $\int_0^a \sqrt{a^2-x^2}\, dx = \frac{\pi a^2}{4}.$

10. $\int_0^a x^2\sqrt{a^2-x^2}\, dx = \frac{\pi a^4}{16}.$

11. $\int_0^a x^2(a^2-x^2)^{\frac{3}{2}} dx = \frac{\pi a^6}{32}.$

12. $\int \frac{dy}{y(a^2+y^2)^{\frac{1}{2}}} = \frac{1}{2a} \log \frac{\sqrt{a^2+y^2}-a}{\sqrt{a^2+y^2}+a} + C.$

13. $\int t^3(1+2t^2)^{\frac{3}{2}} dt = (1+2t^2)^{\frac{5}{2}} \frac{5t^2-1}{70} + C.$

14. $\int u(1+u)^{\frac{3}{2}} du = \frac{2}{35}(1+u)^{\frac{5}{2}}(5u-2) + C.$

15. $\int \frac{\alpha^2 d\alpha}{(a+b\alpha^2)^{\frac{5}{2}}} = \frac{\alpha^3}{3a(a+b\alpha^2)^{\frac{3}{2}}} + C.$

16. $\int \theta^5(1+\theta^2)^{\frac{2}{3}} d\theta = \frac{3}{22}(1+\theta^2)^{\frac{11}{3}} - \frac{3}{8}(1+\theta^2)^{\frac{8}{3}} + \frac{3}{10}(1+\theta^2)^{\frac{5}{3}} + C.$

17. $\int \frac{dx}{x^2(a+x^3)^{\frac{5}{3}}} = -\frac{3x^3+2a}{2a^2x(a+x^3)^{\frac{2}{3}}} + C$

**197. Transformation of trigonometric differentials.**

From Trigonometry

(*A*) $$\sin x = 2 \sin\frac{x}{2}\cos\frac{x}{2},$$ 37, p. 2

(*B*) $$\cos x = \cos^2\frac{x}{2} - \sin^2\frac{x}{2}.$$ 37, p. 2

But $$\sin\frac{x}{2} = \frac{1}{\csc\frac{x}{2}} = \frac{1}{\sqrt{\cot^2\frac{x}{2}+1}} = \frac{\tan\frac{x}{2}}{\sqrt{1+\tan^2\frac{x}{2}}},$$

and $$\cos\frac{x}{2} = \frac{1}{\sec\frac{x}{2}} = \frac{1}{\sqrt{1+\tan^2\frac{x}{2}}}.$$

If we now assume

$$\tan\frac{x}{2} = z, \quad \text{or,} \quad x = 2 \operatorname{arc} \tan z,$$

we get $$\sin\frac{x}{2} = \frac{z}{\sqrt{1+z^2}}, \quad \cos\frac{x}{2} = \frac{1}{\sqrt{1+z^2}}.$$

Substituting in $(A)$ and $(B)$,

$$\sin x = \frac{2z}{1+z^2}, \quad \cos x = \frac{1-z^2}{1+z^2}.$$

Also by differentiating $x = 2 \operatorname{arc} \tan z$ we have $dx = \dfrac{2\,dz}{1+z^2}$.

Since $\sin x$, $\cos x$, and $dx$ are here expressed rationally in terms of $z$, it follows that

*A trigonometric differential involving $\sin x$ and $\cos x$ rationally only can be transformed by means of the substitution*

$$\tan \frac{x}{2} = z,$$

*or, what is the same thing, by the substitutions*

$$\sin x = \frac{2z}{1+z^2}, \quad \cos x = \frac{1-z^2}{1+z^2}, \quad dx = \frac{2\,dz}{1+z^2}$$

*into another differential expression which is rational in $z$.*

It is evident that if a trigonometric differential involves $\tan x$, $\cot x$, $\sec x$, $\csc x$ rationally only, it will be included in the above theorem, since these four functions can be expressed rationally in terms of $\sin x$, or $\cos x$, or both. It follows, therefore, that *any rational trigonometric differential can be integrated.**

**EXAMPLES**

1. $\displaystyle\int \frac{(1+\sin x)\,dx}{\sin x(1+\cos x)} = \frac{1}{4}\tan^2\frac{x}{2} + \tan\frac{x}{2} + \frac{1}{2}\log\tan\frac{x}{2} + C.$

**Solution.** Since this differential is rational in $\sin x$ and $\cos x$, we make the above substitutions at once, giving

$$\int \frac{(1+\sin x)\,dx}{\sin x(1+\cos x)} = \int \frac{\left(1+\dfrac{2z}{1+z^2}\right)\dfrac{2\,dz}{1+z^2}}{\dfrac{2z}{1+z^2}\left(1+\dfrac{1-z^2}{1+z^2}\right)}$$

$$= \int \frac{(1+z^2+2z)\,dz}{z(1+z^2+1-z^2)} = \frac{1}{2}\int (z+2+z^{-1})\,dz$$

$$= \frac{1}{2}\left(\frac{z^2}{2} + 2z + \log z\right) + C$$

$$= \frac{1}{4}\tan^2\frac{x}{2} + \tan\frac{x}{2} + \frac{1}{2}\log\left(\tan\frac{x}{2}\right) + C.$$

* See footnote, p. 341.

2. $\int_0^{\frac{\pi}{2}} \frac{dx}{1+\sin x} = 1.$

3. $\int_{\frac{\pi}{2}}^{\pi} \frac{dx}{1-\cos x} = 1.$

4. $\int_0^{\pi} \frac{dx}{1+\cos a \cos x} = \frac{\pi}{\sin a}.$

5. $\int_0^{\pi} \frac{dy}{3+2\cos y} = \frac{\pi}{\sqrt{5}}.$

6. $\int_0^{\frac{\pi}{2}} \frac{d\alpha}{2+\cos\alpha} = \frac{\pi}{3\sqrt{3}}.$

7. $\int_{\frac{2\pi}{3}}^{\frac{\pi}{2}} \frac{dx}{1+\cos x} = 1 - \sqrt{3}.$

8. $\int \frac{dx}{1-\sin x} = \frac{2}{1-\tan\frac{x}{2}} + C.$

9. $\int \frac{dx}{4-5\sin x} = \frac{1}{3}\log\left(\frac{\tan\frac{x}{2}-2}{2\tan\frac{x}{2}-1}\right) + C.$

10. $\int \frac{dx}{5-3\cos x} = \frac{1}{2}\arctan\left(2\tan\frac{x}{2}\right) + C.$

11. $\int \frac{dx}{5-4\cos 2x} = \frac{1}{3}\arctan(3\tan x) + C.$

12. $\int \frac{dt}{2-\cos t} = \frac{2}{\sqrt{3}}\arctan\left(\sqrt{3}\tan\frac{t}{2}\right) + C.$

13. $\int \frac{dx}{5+4\sin 2x} = \frac{1}{3}\arctan\left(\frac{5\tan x+4}{3}\right) + C.$

14. $\int \frac{\cos x\,dx}{1+\cos x} = 2\arctan\left(\tan\frac{x}{2}\right) - \tan\frac{x}{2} + C = x - \tan\frac{x}{2} + C.$

15. Derive by the method of this article formulas (16) and (17), p. 284.

16. $\int \frac{\sin x\,dx}{1+\sin x} = \frac{2}{1+\tan\frac{x}{2}} + 2\arctan\left(\tan\frac{x}{2}\right) + C = \frac{2}{1+\tan\frac{x}{2}} + x + C.$

**198. Miscellaneous substitutions.** So far the substitutions considered have rationalized the given differential expression. In a great number of cases, however, integrations may be effected by means of substitutions which do not rationalize the given differential, but no general rule can be given, and the experience gained in working out a large number of problems must be our guide.

A very useful substitution is

$$x = \frac{1}{z}, \qquad dx = -\frac{dz}{z^2},$$

called the *reciprocal substitution.* Let us use this substitution in the next example.

ILLUSTRATIVE EXAMPLE 1. Find $\int \frac{\sqrt{a^2-x^2}}{x^4}dx.$

**Solution.** Making the substitution $x = \frac{1}{z}$, $dx = -\frac{dz}{z^2}$, we get

$$\int \frac{\sqrt{a^2-x^2}}{x^4}dx = -\int (a^2z^2-1)^{\frac{1}{2}} z\,dz = -\frac{(a^2z^2-1)^{\frac{3}{2}}}{3a^2} + C = -\frac{(a^2-x^2)^{\frac{3}{2}}}{3a^2x^3} + C.$$

**EXAMPLES**

1. $\int \frac{dx}{x(a^3+x^3)} = \frac{1}{3a^3}\log\frac{x^3}{a^3+x^3}+C.$ Assume $x^3 = z$.

2. $\int \frac{x^2-x}{(x-2)^3}dx = \log(x-2) - \frac{3x-5}{(x-2)^2}+C$ Assume $x-2=z$.

3. $\int \frac{x^3\,dx}{(x+1)^4} = \frac{18x^2+27x+11}{6(x+1)^3}+\log(x+1)+C.$ Assume $x+1=z$.

4. $\int \frac{dx}{(a^2+x^2)^{\frac{3}{2}}} = \frac{x}{a^2\sqrt{a^2+x^2}}+C.$ Assume $x=\frac{1}{z}$.

5. $\int \frac{dx}{x\sqrt{a^2+x^2}} = \frac{1}{a}\log\frac{cx}{a+\sqrt{a^2+x^2}}.$ Assume $x=\frac{a}{z}$.

6. $\int \frac{dx}{x\sqrt{1+x+x^2}} = \log\frac{cx}{2+x+2\sqrt{1+x+x^2}}.$ Assume $x=\frac{1}{z}$.

7. $\int \frac{\sqrt{1+\log x}}{x}dx = \frac{2}{3}(1+\log x)^{\frac{3}{2}}+C.$ Assume $1+\log x = z$.

8. $\int \frac{e^{2x}dx}{(e^x+1)^{\frac{1}{4}}} = \frac{4}{21}(3e^x-4)(e^x+1)^{\frac{3}{4}}+C.$ Assume $e^x+1=z$.

9. $\int \frac{dx}{e^{2x}-2e^x} = \frac{1}{2e^x}-\frac{x}{4}+\frac{1}{4}\log(e^x-2)+C.$ Assume $e^x=z$.

10. $\int \frac{xdx}{(1+x^3)^{\frac{2}{3}}} = \frac{1}{2}\log\frac{1}{(x^3+1)^{\frac{1}{3}}-x} - \frac{1}{\sqrt{3}}\arctan\frac{1}{\sqrt{3}}\left[\frac{2x}{(x^3+1)^{\frac{1}{3}}}+1\right]+C.$
Assume $x^3=\frac{z^3}{1-z^3}$.

11. $\int_{\frac{1}{3}}^{1}\frac{(x-x^3)^{\frac{1}{3}}dx}{x^4} = 6.$ Assume $x=\frac{1}{z}$.

12. $\int_0^{\frac{\pi}{4}}\frac{(\sin\theta+\cos\theta)\,d\theta}{3+\sin 2\theta} = \frac{\log 3}{4}.$ Assume $\sin\theta - \cos\theta = z$

13. $\int_0^1\frac{dx}{e^x+e^{-x}} = \arctan e - \frac{\pi}{4}.$ Assume $e^x=z$.

14. $\int_0^a\frac{dx}{\sqrt{ax-x^2}} = \pi.$ Assume $x = a\sin^2 z$.

15. $\int_0^{\log 5}\frac{e^x\sqrt{e^x-1}}{e^x+3}dx = 4-\pi.$ Assume $e^x-1=z^2$.

16. $\int_0^1\sqrt{2t+t^2}\,dt = \sqrt{3}-\frac{1}{2}\log(2+\sqrt{3}).$ Assume $t+1=z$.

17. $\int_0^{\log 2}\sqrt{e^x-1}\,dx = \frac{4-\pi}{2}.$ Assume $e^x-1=z$.

18. $\int_1^{2+\sqrt{5}}\frac{(x^2+1)\,dx}{x\sqrt{x^4+7x^2+1}} = \log 3.$ Assume $x-\frac{1}{x}=z$.

# CHAPTER XXVII

## INTEGRATION BY PARTS. REDUCTION FORMULAS

**199. Formula for integration by parts.** If $u$ and $v$ are functions of a single independent variable, we have, from the formula for the differentiation of a product (V, p. 34),

$$d(uv) = udv + vdu,$$

or, transposing, $$udv = d(uv) - vdu.$$

Integrating this, we get the inverse formula,

$$(A) \qquad \int udv = uv - \int vdu,$$

called the **formula for integration by parts.** This formula makes the integration of $udv$, which we may not be able to integrate directly, depend on the integration of $dv$ and $vdu$, which may be in such form as to be readily integrable. This method of *integration by parts* is one of the most useful in the Integral Calculus.

To apply this formula in any given case the given differential must be separated into two factors, namely, $u$ and $dv$. No general directions can be given for choosing these factors, except that

(a) *dx is always a part of dv;*

(b) *it must be possible to integrate dv; and*

(c) *when the expression to be integrated is the product of two functions, it is usually best to choose the most complicated looking one that it is possible to integrate as part of dv.*

The following examples will show in detail how the formula is applied:

ILLUSTRATIVE EXAMPLE 1. Find $\int x \cos x dx$.

**Solution.** Let $$u = x \text{ and } dv = \cos x dx;$$

then $$du = dx \text{ and } v = \int \cos x dx = \sin x.$$

Substituting in (A),

$$\int \overbrace{x}^{u}\, \overbrace{\cos x dx}^{dv} = \overbrace{x}^{u}\, \overbrace{\sin x}^{v} - \int \overbrace{\sin x}^{v}\, \overbrace{dx}^{du}$$
$$= x \sin x + \cos x + C.$$

ILLUSTRATIVE EXAMPLE 2. Find $\int x\log x\,dx$.

**Solution.** Let $$u=\log x \quad\text{and}\quad dv=x\,dx;$$

then $$du=\frac{dx}{x}\quad\text{and}\quad v=\int x\,dx=\frac{x^2}{2}.$$

Substituting in (*A*),

$$\int x\log x\,dx=\log x\cdot\frac{x^2}{2}-\int\frac{x^2}{2}\cdot\frac{dx}{x}$$
$$=\frac{x^2}{2}\log x-\frac{x^2}{4}+C$$

ILLUSTRATIVE EXAMPLE 3. Find $\int xe^{ax}dx$.

**Solution.** Let $$u=e^{ax}\quad\text{and}\quad dv=x\,dx;$$

then $$du=e^{ax}\cdot a\,dx\quad\text{and}\quad v=\int x\,dx=\frac{x^2}{2}.$$

Substituting in (*A*),

$$\int xe^{ax}dx=e^{ax}\cdot\frac{x^2}{2}-\int\frac{x^2}{2}e^{ax}a\,dx$$
$$=\frac{x^2e^{ax}}{2}-\frac{a}{2}\int x^2e^{ax}dx.$$

But $x^2e^{ax}dx$ is not as simple to integrate as $xe^{ax}dx$, which fact indicates that we did not choose our factors suitably. Instead,

let $$u=x\quad\text{and}\quad dv=e^{ax}dx;$$

then $$du=dx\quad\text{and}\quad v=\int e^{ax}dx=\frac{e^{ax}}{a}.$$

Substituting in (*A*),

$$\int xe^{ax}dx=x\cdot\frac{e^{ax}}{a}-\int\frac{e^{ax}}{a}dx$$
$$=\frac{xe^{ax}}{a}-\frac{e^{ax}}{a^2}+C=\frac{e^{ax}}{a}\left(x-\frac{1}{a}\right)+C.$$

It may be necessary to apply the formula for integration by parts more than once, as in the following example:

ILLUSTRATIVE EXAMPLE 4. Find $\int x^2e^{ax}dx$.

**Solution.** Let $$u=x^2\quad\text{and}\quad dv=e^{ax}dx;$$

then $$du=2x\,dx\quad\text{and}\quad v=\int e^{ax}dx=\frac{e^{ax}}{a}.$$

Substituting in (*A*),

$$\int x^2e^{ax}dx=x^2\cdot\frac{e^{ax}}{a}-\int\frac{e^{ax}}{a}\cdot 2x\,dx$$

$$(B)\qquad =\frac{x^2e^{ax}}{a}-\frac{2}{a}\int xe^{ax}dx.$$

The integral in the last term may be found by applying formula $(A)$ again, which gives

$$\int xe^{ax}dx = \frac{e^{ax}}{a}\left(x - \frac{1}{a}\right).$$

Substituting this result in $(B)$, we get

$$\int x^2e^{ax}dx = \frac{x^2e^{ax}}{a} - \frac{2\,e^{ax}}{a^2}\left(x - \frac{1}{a}\right) + C = \frac{e^{ax}}{a}\left(x^2 - \frac{2\,x}{a} + \frac{2}{a^2}\right) + C.$$

Among the most important applications of the method of integration by parts is the integration of

(a) *differentials involving products,*
(b) *differentials involving logarithms,*
(c) *differentials involving inverse circular functions.*

EXAMPLES

1. $\int x^2 \log x dx = \frac{x^3}{3}\left(\log x - \frac{1}{3}\right) + C.$

2. $\int \alpha \sin \alpha d\alpha = - \alpha \cos \alpha + \sin \alpha + C.$

3. $\int \text{arc} \sin x dx = x \,\text{arc} \sin x + \sqrt{1 - x^2} + C.$

HINT. Let $u = \text{arc} \sin x$ and $dv = dx$, etc.

4. $\int \log x dx = x(\log x - 1) + C.$

5. $\int \text{arc} \tan x dx = x \,\text{arc} \tan x - \log(1 + x^2)^{\frac{1}{2}} + C.$

6. $\int x^n \log x dx = \frac{x^{n+1}}{n+1}\left(\log x - \frac{1}{n+1}\right) + C. \quad n \neq -1.$

7. $\int x \,\text{arc} \tan x dx = \frac{x^2+1}{2} \text{arc} \tan x - \frac{x}{2} + C.$

8 $\int \text{arc} \cot y dy = y \,\text{arc} \cot y + \frac{1}{2} \log(1 + y^2) + C.$

9. $\int xa^x dx = a^x\left[\frac{x}{\log a} - \frac{1}{\log^2 a}\right] + C. \quad a > 0 \text{ and } \neq 1.$

10. $\int t^2a^t dt = a^t\left[\frac{t^2}{\log a} - \frac{2\,t}{\log^2 a} + \frac{2}{\log^3 a}\right] + C. \quad a > 0 \text{ and } \neq 1.$

11. $\int \cos\theta \log \sin \theta d\theta = \sin\theta(\log \sin \theta - 1) + C.$

12. $\int x^2e^x dx = e^x(x^2 - 2\,x + 2) + C.$

13. $\int x \sin x \cos x dx = \frac{1}{8} \sin 2\,x - \frac{1}{4} x \cos 2\,x + C.$

14. $\int x^2e^{-x} dx = e^{-x}(2 - 2\,x - x^2) + C.$

15. $\int \text{arc} \tan \sqrt{x} dx = x \,\text{arc} \tan \sqrt{x} - \sqrt{x} + \text{arc} \tan \sqrt{x} + C.$

16. $\int xa^x dx = \frac{a^x}{\log a}\left[x - \frac{1}{\log a}\right] + C.$

17. $\int_0^1 x \log x dx = -\frac{1}{4}.$

18. $\int_0^1 \log y dy = -1.$

19. $\int_0^{\frac{\pi}{2}} \alpha^2 \sin \alpha d\alpha = \pi - 2.$

20. $\int x^3 \log x dx = \frac{x^4}{4}\left(\log x - \frac{1}{4}\right) + C.$

21. $\int_0^1 \text{arc} \sin z dz = \frac{\pi}{2} - 1.$

22. $\int_0^1 \text{arc} \tan \theta d\theta = \frac{\pi}{4} - \log \sqrt{2}.$

23. $\int_0^1 s^2 \log s ds = -\frac{1}{9}.$

24. $\int x^3 e^{ax} dx = \frac{e^{ax}}{a}\left(x^3 - \frac{3x^2}{a} + \frac{6x}{a^2} - \frac{6}{a^3}\right) + C.$

25. $\int \phi^2 \sin \phi d\phi = 2 \cos \phi + 2 \phi \sin \phi - \phi^2 \cos \phi + C.$

26. $\int (\log x)^2 dx = x[\log^2 x - 2 \log x + 2] + C.$

27. $\int \alpha \tan^2 \alpha d\alpha = \alpha \tan \alpha - \frac{\alpha^2}{2} + \log \cos \alpha + C.$

28. $\int \frac{\log x dx}{(x+1)^2} = \frac{x}{x+1} \log x - \log(x+1) + C.$

HINT. Let $u = \log x$ and $dv = \frac{dx}{(x+1)^2}$, etc.

29. $\int x^2 \text{arc} \sin x dx = \frac{x^3}{3} \text{arc} \sin x + \frac{x^2 + 2}{9}\sqrt{1 - x^2} + C.$

30. $\int \sec^2 \theta \log \tan \theta d\theta = \tan \theta (\log \tan \theta - 1) + C.$

31. $\int \log(\log x) \frac{dx}{x} = \log x \cdot \log(\log x) - \log x + C.$

32. $\int \frac{\log(x+1) dx}{\sqrt{x+1}} = 2\sqrt{x+1}[\log(x+1) - 2] + C.$

33. $\int x^3 (a - x^2)^{\frac{1}{2}} dx = -\frac{1}{3} x^2 (a - x^2)^{\frac{3}{2}} - \frac{2}{15}(a - x^2)^{\frac{5}{2}} + C.$

HINT. Let $u = x^2$ and $dv = (a - x^2)^{\frac{1}{2}} x dx$, etc.

34. $\int \sqrt{a^2 - x^2} dx = \frac{x}{2}\sqrt{a^2 - x^2} + \frac{a^2}{2} \text{arc} \sin \frac{x}{a} + C.$

35. $\int \frac{x^3 dx}{\sqrt{1 - x^2}} = -\frac{1}{3}(x^2 + 2)(1 - x^2)^{\frac{1}{2}} + C.$

36. $\int \sqrt{a^2 + x^2} dx = \frac{x}{2}\sqrt{a^2 + x^2} + \frac{a^2}{2} \log(x + \sqrt{a^2 + x^2}) + C.$

37. $\int \frac{x^2 dx}{\sqrt{a^2 - x^2}} = -\frac{x}{2}\sqrt{a^2 - x^2} + \frac{a^2}{2} \text{arc} \sin \frac{x}{a} + C.$

38. $\int \frac{(\log x)^2 dx}{x^{\frac{5}{2}}} = -\frac{2}{3x^{\frac{3}{2}}}\left[\log^2 x + \frac{4}{3} \log x + \frac{8}{9}\right] + C.$

**200. Reduction formulas for binomial differentials.** It was shown in § 195, p. 340, that any binomial differential may be reduced to the form

$$x^m (a + bx^n)^p dx,$$

where $p$ is a rational number, $m$ and $n$ are integers, and $n$ is positive. Also in § 196, p. 341, we learned how to integrate such a differential expression in certain cases.

In general we can integrate such an expression by parts, using (**A**), p. 347, if it can be integrated at all. To apply the method of *integration by parts* to every example, however, is rather a long and tedious process. When the binomial differential cannot be integrated readily by any of the methods shown so far, it is customary to employ *reduction formulas* deduced by the method of integration by parts. By means of these reduction formulas the given differential is expressed as the sum of two terms, one of which is not affected by the sign of integration, and the other is an integral of the same form as the original expression, but one which is easier to integrate. The following are the four principal **reduction formulas**:

$$(A)\quad \int x^m(a+bx^n)^p\,dx = \frac{x^{m-n+1}(a+bx^n)^{p+1}}{(np+m+1)b} - \frac{(m-n+1)a}{(np+m+1)b}\int x^{m-n}(a+bx^n)^p\,dx.$$

$$(B)\quad \int x^m(a+bx^n)^p\,dx = \frac{x^{m+1}(a+bx^n)^p}{np+m+1} + \frac{anp}{np+m+1}\int x^m(a+bx^n)^{p-1}\,dx.$$

$$(C)\quad \int x^m(a+bx^n)^p\,dx = \frac{x^{m+1}(a+bx^n)^{p+1}}{(m+1)a} - \frac{(np+n+m+1)b}{(m+1)a}\int x^{m+n}(a+bx^n)^p\,dx.$$

$$(D)\quad \int x^m(a+bx^n)^p\,dx = -\frac{x^{m+1}(a+bx^n)^{p+1}}{n(p+1)a} + \frac{np+n+m+1}{n(p+1)a}\int x^m(a+bx^n)^{p+1}\,dx.$$

While it is not desirable for the student to memorize these formulas, he should know what each one will do and when each one fails. Thus:

*Formula* (**A**) *diminishes $m$ by $n$.* (**A**) *fails when $np+m+1=0$.*
*Formula* (**B**) *diminishes $p$ by 1.* (**B**) *fails when $np+m+1=0$.*
*Formula* (**C**) *increases $m$ by $n$.* (**C**) *fails when $m+1=0$.*
*Formula* (**D**) *increases $p$ by 1.* (**D**) *fails when $p+1=0$.*

I. *To derive formula* (**A**). The formula for integration by parts is

$$(A) \qquad \int u\,dv = uv - \int v\,du. \qquad (A), \text{ p. } 347$$

We may apply this formula in the integration of

$$\int x^m (a+bx^n)^p\,dx$$

by placing $u = x^{m-n+1}$* and $dv = (a+bx^n)^p x^{n-1}dx$;

then $du = (m-n+1)x^{m-n}dx$ and $v = \dfrac{(a+bx^n)^{p+1}}{nb(p+1)}$.

Substituting in $(A)$,

$$(B) \qquad \int x^m (a+bx^n)^p\,dx = \frac{x^{m-n+1}(a+bx^n)^{p+1}}{nb(p+1)} - \frac{m-n+1}{nb(p+1)}\int x^{m-n}(a+bx^n)^{p+1}dx.$$

But $\displaystyle\int x^{m-n}(a+bx^n)^{p+1}dx = \int x^{m-n}(a+bx^n)^p(a+bx^n)\,dx$

$$= a\int x^{m-n}(a+bx^n)^p dx + b\int x^m(a+bx^n)^p\,dx.$$

Substituting this in $(B)$, we get

$$\int x^m(a+bx^n)^p dx = \frac{x^{m-n+1}(a+bx^n)^{p+1}}{nb(p+1)} - \frac{(m-n+1)a}{nb(p+1)}\int x^{m-n}(a+bx^n)^p dx - \frac{m-n+1}{n(p+1)}\int x^m(a+bx^n)^p dx.$$

Transposing the last term to the first member, combining, and solving for $\int x^m(a+bx^n)^p dx$, we obtain

$$(\mathbf{A}) \qquad \int x^m(a+bx^n)^p dx = \frac{x^{m-n+1}(a+bx^n)^{p+1}}{b(np+m+1)} - \frac{a(m-n+1)}{b(np+m+1)}\int x^{m-n}(a+bx^n)^p dx.$$

* In order to integrate $dv$ by (4) it is necessary that $x$ outside the parenthesis shall have the exponent $n-1$. Subtracting $n-1$ from $m$ leaves $m-n+1$ for the exponent of $x$ in $u$.

It is seen by formula (*A*) that the integration of $x^m(a+bx^n)^p dx$ is made to depend upon the integration of another differential of the same form in which $m$ is replaced by $m-n$. By repeated applications of formula (*A*), $m$ may be diminished by any multiple of $n$.

When $np+m+1=0$, formula (*A*) evidently fails (the denominator vanishing). But in that case

$$\frac{m+1}{n}+p=0;$$

hence we can apply the method of § 196, p. 341, and the formula is not needed.

II. *To derive formula* (*B*). Separating the factors, we may write

$$(C)\quad \int x^m(a+bx^n)^p dx = \int x^m(a+bx^n)^{p-1}(a+bx^n)\,dx$$
$$= a\int x^m(a+bx^n)^{p-1}dx$$
$$+ b\int x^{m+n}(a+bx^n)^{p-1}dx.$$

Now let us apply formula (*A*) to the last term of (*C*) by substituting in the formula $m+n$ for $m$, and $p-1$ for $p$. This gives

$$b\int x^{m+n}(a+bx^n)^{p-1}dx = \frac{x^{m+1}(a+bx^n)^p}{np+m+1} - \frac{a(m+1)}{np+m+1}\int x^m(a+bx^n)^{p-1}dx.$$

Substituting this in (*C*), and combining like terms, we get

$$(B)\quad \int x^m(a+bx^n)^p dx = \frac{x^{m+1}(a+bx^n)^p}{np+m+1}$$
$$+\frac{anp}{np+m+1}\int x^m(a+bx^n)^{p-1}dx.$$

Each application of formula (*B*) diminishes $p$ by unity. Formula (*B*) fails for the same case as (*A*).

III. *To derive formula* (**C**). Solving formula (*A*) for

$$\int x^{m-n}(a+bx^n)^p dx,$$

and substituting $m+n$ for $m$, we get

$$(C)\quad \int x^m(a+bx^n)^p dx = \frac{x^{m+1}(a+bx^n)^{p+1}}{a(m+1)}$$
$$-\frac{b(np+n+m+1)}{a(m+1)}\int x^{m+n}(a+bx^n)^p dx.$$

Therefore each time we apply (**C**), $m$ is replaced by $m+n$. When $m+1=0$, formula (**C**) fails, but then the differential expression can be integrated by the method of § 196, p. 341, and the formula is not needed.

IV. *To derive formula* (**D**). Solving formula (**B**) for

$$\int x^m(a+bx^n)^{p-1}dx,$$

and substituting $p+1$ for $p$, we get

$$(D)\quad \int x^m(a+bx^n)^p dx = -\frac{x^{m+1}(a+bx^n)^{p+1}}{an(p+1)} + \frac{np+n+m+1}{an(p+1)}\int x^m(a+bx^n)^{p+1}dx.$$

Each application of (**D**) increases $p$ by unity. Evidently (**D**) fails when $p+1=0$, but then $p=-1$ and the expression is rational.

**EXAMPLES**

1. $\displaystyle\int\frac{x^3dx}{\sqrt{1-x^2}} = -\frac{1}{3}(x^2+2)(1-x^2)^{\frac{1}{2}}+C.$

**Solution.** Here $m=3$, $n=2$, $p=-\frac{1}{2}$, $a=1$, $b=-1$.

We apply reduction formula ($A$) in this case because the integration of the differential would then depend on the integration of $\int x(1-x^2)^{-\frac{1}{2}}dx$, which comes under (4), p. 284. Hence, substituting in ($A$), we obtain

$$\int x^3(1-x^2)^{-\frac{1}{2}}dx = \frac{x^{3-2+1}(1-x^2)^{-\frac{1}{2}+1}}{-1(-1+3+1)} - \frac{1(3-2+1)}{-1(-1+3+1)}\int x^{3-2}(1-x^2)^{-\frac{1}{2}}dx$$
$$= -\tfrac{1}{3}x^2(1-x^2)^{\frac{1}{2}} + \tfrac{2}{3}\int x(1-x^2)^{-\frac{1}{2}}dx$$
$$= -\tfrac{1}{3}x^2(1-x^2)^{\frac{1}{2}} - \tfrac{2}{3}(1-x^2)^{\frac{1}{2}} + C$$
$$= -\tfrac{1}{3}(x^2+2)(1-x^2)^{\frac{1}{2}} + C.$$

2. $\displaystyle\int\frac{x^4dx}{(a^2-x^2)^{\frac{1}{2}}} = -\left(\frac{1}{4}x^3+\frac{3}{8}a^2x\right)\sqrt{a^2-x^2}+\frac{3}{8}a^4\arcsin\frac{x}{a}+C.$

HINT. Apply ($A$) twice.

3. $\displaystyle\int(a^2+x^2)^{\frac{1}{2}}dx = \frac{x}{2}\sqrt{a^2+x^2}+\frac{a^2}{2}\log(x+\sqrt{a^2+x^2})+C.$

HINT. Here $m=0$, $n=2$, $p=\frac{1}{2}$, $a=a^2$, $b=1$. Apply ($B$) once.

4. $\displaystyle\int\frac{dx}{x^3\sqrt{x^2-1}} = \frac{(x^2-1)^{\frac{1}{2}}}{2x^2}+\frac{1}{2}\operatorname{arc\,sec} x+C.$

HINT. Apply ($C$) once.

5. $\displaystyle\int\frac{x^2dx}{\sqrt{a^2-x^2}} = -\frac{x}{2}\sqrt{a^2-x^2}+\frac{a^2}{2}\arcsin\frac{x}{a}+C.$

6. $\int \frac{x^3dx}{\sqrt{a^2+x^2}} = \frac{1}{3}(x^2 - 2a^2)\sqrt{a^2+x^2} + C.$

7. $\int \frac{x^5dx}{\sqrt{1-x^2}} = -\left(\frac{x^4}{5} + \frac{4x^2}{15} + \frac{8}{15}\right)\sqrt{1-x^2} + C.$

8. $\int x^2\sqrt{a^2-x^2}\,dx = \frac{x}{8}(2x^2-a^2)\sqrt{a^2-x^2} + \frac{a^4}{8}\arcsin\frac{x}{a} + C.$

HINT. Apply (*A*) and then (*B*).

9. $\int \frac{dx}{(a^2+x^2)^2} = \frac{x}{2a^2(a^2+x^2)} + \frac{1}{2a^3}\arctan\frac{x}{a} + C.$

HINT. Apply (*D*) once.

10. $\int \frac{dx}{x^3\sqrt{a^2-x^2}} = -\frac{\sqrt{a^2-x^2}}{2a^2x^2} + \frac{1}{2a^3}\log\frac{x}{a+\sqrt{a^2-x^2}} + C.$

11. $\int \frac{x^3dx}{(a^2+x^2)^{\frac{3}{2}}} = \frac{x^2+2a^2}{(a^2+x^2)^{\frac{1}{2}}} + C.$

12. $\int \frac{dx}{(a^2-x^2)^{\frac{5}{2}}} = \frac{(3a^2-2x^2)x}{3a^4(a^2-x^2)^{\frac{3}{2}}} + C.$

13. $\int (x^2+a^2)^{\frac{3}{2}}dx = \frac{x}{8}(2x^2+5a^2)\sqrt{x^2+a^2} + \frac{3a^4}{8}\log(x+\sqrt{x^2+a^2}) + C.$

14. $\int x^2(x^2+a^2)^{\frac{1}{2}}dx = \frac{x}{8}(2x^2+a^2)\sqrt{x^2+a^2} - \frac{a^4}{8}\log(x+\sqrt{x^2+a^2}) + C.$

15. $\int \frac{x^2dx}{\sqrt{2ax-x^2}} = -\frac{x+3a}{2}(2ax-x^2)^{\frac{1}{2}} + \frac{3a^2}{2}\operatorname{arc vers}\frac{x}{a} + C.$

HINT. $\int \frac{x^2dx}{\sqrt{2ax-x^2}} = \int x^{\frac{3}{2}}(2a-x)^{-\frac{1}{2}}dx.$ Apply (*A*) twice.

16. $\int \frac{dx}{x^2(a^2-x^2)^{\frac{1}{2}}} = -\frac{\sqrt{a^2-x^2}}{a^2x} + C.$

17. $\int \frac{y^3dy}{\sqrt{2ry-y^2}} = -\frac{2y^2+5r(y+3r)}{6}\sqrt{2ry-y^2} + \frac{5}{2}r^3\operatorname{arc vers}\frac{y}{r} + C.$

18. $\int \frac{tdt}{\sqrt{2at-t^2}} = -(2at-t^2)^{\frac{1}{2}} + a\operatorname{arc vers}\frac{t}{a} + C.$

19. $\int \frac{ds}{(a^2+s^2)^3} = \frac{s}{4a^2(a^2+s^2)^2} + \frac{3s}{8a^4(a^2+s^2)} + \frac{3}{8a^5}\arctan\frac{s}{a} + C.$

20. $\int \frac{r^8dr}{\sqrt{1-r^3}} = -\frac{2}{45}(3r^6+4r^3+8)\sqrt{1-r^3} + C.$

21. $\int \frac{x^2dx}{(a^2-x^2)^2}.$

22. $\int \frac{dx}{x^2(1+x^2)^{\frac{3}{2}}}.$

23. $\int \frac{x^5dx}{\sqrt{1-x^3}}.$

24. $\int \frac{x^2dx}{\sqrt{1-x^6}}.$

25. $\int t^5\sqrt{a^2+t^3}\,dt.$

26. $\int \frac{z^5dz}{\sqrt{z^4+9}}.$

27. $\int \frac{\sqrt{a^2+x^2}\,dx}{x}.$

28. $\int \frac{(1-x^3)^{\frac{5}{2}}dx}{x}.$

29. $\int \frac{s^7ds}{(a+bs^4)^{\frac{2}{3}}}.$

30. $\int \frac{\alpha^8d\alpha}{\sqrt{3-\alpha^3}}.$

31. $\int \frac{\sqrt{4-y^2}\,dy}{y^4}.$

32. $\int \frac{dt}{t(a^4-t^4)^{\frac{2}{3}}}.$

**201. Reduction formulas for trigonometric differentials.** The method of the last section, which makes the given integral depend on another integral of the same form, is called *successive reduction.*

We shall now apply the same method to trigonometric differentials by deriving and illustrating the use of the following **trigonometric reduction formulas:**

$$(E)\qquad \int \sin^m x \cos^n x dx = \frac{\sin^{m+1} x \cos^{n-1} x}{m+n} + \frac{n-1}{m+n}\int \sin^m x \cos^{n-2} x dx.$$

$$(F)\qquad \int \sin^m x \cos^n x dx = -\frac{\sin^{m-1} x \cos^{n+1} x}{m+n} + \frac{m-1}{m+n}\int \sin^{m-2} x \cos^n x dx.$$

$$(G)\qquad \int \sin^m x \cos^n x dx = -\frac{\sin^{m+1} x \cos^{n+1} x}{n+1} + \frac{m+n+2}{n+1}\int \sin^m x \cos^{n+2} x dx.$$

$$(H)\qquad \int \sin^m x \cos^n x dx = \frac{\sin^{m+1} x \cos^{n+1} x}{m+1} + \frac{m+n+2}{m+1}\int \sin^{m+2} x \cos^n x dx.$$

Here the student should note that

*Formula* (*E*) *diminishes* $n$ *by* 2. (*E*) *fails when* $m+n=0$.
*Formula* (*F*) *diminishes* $m$ *by* 2. (*F*) *fails when* $m+n=0$.
*Formula* (*G*) *increases* $n$ *by* 2. (*G*) *fails when* $n+1=0$.
*Formula* (*H*) *increases* $m$ *by* 2. (*H*) *fails when* $m+1=0$.

To derive these we apply, as before, the formula for integration by parts, namely,

$$(A)\qquad \int u dv = uv - \int v du. \qquad (A), \text{ p. } 347$$

Let $u = \cos^{n-1} x$, and $dv = \sin^m x \cos x dx$;

then $du = -(n-1)\cos^{n-2} x \sin x dx$, and $v = \dfrac{\sin^{m+1} x}{m+1}$.

Substituting in $(A)$, we get

$$(B)\qquad \int \sin^m x \cos^n x\,dx = +\frac{\sin^{m+1}x\cos^{n-1}x}{m+1} + \frac{n-1}{m+1}\int \sin^{m+2}x\cos^{n-2}x\,dx.$$

In the same way, if we

let $\qquad u = \sin^{m-1}x,\quad$ and $\quad dv = \cos^n x \sin x\,dx,$

we obtain

$$(C)\qquad \int \sin^m x \cos^n x\,dx = -\frac{\sin^{m-1}x\cos^{n+1}x}{n+1} + \frac{m-1}{n+1}\int \sin^{m-2}x\cos^{n+2}x\,dx.$$

But $\int \sin^{m+2}x\cos^{n-2}x\,dx = \int \sin^m x(1-\cos^2 x)\cos^{n-2}x\,dx$

$$= \int \sin^m x\cos^{n-2}x\,dx - \int \sin^m x\cos^n x\,dx.$$

Substituting this in $(B)$, combining like terms, and solving for $\int \sin^m x \cos^n x\,dx$, we get

$$(E)\qquad \int \sin^m x \cos^n x\,dx = \frac{\sin^{m+1}x\cos^{n-1}x}{m+n} + \frac{n-1}{m+n}\int \sin^m x\cos^{n-2}x\,dx.$$

Making a similar substitution in $(C)$, we get

$$(F)\qquad \int \sin^m x \cos^n x\,dx = -\frac{\sin^{m-1}x\cos^{n+1}x}{m+n} + \frac{m-1}{m+n}\int \sin^{m-2}x\cos^n x\,dx.$$

Solving formula $(E)$ for the integral on the right-hand side, and increasing $n$ by 2, we get

$$(G)\qquad \int \sin^m x \cos^n x\,dx = -\frac{\sin^{m+1}x\cos^{n+1}x}{n+1} + \frac{m+n+2}{n+1}\int \sin^m x\cos^{n+2}x\,dx.$$

In the same way we get, from formula (*F*),

$$(H)\qquad \int \sin^m x \cos^n x dx = \frac{\sin^{m+1} x \cos^{n+1} x}{m+1} + \frac{m+n+2}{m+1}\int \sin^{m+2} x \cos^n x dx.$$

Formulas (*E*) and (*F*) fail when $m+n=0$, formula (*G*) when $n+1=0$, and formula (*H*) when $m+1=0$. But in such cases we may integrate by methods which have been previously explained.

It is clear that when $m$ and $n$ are integers, the integral

$$\int \sin^m x \cos^n x dx$$

may be made to depend, by using one of the above reduction formulas, upon one of the following integrals:

$$\int dx,\ \int \sin x dx,\ \int \cos x dx,\ \int \sin x \cos x dx,\ \int \frac{dx}{\sin x} = \int \csc x dx,$$
$$\int \frac{dx}{\cos x} = \int \sec x dx,\ \int \frac{dx}{\cos x \sin x},\ \int \tan x dx,\ \int \cot x dx,$$

all of which we have learned how to integrate.

## EXAMPLES

1. $\int \sin^2 x \cos^4 x dx = -\frac{\sin x \cos^5 x}{6} + \frac{\sin x \cos^3 x}{24} + \frac{1}{16}(\sin x \cos x + x) + C.$

**Solution.** First applying formula (*F*), we get

$$(A)\qquad \int \sin^2 x \cos^4 x dx = -\frac{\sin x \cos^5 x}{6} + \frac{1}{6}\int \cos^4 x dx.$$

[Here $m=2$, $n=4$.]

Applying formula (*E*) to the integral in the second member of (*A*), we get

$$(B)\qquad \int \cos^4 x dx = \frac{\sin x \cos^3 x}{4} + \frac{3}{4}\int \cos^2 x dx.$$

[Here $m=0$, $n=4$.]

Applying formula (*E*) to the second member of (*B*) gives

$$(C)\qquad \int \cos^2 x dx = \frac{\sin x \cos x}{2} + \frac{x}{2}.$$

Now substitute the result (*C*) in (*B*), and then this result in (*A*). This gives the answer as above.

2. $\int \sin^4 x \cos^2 x dx = \frac{\cos x}{2}\left(\frac{\sin^5 x}{3} - \frac{\sin^3 x}{12} - \frac{\sin x}{8}\right) + \frac{x}{16} + C.$

3. $\int \frac{dx}{\sin^4 x \cos^2 x} = \tan x - 2\cot x - \frac{1}{3}\cot^3 x + C.$

4. $\int \frac{\cos^4 x dx}{\sin^2 x} = -\frac{\cot x}{2}(3 - \cos^2 x) - \frac{3x}{2} + C.$

5. $\int \sec^3 x dx = \frac{1}{2}\sec x \tan x + \frac{1}{2}\log(\sec x + \tan x) + C.$

6. $\int \csc^3 x dx = -\frac{1}{2}\csc x \cot x + \frac{1}{2}\log(\csc x - \cot x) + C.$

7. $\int \frac{\cos^4 \alpha d\alpha}{\sin^3 \alpha} = -\frac{\cos \alpha}{2\sin^2 \alpha} - \cos \alpha - \frac{3}{2}\log \tan\frac{\alpha}{2} + C.$

8. $\int \sin^6 \alpha d\alpha = -\frac{\cos \alpha}{2}\left(\frac{\sin^5 \alpha}{3} + \frac{5}{12}\sin^3 \alpha + \frac{5}{8}\sin \alpha\right) + \frac{5\alpha}{16} + C.$

9. $\int \csc^5 \theta d\theta = -\frac{\cos \theta}{4}\left(\frac{1}{\sin^4 \theta} + \frac{3}{2\sin^2 \theta}\right) + \frac{3}{8}\log \tan\frac{\theta}{2} + C.$

10. $\int \sec^7 \phi d\phi = \frac{\sin \phi}{2\cos^2 \phi}\left(\frac{1}{3\cos^4 \phi} + \frac{5}{12\cos^2 \phi} + \frac{5}{8}\right) + \frac{5}{16}\log(\sec \phi + \tan \phi) + C.$

11. $\int \cos^8 t dt = \frac{\sin t}{8}\left(\cos^7 t + \frac{7}{6}\cos^5 t + \frac{35}{24}\cos^3 t + \frac{35}{16}\cos t\right) + \frac{35t}{128} + C.$

12. $\int \frac{dy}{\sin^4 y \cos^3 y} = -\frac{1}{\cos^2 y}\left(\frac{1}{3\sin^3 y} + \frac{5}{3\sin y} - \frac{5}{2}\sin y\right) + \frac{5}{2}\log(\sec y + \tan y) + C.$

13. $\int_0^{\frac{\pi}{2}} \sin^6 x \cos^4 x dx = \frac{3\pi}{512}.$

14. $\int_0^{\frac{\pi}{2}} \sin^8 x dx = \frac{35\pi}{256}.$

15. $\int_0^{\pi} \sin^6 \theta d\theta = \frac{5\pi}{16}.$

16. $\int_0^{\frac{\pi}{2}} \cos^8 \alpha d\alpha = \frac{35\pi}{256}.$

17. $\int_0^{\pi} \sin^4 x dx = \frac{3\pi}{8}.$

18. $\int_0^{\frac{\pi}{2}} \cos^4 t dt = \frac{3\pi}{16}.$

**202. To find $\int e^{ax} \sin nx dx$ and $\int e^{ax} \cos nx dx$.**

Integrate $e^{ax} \sin nx dx$ by parts,

letting $u = e^{ax}$, and $dv = \sin nx dx$;

then $du = ae^{ax} dx$, and $v = -\frac{\cos nx}{n}$.

Substituting in formula (*A*), p. 347, namely,

$$\int u dv = uv - \int v du,$$

we get

$$(A) \qquad \int e^{ax} \sin nx dx = -\frac{e^{ax}\cos nx}{n} + \frac{a}{n}\int e^{ax}\cos nx dx.$$

Integrate $e^{ax} \sin nx dx$ again by parts,

letting $u = \sin nx$, and $dv = e^{ax} dx$;

then $du = n \cos nx dx$, and $v = \frac{e^{ax}}{a}$.

Substituting in (*A*), p. 347, we get

$$(B) \qquad \int e^{ax} \sin nx dx = \frac{e^{ax} \sin nx}{a} - \frac{n}{a} \int e^{ax} \cos nx dx.$$

Eliminating $\int e^{ax} \cos nx dx$ between (*A*) and (*B*), we have

$$(a^2 + n^2) \int e^{ax} \sin nx dx = e^{ax} (a \sin nx - n \cos nx),$$

or

$$\int e^{ax} \sin nx dx = \frac{e^{ax}(a \sin nx - n \cos nx)}{a^2 + n^2} + C.$$

Similarly, we may obtain

$$\int e^{ax} \cos nx dx = \frac{e^{ax}(n \sin nx + a \cos nx)}{a^2 + n^2} + C.$$

In working out the examples which follow, the student is advised not to use the above results as formulas, but to follow the method by which they were obtained.

**EXAMPLES**

1. $\int e^x \sin x dx = \frac{e^x}{2}(\sin x - \cos x) + C.$

2. $\int e^x \cos x dx = \frac{e^x}{2}(\sin x + \cos x) + C.$

3. $\int e^{2x} \cos 3x dx = \frac{e^{2x}}{13}(3 \sin 3x + 2 \cos 3x) + C.$

4. $\int \frac{\sin x dx}{e^x} = -\frac{\sin x + \cos x}{2 e^x} + C.$

5. $\int \frac{\cos 2x dx}{e^{3x}} = \frac{1}{13 e^{3x}}(2 \sin 2x - 3 \cos 2x) + C.$

6. $\int e^x \sin^2 x dx = \frac{e^x}{2}\left(1 - \frac{2 \sin 2x + \cos 2x}{5}\right) + C.$

7. $\int e^\alpha \cos^2 \alpha d\alpha = \frac{e^\alpha}{2}\left(1 + \frac{2 \sin 2\alpha + \cos 2\alpha}{5}\right) + C.$

8. $\int e^{\frac{x}{2}} \cos \frac{x}{2} dx = e^{\frac{x}{2}}\left(\sin \frac{x}{2} + \cos \frac{x}{2}\right) + C.$

9. $\int e^{a\alpha}(\sin a\alpha + \cos a\alpha) d\alpha = \frac{e^{a\alpha} \sin a\alpha}{a} + C.$

10. $\int e^{3x}(\sin 2x - \cos 2x) dx = \frac{e^{3x}}{13}(\sin 2x - 5 \cos 2x) + C.$

11. $\int_0^\infty e^{-x} \sin x dx = \frac{1}{2}.$ $\qquad \int_0^\infty e^{-3x} \cos 2x dx = \frac{3}{13}.$

## CHAPTER XXVIII

### INTEGRATION A PROCESS OF SUMMATION

**203. Introduction.** Thus far we have defined integration as the *inverse of differentiation.* In a great many of the applications of the Integral Calculus, however, it is preferable to define integration as a *process of summation.* In fact, the Integral Calculus was invented in the attempt to calculate the area bounded by curves, by supposing the given area to be divided into an "infinite number of infinitesimal parts called *elements*, the sum of all these elements being the area required." Historically, the integral sign is merely the long *S*, used by early writers to indicate "sum."

This new definition, as amplified in the next section, is of fundamental importance, and it is essential that the student should thoroughly understand what is meant in order to be able to apply the Integral Calculus to practical problems.

**204. The fundamental theorem of Integral Calculus.** If $\phi(x)$ is the derivative of $f(x)$, then it has been shown in § 174, p. 315, that the value of the definite integral

$$(A) \qquad \int_a^b \phi(x)\,dx = f(b) - f(a)$$

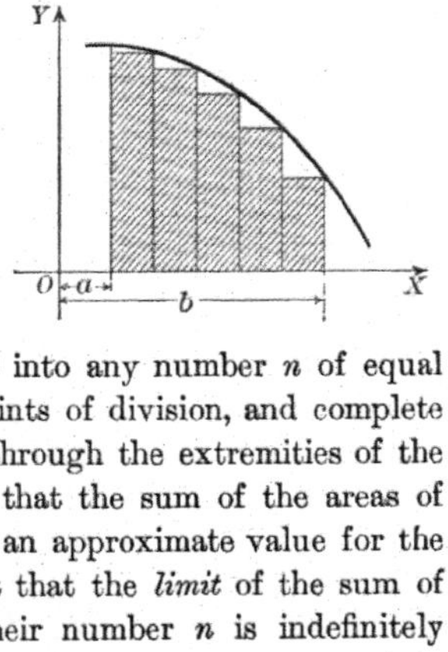


gives the area bounded by the curve $y = \phi(x)$, the $X$-axis, and the ordinates erected at $x = a$ and $x = b$.

Now let us make the following construction in connection with this area. Divide the interval from $x = a$ to $x = b$ into any number $n$ of equal subintervals, erect ordinates at these points of division, and complete rectangles by drawing horizontal lines through the extremities of the ordinates, as in the figure. It is clear that the sum of the areas of these $n$ rectangles (the shaded area) is an approximate value for the area in question. It is further evident that the *limit* of the sum of the areas of these rectangles when their number $n$ is indefinitely increased, will *equal* the area under the curve.

Let us now carry through the following more general construction. Divide the interval into $n$ subintervals, *not necessarily equal*, and erect ordinates at the points of division. Choose a point within each subdivision *in any manner*,* erect ordinates at these points, and through their extremities draw horizontal lines to form rectangles, as in the figure. Then, as before, the sum of the areas of these $n$ rectangles (the shaded area) equals approximately the area under the curve; and the *limit of this sum* as $n$ increases without limit, and each subinterval approaches zero as a limit, is precisely the area under the curve. These considerations show that the definite integral $(A)$ may be regarded as the *limit of a sum*. Let us now formulate this result.

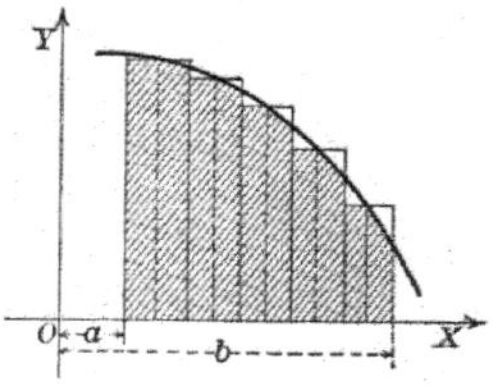


(1) Denote the lengths of the successive subintervals by

$$\Delta x_1, \ \Delta x_2, \ \Delta x_3, \ \cdots, \ \Delta x_n.$$

(2) Denote the abscissas of the points chosen in the subintervals by

$$x_1, \ x_2, \ x_3, \ \cdots, \ x_n.$$

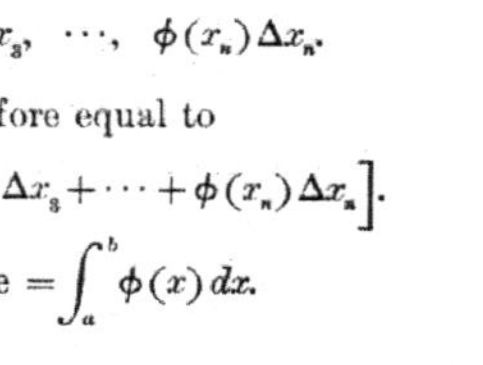


Then the ordinates of the curve at these points are

$$\phi(x_1), \ \phi(x_2), \ \phi(x_3), \ \cdots, \ \phi(x_n).$$

(3) The areas of the successive rectangles are obviously

$$\phi(x_1)\Delta x_1, \ \phi(x_2)\Delta x_2, \ \phi(x_3)\Delta x_3, \ \cdots, \ \phi(x_n)\Delta x_n.$$

(4) The area under the curve is therefore equal to

$$\lim_{n=\infty}\left[\phi(x_1)\Delta x_1+\phi(x_2)\Delta x_2+\phi(x_3)\Delta x_3+\cdots+\phi(x_n)\Delta x_n\right].$$

But from $(A)$ the area under the curve $=\int_a^b \phi(x)\,dx.$
Therefore our discussion gives

$$(B) \quad \int_a^b \phi(x)\,dx = \lim_{n=\infty}\left[\phi(x_1)\Delta x_1+\phi(x_2)\Delta x_2+\cdots+\phi(x_n)\Delta x_n\right].$$

* This construction includes the previous one as a special case, namely, when the point is chosen at one extremity of a subinterval.

This equation has been derived by making use of the notion of area. Intuition has aided us in establishing the result. Let us now regard $(B)$ *simply as a theorem in analysis*, which may then be stated as follows:

### Fundamental Theorem of the Integral Calculus

*Let $\phi(x)$ be continuous for the interval $x=a$ to $x=b$. Let this interval be divided into n subintervals whose lengths are $\Delta x_1, \Delta x_2, \cdots, \Delta x_n$, and points be chosen, one in each subinterval, their abscissas being $x_1, x_2, \cdots, x_n$ respectively. Consider the sum*

$$(C) \qquad \phi(x_1)\Delta x_1 + \phi(x_2)\Delta x_2 + \cdots + \phi(x_n)\Delta x_n = \sum_{i=1}^{n}\phi(x_i)\Delta x_i.$$

*Then the limiting value of this sum when n increases without limit, and each subinterval approaches zero as a limit, equals the value of the definite integral*

$$\int_a^b \phi(x)\,dx.$$

Equation $(B)$ may be abbreviated as follows:

$$(D) \qquad \int_a^b \phi(x)\,dx = \underset{n=\infty}{\text{limit}} \sum_{i=1}^{n}\phi(x_i)\Delta x_i.$$

The importance of this theorem results from the fact that *we are able to calculate by integration a magnitude which is the limit of a sum of the form* $(C)$.

It may be remarked that each term in the sum $(C)$ is a *differential* expression, since the lengths $\Delta x_1, \Delta x_2, \cdots, \Delta x_n$ approach zero as a limit. Each term is also called an element of the magnitude to be calculated.

The following rule will be of service in applying this theorem to practical problems.

### Fundamental Theorem. Rule

First Step. *Divide the required magnitude into similar parts such that it is clear that the desired result will be found by taking the limit of a sum of such parts.*

Second Step. *Find expressions for the magnitudes of these parts such that their sum will be of the form* $(C)$.

Third Step. *Having chosen the proper limits $x=a$ and $x=b$, we apply the Fundamental Theorem*

$$\underset{n=\infty}{\text{limit}} \sum_{i=1}^{n}\phi(x_i)\Delta x_i = \int_a^b \phi(x)\,dx$$

*and integrate.*

**205. Analytical proof of the Fundamental Theorem.** As in the last section, divide the interval from $x = a$ to $x = b$ into any number $n$ of subintervals, not necessarily equal, and denote the abscissas of these points of division by $b_1, b_2, \cdots, b_{n-1}$, and the lengths of the subintervals by $\Delta x_1, \Delta x_2, \cdots, \Delta x_n$. Now, however, we let $x_1', x_2', \cdots, x_n'$ denote abscissas, one in each interval, determined by the Theorem of Mean Value (44), p. 165, erect ordinates at these points, and through their extremities draw horizontal lines to form rectangles, as in the figure. Note that here $\phi(x)$ takes the place of $\phi'(x)$. Applying (44) to the first interval ($a = a$, $b = b_1$, and $x_1'$ lies between $a$ and $b_1$), we have

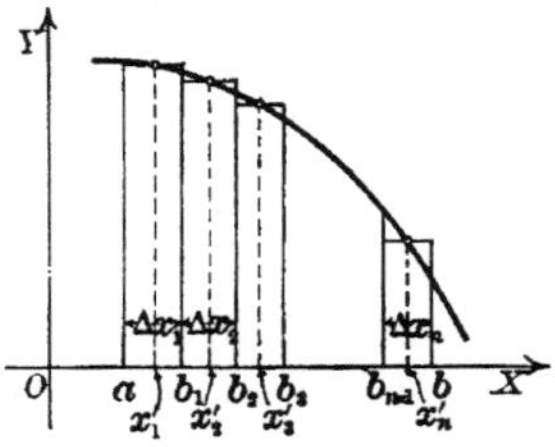


$$\frac{f(b_1) - f(a)}{b_1 - a} = \phi(x_1'),$$

or, since $$b_1 - a = \Delta x_1,$$

$$f(b_1) - f(a) = \phi(x_1')\,\Delta x_1.$$

Also $$f(b_2) - f(b_1) = \phi(x_2')\,\Delta x_2, \text{ for the second interval,}$$

$$f(b_3) - f(b_2) = \phi(x_3')\,\Delta x_3, \text{ for the third interval,}$$

$$\cdot \quad \cdot \quad \cdot \quad \cdot \quad \cdot \quad \cdot \quad \cdot \quad \text{etc.,}$$

$$f(b) - f(b_{n-1}) = \phi(x_n')\,\Delta x_n, \text{ for the } n\text{th interval.}$$

Adding these, we get

(E) $$f(b) - f(a) = \phi(x_1')\,\Delta x_1 + \phi(x_2')\,\Delta x_2 + \cdots + \phi(x_n')\,\Delta x_n.$$

But $$\phi(x_1') \cdot \Delta x_1 = \text{area of the first rectangle,}$$

$$\phi(x_2') \cdot \Delta x_2 = \text{area of the second rectangle, etc.}$$

Hence the sum on the right-hand side of (E) equals the sum of the areas of the rectangles. But from (A), p. 361, the left-hand side of (E) equals the area between the curve $y = \phi(x)$, the axis of $X$, and the ordinates at $x = a$ and $x = b$. Then the sum

(F) $$\sum_{i=1}^{n} \phi(x_i')\,\Delta x_i$$

equals this area. And while the corresponding sum

(G) $$\sum_{i=1}^{n} \phi(x_i)\,\Delta x_i$$

[Where $x_i$ is *any* abscissa of the subinterval $\Delta_i$.]

(formed as in last section) does not also give the area, nevertheless we may show that the two sums (F) and (G) approach equality when $n$ increases without limit and each subinterval approaches zero as a limit. For the difference $\phi(x_i') - \phi(x_i)$ does not exceed in numerical value the difference of the greatest and smallest ordinates in $\Delta x_i$. And furthermore it is always possible* to make all these differences less in numerical value than any assignable positive number $\epsilon$, however small, by continuing the process of subdivision far enough, i.e. by choosing $n$ sufficiently large. Hence for such a choice of $n$ the difference of the sums (F) and (G) is less in numerical value than $\epsilon(b - a)$,

* That such is the case is shown in advanced works on the Calculus.

i.e. less than any assignable positive quantity, however small. Accordingly as $n$ increases without limit, the sums $(F)$ and $(G)$ approach equality, and since $(F)$ is always equal to the area, the fundamental result follows that

$$\int_a^b \phi(x)\,dx = \lim_{n=\infty} \sum_{i=1}^{n} \phi(x_i)\,\Delta x_i,$$

in which the interval $[a, b]$ is subdivided in any manner whatever, and $x_i$ is any abscissa in the corresponding subinterval.

**206. Areas of plane curves. Rectangular coördinates.** As already explained, the area between a curve, the axis of $X$, and the ordinates $x = a$ and $x = b$ is given by the formula

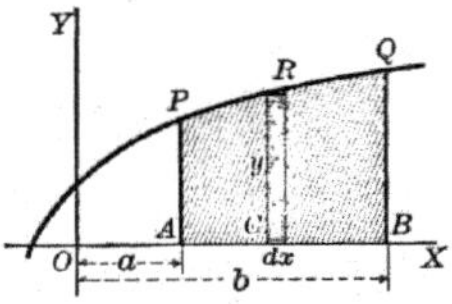


$$(A) \qquad \text{area} = \int_a^b y\,dx,$$

the value of $y$ in terms of $x$ being substituted from the equation of the curve.

Equation $(A)$ is readily memorized by observing that $ydx$ represents the area of a rectangle (as $CR$) of base $dx$ and altitude $y$. It is convenient to think of the required area $ABQP$ as the limit of the sum of all such rectangles (strips) between the ordinates $AP$ and $BQ$.

Let us now apply the Fundamental Theorem, p. 363, to the calculation of the area bounded by the curve $x = \phi(y)$, ($AB$ in figure), the axis of $Y$, and the horizontal lines $y = c$ and $y = d$.

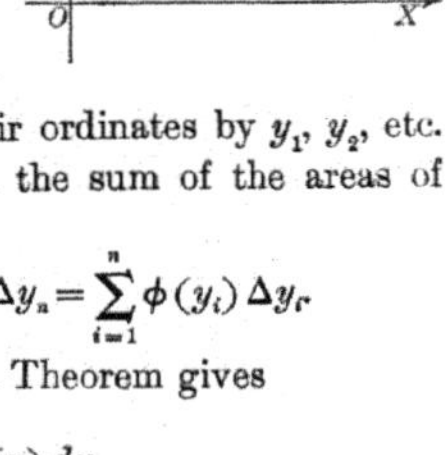


First Step. Construct the $n$ rectangles as in the figure. The required area is clearly the limit of the sum of the areas of these rectangles as their number increases without limit and the altitude of each one approaches zero as a limit.

Second Step. Denote the altitudes by $\Delta y_1$, $\Delta y_2$, etc. Take the point in each interval at the upper extremity and denote their ordinates by $y_1$, $y_2$, etc. Then the bases are $\phi(y_1)$, $\phi(y_2)$, etc., and the sum of the areas of the rectangles is

$$\phi(y_1)\,\Delta y_1 + \phi(y_2)\,\Delta y_2 + \cdots + \phi(y_n)\,\Delta y_n = \sum_{i=1}^{n} \phi(y_i)\,\Delta y_i.$$

Third Step. Applying the Fundamental Theorem gives

$$\lim_{n=\infty} \sum_{i=1}^{n} \phi(y_i)\,\Delta y_i = \int_c^d \phi(y)\,dy.$$

Hence the area between a curve, the axis of $Y$, and the horizontal lines $y = c$ and $y = d$ is given by the formula

$$(B) \qquad \text{area} = \int_c^d x\,dy,$$

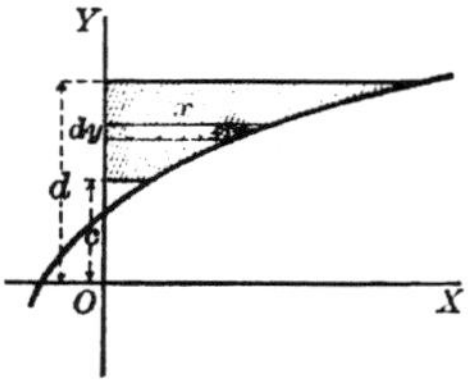


the value of $x$ in terms of $y$ being substituted from the equation of the curve. Formula $(B)$ is remembered as indicating the limit of the sum of all horizontal strips (rectangles) within the required area, $x$ and $dy$ being the base and altitude of any strip.

Illustrative Example 1. Find the area included between the semicubical parabola $y^2 = x^3$ and the line $x = 4$.

**Solution.** Let us first find the area $OMP$, half of the required area $OPP'$. For the upper branch of the curve $y = \sqrt{x^3}$, and summing up all the strips between the limits $x = 0$ and $x = 4$, we get, by substituting in $(A)$,

$$\text{area } OMP = \int_0^4 y\,dx = \int_0^4 x^{\frac{3}{2}}dx = \tfrac{64}{5} = 12\tfrac{4}{5}.$$

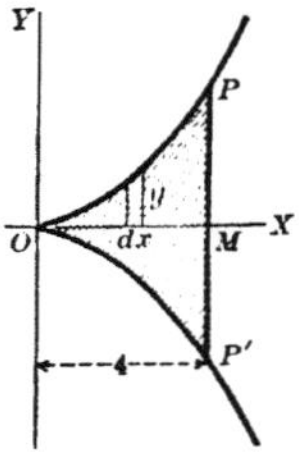


Hence area $OPP' = 2 \cdot \frac{64}{5} = 25\frac{3}{5}$.

If the unit of length is one inch, the area of $OPP'$ is $25\frac{3}{5}$ square inches.

Note. For the lower branch $y = -x^{\frac{3}{2}}$; hence

$$\text{area } OMP' = \int_0^4 (-x^{\frac{3}{2}})\,dx = -12\tfrac{4}{5}.$$

*This area lies below the axis of $x$ and has a negative sign* because the ordinates are negative.

In finding the area $OMP$ above, *the result was positive* because the ordinates were positive, *the area lying above the axis of $x$.*

The above result, $25\frac{3}{5}$, was the total area regardless of sign. As we shall illustrate in the next example, it is important to note the sign of the area when the curve crosses the axis of $X$ within the limits of integration.

Illustrative Example 2. Find the area of one arch of the sine curve $y = \sin x$.

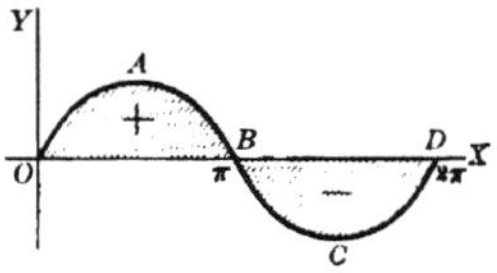


**Solution.** Placing $y = 0$ and solving for $x$, we find

$$x = 0, \quad \pi, \quad 2\pi, \text{ etc.}$$

Substituting in $(A)$, p. 365,

$$\text{area } OAB = \int_a^b y\,dx = \int_0^{\pi} \sin x\,dx = 2.$$

Also

$$\text{area } BCD = \int_a^b y\,dx = \int_{\pi}^{2\pi} \sin x\,dx = -2,$$

and

$$\text{area } OABCD = \int_a^b y\,dx = \int_0^{2\pi} \sin x\,dx = 0.$$

This last result takes into account the signs of the two separate areas composing the whole. The total area regardless of these signs equals 4.

ILLUSTRATIVE EXAMPLE 3. Find the area included between the parabola $x^2 = 4\,ay$ and the witch

$$y = \frac{8\,a^3}{x^2 + 4\,a^2}.$$

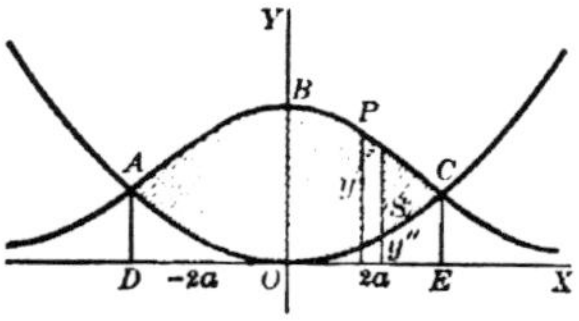


**Solution.** To determine the limits of integration we solve the equations simultaneously to find where the curves intersect. The coördinates of $A$ are found to be $(-2\,a, a)$, and of $C$ $(2\,a, a)$.

It is seen from the figure that

$$\text{area } AOCB = \text{area } DECBA - \text{area } DECOA.$$

But $$\text{area } DECBA = 2 \times \text{area } OECB = 2\int_0^{2a} \frac{8\,a^3dx}{x^2 + 4\,a^2} = 2\,\pi a^2,$$

and $$\text{area } DECOA = 2 \times \text{area } OEC = 2\int_0^{2a} \frac{x^2}{4\,a}dx = \frac{4\,a^2}{3}.$$

Hence $$\text{area } AOCB = 2\,\pi a^2 - \frac{4\,a^2}{3} = 2\,a^2(\pi - \tfrac{2}{3}).\ \ \textit{Ans.}$$

Another method is to consider the strip $PS$ as an element of the area. If $y'$ is the ordinate corresponding to the witch, and $y''$ to the parabola, the differential expression for the area of the strip $PS$ equals $(y' - y'')\,dx$. Substituting the values of $y'$ and $y''$ in terms of $x$ from the given equations, we get

$$\begin{aligned}\text{area } AOCB &= 2 \times \text{area } OCB\\ &= 2\int_0^{2a} (y' - y'')\,dx\\ &= 2\int_0^{2a}\left(\frac{8\,a^3}{x^2 + 4\,a^2} - \frac{x^2}{4\,a}\right)dx\\ &= 2\,a^2(\pi - \tfrac{2}{3}).\end{aligned}$$

ILLUSTRATIVE EXAMPLE 4. Find the area of the ellipse $\frac{x^2}{a^2} + \frac{y^2}{b^2} = 1$

**Solution.** To find the area of the quadrant $OAB$, the limits are $x = 0$, $x = a$; and

$$y = \frac{b}{a}\sqrt{a^2 - x^2}.$$

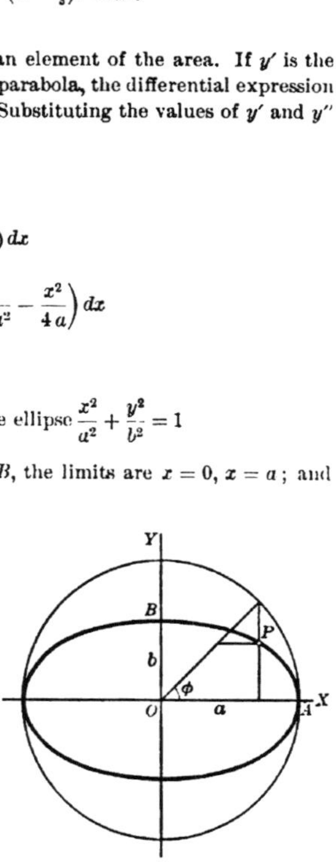


Hence, substituting in $(A)$, p. 365,

$$\begin{aligned}\text{area } OAB &= \frac{b}{a}\int_0^a (a^2 - x^2)^{\frac{1}{2}}\,dx\\ &= \left[\frac{bx}{2\,a}(a^2 - x^2)^{\frac{1}{2}} + \frac{ab}{2}\text{arc sin}\frac{x}{a}\right]_0^a\\ &= \frac{\pi ab}{4}.\end{aligned}$$

Therefore the entire area of the ellipse equals $\pi ab$.

**207. Area when equation of the curve is given in parametric form.** Let the equation of the curve be given in the parametric form

$$x = f(t), \qquad y = \phi(t).$$

We then have $\quad y = \phi(t),$ and $\ dx = f'(t)\,dt,$

which substituted* in (*A*), p. 365, gives

$$(A) \qquad \text{area} = \int_{t_1}^{t_2} \phi(t) f'(t)\,dt,$$

where $\qquad t = t_1$ when $x = a,$ and $t = t_2$ when $x = b.$

We may employ this formula (*A*) when finding the area under a curve given in parametric form. Or we may find $y$ and $dx$ from the parametric equations of the curve in terms of $t$ and $dt$ and then substitute the results directly in (*A*), p. 365.

Thus in finding the area of the ellipse in Illustrative Example 4, p. 367, it would have been simpler to use the parametric equations of the ellipse

$$x = a \cos \phi, \qquad y = b \sin \phi,$$

where the eccentric angle $\phi$ is the parameter (§ 66, p. 81).

Here $\qquad y = b \sin \phi,$ and $dx = -a \sin \phi d\phi.$

When $\qquad x = 0, \ \phi = \frac{\pi}{2};$ and when $x = a, \ \phi = 0.$

Substituting these in (*A*), above, we get

$$\text{area } OAB = \int_0^a y dx = -\int_{\frac{\pi}{2}}^{0} ab \sin^2 \phi d\phi = \frac{\pi ab}{4}.$$

Hence the entire area equals $\pi ab$. *Ans.*

**EXAMPLES**

1. Find the area bounded by the line $y = 5x$, the axis of $X$, and the ordinate $x = 2$. *Ans.* 10.

2. Find the area bounded by the parabola $y^2 = 4x$, the axis of $Y$, and the lines $y = 4$ and $y = 6$. *Ans.* $12\frac{2}{3}$.

3. Find the area of the circle $x^2 + y^2 = r^2$. *Ans.* $\pi r^2$.

4. Find the area bounded by $y^2 = 9x$ and $y = 3x$. *Ans.* $\frac{1}{2}$.

5. Find the area bounded by the coördinate axes and the curve $y = e^x$. *Ans.* 1.

6. Find the area bounded by the curve $y = \log x$, the axis of $y$, and the lines $y = 0$ and $y = 2$. *Ans.* $e^2 - 1$.

7. Find the entire area of the curve $x^{\frac{2}{3}} + y^{\frac{2}{3}} = a^{\frac{2}{3}}$. *Ans.* $\frac{3}{8}\pi a^2$.

8. Find the area between the catenary $y = \frac{a}{2}[e^{\frac{x}{a}} + e^{-\frac{x}{a}}]$, the axis of $Y$, the axis of $X$, and the line $x = a$. *Ans.* $\frac{a^2}{2e}[e^2 - 1]$.

* For a rigorous proof of this substitution the student is referred to more advanced treatises on the Calculus.

9. Find the area between the curve $y = \log x$, the axis of $X$, and the ordinates $x = 1$ and $x = a$. *Ans.* $a(\log a - 1) + 1$.

10. Find the entire area of the curve

$$\left(\frac{x}{a}\right)^2 + \left(\frac{y}{b}\right)^{\frac{2}{3}} = 1. \qquad \textit{Ans. } \frac{3\pi ab}{4}.$$

11. Find the entire area of the curve $a^2y^2 = x^3(2a - x)$. *Ans.* $\pi a^2$.

12. Find the area bounded by the curves

$$x(y - e^x) = \sin x, \quad \text{and} \quad 2xy = 2\sin x + x^3,$$

the axis of $Y$, and the ordinate $x = 1$. *Ans.* $\int_0^1 (e^x - \frac{1}{2}x^2)\,dx = e - \frac{7}{6} = 1.55 + \cdots$.

13. Find the area between the witch $y = \frac{8a^3}{x^2 + 4a^2}$ and the axis of $X$, its asymptote. *Ans.* $4\pi a^2$.

14. Find the area between the cissoid $y^2 = \frac{x^3}{2a - x}$ and its asymptote, the line $x = 2a$. *Ans.* $3\pi a^2$.

15. Find the area bounded by $y = x^3$, $y = 8$, and the axis of $Y$. *Ans.* 12.

16. Find the area included between the two parabolas $y^2 = 2px$ and $x^2 = 2py$. *Ans.* $\frac{4p^2}{3}$.

17. Find the area included between the parabola $y^2 = 2x$ and the circle $y^2 = 4x - x^2$, and lying outside of the parabola. *Ans.* 0.475.

18. Find the area bounded by $y = x^2$, $y = x$, $y = 2x$. *Ans.* $\frac{7}{6}$.

19. Find an expression for the area bounded by the equilateral hyperbola $x^2 - y^2 = a^2$, the axis of $X$, and a line drawn from the origin to any point $(x, y)$. *Ans.* $\frac{a^2}{2}\log\frac{x + y}{a}$.

20. Find by integration the area of the triangle bounded by the axis of $Y$ and the lines $2x + y + 8 = 0$ and $y = -4$. *Ans.* 4.

21. Find the area of the circle

$$\begin{cases} x = r\cos\theta, \\ y = r\sin\theta, \end{cases}$$

$\theta$ being the parameter. *Ans.* $\pi r^2$.

22. Find the area of the ellipse

$$x = a\cos\phi,$$
$$y = b\sin\phi,$$

where the eccentric angle $\phi$ is the parameter. *Ans.* $\pi ab$.

23. Find the area of the cardioid

$$x = a(2\cos t - \cos 2t),$$
$$y = a(2\sin t - \sin 2t). \qquad \textit{Ans. } \tfrac{3}{2}\pi a^2.$$

24. Find the area of one arch of the cycloid

$$\begin{cases} x = a(\theta - \sin\theta), \\ y = a(1 - \cos\theta), \end{cases}$$

$\theta$ being the parameter.

HINT. Since $x$ varies from 0 to $2\pi a$, $\theta$ varies from 0 to $2\pi$.

*Ans.* $3\pi a^2$; that is, three times the area of the generating circle.

**25.** The locus of $A$ in the figure, p. 82, is called the "companion to the cycloid." Its equations are

$$x = a\theta,$$
$$y = a(1 - \cos\theta).$$

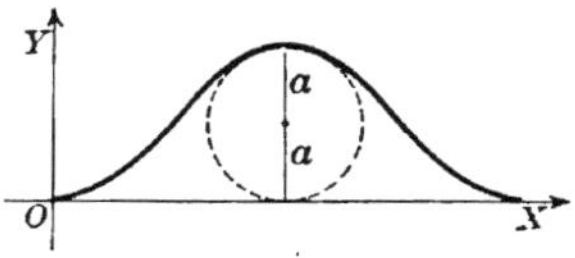


Find the area of one arch. *Ans.* $2\pi a^2$.

**26.** Find the area of the hypocycloid

$$\begin{cases} x = a\cos^3\theta, \\ y = a\sin^3\theta, \end{cases}$$

$\theta$ being the parameter. *Ans.* $\frac{8\pi a^2}{8}$; that is, three eighths of the area of the circumscribing circle.

**27.** Find the area of the loop of the folium of Descartes $x^3 + y^3 = 3axy$.

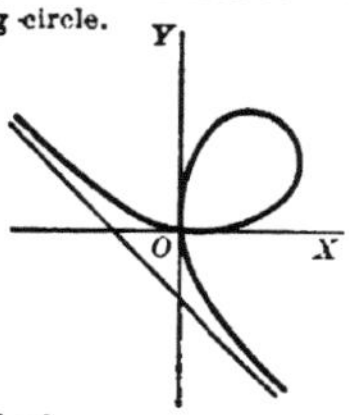


HINT. Let $y = tx$; then $x = \frac{3at}{1+t^3}$,

$$y = \frac{3at^2}{1+t^3}, \quad \text{and} \quad dx = \frac{1-2t^3}{(1+t^3)^2}3a\,dt.$$

The limits for $t$ are 0 and $\infty$.

**28.** Find by integration the areas bounded by the following loci:

(a) $(y - x)^2 = x^3$, $y = 0$. *Ans.* $\frac{1}{10}$.
(b) $(x - y^2)^2 = y^5$, $x = 0$. $\frac{1}{21}$.
(c) $a^2y = x(x^2 - a^2)$, $y = 0$. $\frac{1}{2}a^2$.
(d) $x(1 + y^2) = 1$, $x = 0$. $\pi$.
(e) $y = x(1 - x^2)$, $y = 0$. $\frac{1}{2}$.
(f) $x = y^2(y - 1)$, $x = 0$. $\frac{1}{12}$.
(g) $y^2 = x^4(2x + 1)$. Area of loop. $\frac{4}{105}$.
(h) $y^2 = x^2(2x + 1)$. Area of loop. $\frac{2}{15}$.
(i) $y = x + 4$, $y = 2x + 4$, $y = 0$.
(j) $y = x^2 + 5$, $y = 0$, $x = 0$, $x = 3$.
(k) $y = 2x^3$, $x = 0$, $y = 2$, $y = 4$.
(l) $x^2 = y + 9$, $y = 0$.
(m) $y^2 - 4 + x = 0$, $x = 0$.
(n) $xy = x^2 - 1$, $y = 0$, $x = \frac{1}{2}$, $x = 1$.
(o) $xy = 4$, $y = 1$, $y = 5$.
(p) $x = 10^y$, $y = \frac{1}{2}$, $y = 2$.

**208. Areas of plane curves. Polar coördinates.** Let it be required to find the area bounded by a curve and two of its radii vectors. For this purpose we employ polar coördinates. Assume the equation of the curve to be

$$\rho = f(\theta),$$

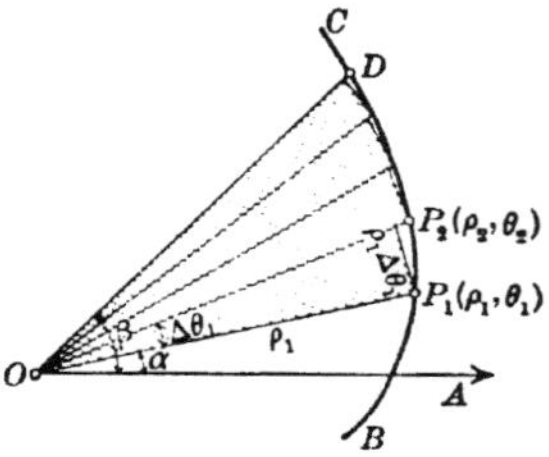


and let $OP$ and $OD$ be the two radii. Denote by $\alpha$ and $\beta$ the angles which the radii make with the polar axis. Apply the Fundamental Theorem, p. 363.

FIRST STEP. The required area is clearly the limit of the sum of circular sectors constructed as in the figure.

SECOND STEP. Let the angles of the successive sectors be $\Delta\theta_1$, $\Delta\theta_2$, etc., and their radii $\rho_1$, $\rho_2$, etc. Then the sum of the areas of the sectors * is

$$\tfrac{1}{2}\rho_1^2\Delta\theta_1 + \tfrac{1}{2}\rho_2^2\Delta\theta_2 + \cdots + \tfrac{1}{2}\rho_n^2\Delta\theta_n = \sum_{i=1}^{n}\tfrac{1}{2}\rho_i^2\Delta\theta_i.$$

* The area of a circular sector = $\frac{1}{2}$ radius × arc. Hence the area of first sector = $\frac{1}{2}\rho_1 \cdot \rho_1\Delta\theta_1 = \frac{1}{2}\rho_1^2\Delta\theta_1$, etc.

THIRD STEP. Applying the Fundamental Theorem,

$$\lim_{n=\infty} \sum_{i=1}^{n} \tfrac{1}{2} \rho_i^2 \Delta\theta_i = \int_{\alpha}^{\beta} \tfrac{1}{2} \rho^2 d\theta.$$

Hence the area swept over by the radius vector of the curve in moving from the position $OP_1$ to the position $OD$ is given by the formula

$$\text{(A)} \qquad \textbf{area} = \tfrac{1}{2} \int_{\alpha}^{\beta} \rho^2 d\theta,$$

the value of $\rho$ in terms of $\theta$ being substituted from the equation of the curve.

ILLUSTRATIVE EXAMPLE 1. Find the entire area of the lemniscate $\rho^2 = a^2 \cos 2\theta$.

**Solution.** Since the figure is symmetrical with respect to both $OX$ and $OY$, the whole area = 4 times the area of $OAB$.

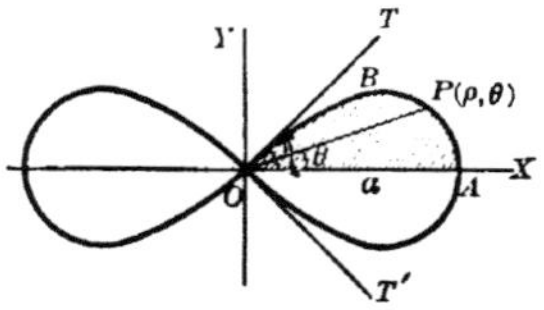


Since $\rho = 0$ when $\theta = \frac{\pi}{4}$, we see that if $\theta$ varies from 0 to $\frac{\pi}{4}$, the radius vector $OP$ sweeps over the area $OAB$. Hence, substituting in $(A)$,

$$\text{entire area} = 4 \times \text{area } OAB = 4 \cdot \frac{1}{2} \int_{\alpha}^{\beta} \rho^2 d\theta$$

$$= 2a^2 \int_0^{\frac{\pi}{4}} \cos 2\theta d\theta = a^2;$$

that is, the area of both loops equals the area of a square constructed on $OA$ as one side.

## EXAMPLES

1. Find the area swept over in one revolution by the radius vector of the spiral of Archimedes, $\rho = a\theta$, starting with $\theta = 0$. How much additional area is swept over in the second revolution?

*Ans.* $\frac{4\pi^3 a^2}{3}$; $8\pi^3 a^2$.

2. Find the area of one loop of the curve $\rho = a \cos 2\theta$. *Ans.* $\frac{\pi a^2}{8}$.

3. Show that the entire area of the curve $\rho = a \sin 2\theta$ equals one half the area of the circumscribed circle.

4. Find the entire area of the cardioid $\rho = a(1 - \cos\theta)$.

*Ans.* $\frac{3\pi a^2}{2}$; that is, six times the area of the generating circle.

5. Find the area of the circle $\rho = a \cos\theta$. *Ans.* $\frac{\pi a^2}{4}$.

6. Prove that the area of the three loops of $\rho = a \sin 3\theta$ equals one fourth of the area of the circumscribed circle.

7. Prove that the area generated by the radius vector of the spiral $\rho = e^{\theta}$ equals one fourth of the area of the square described on the radius vector.

8. Find the area of that part of the parabola $\rho = a \sec^2 \frac{\theta}{2}$ which is intercepted between the curve and the latus rectum.

*Ans.* $\frac{8a^2}{3}$.

9. Show that the area bounded by any two radii vectors of the hyperbolic spiral $\rho\theta = a$ is proportional to the difference between the lengths of these radii.

10. Find the area of the ellipse $\rho^2 = \dfrac{a^2b^2}{a^2 \sin^2\theta + b^2 \cos^2\theta}$. *Ans.* $\pi ab$.

11. Find the entire area of the curve $\rho = a(\sin 2\theta + \cos 2\theta)$. *Ans.* $\pi a^2$.

12. Find the area of one loop of the curve $\rho^2 \cos\theta = a^2 \sin 3\theta$. *Ans.* $\dfrac{3a^2}{4} - \dfrac{a^2}{2}\log 2$.

13. Find the area below $OX$ within the curve $\rho = a \sin^3 \dfrac{\theta}{3}$. *Ans.* $(10\pi + 27\sqrt{3})\dfrac{a^2}{64}$.

14. Find the area bounded by $\rho^2 = a^2 \sin 4\theta$. *Ans.* $a^2$.

15. Find the area bounded by the following curves and the given radii vectors:

(a) $\rho = \tan\theta,\ \theta = 0,\ \theta = \dfrac{\pi}{4}$.

(b) $\rho = e^{\frac{1}{2}\theta},\ \theta = \dfrac{\pi}{4},\ \theta = \dfrac{\pi}{2}$.

(c) $\rho = a^2 \sec^2 \dfrac{\theta}{2},\ \theta = \dfrac{\pi}{3},\ \theta = \dfrac{2\pi}{3}$.

(d) $\rho = \sec\theta + \tan\theta,\ \theta = 0,\ \theta = \dfrac{\pi}{4}$.

(e) $\rho = \sin\dfrac{\theta}{2} + \cos\dfrac{\theta}{2},\ \theta = 0,\ \theta = \dfrac{\pi}{4}$.

(f) $\rho = a\sin\theta + b\cos\theta,\ \theta = 0,\ \theta = \dfrac{\pi}{2}$.

16. Find the area inclosed by each of the following curves:

(a) $\rho^2 = 4\sin 2\theta$.

(b) $\rho = a\cos 3\theta$.

(c) $\rho = 8\sin 4\theta$.

(d) $\rho = 1 + 2\cos\theta$.

(e) $\rho = 3 + \cos\theta$.

(f) $\rho = 2 - \sin\theta$.

(g) $\rho^2 = a^2(1 - \cos\theta)$.

(h) $\rho = a(1 + \sin\theta)$.

(i) $\rho = a\cos 5\theta$.

**209. Length of a curve.** By the *length of a straight line* we commonly mean the number of times we can superpose upon it another straight line employed as a unit of length, as when the carpenter measures the length of a board by making end-to-end applications of his foot rule.

Since it is impossible to make a straight line coincide with an arc of a curve, we cannot measure curves in the same manner as we measure straight lines. We proceed then as follows:

Divide the curve (as $AB$) into any number of parts in any manner whatever (as at $C$, $D$, $E$) and connect the adjacent points of division, forming chords (as $AC$, $CD$, $DE$, $EB$).

*The length of the curve is defined as the limit of the sum of the chords as the number of points of division increases without limit in such a way that at the same time each chord separately approaches zero as a limit.*

Since this limit will also be the measure of the length of some straight line, the finding of the length of a curve is also called "the rectification of the curve."

The student has already made use of this definition for the length of a curve in his Geometry. Thus the circumference of a circle is defined as the limit of the perimeter of the inscribed (or circumscribed) regular polygon when the number of sides increases without limit.

The method of the next section for finding the length of a plane curve is based on the above definition, and the student should note very carefully how it is applied.

**210. Lengths of plane curves. Rectangular coördinates.** We shall now proceed to express, in analytical form, the definition of the last section, making use of the Fundamental Theorem.

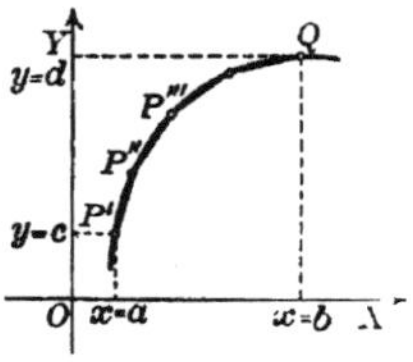


Given the curve
$$y=f(x),$$
and the points $P'(a, c)$, $Q(b, d)$ on it; to find the length of the arc $P'Q$.

FIRST STEP. Take any number $n$ of points on the curve between $P'$ and $Q$ and draw the chords joining the adjacent points, as in the figure. The required length of arc $P'Q$ is evidently the limit of the sum of the lengths of such chords.

SECOND STEP. Consider any one of these chords, $P'P''$ for example, and let the coördinates of $P'$ and $P''$ be

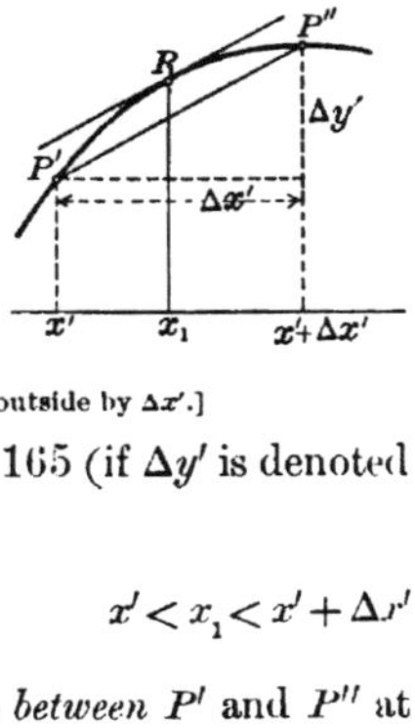


$$P'(x', y') \text{ and } P''(x'+\Delta x', y'+\Delta y').$$

Then, as in § 90, p. 134,

$$P'P''=\sqrt{(\Delta x')^2+(\Delta y')^2},$$

or,
$$P'P''=\left[1+\left(\frac{\Delta y'}{\Delta x'}\right)^2\right]^{\frac{1}{2}}\Delta x'.$$

[Dividing inside the radical by $(\Delta x')^2$ and multiplying outside by $\Delta x'$.]

But from the Theorem of Mean Value, (**44**), p. 165 (if $\Delta y'$ is denoted by $f(b)-f(a)$ and $\Delta x'$ by $b-a$), we get

$$\frac{\Delta y'}{\Delta x'}=f'(x_1), \qquad x'<x_1<x'+\Delta x'$$

$x_1$ being the abscissa of a point $P_1$ on the curve *between* $P'$ and $P''$ at which the tangent is parallel to the chord.

Substituting, $P'P''=[1+f'(x_1)^2]^{\frac{1}{2}}\Delta x'$ = length of first chord.

Similarly, $P''P'''=[1+f'(x_2)^2]^{\frac{1}{2}}\Delta x''$ = length of second chord,

. . . . . . . . . . . . .

$P^{(n)}Q=[1+f'(x_n)^2]^{\frac{1}{2}}\Delta x^{(n)}$ = length of $n$th chord.

The length of the inscribed broken line joining $P'$ and $Q$ (sum of the chords) is then the sum of these expressions, namely,

$$[1+f'(x_1)^2]^{\frac{1}{2}}\Delta x'+[1+f'(x_2)^2]^{\frac{1}{2}}\Delta x''+\cdots+[1+f'(x_n)^2]^{\frac{1}{2}}\Delta x^{(n)}$$
$$=\sum_{i=1}^{n}[1+f'(x_i)^2]^{\frac{1}{2}}\Delta x^{(i)}.$$

THIRD STEP. Applying the Fundamental Theorem,

$$\underset{n=\infty}{\text{limit}} \sum_{i=1}^{n} [1+f'(x_i)^2]^{\frac{1}{2}} \Delta x^{(i)} = \int_a^b [1+f'(x)^2]^{\frac{1}{2}} dx.$$

Hence, denoting the length of arc $P'Q$ by $s$, we have the **formula for the length of the arc**

$$s = \int_a^b [1+f'(x)^2]^{\frac{1}{2}} dx, \text{ or}$$

$$(A) \qquad s = \int_a^b \left[1 + \left(\frac{dy}{dx}\right)^2\right]^{\frac{1}{2}} dx,$$

where $\frac{dy}{dx}$ must be found in terms of $x$ from the equation of the given curve.

Sometimes it is more convenient to use $y$ as the independent variable. To derive a formula to cover this case, we know from (**35**), p. 148, that

$$\frac{dy}{dx} = \frac{1}{\frac{dx}{dy}}; \text{ hence } dx = \frac{dx}{dy} dy.$$

Substituting this value of $dx$ in (*A*), and noting that the corresponding $y$ limits are $c$ and $d$, we get* the **formula for the length of the arc,**

$$(B) \qquad s = \int_c^d \left[\left(\frac{dx}{dy}\right)^2 + 1\right]^{\frac{1}{2}} dy,$$

where $\frac{dx}{dy}$ in terms of $y$ must be found from the equation of the given curve.

ILLUSTRATIVE EXAMPLE 1. Find the length of the circle $x^2 + y^2 = r^2$.

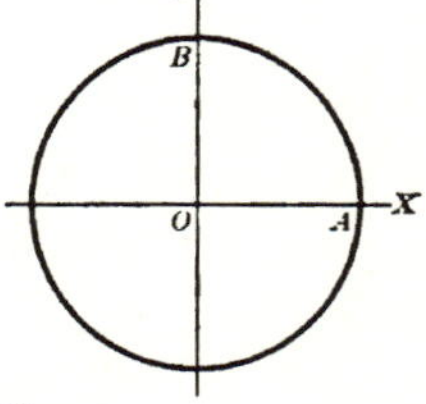


**Solution.** Differentiating, $\frac{dy}{dx} = -\frac{x}{y}$.
Substituting in (*A*),

$$\text{arc } BA = \int_0^r \left[1 + \frac{x^2}{y^2}\right]^{\frac{1}{2}} dx$$

$$= \int_0^r \left[\frac{y^2 + x^2}{y^2}\right]^{\frac{1}{2}} dx = \int_0^r \left[\frac{r^2}{r^2 - x^2}\right]^{\frac{1}{2}} dx.$$

[Substituting $y^2 = r^2 - x^2$ from the equation of the circle in order to get everything in terms of $x$.]

$$\therefore \text{ arc } BA = r\int_0^r \frac{dx}{\sqrt{r^2 - x^2}} = \left[r \text{ arc sin} \frac{x}{r}\right]_0^r = \frac{\pi r}{2}.$$

Hence the total length equals $2\pi r$. *Ans.*

* $s = \int_c^d \left[1 + \left(\frac{dy}{dx}\right)^2\right]^{\frac{1}{2}} \frac{dx}{dy} dy = \int_c^d \left[\left(\frac{dx}{dy}\right)^2 + \left(\frac{dy}{dx}\right)^2 \left(\frac{dx}{dy}\right)^2\right]^{\frac{1}{2}} dy = \int_c^d \left[\left(\frac{dx}{dy}\right)^2 + 1\right]^{\frac{1}{2}} dy.$

**EXAMPLES**

**1.** Find the length of the arc of the semicubical parabola $ay^2 = x^3$ from the origin to the ordinate $x = 5a$.

*Ans.* $\frac{335a}{27}$.

**2.** Find the entire length of the hypocycloid $x^{\frac{2}{3}} + y^{\frac{2}{3}} = a^{\frac{2}{3}}$. *Ans.* $6a$.

**3.** Rectify the catenary $y = \frac{a}{2}(e^{\frac{x}{a}} + e^{-\frac{x}{a}})$ from $x = 0$ to the point $(x, y)$.

*Ans.* $\frac{a}{2}(e^{\frac{x}{a}} - e^{-\frac{x}{a}})$.

**4.** Find the length of one complete arch of the cycloid

$$x = r \operatorname{arc vers} \frac{y}{r} - \sqrt{2ry - y^2}.$$

*Ans.* $8r$.

HINT. Use $(B)$. Here $\frac{dx}{dy} = \frac{y}{\sqrt{2ry - y^2}}$.

**5.** Find the length of the arc of the parabola $y^2 = 2px$ from the vertex to one extremity of the latus rectum.

*Ans.* $\frac{p\sqrt{2}}{2} + \frac{p}{2}\log(1 + \sqrt{2})$.

**6.** Rectify the curve $9ay^2 = x(x - 3a)^2$ from $x = 0$ to $x = 3a$. *Ans.* $2a\sqrt{3}$.

**7.** Find the length in one quadrant of the curve $\left(\frac{x}{a}\right)^{\frac{2}{3}} + \left(\frac{y}{b}\right)^{\frac{2}{3}} = 1$.

*Ans.* $\frac{a^2 + ab + b^2}{a + b}$.

**8.** Find the length between $x = a$ and $x = b$ of the curve $e^y = \frac{e^x + 1}{e^x - 1}$.

*Ans.* $\log \frac{e^{2b} - 1}{e^{2a} - 1} + a - b$.

**9.** The equations of the involute of a circle are

$$\begin{cases} x = a(\cos\theta + \theta\sin\theta), \\ y = a(\sin\theta - \theta\cos\theta). \end{cases}$$

Find the length of the arc from $\theta = 0$ to $\theta = \theta_1$. *Ans.* $\frac{1}{2}a\theta_1^2$.

**10.** Find the length of arc of curve $\begin{cases} x = e^\theta \sin\theta \\ y = e^\theta \cos\theta \end{cases}$ from $\theta = 0$ to $\theta = \frac{\pi}{2}$.

*Ans.* $\sqrt{2}(e^{\frac{\pi}{2}} - 1)$.

**11.** Find the lengths of arcs in the following curves:

(a) $y = \log\frac{e^x - 1}{e^x + 1}$; $x = 1$, $x = 2$.

(b) $y = \log(1 - x^2)$; $x = 0$, $x = \frac{1}{2}$.

(c) $y = \frac{x^2}{4} - \frac{1}{2}\log x$; $x = 1$, $x = 2$.

(d) $y = \log x$; $x = 1$, $x = 4$.

(e) $y = \log\sec x$; $x = 0$, $x = \frac{\pi}{3}$.

(f) $y = \log\csc x$; $x = \frac{\pi}{6}$, $x = \frac{\pi}{2}$.

**211. Lengths of plane curves. Polar coördinates.** Formulas $(A)$ and $(B)$ of the last section for finding the lengths of curves whose equations are given in rectangular coördinates involved the differential expressions

$$\left[1 + \left(\frac{dy}{dx}\right)^2\right]^{\frac{1}{2}} dx \quad \text{and} \quad \left[\left(\frac{dx}{dy}\right)^2 + 1\right]^{\frac{1}{2}} dy.$$

In each case, if we introduce the differential of the independent variable inside the radical, they reduce to the form

$$[dx^2 + dy^2]^{\frac{1}{2}}.$$

Let us now transform this expression into polar coördinates by means of the substitutions $x = \rho\cos\theta,\quad y = \rho\sin\theta.$

Then
$$dx = -\rho\sin\theta d\theta + \cos\theta d\rho,$$
$$dy = \rho\cos\theta d\theta + \sin\theta d\rho,$$
and we have
$$[dx^2 + dy^2]^{\frac{1}{2}} = [(-\rho\sin\theta d\theta + \cos\theta d\rho)^2 + (\rho\cos\theta d\theta + \sin\theta d\rho)^2]^{\frac{1}{2}}$$
$$= [\rho^2 d\theta^2 + d\rho^2]^{\frac{1}{2}}.$$

If the equation of the curve is
$$\rho = f(\theta),$$
then
$$d\rho = f'(\theta)\,d\theta = \frac{d\rho}{d\theta}\,d\theta.$$

Substituting this in the above differential expression, we get
$$\left[\rho^2 + \left(\frac{d\rho}{d\theta}\right)^2\right]^{\frac{1}{2}} d\theta.$$

If then $\alpha$ and $\beta$ are the limits of the independent variable $\theta$ corresponding to the limits in (*A*) and (*B*), p. 374, we get the **formula for the length of the arc,**

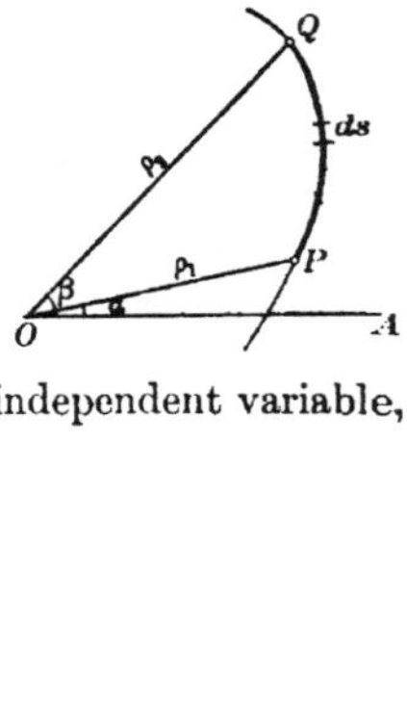


$$(A)\qquad s = \int_{\alpha}^{\beta}\left[\rho^2 + \left(\frac{d\rho}{d\theta}\right)^2\right]^{\frac{1}{2}} d\theta,$$

where $\rho$ and $\frac{d\rho}{d\theta}$ in terms of $\theta$ must be substituted from the equation of the given curve.

In case it is more convenient to use $\rho$ as the independent variable, and the equation is in the form
$$\theta = \phi(\rho),$$
then
$$d\theta = \phi'(\rho)\,d\rho = \frac{d\theta}{d\rho}\,d\rho.$$

Substituting this in $[\rho^2 d\theta^2 + d\rho^2]^{\frac{1}{2}}$

gives
$$\left[\rho^2\left(\frac{d\theta}{d\rho}\right)^2 + 1\right]^{\frac{1}{2}} d\rho.$$

Hence if $\rho_1$ and $\rho_2$ are the corresponding limits of the independent variable $\rho$, we get the **formula for the length of the arc,**

$$(B)\qquad s = \int_{\rho_1}^{\rho_2}\left[\rho^2\left(\frac{d\theta}{d\rho}\right)^2 + 1\right]^{\frac{1}{2}} d\rho,$$

where $\frac{d\theta}{d\rho}$ in terms of $\rho$ must be substituted from the equation of the given curve.

ILLUSTRATIVE EXAMPLE 1. Find the perimeter of the cardioid $\rho = a(1 + \cos\theta)$.

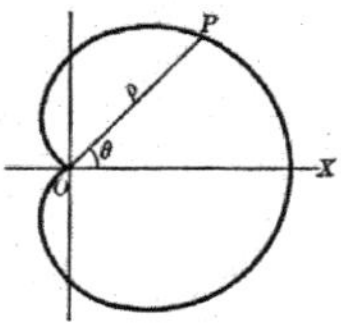


**Solution.** Here $\frac{d\rho}{d\theta} = -a\sin\theta$.

If we let $\theta$ vary from 0 to $\pi$, the point $P$ will generate one half of the curve. Substituting in $(A)$, p. 376,

$$\frac{s}{2} = \int_0^\pi [a^2(1+\cos\theta)^2 + a^2\sin^2\theta]^{\frac{1}{2}}\,d\theta$$

$$= a\int_0^\pi (2 + 2\cos\theta)^{\frac{1}{2}}\,d\theta = 2a\int_0^\pi \cos\frac{\theta}{2}\,d\theta = 4a.$$

$\therefore s = 8a.$ *Ans.*

**EXAMPLES**

1. Find the length of the spiral of Archimedes, $\rho = a\theta$, from the origin to the end of the first revolution. *Ans.* $\pi a\sqrt{1+4\pi^2} + \frac{a}{2}\log(2\pi + \sqrt{1+4\pi^2})$.

2. Rectify the spiral $\rho = e^{a\theta}$ from the origin to the point $(\rho, \theta)$. *Ans.* $\frac{\rho}{a}\sqrt{a^2+1}$.

HINT. Use $(B)$.

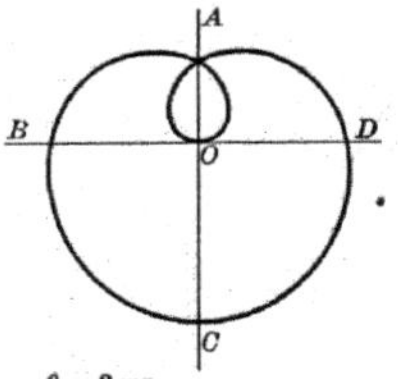


3. Find the length of the curve $\rho = a\sec^2\frac{\theta}{2}$ from $\theta = 0$ to $\theta = \frac{\pi}{2}$. *Ans.* $[\sqrt{2} + \log(\sqrt{2}+1)]a$.

4. Find the circumference of the circle $\rho = 2r\sin\theta$. *Ans.* $2\pi r$.

5. Find the length of the hyperbolic spiral $\rho\theta = a$ from $(\rho_1, \theta_1)$ to $(\rho_2, \theta_2)$.

*Ans.* $\sqrt{a^2+\rho_1^2} - \sqrt{a^2+\rho_2^2} + a\log\frac{\rho_1\left(a+\sqrt{a^2+\rho_2^2}\right)}{\rho_2\left(a+\sqrt{a^2+\rho_1^2}\right)}$.

6. Show that the entire length of the curve $\rho = a\sin^3\frac{\theta}{3}$ is $\frac{3\pi a}{2}$. Show that $OA$, $AB$, $BC$ are in arithmetical progression.

7. Find the length of arc of the cissoid $\rho = 2a\tan\theta\sin\theta$ from $\theta = 0$ to $\theta = \frac{\pi}{4}$.

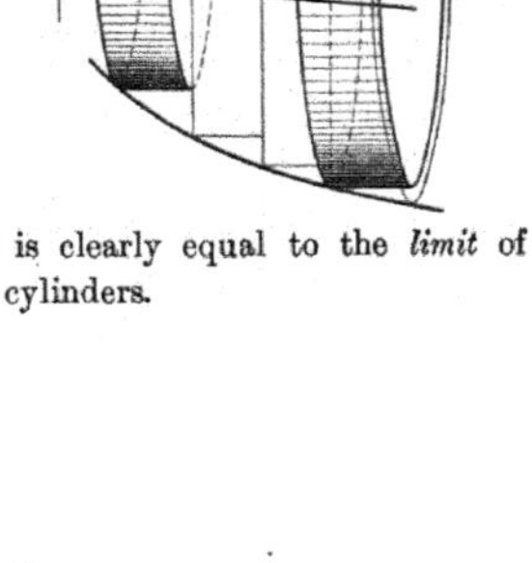


**212. Volumes of solids of revolution.** Let $V$ denote the volume of the solid generated by revolving the plane surface $ABCD$ about the axis of $X$, the equation of the plane curve $DC$ being

$$y = f(x).$$

FIRST STEP. Construct rectangles within the plane area $ABCD$ as in the figure. When this area is revolved about the axis of $X$, each rectangle generates a cylinder of revolution. The required volume is clearly equal to the *limit* of the sum of the volumes of these cylinders.

SECOND STEP. Denote the bases of the rectangles by $\Delta x_1$, $\Delta x_2$, etc., and the corresponding altitudes by $y_1$, $y_2$, etc. Then the volume of the cylinder generated by the rectangle $AEFD$ will be $\pi y_1^2 \Delta x_1$, and the sum of the volumes of all such cylinders is

$$\pi y_1^2 \Delta x_1 + \pi y_2^2 \Delta x_2 + \cdots + \pi y_n \Delta x_n = \sum_{i=1}^{n} \pi y_i \Delta x_i.$$

THIRD STEP. Applying the Fundamental Theorem (using limits $OA = a$ and $OB = b$), $\lim_{n=\infty} \sum_{i=1}^{n} \pi y_i^2 \Delta x_i = \int_a^b \pi y^2 dx.$

Hence the volume generated by revolving, about the axis of $X$, the area bounded by the curve, the axis of $X$, and the ordinates $x = a$ and $x = b$ is given by the formula

$$(A) \qquad V_x = \pi \int_a^b y^2 dx,$$

where the value of $y$ in terms of $x$ must be substituted from the equation of the given curve.

This formula is easily remembered if we consider a slice or disk of the solid between two planes perpendicular to the axis of revolution as an element of the volume, and regard it as a cylinder of infinitesimal altitude $dx$ and with a base of area $\pi y^2$, and hence of volume $\pi y^2 dx$.

Similarly, when $OY$ is the axis of revolution we use the formula

$$(B) \qquad V_y = \pi \int_c^d x^2 dy,$$

where the value of $x$ in terms of $y$ must be substituted from the equation of the given curve.

ILLUSTRATIVE EXAMPLE 1. Find the volume generated by revolving the ellipse $\frac{x^2}{a^2} + \frac{y^2}{b^2} = 1$ about the axis of $X$.

**Solution.** Since $y^2 = \frac{b^2}{a^2}(a^2 - x^2)$, and the required volume is twice the volume generated by $OAB$, we get, substituting in $(A)$,

$$\frac{V_x}{2} = \pi \int_0^a y^2 dx = \pi \int_0^a \frac{b^2}{a^2}(a^2 - x^2)\, dx$$

$$= \frac{2\pi a b^2}{3}.$$

$$\therefore V_x = \frac{4\pi a b^2}{3}.$$

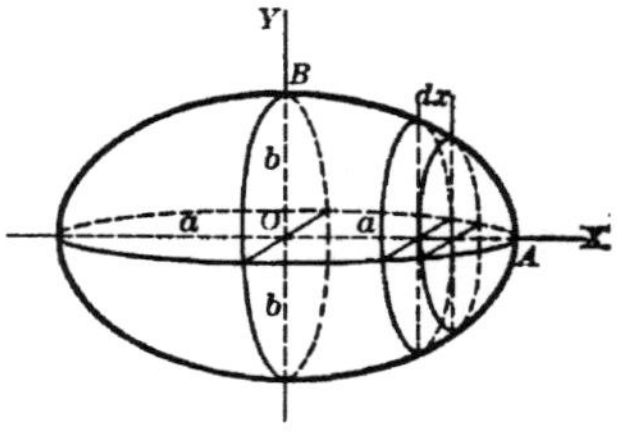


To verify this result, let $b = a$. Then $V_x = \frac{4\pi a^3}{3}$, the volume of a sphere, which is only a special case of the ellipsoid. When the ellipse is revolved about its major axis, the solid generated is called a prolate spheroid; when about its minor axis, an oblate spheroid.

**EXAMPLES**

1. Find the volume of the sphere generated by revolving the circle $x^2 + y^2 = r^2$ about a diameter. *Ans.* $\frac{4}{3}\pi r^3$.

2. Find by integration the volume of the right cone generated by revolving the triangle whose vertices are (0, 0), (*a*, 0), (*a*, *b*) about *OX*. Also find the volume generated by revolving this triangle about *OY*. Verify your results geometrically.

3. Find the volume of the torus (ring) generated by revolving the circle $x^2 + (y - b)^2 = a^2$ about *OX*. *Ans.* $2\pi^2a^2b$.

4. Find by integration the volume of the right cylinder generated by revolving the area bounded by $x = 0$, $y = 0$, $x = 6$, $y = 4$ (a) about *OX*; (b) about *OY*. Verify your results geometrically.

5. Find by integration the volume of the truncated cone generated by revolving the area bounded by $y = 6 - x$, $y = 0$, $x = 0$, $x = 4$ about *OX*. Verify geometrically.

6. Find the volume of the paraboloid of revolution generated by revolving the arc of the parabola $y^2 = 4ax$ between the origin and the point $(x_1, y_1)$ about its axis.

*Ans.* $2\pi a x_1^2 = \frac{\pi y_1^2 x_1}{2}$; i.e. one half of the volume of the circumscribing cylinder.

7. Find the volume generated by revolving the arc in Ex. 6 about the axis of *Y*.

*Ans.* $\frac{\pi y_1^5}{80a^2} = \frac{1}{5}\pi x_1^2 y_1$; i.e. one fifth of the cylinder of altitude $y_1$ and radius of base $x_1$.

8. Find by integration the volume of the cone generated by revolving about *OX* that part of the line $4x - 5y + 3 = 0$ which is intercepted between the coördinate axes. *Ans.* $\frac{9\pi}{100}$.

9 Find the volume generated by revolving about *OX* the curve

$$(x - 4a)y^2 = ax(x - 3a)$$

between the limits $x = 0$ and $x = 3a$. *Ans.* $\frac{\pi a^3}{2}(15 - 16 \log 2)$.

10. Find the volume generated by revolving about *OX* the areas bounded by the following loci:

(a) The hypocycloid $x^{\frac{2}{3}} + y^{\frac{2}{3}} = a^{\frac{2}{3}}$. *Ans.* $\frac{32\pi a^3}{105}$.

(b) The parabola $x^{\frac{1}{2}} + y^{\frac{1}{2}} = a^{\frac{1}{2}}$, $x = 0$, $y = 0$. $\frac{\pi a^3}{15}$.

(c) One arch of $y = \sin x$. $\frac{\pi^2}{2}$.

(d) The parabola $y^2 = 4x$, $x = 4$. $32\pi$.

(e) $y = xe^x$, $x = 1$, $y = 0$. $\frac{\pi}{4}(e^2 - 1)$.

(f) $y^2 = 9x$, $y = 3x$. $\frac{3\pi}{2}$.

(g) The witch $y = \frac{8a^3}{x^2 + 4a^2}$, $y = 0$. $4\pi^2a^3$.

(h) $y^2(4 + x^2) = 1$, $y = 0$, $x = 0$, $x = \infty$.

(i) $y = x^3$, $y = 0$, $x = 1$.

(j) $y^2(6 - x) = x^2$, $y = 0$, $x = 0$, $x = 4$.

(k) $4y^2 = x^3$, $x = 4$.

(l) $y(1 + x^2) = x$, $y = 0$, $x = 0$, $x = 8$.

(m) $y(x - 2)^2 = 1$, $y = 0$, $x = 3$, $x = 4$.

(n) $y^2 = (x + 2)^3$, $y = 0$, $x = -1$, $x = 0$

(o) $(x - 1)y = 2$, $y = 0$, $x = 2$, $x = 5$.

**11.** Find the volume generated by revolving the areas bounded by the following loci:

| | | About $OX$ | About $OY$ |
|---|---|---|---|
| (a) $y = e^x,\ x = 0,\ y = 0.$ | *Ans.* | $\frac{\pi}{2}$. | $2\pi$. |
| (b) $y = x^3,\ x = 2,\ y = 0.$ | | $\frac{128\pi}{7}$. | $\frac{64\pi}{5}$. |
| (c) $ay^2 = x^3,\ y = 0,\ x = a.$ | | $\frac{1}{4}\pi a^3$. | $\frac{4}{7}\pi a^3$. |
| (d) $\frac{x^2}{16} + \frac{y^2}{9} = 1.$ | | $48\pi$. | $64\pi$. |
| (e) $\left(\frac{x}{a}\right)^2 + \left(\frac{y}{b}\right)^{\frac{2}{3}} = 1.$ | | $\frac{32}{35}\pi ab^2$. | $\frac{4}{5}\pi a^2 b$ |

(f) $y^2 = 9x,\ y = 0,\ x = 9.$

(g) $y^2 = 4 - x,\ x = 0.$

(h) $y^2 = x + 9,\ x = 0.$

(i) $x^2 = 1 + y,\ y = 0.$

(j) $x^2 = 16 - y,\ y = 0.$

(k) $x^2 + 9y^2 = 36.$

(l) $y = 2x,\ y = 0,\ x = 3.$

(m) $y = x + 2,\ y = 0,\ x = 0,\ x = 3.$

**12.** Find the volume generated by revolving one arch of the cycloid

$$x = r \operatorname{arc\,vers} \frac{y}{r} - \sqrt{2ry - y^2}$$

about $OX$, its base.

HINT. Substitute $dx = \frac{y\,dy}{\sqrt{2ry - y^2}}$, and limits $y = 0,\ y = 2r$, in $(A)$, p. 378. *Ans.* $5\pi^2 r^3$.

**13.** Find the volume generated by revolving the catenary $y = \frac{a}{2}\left(e^{\frac{x}{a}} + e^{-\frac{x}{a}}\right)$ about the axis of $X$ from $x = 0$ to $x = b$.

*Ans.* $\frac{\pi a^3}{8}\left(e^{\frac{2b}{a}} - e^{-\frac{2b}{a}}\right) + \frac{\pi a^2 b}{2}$.

**14.** Find the volume of the solid generated by revolving the cissoid $y^2 = \frac{x^3}{2a - x}$ about its asymptote $x = 2a$. *Ans.* $2\pi^2 a^3$.

**15.** Given the slope of tangent to the tractrix $\frac{dy}{dx} = -\frac{y}{\sqrt{a^2 - y^2}}$, find the solid generated by revolving it about $OX$. *Ans.* $\frac{2}{3}\pi a^3$.

**16.** Show that the volume of a conical cap of height $a$ cut from the solid generated by revolving the rectangular hyperbola $x^2 - y^2 = a^2$ about $OX$ equals the volume of a sphere of radius $a$.

**17.** Using the parametric equations of the hypocycloid

$$\begin{cases} x = a\cos^3\theta, \\ y = a\sin^3\theta, \end{cases}$$

find the volume of the solid generated by revolving it about $OX$. *Ans.* $\frac{32\pi a^3}{105}$.

**18.** Find the volume generated by revolving one arch of the cycloid

$$\begin{cases} x = a(\theta - \sin\theta), \\ y = a(1 - \cos\theta), \end{cases}$$

about its base $OX$. *Ans.* $5\pi^2 a^3$.

Show that if the arch be revolved about $OY$, the volume generated is $6\pi^3 a^3$.

**19.** Show that the volume of the *egg* generated by revolving the curve

$$x^2y^2 + (x - a)(x - b) = 0, \qquad (a < b)$$

about $OX$ is

$$\pi\left\{(a + b)\log\frac{b}{a} - 2(b - a)\right\}.$$

**20.** Find the volume generated by revolving the curve $x^4 - a^2x^2 + a^2y^2 = 0$ about $OX$. *Ans.* $\frac{4\pi a^3}{15}$.

**213. Areas of surfaces of revolution.** A surface of revolution is generated by revolving the arc $CD$ of the curve

$$y = f(x)$$

about the axis of $X$.

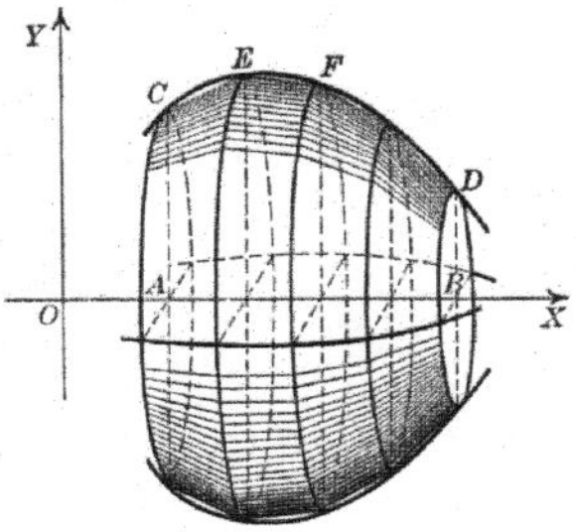


It is desired to measure this surface by making use of the Fundamental Theorem.

First Step. As before, divide the interval $AB$ into subintervals $\Delta x_1$, $\Delta x_2$, etc., and erect ordinates at the points of division. Draw the chords $CE$, $EF$, etc., of the curve. When the curve is revolved, each chord generates the lateral surface of a frustum of a cone of revolution. The required surface of revolution is defined as the limit of the sum of the lateral surfaces of these frustums.

Second Step. For the sake of clearness let us draw the first frustum on a larger scale. Let $M$ be the middle point of the chord $CE$. Then

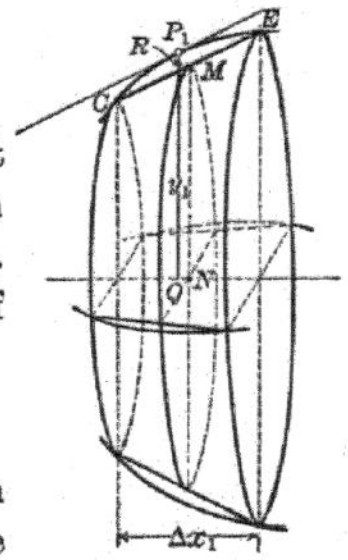


$$(A) \qquad \text{lateral area} = 2\,\pi NM \cdot CE.^*$$

In order to apply the Fundamental Theorem it is necessary to express this product as a function of the abscissa of some point in the interval $\Delta x_1$. As in § 210, p. 373, we get, using the Theorem of Mean Value, the length of chord

$$(B) \qquad CE = [1 + f'(x_1)^2]^{\frac{1}{2}}\Delta x_1,$$

where $x_1$ is the abscissa of the point $P_1(x_1, y_1)$ on the arc $CE$, where the tangent is parallel to the chord $CE$. Let the horizontal line through $M$ intersect $QP_1$ at $R$, and denote $RP_1$ by $\epsilon_1$.† Then

$$(C) \qquad NM = y_1 - \epsilon_1.$$

Substituting $(B)$ and $(C)$ in $(A)$, we get

$$2\,\pi(y_1 - \epsilon_1)[1 + f'(x_1)^2]^{\frac{1}{2}}\Delta x_1 = \text{lateral area of first frustum.}$$

* The lateral area of the frustum of a cone of revolution is equal to the circumference of the middle section multiplied by the slant height.

† The student will observe that as $\Delta x_1$ approaches zero as a limit, $\epsilon_1$ also approaches the limit zero.

Similarly,

$$2\pi(y_2-\epsilon_2)[1+f'(x_2)^2]^{\frac{1}{2}}\Delta x_2 = \text{lateral area of second frustum},$$

$$\cdots\cdots\cdots$$

$$2\pi(y_n-\epsilon_n)[1+f'(x_n)^2]^{\frac{1}{2}}\Delta x_n = \text{lateral area of last frustum}.$$

Hence

$$\sum_{i=1}^{n} 2\pi(y_i-\epsilon_i)[1+f'(x_i)^2]^{\frac{1}{2}}\Delta x_i = \text{sum of lateral areas of frustums}.$$

This may be written

$$(D)\qquad \sum_{i=1}^{n} 2\pi y_i[1+f'(x_i)^2]^{\frac{1}{2}}\Delta x_i - \sum_{i=1}^{n}\epsilon_i[1+f'(x_i)^2]^{\frac{1}{2}}\Delta x_i.$$

THIRD STEP. Applying the Fundamental Theorem to the first sum (using the limits $OA = a$ and $OB = b$), we get

$$\underset{n=\infty}{\text{limit}} \sum_{i=1}^{n} 2\pi y_i[1+f'(x_i)^2]^{\frac{1}{2}}\Delta x_i = \int_a^b 2\pi y[1+f'(x)^2]^{\frac{1}{2}}dx.$$

The limit of the second sum of $(D)$ for $n \doteq \infty$ is zero.* Hence the area of the surface of revolution generated by revolving the arc $CD$ about $OX$ is given by the formula

$$(E)\qquad S_x = 2\pi\int_a^b y\left[1+\left(\frac{dy}{dx}\right)^2\right]^{\frac{1}{2}}dx,$$

where $y$ and $\frac{dy}{dx}$ in terms of $x$ must be substituted from the equation of the revolved curve, and $S$ denotes the required area. Or we may write the formula in the form

$$S = 2\pi\int_a^b y\,ds,$$

remembering that

$$ds = (dx^2+dy^2)^{\frac{1}{2}} = \left[1+\left(\frac{dy}{dx}\right)^2\right]^{\frac{1}{2}}dx. \qquad (27),\ \text{p. } 135$$

This formula is easily remembered if we consider a narrow band of the surface included between two planes perpendicular to the axis of revolution as the element of area, and regard it as the convex surface

* This is easily seen as follows. Denote the second sum by $S_n$. If $\epsilon$ equals the largest of the positive numbers $|\epsilon_1|, |\epsilon_2|, \cdots, |\epsilon_n|$, then

$$S_n \leqq \epsilon\sum_{i=1}^{n}[1+f'(x_i)^2]^{\frac{1}{2}}\Delta x_i.$$

The sum on the right is, by $(B)$, p. 381, equal to the sum of the chords $CE$, $EF$, etc. Let this sum be $l_n$. Then $S_n \leqq \epsilon l_n$. Since $\underset{n=\infty}{\text{limit}}\,\epsilon = 0$, $S_n$ is an infinitesimal, and therefore $\underset{n=\infty}{\text{limit}}\,S_n = 0$.

of a frustum of a cone of revolution of infinitesimal slant height $ds$, and with a middle section whose circumference equals $2\pi y$, hence of area $2\pi y ds$.

Similarly, when $OY$ is the axis of revolution we use the formula

$$(F) \qquad S_y = 2\pi\int_c^d x\left[1+\left(\frac{dx}{dy}\right)^2\right]^{\frac{1}{2}}dy,$$

where the value of $x$ and $\frac{dx}{dy}$ in terms of $y$ must be substituted from the equation of the given curve.

ILLUSTRATIVE EXAMPLE 1. Find the area of the surface of revolution generated by revolving the hypocycloid $x^{\frac{2}{3}} + y^{\frac{2}{3}} = a^{\frac{2}{3}}$ about the axis of $X$.

**Solution.** Here $\frac{dy}{dx} = -\frac{y^{\frac{1}{3}}}{x^{\frac{1}{3}}}$, $y = (a^{\frac{2}{3}} - x^{\frac{2}{3}})^{\frac{3}{2}}$.

Substituting in ($E$), p. 382, noting that the arc $BA$ generates only one half of the surface, we get

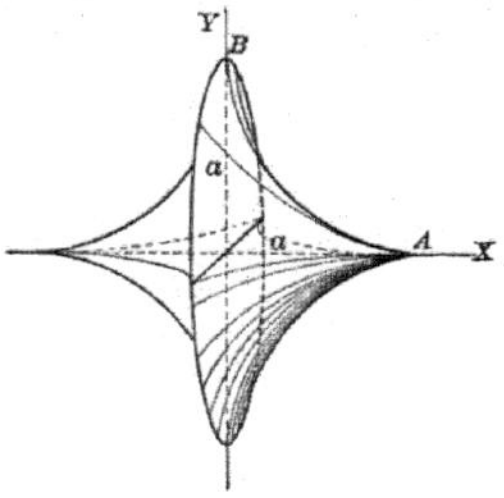


$$\frac{S_x}{2} = 2\pi\int_0^a (a^{\frac{2}{3}} - x^{\frac{2}{3}})^{\frac{3}{2}}\left[1+\frac{y^{\frac{2}{3}}}{x^{\frac{2}{3}}}\right]^{\frac{1}{2}}dx$$

$$= 2\pi\int_0^a (a^{\frac{2}{3}} - x^{\frac{2}{3}})^{\frac{3}{2}}\left(\frac{a^{\frac{2}{3}}}{x^{\frac{2}{3}}}\right)^{\frac{1}{2}}dx$$

$$= 2\pi a^{\frac{1}{3}}\int_0^a (a^{\frac{2}{3}} - x^{\frac{2}{3}})^{\frac{3}{2}}x^{-\frac{1}{3}}\,dx$$

$$= \frac{6\pi a^2}{5}.$$

$$\therefore S_x = \frac{12\pi a^2}{5}.$$

## EXAMPLES

1. Find the area of the surface of the sphere generated by revolving the circle $x^2 + y^2 = r^2$ about a diameter. *Ans.* $4\pi r^2$.

2. Find the area of the surface generated by revolving the parabola $y^2 = 4ax$ about $OX$, from the origin to the point where $x = 3a$. *Ans.* $\frac{56}{3}\pi a^2$.

3. Find by integration the area of the surface of the cone generated by revolving about $OX$ the line joining the origin to the point $(a, b)$. *Ans.* $\pi b\sqrt{a^2+b^2}$.

4. Find by integration the area of the surface of the cone generated by revolving the line $y = 2x$ from $x = 0$ to $x = 2$ (a) about $OX$; (b) about $OY$. Verify your results geometrically.

5. Find by integration the lateral area of the cylinder generated by revolving the line $x = 4$ about $OY$ from $y = 0$ to $y = 6$, and verify your result geometrically.

6. Find by integration the lateral area of the frustum of a cone of revolution generated by revolving the line $2y = x - 4$ about $OX$ from $x = 0$ to $x = 3$, and verify your results geometrically.

**7.** Parabolic mirrors and reflectors have the shape of a paraboloid of revolution. Find the area of the reflecting surface of such a mirror 2 feet deep and 6 feet wide. *Ans.* $\frac{49}{4}\pi$.

This equals the area of a circle 7 feet in diameter.

**8.** Find the surface of the torus (ring) generated by revolving the circle $x^2 + (y - b)^2 = a^2$ about $OX$. *Ans.* $4\pi^2 ab$.

HINT. Using the positive value of $\sqrt{a^2 - x^2}$ gives the outside surface, and the negative value the inside surface.

**9.** Find the surface generated by revolving an arch of the cycloid

$$x = r \text{ arc vers}\frac{y}{r} - \sqrt{2\,ry - y^2}$$

about its base. *Ans.* $\frac{64\,\pi r^2}{3}$.

**10.** Find the area of the surface of revolution generated by revolving each of the following curves about $OX$:

(a) $y = x^3$, from $x = 0$ to $x = 2$. *Ans.* $\frac{\pi}{27}[(145)^{\frac{3}{2}} - 1]$.

(b) $y = e^{-x}$, from $x = 0$ to $x = \infty$. $\pi\left[\sqrt{2} + \log(1 + \sqrt{2})\right]$.

(c) The loop of $9\,ay^2 = x(3\,a - x)^2$. $3\,\pi a^2$.

(d) $6\,a^2xy = x^4 + 3\,a^4$, from $x = a$ to $x = 2\,a$. $\frac{47}{16}\pi a^2$.

(e) The loop of $8\,a^2y^2 = a^2x^2 - x^4$. $\frac{\pi a^2}{4}$.

(f) $y^2 + 4\,x = 2\log y$, from $y = 1$ to $y = 2$. $\frac{10}{3}\pi$.

(g) $y = e^x$, from $x = -\infty$ to $x = 0$. $\pi\left[\sqrt{2} + \log(1 + \sqrt{2})\right]$.

(h) The cycloid $\begin{cases} x = a(\theta - \sin\theta), \\ y = a(1 - \cos\theta). \end{cases}$ $\frac{64\,\pi a^2}{3}$.

(i) The cardioid $\begin{cases} x = a(2\cos\theta - \cos 2\,\theta), \\ y = a(2\sin\theta - \sin 2\,\theta). \end{cases}$ $\frac{128\,\pi a^2}{5}$.

(j) $y + 2\,x = 4$, from $x = 0$ to $x = 2$.

(k) $3\,y - 2\,x = 6$, from $x = 0$ to $x = 2$.

(l) $y = x^3$, from $x = 0$ to $x = 1$.

(m) $x^2 + 4\,y^2 = 16$.

(n) $9\,x^2 + y^2 = 36$.

(o) $y^2 = 9\,x$, from $x = 0$ to $x = 1$.

**11.** Find the area of the surface of revolution generated by revolving each of the following curves about $OY$:

(a) $x + 2\,y = 6$, from $y = 0$ to $y = 3$.

(b) $3\,x + 2\,y = 12$, from $y = 0$ to $y = 4$.

(c) $x^2 = 4\,y$, from $y = 0$ to $y = 3$.

(d) $x^2 + 16\,y^2 = 16$.

(e) $4\,x^2 + y^2 = 100$.

(f) $3\,x = y^3$, from $y = 0$ to $y = 1$.

(g) $x = y^3$, from $y = 0$ to $y = 3$. *Ans.* $\frac{\pi}{27}[(730)^{\frac{3}{2}} - 1]$.

(h) $6\,a^2xy = x^4 - 3\,a^4$, from $x = a$ to $x = 3\,a$. $(20 + \log 3)\,\pi a^2$.

(i) $4\,y = x^2 - 2\log x$, from $x = 1$ to $x = 4$. $24\,\pi$.

(j) $2\,y = x\sqrt{x^2 - 1} + \log\left(x - \sqrt{x^2 - 1}\right)$, from $x = 2$ to $x = 5$. $78\,\pi$.

**12. Find the area of the surface of revolution generated by revolving each of the following curves:**

| | *About OX* | *About OY* |
|---|---|---|
| (a) The ellipse $\frac{x^2}{a^2}+\frac{y^2}{b^2}=1$. <br> HINT. $e$ = eccentricity of ellipse $=\frac{\sqrt{a^2-b^2}}{a}$. | $2\pi b^2+\frac{2\pi ab}{e}\arcsin e$. | $2\pi a^2+\frac{\pi b^2}{e}\log\frac{1+e}{1-e}$. |
| (b) The catenary $y=\frac{a}{2}(e^{\frac{x}{a}}+e^{-\frac{x}{a}})$, from $x=0$ to $x=a$. | $\frac{\pi a^2}{4}(e^2+4-e^{-2})$. | $2\pi a^2(1-e^{-1})$. |
| (c) $x^4+3=6xy$, from $x=1$ to $x=2$. | $\frac{47}{16}\pi$. | $\pi(\frac{15}{4}+\log 2)$. |
| (d) $\begin{cases} x=e^{\theta}\sin\theta, \\ y=e^{\theta}\cos\theta, \text{ from } \theta=0 \text{ to } \frac{\pi}{2}. \end{cases}$ | $\frac{2\sqrt{2}\pi}{5}(e^{\pi}-2)$. | $\frac{4\pi}{5}(2e^{\pi}+1)$. |
| (e) $3x^2+4y^2=3a^2$. | $\left(\frac{3}{2}+\frac{\pi}{\sqrt{3}}\right)\pi a^2$. | $(4+3\log 3)\frac{\pi a^2}{2}$. |
| (f) $x+y=4$, from $x=0$ to $x=4$. | | |
| (g) $y=2x+4$, from $y=4$ to $y=8$. | | |
| (h) $x^2+2y^2=16$. | | |

**214. Miscellaneous applications.** In § 212 it was shown how to calculate the volume of a solid of revolution by means of a single integration. Evidently we may consider a solid of revolution as generated by a moving circle of varying radius whose center lies on the axis of revolution and whose plane is perpendicular to it. Thus in the figure the circle $ACBD$, whose plane is perpendicular to $OX$, may be supposed to generate the solid of revolution $O-EGFH$, while its center moves from $O$ to $N$, the radius $MC(=y)$ varying continuously with $OM(=x)$ in a manner determined by the equation of the plane curve that is being revolved.

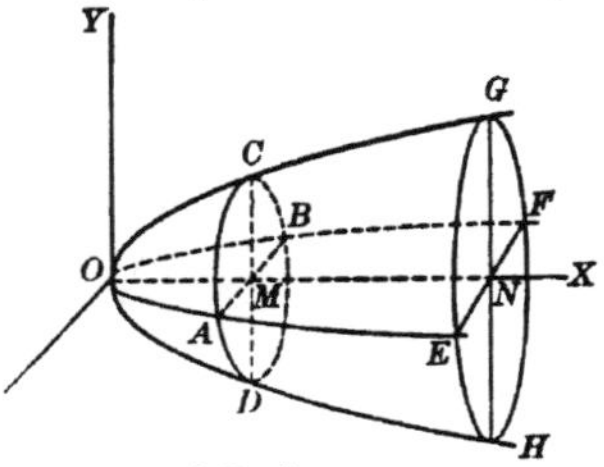


We will now show how this idea may be extended to the calculation of volumes that are not solids of revolution when it is possible to express the area of parallel plane sections of the solid as a function of their distances from a fixed point.

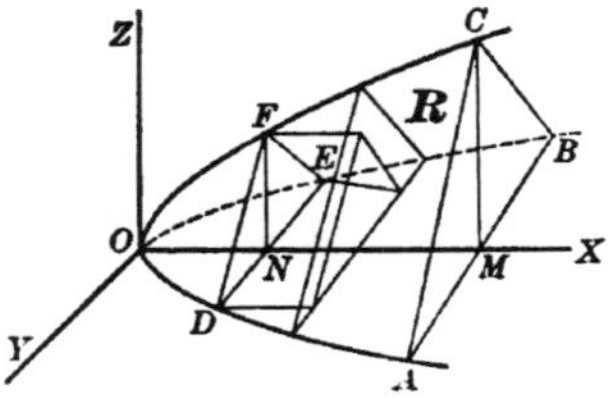


Suppose we divide the solid shown in our figure into $n$ slices by sections perpendicular to $OX$ and take the origin as our fixed point.

Let $FDE$ be one face of such a slice. Construct a right prism upon $FDE$ as a base, the second base lying in the other face of the slice.

Since, by hypothesis, the area of $FDE$ is a function of $ON$, or $x$, let $f(x)$ = area of $FDE$ = area of base of prism, and let $\Delta x$ = altitude of prism.

Hence $f(x)\Delta x$ = volume of prism, and $\sum_{i=1}^{n} f(x_i)\Delta x_i$ = sum of volumes of all such prisms. It is evident that the required volume is the limit of this sum; hence, by the Fundamental Theorem,

$$\lim_{n=\infty} \sum_{i=1}^{n} f(x_i)\,\Delta x_i = \int f(x)\,dx,$$

and we have the formula

$$(A) \qquad \boldsymbol{V = \int f(x)\,dx,}$$

where $f(x)$ is the area of a section of the solid perpendicular to $OX$ expressed in terms of its distance ($=x$) from the origin, the $x$-limits being chosen so as to extend over the entire region $R$ occupied by the solid.

Evidently the solid $O-ABC$ may be considered as being generated by the continuously varying plane section $DEF$ as $ON(=x)$ varies from zero to $OM$. The following examples will further illustrate this principle.

ILLUSTRATIVE EXAMPLE 1. Calculate the volume of the ellipsoid

$$\frac{x^2}{a^2} + \frac{y^2}{b^2} + \frac{z^2}{c^2} = 1$$

by means of a single integration.

**Solution.** Consider a section of the ellipsoid perpendicular to $OX$, as $ABCD$, with semiaxes $b'$ and $c'$. The equation of the ellipse $HEJG$ in the $XOY$-plane is

$$\frac{x^2}{a^2} + \frac{y^2}{b^2} = 1.$$

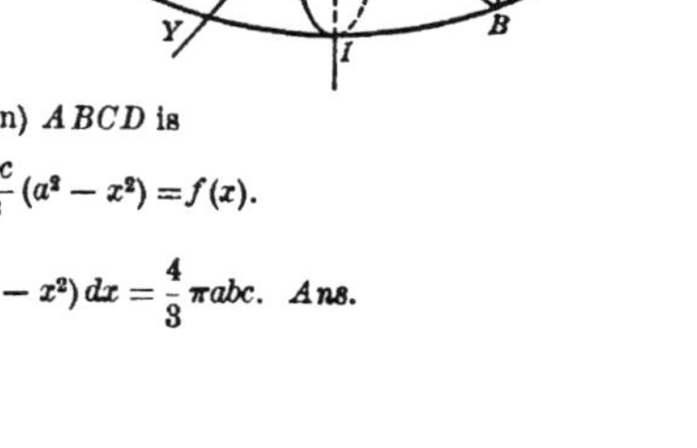


Solving this for $y\,(=b')$ in terms of $x\,(=OM)$ gives

$$b' = \frac{b}{a}\sqrt{a^2 - x^2}.$$

Similarly, from the equation of the ellipse $EFGI$ in the $XOZ$-plane we get

$$c' = \frac{c}{a}\sqrt{a^2 - x^2}.$$

Hence the area of the ellipse (section) $ABCD$ is

$$\pi b'c' = \frac{\pi bc}{a^2}(a^2 - x^2) = f(x).$$

Substituting in $(A)$,

$$V = \frac{\pi bc}{a^2}\int_{-a}^{+a}(a^2 - x^2)\,dx = \frac{4}{3}\pi abc. \quad Ans.$$

We may then think of the ellipsoid as being generated by a variable ellipse $ABCD$ moving from $G$ to $E$, its center always on $OX$ and its plane perpendicular to $OX$.

ILLUSTRATIVE EXAMPLE 2. Find the volume of a right conoid with circular base, the radius of base being $r$ and altitude $a$.

**Solution.** Placing the conoid as shown in the figure, consider a section $PQR$ perpendicular to $OX$. This section is an isosceles triangle; and since

$$RM = \sqrt{2rx - x^2}$$

(found by solving $x^2 + y^2 = 2rx$, the equation of the circle $ORAQ$, for $y$) and $MP = a$,

the area of the section is

$$a\sqrt{2rx - x^2} = f(x).$$

Substituting in $(A)$, p. 386,

$$V = a\int_0^{2r}\sqrt{2rx - x^2}dx = \frac{\pi r^2 a}{2}. \quad Ans.$$

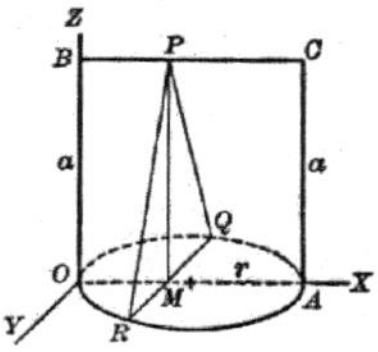


This is one half the volume of the cylinder of the same base and altitude.

We will now take up the study of *fluid pressure* and learn how to calculate the pressure of a fluid on a vertical wall.

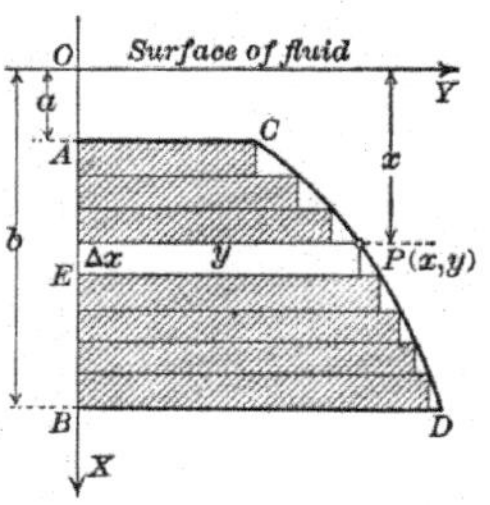


Let $ABCD$ represent part of the area of the vertical surface of one wall of a reservoir. It is desired to determine the total fluid pressure on this area. Draw the axes as in the figure, the $Y$-axis lying in the surface of the fluid. Divide $AB$ into $n$ subintervals and construct horizontal rectangles within the area. Then the area of one rectangle (as $EP$) is $y\Delta x$. If this rectangle was horizontal at the depth $x$, the fluid pressure on it would be

$$Wxy\Delta x,$$

[The pressure of a fluid on any given horizontal surface equals the weight of a column of the fluid standing on that surface as a base and of height equal to the distance of this surface below the surface of the fluid.]

where $W$ = the weight of a unit volume of the fluid. Since fluid pressure is the same in all directions, it follows that $Wxy\Delta x$ will be approximately the pressure on the rectangle $EP$ in its vertical position. Hence the sum

$$\sum_{i=1}^{n} Wx_i y_i \Delta x_i$$

represents approximately the pressure on all the rectangles. The pressure on the area $ABCD$ is evidently the limit of this sum. Hence, by the Fundamental Theorem,

$$\lim_{n=\infty} \sum_{i=1}^{n} Wx_i y_i \Delta x_i = \int Wxy\,dx.$$

Hence the fluid pressure on a vertical submerged surface bounded by a curve, the axis of $X$, and the two horizontal lines $x = a$ and $x = b$ is given by the formula

**(B)** $$\textbf{fluid pressure} = W\int_a^b yx\,dx,$$

where the value of $y$ in terms of $x$ must be substituted from the equation of the given curve.

We shall assume 62 lb. ($= W$) as the weight of a cubic foot of water.

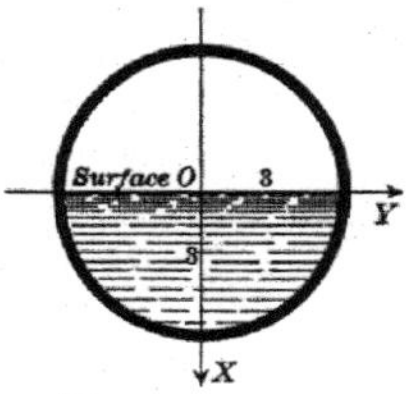


ILLUSTRATIVE EXAMPLE 3. A circular water main 6 ft. in diameter is half full of water. Find the pressure on the gate that closes the main.

**Solution.** The equation of the circle is $x^2 + y^2 = 9$.

Hence $$y = \sqrt{9 - x^2},$$
$$W = 62,$$

and the limits are from $x = 0$ to $x = 3$. Substituting in (B), we get the pressure on the right of the axis of $X$ to be

$$\text{pressure} = 62\int_0^3 \sqrt{9 - x^2}\cdot x\,dx = \left[-\tfrac{62}{3}(9 - x^2)^{\frac{3}{2}}\right]_0^3 = 558.$$

Hence the total pressure $= 2 \times 558 = 1116$ lb. *Ans.*

Let us now consider the problem of finding the work done in emptying reservoirs of the form of solids of revolution with their axes vertical. It is convenient to assume the axis of $X$ of the revolved curve as vertical, and the axis of $Y$ on a level with the top of the reservoir.

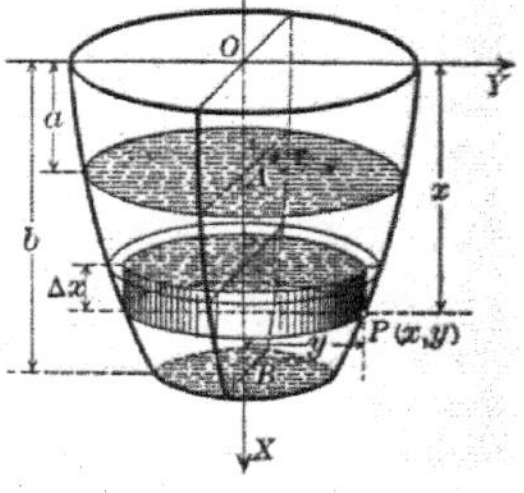


Consider a reservoir such as the one shown; we wish to calculate the work done in emptying it of a fluid from the depth $a$ to the depth $b$.

Divide $AB$ into $n$ subintervals, pass planes perpendicular to the axis of revolution through these points of division, and construct cylinders of revolution, as in § 212, p. 377. The volume of any such cylinder will be $\pi y^2 \Delta x$ and its weight $W\pi y^2 \Delta x$, where $W =$ weight of a cubic unit of the fluid. The work done in lifting this cylinder of the fluid out of the reservoir (through the height $x$) will be

$$W\pi y^2 x\Delta x.$$

[Work done in lifting equals the weight multiplied by the vertical height.]

The work done in lifting all such cylinders to the top is the sum

$$\sum_{i=1}^{n} W\pi y_i^2 x_i \Delta x_i.$$

The work done in emptying that part of the reservoir will evidently be the limit of this sum. Hence, by the Fundamental Theorem,

$$\lim_{n=\infty} \sum_{i=1}^{n} W\pi y_i^2 x_i \Delta x_i = \int W\pi y^2 x dx.$$

Therefore the work done in emptying a reservoir in the form of a solid of revolution from the depth $a$ to the depth $b$ is given by the formula

$$(C) \qquad \textbf{work} = W\pi \int_a^b y^2 x dx,$$

where the value of $y$ in terms of $x$ must be substituted from the equation of the revolved curve.

#### ILLUSTRATIVE EXAMPLES

1. Calculate the work done in pumping out the water filling a hemispherical reservoir 10 feet deep.

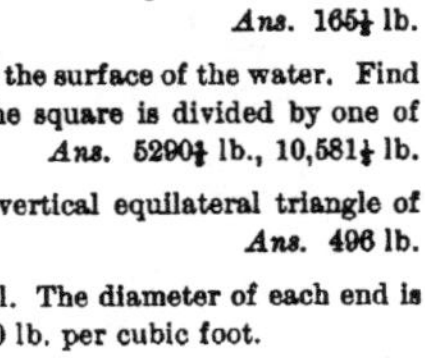


**Solution.** The equation of the circle is $x^2 + y^2 = 100$.

Hence $y^2 = 100 - x^2$,

$W = 62$,

and the limits are from $x = 0$ to $x = 10$.

Substituting in $(C)$, we get

$$\text{work} = 62\pi \int_0^{10} (100 - x^2) x dx = 155{,}000\,\pi \text{ ft. lb.}$$

2. A trough 2 ft. deep and 2 ft. broad at the top has semielliptical ends. If it is full of water, find the pressure on one end. *Ans.* $165\frac{1}{3}$ lb.

3. A floodgate 8 ft. square has its top just even with the surface of the water. Find the pressure on each of the two portions into which the square is divided by one of its diagonals. *Ans.* $5290\frac{2}{3}$ lb., $10{,}581\frac{1}{3}$ lb.

4. Find the pressure on one face of a submerged vertical equilateral triangle of side 4 ft., one side lying in the surface of the water. *Ans.* 496 lb.

5. A horizontal cylindrical oil tank is half full of oil. The diameter of each end is 4 ft. Find the pressure on one end if the oil weighs 50 lb. per cubic foot. *Ans.* $266\frac{2}{3}$ lb.

**6.** Find the work done in pumping out a semielliptical reservoir filled with water. The top is a circle of diameter 6 ft. and the depth is 5 ft. *Ans.* $3487\frac{1}{2}\pi$ ft. lb.

7. Find the pressure on the surface of the reservoir in Example 1.

8. Find the pressure on the surface of the reservoir in Example 6.

**9.** A conical reservoir 12 ft. deep is filled with a liquid weighing 80 lb. per cubic foot. The top of the reservoir is a circle 8 ft. in diameter. Find the energy expended in pumping it out. *Ans.* $15{,}360\,\pi$ ft. lb.

10. The cross section of a trough is a parabola with vertex downward, the latus rectum lying in the surface and being 4 feet long. Find the pressure on one end of the trough when it is full of a liquid weighing $62\frac{1}{2}$ lb. per cubic foot. *Ans.* 66 lb.

11. Find the pressure on a sphere 6 feet in diameter which is immersed in water, its center being 10 feet below the surface of the water.

HINT. Pressure $= 2\pi w\int_{-3}^{3} y(10+x)ds$, and $ds = \frac{3}{y}dx$.

*Ans.* $22320\,\pi$ lb.

12. A board in the form of a parabolic segment by a chord perpendicular to the axis is immersed in water. The vertex is at the surface and the axis is vertical. It is 20 feet deep and 12 feet broad. Find the pressure in tons. *Ans.* 59.52.

13. How far must the board in Example 12 be sunk to double the pressure?

*Ans.* 12 feet.

14. A water tank is in the form of a hemisphere 24 feet in diameter, surmounted by a cylinder of the same diameter and 10 feet high. Find the work done in pumping it out when filled within 2 feet of the top.

15. The center of a square moves along a diameter of a given circle of radius $a$, the plane of the square being perpendicular to that of the circle, and its magnitude varying in such a way that two opposite vertices move on the circumference of the circle. Find the volume of the solid generated. *Ans.* $\frac{8}{3}a^3$.

16. A woodman fells a tree 2 ft. in diameter, cutting halfway through on each side. The lower face of each cut is horizontal, and the upper face makes an angle of 60° with the lower. How much wood does he cut out?

*Ans.* $\frac{4}{\sqrt{3}}$ cu. ft.

17. A variable equilateral triangle moves with its plane perpendicular to the $x$-axis and the ends of its base on the points on the curves $y^2 = 16\,ax$ and $y^2 = 4\,ax$ respectively above the $x$-axis. Find the volume generated by the triangle as it moves from the origin to the points whose abscissa is $a$.

*Ans.* $\frac{\sqrt{3}}{2}a^3$.

18. A rectangle moves from a fixed point, one side being always equal to the distance from this point, and the other equal to the square of this distance. What is the volume generated while the rectangle moves a distance of 2 ft.? *Ans.* 4 cu. ft.

19. On the double ordinates of the ellipse $\frac{x^2}{a^2} + \frac{y^2}{b^2} = 1$, isosceles triangles of vertical angle 90° are described in planes perpendicular to that of the ellipse. Find the volume of the solid generated by supposing such a variable triangle moving from one extremity to the other of the major axis of the ellipse.

*Ans.* $\frac{4\,ab^2}{3}$.

**20.** Determine the amount of attraction exerted by a thin, straight, homogeneous rod of uniform thickness, of length $l$, and of mass $M$, upon a material point $P$ of mass $m$ situated at a distance of $a$ from one end of the rod in its line of direction.

**Solution.*** Suppose the rod to be divided into equal infinitesimal portions (elements) of length $dx$.

$$\frac{M}{l} = \text{mass of a unit length of rod};$$

hence $$\frac{M}{l}dx = \text{mass of any element.}$$

Newton's Law for measuring the attraction between any two masses is

$$\text{force of attraction} = \frac{\text{product of masses}}{(\text{distance between them})^2};$$

therefore the force of attraction between the particle at $P$ and an element of the rod is

$$\frac{\frac{M}{l}mdx}{(x+a)^2},$$

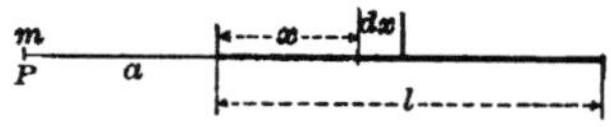


which is then an *element of the force of attraction required*. The total attraction between the particle at $P$ and the rod being the limit of the sum of all such elements between $x = 0$ and $x = l$, we have

$$\text{force of attraction} = \int_0^l \frac{\frac{M}{l}mdx}{(x+a)^2} = \frac{Mm}{l}\int_0^l \frac{dx}{(x+a)^2} = +\frac{Mm}{a(a+l)}. \quad \textit{Ans.}$$

**21.** Determine the amount of attraction in the last example if $P$ lies in the perpendicular bisector of the rod at the distance $a$ from it.

*Ans.* $\dfrac{2\,mM}{a\sqrt{4\,a^2+l^2}}$.

**22.** A vessel in the form of a right circular cone is filled with water. If $h$ is its height and $r$ the radius of base, what time will it require to empty itself through an orifice of area $a$ at the vertex?

**Solution.** Neglecting all hurtful resistances, it is known that the velocity of discharge through an orifice is that acquired by a body falling freely from a height equal to the depth of the water. If then $x$ denote depth of water,

$$v = \sqrt{2\,gx}.$$

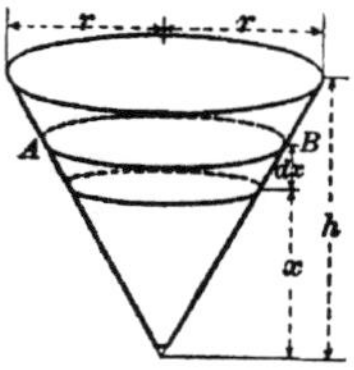


Denote by $dQ$ the volume of water discharged in time $dt$, and by $dx$ the corresponding fall of surface. The volume of water discharged through the orifice in a unit of time is

$$a\sqrt{2\,gx},$$

being measured as a right cylinder of area of base $a$ and altitude $v\left(=\sqrt{2\,gx}\right)$. Therefore in time $dt$,

$$(A) \qquad dQ = a\sqrt{2\,gx}\,dt.$$

Denoting by $S$ the area of surface of water when the depth is $x$, we have, from Geometry,

$$\frac{S}{\pi r^2} = \frac{x^2}{h^2}, \quad \text{or,} \quad S = \frac{\pi r^2 x^2}{h^2}.$$

* The two following examples indicate commonly employed "short methods," the detailed exposition followed in the preceding sections being omitted. The student should however supply this.

But the volume of water discharged in time $dt$ may also be considered as the volume of cylinder $AB$ of area of base $S$ and altitude $dx$; hence

(B) $$dQ = Sdx = \frac{\pi r^2 x^2 dx}{h^2}.$$

Equating (A) and (B) and solving for $dt$,

$$dt = \frac{\pi r^2 x^2 dx}{ah^2\sqrt{2gx}}.$$

Therefore $$t = \int_0^h \frac{\pi r^2 x^2 dx}{ah^2\sqrt{2gx}} = \frac{2\pi r^2\sqrt{h}}{5a\sqrt{2g}}. \quad \textit{Ans.}$$

**23.** A perfect gas in a cylinder expands against a piston head from the volume $v_0$ to the volume $v_1$, the temperature remaining constant. Find the work done.

**Solution.** Let $c$ = area of cross section of cylinder.

If $dv$ = increment of volume,

then $\frac{dv}{c}$ = distance piston head moves while volume takes on the increment $dv$.

By Boyle's Law, $$pv = k (= \text{const.}).$$

$$\therefore p = \frac{k}{v} = \text{pressure on piston head.}$$

Hence element of work done $= \frac{k}{v}\cdot\frac{dv}{c}$ (= pressure × dist.).

$$\therefore \text{total work done} = \int_{v_0}^{v_1}\frac{kdv}{vc} = \frac{k}{c}\int_{v_0}^{v_1}\frac{dv}{v} = \frac{k}{c}\log\frac{v_1}{v_0}.$$

# CHAPTER XXIX

## SUCCESSIVE AND PARTIAL INTEGRATION

**215. Successive integration.** Corresponding to *successive differentiation* in the Differential Calculus we have the inverse process of *successive integration* in the Integral Calculus. We shall illustrate by means of examples the details of this process, and show how problems arise where it is necessary to apply it.

ILLUSTRATIVE EXAMPLE 1. Given $\frac{d^3y}{dx^3} = 6x$, to find $y$.

**Solution.** We may write this

$$\frac{d\left(\frac{d^2y}{dx^2}\right)}{dx} = 6x,$$

or,
$$d\left(\frac{d^2y}{dx^2}\right) = 6xdx.$$

Integrating,
$$\frac{d^2y}{dx^2} = \int 6xdx,$$

or,
$$\frac{d^2y}{dx^2} = 3x^2 + c_1.$$

This may also be written
$$\frac{d\left(\frac{dy}{dx}\right)}{dx} = 3x^2 + c_1,$$

or,
$$d\left(\frac{dy}{dx}\right) = (3x^2 + c_1)\,dx.$$

Integrating again,
$$\frac{dy}{dx} = \int (3x^2 + c_1)\,dx, \text{ or,}$$

(A)
$$\frac{dy}{dx} = x^3 + c_1x + c_2.$$

Again
$$dy = (x^3 + c_1x + c_2)\,dx, \text{ and integrating,}$$

(B)
$$y = \frac{x^4}{4} + \frac{c_1x^2}{2} + c_2x + c_3. \quad Ans.$$

The result (A) is also written in the form

$$\frac{dy}{dx} = \int\int 6\,xdxdx \quad \left(\text{or} = \int\int 6\,xdx^2\right),$$

and is called a *double integral*, while (B) is written in the form

$$y = \int\int\int 6\,xdxdxdx \quad \left(\text{or} = \int\int\int 6\,xdx^3\right),$$

and is called a *triple integral.* In general, a *multiple integral* requires two or more successive integrations. As before, if there are no limits assigned, as in the above example, the integral is indefinite; if there are limits assigned for each successive integration, the integral is definite.

ILLUSTRATIVE EXAMPLE 2. Find the equation of a curve for every point of which the second derivative of the ordinate with respect to the abscissa equals 4.

Solution. Here $\frac{d^2y}{dx^2} = 4$. Integrating as in Illustrative Example 1,

(C) $$\frac{dy}{dx} = 4x + c_1.$$

(D) $$y = 2x^2 + c_1x + c_2. \quad \textit{Ans.}$$

This is the equation of a parabola with its axis parallel to $OY$ and extending upward. By giving the arbitrary constants of integration $c_1$ and $c_2$ all possible values, we obtain all such parabolas.

In order to determine $c_1$ and $c_2$, two more conditions are necessary. Suppose we say (a) that at the point where $x = 2$ the slope of the tangent to the parabola is zero; and (b) that the parabola passes through the point $(2, -1)$.

(a) Substituting $x = 2$ and $\frac{dy}{dx} = 0$ in (C)

gives $$0 = 8 + c_1.$$

Hence $$c_1 = -8,$$

and (D) becomes $$y = 2x^2 - 8x + c_2.$$

(b) The coördinates of $(2, -1)$ must satisfy this equation; therefore

$$-1 = 8 - 16 + c_2, \quad \text{or,} \quad c_2 = +7$$

Therefore the equation of the particular parabola which satisfies all three conditions is

$$y = 2x^2 - 8x + 7.$$

## EXAMPLES

1. Given $\frac{d^3y}{dx^3} = ax^2$, find $y$. *Ans.* $y = \frac{ax^5}{60} + \frac{c_1x^2}{2} + c_2x + c_3.$

2. Given $\frac{d^3y}{dx^3} = 0$, find $y$. $y = \frac{c_1x^2}{2} + c_2x + c_3.$

3. Given $d^3y = \frac{2\,dx^3}{x^3}$, find $y$. $y = \log x + \frac{c_1x^2}{2} + c_2x + c_3.$

4. Given $\frac{d^3\rho}{d\theta^3} = \sin\theta$, find $\rho$. $\rho = \cos\theta + \frac{c_1\theta^2}{2} + c_2\theta + c_3.$

5. Given $\frac{d^3s}{dt^3} = 3t^2 - \frac{1}{t^3}$, find $s$. $s = \frac{t^5}{20} - \frac{1}{2}\log t + \frac{c_1t^2}{2} + c_2t + c_3.$

6. Given $d^2\rho = \sin\phi\cos^2\phi\,d\phi^2$, find $\rho$. $\rho = \frac{\sin^3\phi}{9} - \frac{1}{3}\sin\phi + c_1\phi + c_2.$

7. Determine the equations of all curves having zero curvature.

HINT. $\frac{d^2y}{dx^2} = 0$, from (40), p. 157, since $K = 0$.

*Ans.* $y = c_1x + c_2$, a doubly infinite system of straight lines.

**8.** The acceleration of a moving point is constant and equal to $f$; find the distance (space) traversed.

HINT. $\frac{d^2s}{dt^2}=f.$ *Ans.* $s=\frac{ft^2}{2}+c_1t+c_2.$

**9.** Show in Ex. 8 that $c_1$ stands for the initial velocity and $c_2$ for the initial distance.

**10.** Find the equation of the curve at each point of which the second derivative of the ordinate with respect to the abscissa is four times the abscissa, and which passes through the origin and the point (2, 4). *Ans.* $3y=2x(x^2-1).$

**11.** Given $\frac{d^4y}{dx^4}=x\cos x$, find $y$. *Ans.* $y=x\cos x-4\sin x+\frac{c_1x^3}{6}+\frac{c_2x^2}{2}+c_3x+c_4.$

**12.** Given $\frac{d^3y}{dx^3}=\sin^2 x$; find $y$. *Ans.* $y=\frac{7\cos x}{9}-\frac{\cos^3 x}{27}+\frac{c_1x^2}{2}+c_2x+c_3.$

**216. Partial integration.** Corresponding to *partial differentiation* in the Differential Calculus we have the inverse process of *partial integration* in the Integral Calculus. As may be inferred from the connection, partial integration means that, having given a differential expression involving two or more independent variables, we integrate it, considering first *a single one only* as varying and all the rest constant. Then we integrate the result, considering another one as varying and the others constant, and so on. Such integrals are called *double*, *triple*, etc., according to the number of variables, and are called *multiple integrals.**

Thus the expression

$$u=\iint f(x,y)\,dy\,dx$$

indicates that we wish to find a function $u$ of $x$ and $y$ such that

$$\frac{\partial^2 u}{\partial x\partial y}=f(x,y).$$

In the solution of this problem the only new feature is that the constant of integration has a new form. We shall illustrate this by means of examples. Thus suppose we wish to find $u$, having given

$$\frac{\partial u}{\partial x}=2x+y+3.$$

Integrating this with respect to $x$, considering $y$ as constant, we have

$$u=x^2+xy+3x+\phi,$$

* The integrals of the same name in the last section are special cases of these, namely, when we integrate with respect to the same variable throughout.

where $\phi$ denotes the constant of integration. But since $y$ was regarded as constant during this integration, it may happen that $\phi$ involves $y$ in some way; in fact, $\phi$ will in general be a function of $y$. We shall then indicate this dependence of $\phi$ on $y$ by replacing $\phi$ by the symbol $\phi(y)$. Hence the most general form of $u$ is

$$u = x^2 + xy + 3x + \phi(y),$$

where $\phi(y)$ *denotes an arbitrary function of* $y$.

As another problem let us find

$$(A) \qquad u = \iint (x^2 + y^2)\, dy dx.$$

This means that we wish to find $u$, having given

$$\frac{\partial^2 u}{\partial x \partial y} = x^2 + y^2.$$

Integrating first with respect to $y$, regarding $x$ as constant, we get

$$\frac{\partial u}{\partial x} = x^2 y + \frac{y^3}{3} + \psi(x),$$

where $\psi(x)$ is an arbitrary function of $x$ and is to be regarded as the constant of integration.

Now integrating this result with respect to $x$, regarding $y$ as constant, we have

$$u = \frac{x^3 y}{3} + \frac{x y^3}{3} + \Psi(x) + \Phi(y),$$

where $\Phi(y)$ is the constant of integration, and

$$\Psi(x) = \int \psi(x)\, dx.$$

**217. Definite double integral. Geometric interpretation.** Let $f(x, y)$ be a continuous and single-valued function of $x$ and $y$. Geometrically,

$$(A) \qquad z = f(x, y)$$

is the equation of a surface, as *KL*. Take some area *S* in the *XY*-plane and construct upon *S* as a base the right cylinder whose elements are accordingly parallel to *OZ*. Let this cylinder intersect *KL* in the area *S'*, and now let us find the volume *V* of the solid bounded by *S*, *S'*, and the cylindrical surface. We proceed as follows:

At equal distances apart ($= \Delta x$) in the area *S* draw a set of lines parallel to *OY*, and then a second set parallel to *OX* at equal distances apart ($= \Delta y$). Through these lines pass planes parallel to *YOZ* and

*XOZ* respectively. Then within the areas $S$ and $S'$ we have a network of lines, as in the figure, that in $S$ being composed of rectangles, each of area $\Delta x \cdot \Delta y$. This construction divides the cylinder into a number of vertical columns, such as *MNPQ*, whose upper and lower bases are corresponding portions of the networks in $S'$ and $S$ respectively. As the upper bases of these columns are curvilinear, we of course cannot calculate the volume of the columns directly. Let us replace these columns by prisms whose upper bases are found thus: each column is cut through by a plane parallel to *XY* passed through that vertex of the upper base for which $x$ and $y$ have the least numerical values. Thus the column *MNPQ* is replaced by the right prism *MNPR*, the upper base being in a plane through $P$ parallel to the *XOY*-plane.

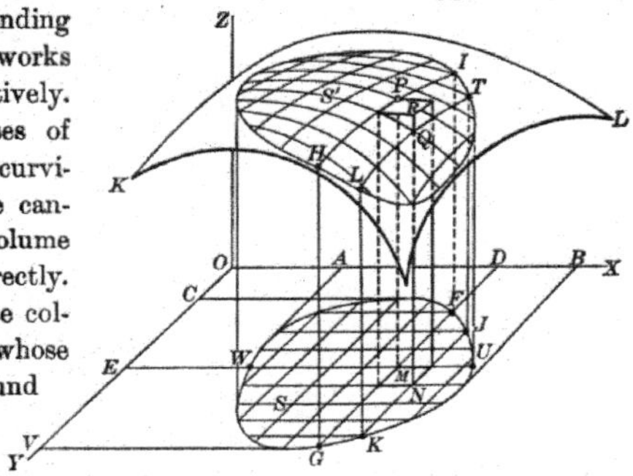


If the coördinates of $P$ are $(x, y, z)$, then $MP = z = f(x, y)$, and therefore

$$(B) \qquad \text{volume of } MNPR = f(x, y)\,\Delta y \cdot \Delta x.$$

Calculating the volume of each of the other prisms formed in the same way by replacing $x$ and $y$ in $(B)$ by corresponding values, and adding the results, we obtain a volume $V'$ approximately equal to $V$; that is,

$$(C) \qquad V' = \sum\sum f(x, y)\,\Delta y \cdot \Delta x;$$

where the double summation sign $\sum\sum$ indicates that there are *two variables* in the quantity to be summed up.

If now in the figure we increase the number of divisions of the network in $S$ indefinitely by letting $\Delta x$ and $\Delta y$ diminish indefinitely, and calculate in each case the double sum $(C)$, then obviously $V'$ will approach $V$ as a limit, and hence we have the fundamental result

$$(D) \qquad V = \lim_{\substack{\Delta y = 0 \\ \Delta x = 0}} \sum\sum f(x, y)\,\Delta y \cdot \Delta x.$$

The required volume may also be found as follows: Consider any one of the successive slices into which the solid is divided by the planes parallel to $YZ$; for example, the slice whose faces are $FIGH$ and $TLJK$. The thickness of this slice is $\Delta x$. Now the values of $z$ along the curve $HI$ are found by writing $x = OD$ in the equation $z = f(x, y)$; that is, along $HI$

$$z = f(OD, y).$$

Hence the area $FIGH = \int_{DF}^{DG} f(OD, y)\,dy.$

The volume of the slice under discussion is approximately equal to that of a prism whose base is $FIGH$ and altitude $\Delta x$; that is, equal to

$$\Delta x \cdot \text{area } FIGH = \Delta x \int_{DF}^{DG} f(OD, y)dy.$$

The required volume of the whole solid is evidently the limit of the sum of all prisms constructed in like manner, as $x (= OD)$ varies from $OA$ to $OB$; that is,

$$(E) \qquad V = \int_{OA}^{OB} dx \int_{DF}^{DG} f(x, y)\,dy.$$

Similarly, it may be shown that

$$(F) \qquad V = \int_{OC}^{OV} dy \int_{EW}^{EU} f(x, y)\,dx.$$

The integrals $(E)$ and $(F)$ are also written in the more compact form

$$\int_{OA}^{OB}\int_{DF}^{DG} f(x, y)dydx \quad \text{and} \quad \int_{OC}^{OV}\int_{EW}^{EU} f(x, y)\,dxdy.$$

In $(E)$ the limits $DF$ and $DG$ are functions of $x$, since they are found by solving the equation of the boundary curve of the base of the solid for $y$.

Similarly, in $(F)$ the limits $EW$ and $EU$ are functions of $y$. Now comparing $(D)$, $(E)$, and $(F)$ gives the result

$$(G) \quad V = \underset{\substack{\Delta y = 0 \\ \Delta x = 0}}{\text{limit}} \sum\sum f(x, y)\,\Delta y \cdot \Delta x = \int_{a_2}^{a_1}\int_{u_2}^{u_1} f(x, y)\,dydx$$

$$= \int_{b_2}^{b_1}\int_{v_2}^{v_1} f(x, y)\,dxdy,$$

where $v_1$ and $v_2$ are, in general, functions of $y$, and $u_1$ and $u_2$ functions of $x$, the second integral sign applying to the first differential and being calculated first.

Our result may be stated in the following form:

*The definite double integral*

$$\int_{a_2}^{a_1}\int_{u_2}^{u_1} f(x, y)\, dydx$$

*may be interpreted as that portion of the volume of a truncated right cylinder which is included between the plane XOY and the surface*

$$z = f(x, y),$$

*the base of the cylinder being the area bounded by the curves*

$$y = u_1, \quad y = u_2, \qquad x = a_1, \quad x = a_2.$$

Similarly for the second integral.

It is instructive to look upon the above process of finding the volume of the solid as follows:

Consider a column of infinitesimal base $dydx$ and altitude $z$ as an element of the volume. Summing up all such elements from $y = DF$ to $y = DG$, $x$ in the meanwhile being constant (say $= OD$), gives the volume of an indefinitely thin slice having $FGHI$ as one face. The volume of the whole solid is then found by summing up all such slices from $x = OA$ to $x = OB$.

In partial integration involving two variables the order of integration denotes that the limits on the inside integral sign correspond to the variable whose differential is written inside, the differentials of the variables and their corresponding limits on the integral signs being written in the reverse order.

**Illustrative Example 1.** Find the value of the definite double integral

$$\int_0^a \int_0^{\sqrt{a^2 - x^2}} (x + y)\, dydx.$$

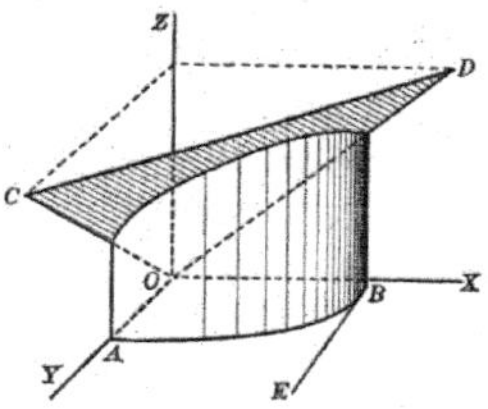


**Solution.** $\int_0^a \int_0^{\sqrt{a^2 - x^2}} (x + y)\, dydx$

$$= \int_0^a \left[\int_0^{\sqrt{a^2 - x^2}} (x + y)\, dy\right] dx$$

$$= \int_0^a \left[xy + \frac{y^2}{2}\right]_0^{\sqrt{a^2 - x^2}} dx$$

$$= \int_0^a \left(x\sqrt{a^2 - x^2} + \frac{a^2 - x^2}{2}\right) dx$$

$$= \frac{2a^3}{3}. \quad \textit{Ans.}$$

Interpreting this result geometrically, it means that we have found the volume of the solid of cylindrical shape standing on $OAB$ as base and bounded at the top by the surface (plane) $z = x + y$.

The attention of the student is now particularly called to the manner in which the limits do bound the base $OAB$, which corresponds to the area $S$ in the figure, p. 397. Our solid here stands on a base in the $XY$-plane bounded by

$$\left.\begin{array}{l} y = 0 \text{ (line } OB) \\ y = \sqrt{a^2 - x^2} \text{ (quadrant of circle } AB) \end{array}\right\} \text{from } y \text{ limits;}$$

$$\left.\begin{array}{l} x = 0 \text{ (line } OA) \\ x = a \text{ (line } BE) \end{array}\right\} \text{from } x \text{ limits.}$$

**218. Value of a definite double integral over a region $S$.** In the last section we represented the definite double integral as a volume. This does not necessarily mean that every definite double integral is a volume, for the physical interpretation of the result depends on the nature of the quantities represented by $x$, $y$, $z$. Thus, if $x$, $y$, $z$ are simply considered as the coördinates of a point in space, and nothing more, then the result is indeed a volume. In order to give the definite double integral in question an interpretation not necessarily involving the geometrical concept of volume, we observe at once that the variable $z$ does not occur explicitly in the integral, and therefore we may confine ourselves to the $XY$-plane. In fact, let us consider simply a region $S$ in the $XY$-plane, and a given function $f(x, y)$. Then, drawing a network as before, calculate the value of

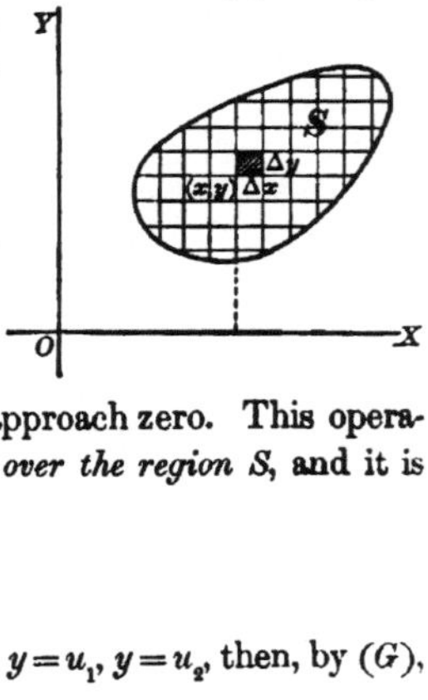


$$f(x, y)\,\Delta y\Delta x$$

*for each point* of the network*, and sum up, finding in this way

$$\sum\sum f(x, y)\,\Delta y\Delta x,$$

and finally pass to the limit as $\Delta x$ and $\Delta y$ approach zero. This operation we call *integrating the function $f(x, y)$ over the region $S$*, and it is denoted by the symbol

$$\iint_S f(x, y)\,dydx.$$

If $S$ is bounded by the curves $x = a_1$, $x = a_2$, $y = u_1$, $y = u_2$, then, by $(G)$,

$$\iint_S f(x, y)\,dydx = \int_{a_1}^{a_2}\int_{u_1}^{u_2} f(x, y)\,dydx.$$

* More generally, divide the interval on $OX$ into subintervals $\Delta x_1, \Delta x_2, \cdots, \Delta x_n$, and on $OY$ into $\Delta y_1, \Delta y_2, \cdots, \Delta y_m$. Draw the network, and in each rectangle $\Delta x_i \Delta y_k$ (not necessarily a corner) choose a point $x_i$, $y_k$. Then it is clear *intuitionally* that

$$\iint_S f(x, y)\,dxdy = \lim_{\substack{m=\infty \\ n=\infty}} \sum_{i=1}^{m}\sum_{k=1}^{n} f(x_i, y_k)\,\Delta x_i \Delta y_k.$$

We may state our result as follows:

**Theorem.** *To integrate a given function $f(x, y)$ over a given region $S$ in the $XOY$-plane means to calculate the value of*

$$\underset{\substack{\Delta x = 0\\ \Delta y = 0}}{\text{limit}} \sum\sum f(x, y)\Delta y \Delta x,$$

*as explained above, and the result is equal to the definite double integral*

$$\int_{a_1}^{a_2}\int_{u_1}^{u_2} f(x, y)\, dydx, \quad \text{or,} \quad \int_{b_1}^{b_2}\int_{v_1}^{v_2} f(x, y)\, dxdy,$$

*the limits being chosen so that the entire region $S$ is covered. This process is indicated briefly by*

$$\iint\limits_S f(x, y)\, dydx.$$

In what follows we shall show how the area of the region itself and its moment of inertia may be calculated in this way.

Before attempting to apply partial integration to practical problems it is best that the student should acquire by practice some facility in evaluating definite multiple integrals.

ILLUSTRATIVE EXAMPLE 1. Verify $\int_b^{2b}\int_0^a (a - y)\, x^2 dydx = \frac{7\, a^2 b^3}{6}.$

**Solution.** $\int_b^{2b}\int_0^a (a - y)\, x^2 dydx = \int_b^{2b}\left[ay - \frac{y^2}{2}\right]_0^a x^2 dx = \int_b^{2b}\frac{a^2}{2}x^2 dx = \frac{7\, a^2 b^3}{6}.$

ILLUSTRATIVE EXAMPLE 2. Verify $\int_0^a\int_{-\sqrt{a^2 - x^2}}^{\sqrt{a^2 - x^2}} x dydx = \frac{2\, a^3}{3}.$

**Solution.** $\int_0^a\int_{-\sqrt{a^2 - x^2}}^{\sqrt{a^2 - x^2}} x dydx = \int_0^a \left[xy\right]_{-\sqrt{a^2 - x^2}}^{\sqrt{a^2 - x^2}} dx$

$$= \int_0^a 2\, x\sqrt{a^2 - x^2}\, dx = \left[-\frac{2}{3}\left(a^2 - x^2\right)^{\frac{3}{2}}\right]_0^a = \frac{2}{3}a^3.$$

In partial integration involving three variables the order of integration is denoted in the same way as for two variables; that is, the order of the limits on the integral signs, reading from the inside to the left, is the same as the order of the corresponding variables whose differentials are read from the inside to the right.

ILLUSTRATIVE EXAMPLE 3. Verify $\int_2^3\int_1^2\int_2^5 xy^2 dzdydx = \frac{35}{2}.$

**Solution.** $\int_2^3\int_1^2\int_2^5 xy^2 dzdydx = \int_2^3\int_1^2\left[\int_2^5 xy^2 dz\right]dydx = \int_2^3\int_1^2\left[xy^2 z\right]_2^5 dydx$

$$= 3\int_2^3\int_1^2 xy^2 dydx = 3\int_2^3\left[\int_1^2 xy^2 dy\right]dx$$

$$= 3\int_2^3\left[\frac{xy^3}{3}\right]_1^2 dx = 7\int_2^3 xdx = \frac{35}{2}.$$

**EXAMPLES**

Verify the following:

1. $\int_0^a \int_0^b xy(x-y)\,dy\,dx = \frac{a^2b^2}{6}(a-b).$

2. $\int_{\frac{b}{2}}^{b} \int_0^{\frac{r}{b}} r\,d\theta\,dr = \frac{7b^2}{24}.$

3. $\int_0^a \int_{y-a}^{2y} xy\,dx\,dy = \frac{11a^4}{24}.$

4. $\int_b^a \int_0^b \int_a^{2a} x^2y^2z\,dz\,dy\,dx = \frac{a^2b^3}{6}(a^3-b^3).$

5. $\int_0^{2a} \int_0^{\sqrt{2ax-x^2}} \int_0^{\frac{x^2+y^2}{a}} dz\,dy\,dx = \frac{3\pi a^3}{4}.$

6. $\int_0^{\pi} \int_0^{a(1+\cos\theta)} r^2 \sin\theta\,dr\,d\theta = \frac{4a^3}{3}.$

7. $\int_0^b \int_t^{10t} \sqrt{st-t^2}\,ds\,dt = 6b^3.$

8. $\int_a^{2a} \int_v^{\frac{v^2}{a}} (w+2v)\,dw\,dv = \frac{143a^3}{30}.$

9. $\int_0^1 \int_0^{v^2} e^{\frac{w}{v}}\,dw\,dv = \frac{1}{2}.$

10. $\int_0^a \int_0^{\sqrt{x}} dy\,dx = \frac{2}{3}a^{\frac{3}{2}}.$

11. $\int_0^a \int_{2x}^{x^{\frac{3}{2}}} xy^2\,dy\,dx = \frac{2}{3}a^5\left(\frac{a^{\frac{3}{2}}}{13} - \frac{4}{5}\right)$

12. $\int_0^a \int_0^x \int_0^y x^3y^2z\,dz\,dy\,dx = \frac{a^9}{90}.$

13. $2a\int_0^a \int_0^{\sqrt{a^2-ax}} \frac{dz\,dx}{\sqrt{ax-x^2}} = 4a^2$

14. $\int_{-a}^{a} \int_0^{\frac{y^2}{a}} (x+y)\,dx\,dy = \frac{a^3}{5}.$

15. $\int_0^{\pi} \int_0^{a\cos\theta} \rho\sin\theta\,d\rho\,d\theta = \frac{a^2}{3}.$

16. $\int_0^{2a} \int_0^x (x^2+y^2)\,dy\,dx = \frac{16a^4}{3}.$

17. $\int_0^a \int_{\frac{x^2}{a}}^{x} \frac{x\,dx\,dy}{x^2+y^2} = \frac{a}{2}\log 2.$

18. $\int_0^{\frac{\pi}{2}} \int_{a\cos\theta}^{a} \rho^4\,d\rho\,d\theta = \left(\pi - \frac{16}{15}\right)\frac{a^5}{10}.$

19. $\int_b^a \int_\beta^\alpha r^2 \sin\theta\,d\theta\,dr = \frac{a^3-b^3}{3}(\cos\beta - \cos\alpha).$

20. $\int_0^1 \int_0^x \int_0^{x+y} e^{x+y+z}\,dz\,dy\,dx = \frac{e^4-3}{8} - \frac{3e^2}{4} + e.$

21. $\int_0^a \int_0^b \int_0^c (x^2+y^2+z^2)\,dz\,dy\,dx = \frac{abc}{3}(a^2+b^2+c^2).$

22. $\int_0^b \int_y^{10y} \sqrt{xy-y^2}\,dx\,dy = 6b^3.$

23. $\int_1^2 \int_0^z \int_0^{x\sqrt{3}} \frac{x\,dx\,dy\,dz}{x^2+y^2} = \frac{\pi}{2}.$

**219. Plane area as a definite double integral. Rectangular coördinates.** As a simple application of the theorem of the last section (p. 401), we shall now determine the area of the region $S$ itself in the $XOY$-plane by double integration.*

* Some of the examples that will be given in this and the following articles may be solved by means of a single integration by methods already explained. The only reason in such cases for using successive integration is to familiarize the student with a new method for solution which is sometimes the only one possible.

As before, draw lines parallel to $OX$ and $OY$ at distances $\Delta x$ and $\Delta y$ respectively. Now take any one of the rectangles formed in this way, then

$$\text{element of area} = \text{area of rectangle } PQ = \Delta y \cdot \Delta x,$$

the coördinates of $P$ being $(x, y)$.

Denoting by $A$ the entire area of region $S$, we have, using the notion of a double summation,

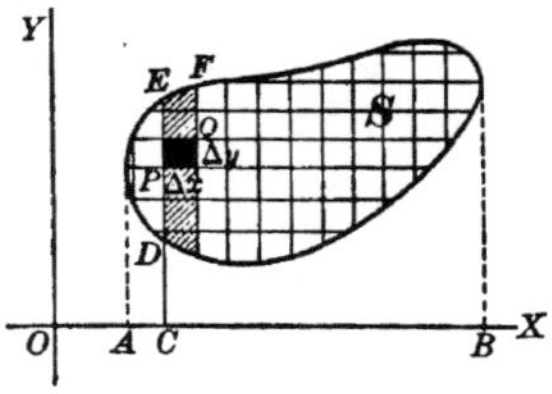


$$(A) \qquad A = \lim_{\substack{\Delta x = 0 \\ \Delta y = 0}} \sum\sum \Delta y \cdot \Delta x.$$

We calculate this by the theorem on p. 401, setting $f(x, y) = 1$, and get

$$(B) \qquad A = \int_{OA}^{OB} \int_{CD}^{CE} dy dx,$$

where $CD$ and $CE$ are, in general, functions of $x$, and $OA$ and $OB$ are constants giving the extreme values of $x$, all four of these quantities being determined from the equations of the curve or curves which bound the region $S$.

It is instructive to interpret this double integral geometrically by referring to our figure. When we integrate first with respect to $y$, keeping $x$ $(= OC)$ constant, we are summing up all the elements in a vertical strip (as $DF$). Then integrating the result with respect to $x$ means that we are summing up all such vertical strips included in the region, and this obviously gives the entire area of the region $S$.

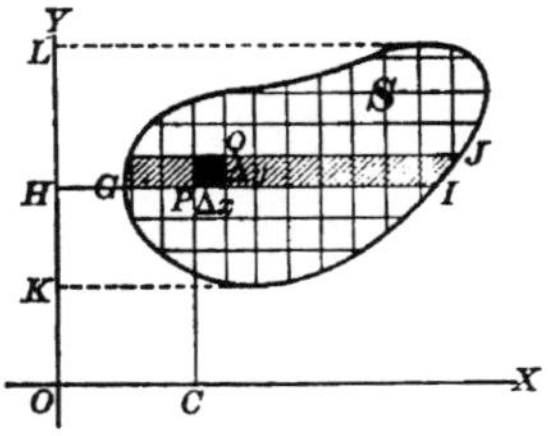


Or, if we change the order of integration, we have

$$(C) \qquad A = \int_{OK}^{OL} \int_{HG}^{HI} dx dy,$$

where $HG$ and $HI$ are, in general, functions of $y$, and $OK$ and $OL$ are constants giving the extreme values of $y$, all four of these quantities being determined from the equations of the curve or curves which bound the region $S$. Geometrically, this means that we now commence by summing up all the elements in a horizontal strip (as $GJ$), and then find the entire area by summing up all such strips within the region.

Corresponding to the two orders of summation (integration), the following notation and figures are sometimes used:

(*D*) $$A = \iint\limits_{S} dydx, \quad A = \iint\limits_{S} dxdy.$$

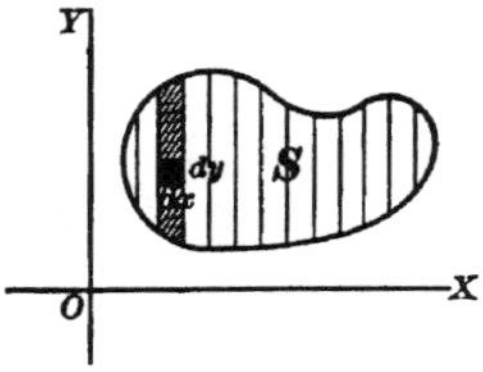


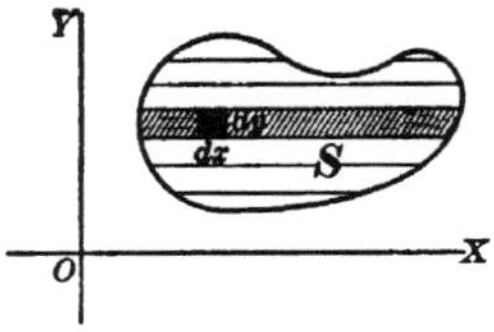


Referring to the result stated on p. 401, we may say:

*The area of any region is the value of the double integral of the function $f(x, y) = 1$ taken over that region.*

Or, also, from § 217, p. 396,

*The area equals numerically the volume of a right cylinder of unit height erected on the base S.*

Illustrative Example 1. Calculate the area of the circle $x^2 + y^2 = r^2$ by double integration.

**Solution.** Summing up first the elements in a vertical strip, we have from (*B*), p. 403,

$$A = \int_{OB}^{OA} \int_{MS}^{MR} dydx.$$

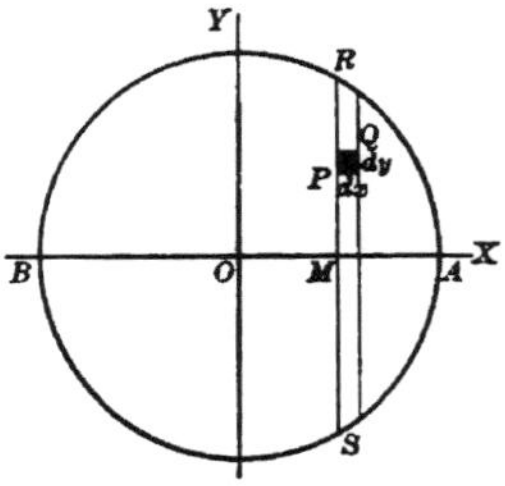


From the equation of the boundary curve (circle) we get

$$MR = \sqrt{r^2 - x^2}, \quad MS = -\sqrt{r^2 - x^2},$$
$$OB = -r, \quad OA = r.$$

Hence $$A = \int_{-r}^{+r} \int_{-\sqrt{r^2-x^2}}^{+\sqrt{r^2-x^2}} dydx$$
$$= 2\int_{-r}^{r} \sqrt{r^2 - x^2}dx = \pi r^2. \quad \textit{Ans.}$$

When the region whose area we wish to find is symmetrical with respect to one or both of the coördinate axes, it sometimes saves us labor to calculate the area of only a part at first. In the above example we may choose our limits so as to cover only one quadrant of the circle, and then multiply the result by 4. Thus

$$\frac{A}{4} = \int_0^r \int_0^{\sqrt{r^2-x^2}} dydx = \int_0^r \sqrt{r^2 - x^2}dx = \frac{\pi r^2}{4}.$$

$$\therefore A = \pi r^2. \quad \textit{Ans.}$$

ILLUSTRATIVE EXAMPLE 2. Calculate that portion of the area which lies above $OX$ bounded by the semicubical parabola $y^2 = x^3$ and the straight line $y = x$.

**Solution.** Summing up first the elements in a horizontal strip, we have from $(C)$, p. 403,

$$A = \int_0^{OD}\int_{AB}^{AC} dxdy.$$

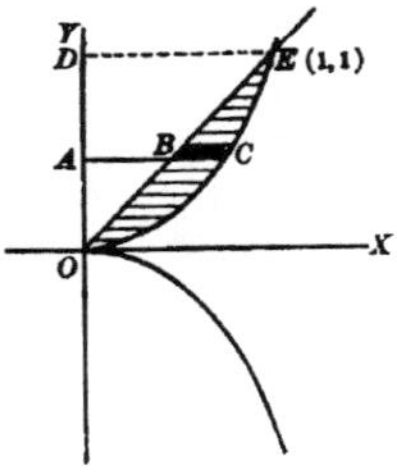


From the equation of the line, $AB = y$, and from the equation of the curve, $AC = y^{\frac{2}{3}}$, solving each one for $x$. To determine $OD$, solve the two equations simultaneously to find the point of intersection $E$. This gives the point $(1, 1)$; hence $OD = 1$. Therefore

$$A = \int_0^1\int_y^{y^{\frac{2}{3}}} dxdy = \int_0^1 (y^{\frac{2}{3}} - y)\,dy = \left[\frac{3y^{\frac{5}{3}}}{5} - \frac{y^2}{2}\right]_0^1$$
$$= \tfrac{3}{5} - \tfrac{1}{2} = \tfrac{1}{10}. \quad Ans.$$

## EXAMPLES

1. Find by double integration the area between the straight line and a parabola with its axis along $OX$, each of which joins the origin and the point $(a, b)$.
*Ans.* $\int_0^a\int_{\frac{bx}{a}}^{b\sqrt{\frac{x}{a}}} dydx = \frac{ab}{6}$.

2. Find by double integration the area between the two parabolas $3y^2 = 25x$ and $5x^2 = 9y$.
*Ans.* 5.

3. Required the area in the first quadrant which lies between the parabola $y^2 = ax$ and the circle $y^2 = 2ax - x^2$.
*Ans.* $\frac{\pi a^2}{4} - \frac{2a^2}{3}$.

4. Solve Problems 2 and 3 by first summing up all the elements in a horizontal strip, and then summing up all such strips.

*Ans.* Ex. 2, $\int_0^5\int_{\frac{3y^2}{25}}^{3\sqrt{\frac{y}{5}}} dxdy = 5$. Ex. 3, $\int_0^a\int_{a-\sqrt{a^2-y^2}}^{\frac{y^2}{a}} dxdy = \frac{\pi a^2}{4} - \frac{2a^2}{3}$.

5. Find by double integration the areas bounded by the following loci:

(a) $x^{\frac{1}{2}} + y^{\frac{1}{2}} = a^{\frac{1}{2}}$, $x + y = a$. *Ans.* $\frac{a^2}{3}$.

(b) $y^2 = 9 + x$, $y^2 = 9 - 3x$. 48.

(c) $y = \sin x$, $y = \cos x$, $x = 0$. $\sqrt{2} - 1$.

(d) $y = \frac{8a^3}{x^2 + 4a^2}$, $2y = x$, $x = 0$. $a^2(\pi - 1)$.

(e) $x^{\frac{2}{3}} + y^{\frac{2}{3}} = a^{\frac{2}{3}}$, $x + y = a$. $\frac{a^2}{2} - \frac{3\pi a^2}{32}$.

(f) $y^2 = x + 4$, $y^2 = 4 - 2x$.

(g) $y^2 = 4a^2 - x^2$, $y^2 = 4a^2 - 4ax$.

(h) $x^2 + y^2 = 25$, $27y^2 = 16x^3$.

(i) $4y^2 = x^3$, $y = x$.

(j) $y^2 = ax$, $y^2 = \frac{x^3}{2a - x}$.

(k) $x^2 - y^2 = 14$, $x^2 + y^2 = 36$.

**220. Plane area as a definite double integral. Polar coördinates.** Suppose the equations of the curve or curves which bound the region $S$ are given in polar coördinates. Then the region may be divided into checks bounded by radial lines drawn from the origin, and concentric circles drawn with centers at the origin. Let $PS = \Delta\rho$ and angle $POR = \Delta\theta$. Then arc $PR = \rho\Delta\theta$, and the area of the shaded check, considered as a rectangle, is $\rho\Delta\theta\cdot\Delta\rho$. The sum of the areas of all such checks in the region will be

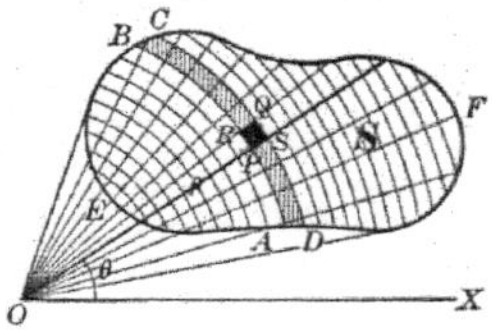


$$\sum \rho\Delta\rho\Delta\theta.$$

Since the required area is evidently the limit of this sum, we have the formula

$$(A) \qquad A = \iint_S \rho d\rho d\theta.$$

Here, again, the summation (integration) may be effected in two ways.

When we integrate first with respect to $\theta$, keeping $\rho$ constant, it means that we sum up all the elements (checks) in a segment of a circular ring (as $ABCD$), and next integrating with respect to $\rho$, that we sum up all such rings within the entire region. Our limits then appear as follows:

$$(B) \qquad A = \int_{OE}^{OF}\int_{\text{angle } XOA}^{\text{angle } XOB} \rho d\theta d\rho,$$

the angles $XOA$ and $XOB$ being, in general, functions of $\rho$, and $OE$ and $OF$ constants giving the extreme values of $\rho$.

Suppose we now reverse the order of integration. Integrating first with respect to $\rho$, keeping $\theta$ constant, means that we sum up all the elements (checks) in a wedge-shaped strip (as $GKLH$). Then integrating with respect to $\theta$, we sum up all such strips within the region $S$. Here

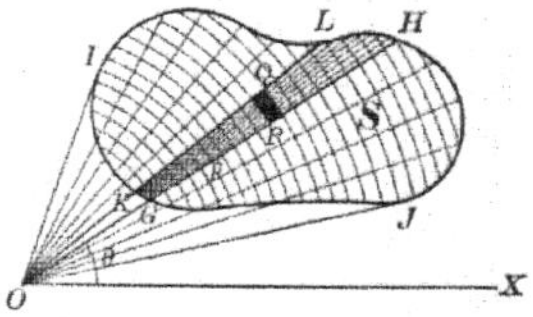


$$(C) \qquad A = \int_{\text{angle } XOJ}^{\text{angle } XOI}\int_{OG}^{OH} \rho d\rho d\theta,$$

$OH$ and $OG$ being, in general, functions of $\theta$, and the angles $XOJ$ and $XOI$ being constants giving the extreme values of $\theta$.

Corresponding to the two orders of summation (integration), the following notation and figures may be conveniently employed:

(D) $$A = \iint\limits_S \rho d\rho d\theta, \qquad A = \iint\limits_S \rho d\theta d\rho.$$

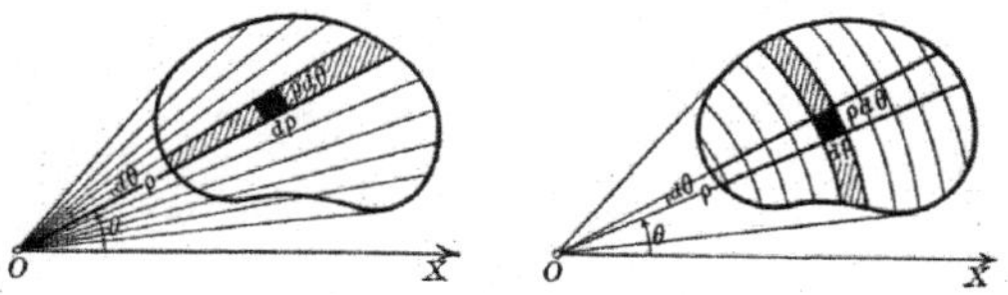

These are easily remembered if we think of the elements (checks) as being rectangles with dimensions $\rho d\theta$ and $d\rho$, and hence of area $\rho d\theta d\rho$.

ILLUSTRATIVE EXAMPLE 1. Find the area of the circle $\rho = 2r\cos\theta$ by double integration.

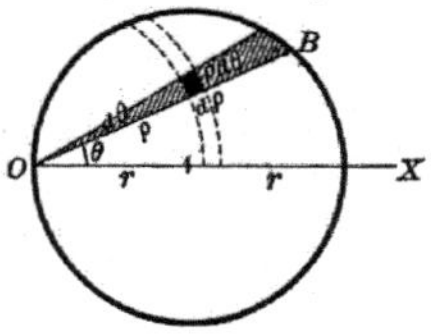

**Solution.** Summing up all the elements in a sector (as $OB$), the limits are 0 and $2r\cos\theta$; and summing up all such sectors, the limits are 0 and $\frac{\pi}{2}$ for the semicircle $OXB$. Substituting in (D),

$$\frac{A}{2} = \int_0^{\frac{\pi}{2}}\int_0^{2r\cos\theta} \rho d\rho d\theta = \frac{\pi r^2}{2}, \text{ or, } A = \pi r^2. \textit{ Ans.}$$

## EXAMPLES

1. In the above example find the area by integrating first with respect to $\theta$.

2. Find by double integration the entire areas in Examples 1–16, pp. 368, 369.

3. Find by double integration the area of that part of the parabola $\rho = a\sec^2\frac{\theta}{2}$ intercepted between the curve and its latus rectum.

$$\textit{Ans. } 2\int_0^{\frac{\pi}{2}}\int_0^{a\sec^2\frac{\theta}{2}} \rho d\rho d\theta = \frac{8a^2}{3}.$$

4. Find by double integration the area between the two circles $\rho = a\cos\theta$, $\rho = b\cos\theta$, $b > a$; integrating first with respect to $\rho$.

$$\textit{Ans. } 2\int_0^{\frac{\pi}{2}}\int_{a\cos\theta}^{b\cos\theta} \rho d\rho d\theta = \frac{\pi}{4}(b^2 - a^2).$$

5. Solve the last problem by first integrating with respect to $\theta$.

6. Find by double integration the area bounded by the following loci:

(a) $\rho = 6\sin\theta$, $\rho = 12\sin\theta$. *Ans.* $27\pi$.

(b) $\rho\cos\theta = 4$, $\rho = 8$. $\frac{64\pi}{3} - 16\sqrt{3}$.

(c) $\rho = a\sec^2\frac{\theta}{2}$, $\rho = 2a$. $2a^2\pi - \frac{8a^2}{3}$.

(d) $\rho = a(1 + \cos\theta)$, $\rho = 2a\cos\theta$. $\frac{\pi a^2}{2}$.

(e) $\rho\sin\theta = 5$, $\rho = 10$.

(f) $\rho = 8\cos\theta$, $\rho\cos\theta = 2$.

(g) $\rho = 2\cos\theta$, $\rho = 8\cos\theta$.

**221. Moment of area.** Consider an element of the area of the region $S$, as $PQ$, the coördinates of $P$ being $(x, y)$. Multiplying the area of this element $(=\Delta y \Delta x)$ by the distance of $P$ from the $Y$-axis $(=x)$, we get the product

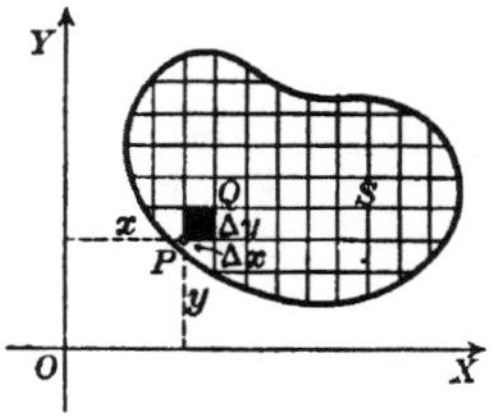


(A) $$x\Delta y\Delta x,$$

which is called the *moment* of the element $PQ$ with respect to the $Y$-axis. Form a similar product for every element within the region and add all such products by a double summation. Then the limit of this sum, namely,

(B) $$\lim_{\substack{\Delta y = 0 \\ \Delta x = 0}} \sum\sum x\Delta y\Delta x = \iint x\,dy\,dx,$$

*defines the **moment of area** of the region $S$ with respect to the $Y$-axis.*

Denoting this moment by $M_y$, we get

(C) $$M_y = \iint x\,dy\,dx,$$

the limits of integration being determined in the same way as for finding the area.

In the same manner, if we denote the moment of area with respect to the $X$-axis by $M_x$, we get

(D) $$M_x = \iint y\,dy\,dx,^*$$

the limits being the same as for (C).

**222. Center of area.** This is defined as the point $(\bar{x}, \bar{y})$ given by the formulas

(E) $$\bar{x} = \frac{M_y}{\text{area}}, \qquad \bar{y} = \frac{M_x}{\text{area}}, \text{ or,}$$

(F) $$\bar{x} = \frac{\iint x\,dy\,dx}{\iint dy\,dx}, \quad \bar{y} = \frac{\iint y\,dy\,dx}{\iint dy\,dx}.$$

From (E), $\quad$ $\text{area} \cdot \bar{x} = M_y$ and $\text{area} \cdot \bar{y} = M_x.$

Hence, *if we suppose the area of a region to be concentrated at $(\bar{x}, \bar{y})$, the moments of area with respect to the coördinate axes remain unchanged.*

* From the result on p. 401 we may say that $M_x$ is the value of the double integral of the function $f(x, y) = y$ taken over the region. Similarly, $M_y$ is the value when $f(x, y) = x$.

The center of area of a thin homogeneous plate or lamina is the same as its *center of mass* (or *center of gravity*).*

If a coördinate axis is an axis of symmetry of the area, it is evident that the corresponding coördinate of the center of area will be zero.

In polar coördinates $x = \rho\cos\theta$, $y = \rho\sin\theta$, and element of area $= \rho\Delta\rho\Delta\theta$ replaces $\Delta y\Delta x$. Hence formulas (*F*) become

$$(G)\qquad \bar{x} = \frac{\displaystyle\iint \rho^2\cos\theta\, d\rho d\theta}{\displaystyle\iint \rho\, d\rho d\theta}, \quad \bar{y} = \frac{\displaystyle\iint \rho^2\sin\theta\, d\rho d\theta}{\displaystyle\iint \rho\, d\rho d\theta},$$

the limits being the same throughout and determined (as before) in the same way as for finding the area.

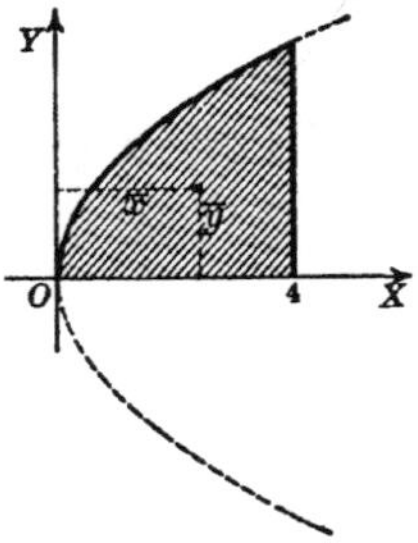


ILLUSTRATIVE EXAMPLE 1. Find the center of the area bounded by $y^2 = 4x$, $x = 4$, $y = 0$, and lying above $OX$.

**Solution.** By (*C*), p. 408, $M_y = \int_0^4\int_0^{2x^{\frac{1}{2}}} x\,dydx = \frac{128}{5}$.

By (*D*), p. 408, $M_x = \int_0^4\int_0^{2x^{\frac{1}{2}}} y\,dydx = 16$.

$$\text{Area} = \int_0^4\int_0^{2x^{\frac{1}{2}}} dydx = \frac{32}{3}.$$

Substituting in (*E*), p. 408,

$\bar{x} = \frac{128}{5} \div \frac{32}{3} = \frac{12}{5}$, and $\bar{y} = 16 \div \frac{32}{3} = \frac{3}{2}$. *Ans.*

**EXAMPLES**

1. Find the centers of the areas bounded by the following loci:

(a) The quadrant of a circle. *Ans.* $\bar{x} = \frac{4r}{3\pi} = \bar{y}$.

(b) The quadrant of an ellipse. $\bar{x} = \frac{4a}{3\pi}$, $\bar{y} = \frac{4b}{3\pi}$.

(c) $y = \sin x$, $y = 0$, from $x = 0$ to $x = \pi$. $\bar{x} = \frac{\pi}{2}$, $\bar{y} = \frac{\pi}{8}$.

(d) A quadrant of $x^{\frac{2}{3}} + y^{\frac{2}{3}} = a^{\frac{2}{3}}$. $\bar{x} = \frac{256a}{315\pi} = \bar{y}$.

(e) $y^2 = 4ax$, $x = h$. $\bar{x} = \frac{3}{5}h$, $\bar{y} = 0$.

(f) $y = 2x$, $y = 0$, $x = 3$. $\bar{x} = 2 = \bar{y}$.

(g) $y^2 = 8x$, $y = 0$, $y + x = 6$. $\bar{x} = 2.48$, $\bar{y} = 1.4$.

(h) $(2a - x)y^2 = x^3$, $x = 2a$. $\bar{x} = \frac{5a}{3}$, $\bar{y} = 0$.

(i) $y^2(a^2 - x^2) = a^4$, $x = 0$, $x = a$. $\bar{x} = \frac{2a}{\pi}$, $\bar{y} = 0$.

(j) $x^{\frac{1}{2}} + y^{\frac{1}{2}} = a^{\frac{1}{2}}$, $x = 0$, $y = 0$. $\bar{x} = \frac{a}{5} = \bar{y}$.

(k) Cycloid $x = a(\theta - \sin\theta)$, $y = a(1 - \cos\theta)$. $\bar{x} = a\pi$, $\bar{y} = \frac{5a}{6}$.

* If the plate is supported loosely on a horizontal axis through its center of gravity, there will be no tendency to rotate, whatever the position of the plate may be.

2. Find the centers of the areas bounded by the following curves:

(a) One loop of $\rho^2 = a^2 \cos 2\theta$. *Ans.* $\bar{x} = \frac{\pi\sqrt{2}a}{8}$, $\bar{y} = 0$.

(b) One loop of $\rho = a \sin 2\theta$. $\bar{x} = \frac{128a}{105\pi} = \bar{y}$.

(c) Cardioid $\rho = a(1 + \cos\theta)$. $\bar{x} = \frac{5a}{6}$, $\bar{y} = 0$.

(d) $\rho = 6 \sin\theta$, $\rho = 12 \sin\theta$.

(e) $\rho \cos\theta = 4$, $\rho = 8$.

(f) $\rho = 8 \cos\theta$, $\rho \cos\theta = 2$.

(g) $\rho = 2 \cos\theta$, $\rho = 8 \cos\theta$.

**223. Moment of inertia of plane areas.** Consider an element of the area of the region $S$, as $PQ$, the coördinates of $P$ being $(x, y)$. Multiplying the area of this element $(= \Delta y \Delta x)$ by the *square* of the distance $(= x)$ of $P$ from the $Y$-axis, we get the product

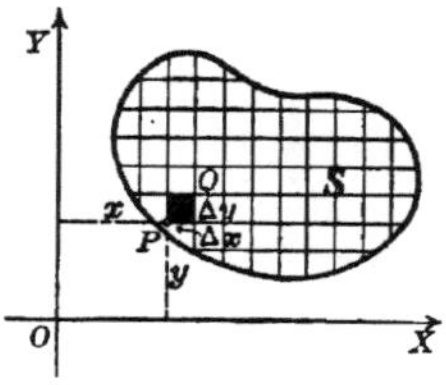


(A) $$x^2 \Delta y \Delta x,$$

which is called the *moment of inertia** of the element $PQ$ with respect to the $Y$-axis. Form a similar product for every element within the region and add all such products by a double summation. Then the limit of this sum, namely

(B) $$\lim_{\substack{\Delta x = 0 \\ \Delta y = 0}} \sum\sum x^2 \Delta y \Delta x = \iint x^2 dy dx,$$

*defines the moment of inertia of the area of S with respect to the Y-axis.*

Denoting this moment by $I_y$, we get

(C) $$I_y = \iint x^2 dy dx,$$

the limits of integration being determined in the same way as for finding the area.

In the same manner, if we denote the *moment of inertia of the area with respect to the X-axis by* $I_x$, we get

(D) $$I_x = \iint y^2 dy dx;$$

the limits being the same as for (C).

**224. Polar moment of inertia. Rectangular coördinates.** Consider an element of the area of region $S$, as $PQ$. If the coördinates of $P$ are $(x, y)$, the distance of $P$ from $O$ is $\sqrt{x^2 + y^2}$. Multiplying the area of

* Because the element of area is multiplied by the *square* of its distance from the $Y$-axis it is sometimes called the *second moment*, to conform with the definition of moment of area (§ 221, p. 408).

element ($=\Delta y\Delta x$) by the square of the distance of $P$ from the origin, we have the product

$$(x^2+y^2)\Delta y\Delta x,$$

which is called the *polar moment of inertia* of the element $PQ$ with respect to the origin. The value of the double sum

$$(E)\qquad \underset{\substack{\Delta y=0\\ \Delta x=0}}{\text{limit}}\sum\sum(x^2+y^2)\Delta y\Delta x=\iint(x^2+y^2)\,dydx$$

*defines the polar moment of inertia of the area within the region S with respect to the origin.*

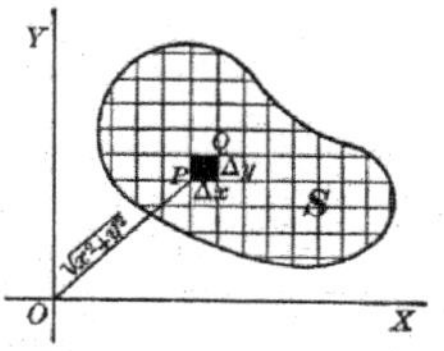


Denoting this moment of inertia by $I_0$, we get

$$(F)\qquad I_0=\iint(x^2+y^2)\,dydx,\text{*}$$

the limits of integration being determined in the same way as for finding the area.

From $(F)$,

$$I_0=\iint(x^2+y^2)\,dydx=\iint x^2dydx+\iint y^2dydx.$$

By comparison with $(C)$ and $(D)$ we get

$$(G)\qquad I_0=I_x+I_y,$$

and hence the

**Theorem.** *The polar moment of inertia of a plane area with respect to any point equals the sum of its moments of inertia with respect to any two perpendicular axes through that point.*

**225. Polar moment of inertia. Polar coördinates.** Since the element of area is now $\rho\Delta\rho\Delta\theta$, and $x^2+y^2=\rho^2$, we get, by substitution in $(E)$,

$$(H)\qquad I_0=\iint\rho^3d\rho d\theta,$$

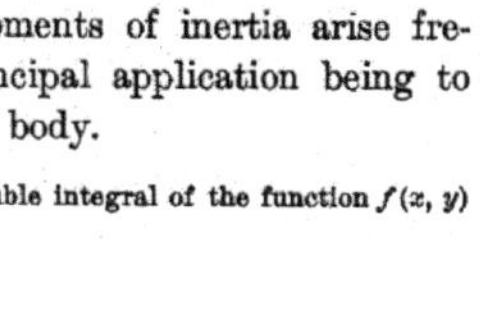


the limits of integration being the same as for finding the area.

Since the element of area ($=\Delta y\Delta x=\rho\Delta\rho\Delta\theta$) is essentially positive and $x^2$, $y^2$, $\rho^2$ are always positive, it follows that moment of inertia is never zero, but always a positive number. Moments of inertia arise frequently in engineering problems, the principal application being to the calculation of the energy of a rotating body.

* We may then say that $I_0$ is the value of the double integral of the function $f(x, y)=x^2+y^2$ over the area.

### EXAMPLES

1. Find $I_0$ over the area bounded by the lines $x = a$, $y = 0$, $y = \frac{b}{a}x$.

**Solution.** These lines bound a triangle $OAB$. Summing up all the elements in a vertical strip (as $PQ$), the $y$-limits are zero and $\frac{b}{a}x$ (found from the equation of the line $OB$). Summing up all such strips within the region (triangle), the $x$-limits are zero and $a$ (= $OA$). Hence, by ($F$),

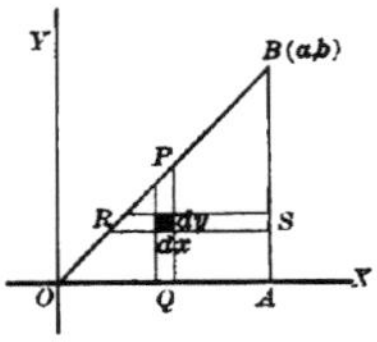


$$I_0 = \int_0^a \int_0^{\frac{b}{a}x} (x^2 + y^2)\, dydx = ab\left(\frac{a^2}{4} + \frac{b^2}{12}\right). \quad Ans.$$

If we suppose the triangle to be composed of horizontal strips (as $RS$),

$$I_0 = \int_0^b \int_{\frac{ay}{b}}^a (x^2 + y^2)\, dxdy = ab\left(\frac{a^2}{4} + \frac{b^2}{12}\right). \quad Ans.$$

2. Find $I_0$ over the rectangle bounded by the lines $x = a$, $y = b$, and the coördinate axes.

$$Ans. \int_0^a \int_0^b (x^2 + y^2)\, dydx = \frac{a^3b + ab^3}{3}.$$

3. Find $I_0$ over the right triangle formed by the coördinate axes and the line joining the points $(a, 0)$, $(0, b)$.

$$Ans. \int_0^a \int_0^{\frac{b(a-x)}{a}} (x^2 + y^2)\, dydx = \frac{ab(a^2 + b^2)}{12}.$$

4. Find $I_x$ for the region within the circle $x^2 + y^2 = r^2$. $\quad Ans. \frac{\pi r^4}{4}$.

5. Find $I_y$ for the ellipse $\frac{x^2}{a^2} + \frac{y^2}{b^2} = 1$. $\quad Ans. \frac{\pi a^3 b}{4}$.

6. Find $I_0$ over the region between the straight line and a parabola with axis along $OX$, each of which joins the origin and the point $(a, b)$.

$$Ans. \int_0^a \int_{\frac{bx}{a}}^{b\sqrt{\frac{x}{a}}} (x^2 + y^2)\, dydx = \frac{ab}{4}\left(\frac{a^2}{7} + \frac{b^2}{5}\right).$$

7. Find $I_0$ over the region bounded by the parabola $y^2 = 4\,ax$, the line $x + y - 3\,a = 0$, and $OX$.

$$Ans. \int_0^a \int_0^{2\sqrt{ax}} (x^2 + y^2)\, dydx + \int_a^{3a} \int_0^{3a-x} (x^2 + y^2)\, dydx = \frac{314\,a^4}{35},$$

$$\text{or, } \int_0^{2a} \int_{y^2}^{3a-y} (x^2 + y^2)\, dxdy = \frac{314\,a^4}{35}.$$

8. Find $I_0$ over the region bounded by the circle $\rho = 2\,r\cos\theta$.

**Solution.** Summing up the elements in the triangular-shaped strip $OP$, the $\rho$-limits are zero and $2\,r\cos\theta$ (found from the equation of the circle). Summing up all such strips, the $\theta$-limits are $-\frac{\pi}{2}$ and $\frac{\pi}{2}$. Hence, by ($H$),

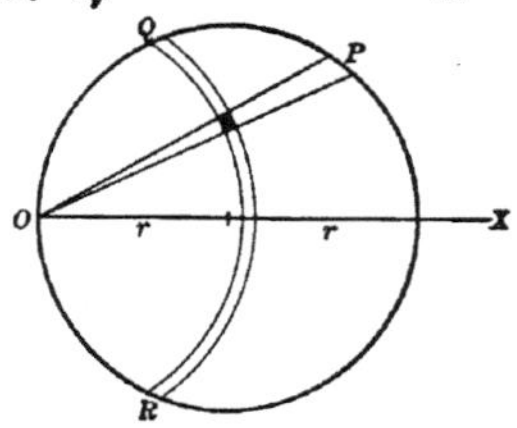


$$I_0 = \int_{-\frac{\pi}{2}}^{\frac{\pi}{2}} \int_0^{2r\cos\theta} \rho^3 d\rho d\theta = \frac{3\,\pi r^4}{2}. \quad Ans.$$

Summing up first the elements in a circular strip (as $QR$), we have

$$I_0 = \int_0^{2r} \int_{-\arccos\frac{\rho}{2r}}^{\arccos\frac{\rho}{2r}} \rho^3 d\theta d\rho = \frac{3\,\pi r^4}{2}. \quad Ans.$$

**9.** Find $I_0$ over the area bounded by the parabola $\rho = a\sec^2\frac{\theta}{2}$, its latus rectum, and the initial line $OX$.

$$Ans. \int_0^{\frac{\pi}{2}}\int_0^{a\sec^2\frac{\theta}{2}} \rho^3 d\rho d\theta = \frac{48\,a^4}{35}.$$

**10.** Find $I_0$ over the entire area of the cardioid $\rho = a(1-\cos\theta)$.

$$Ans. \ 2\int_0^{\pi}\int_0^{a(1-\cos\theta)} \rho^3 d\rho d\theta = \frac{35\,\pi a^4}{16}.$$

**11.** Find $I_0$ for the lemniscate $\rho^2 = a^2\cos 2\theta$. *Ans.* $\frac{\pi a^4}{8}$.

**12.** Find $I_x$ and $I_y$ for area bounded by $y^2 = 4\,ax$, $y = 0$, $x = x_1$.

$$Ans. \ I_x = \frac{2\,x_1 y_1^3}{15},\ I_y = \frac{2\,x_1^3 y_1}{7}.$$

**13.** Find the moment of inertia of the area of a right triangle with respect to the vertex of the right angle. $a$ and $b$ are the lengths of the perpendicular sides.

$$Ans. \ \frac{ab}{12}(a^2+b^2).$$

**14.** Find $I_y$ for the area bounded by $y^2 = 4\,ax$, $x + y = 3\,a$, $y = 0$. *Ans.* $I_y = \frac{46}{7}a^4$.

**15.** Find the moment of inertia of a rectangle whose sides are $2\,a$, $2\,b$, about an axis through its center parallel to the side $2\,b$; to the side $2\,a$. *Ans.* $\frac{a^3b}{3}$; $\frac{ab^3}{3}$.

**16.** Find $I_x$ for $x^{\frac{2}{3}} + y^{\frac{2}{3}} = a^{\frac{2}{3}}$. *Ans.* $\frac{21}{512}\pi a^4$.

**17.** Find $I_0$ over the area of one loop of $\rho = a\cos 2\,\theta$.

**226. General method for finding the areas of surfaces.** The method given in § 213 for finding the area of a surface applied only to surfaces of revolution. We shall now give a more general method. Let

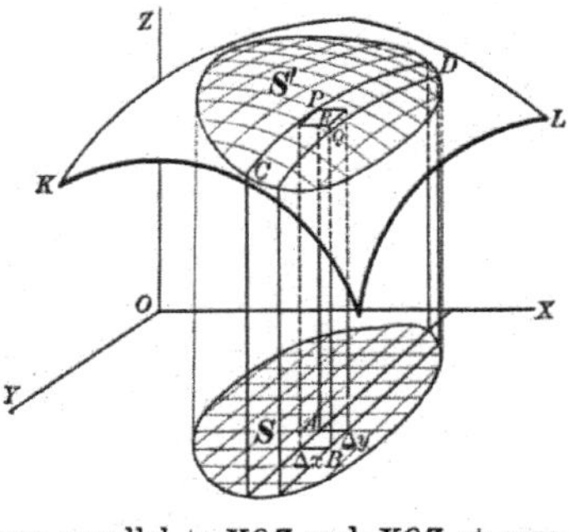


$$(A) \qquad z = f(x, y)$$

be the equation of the surface $KL$ in the figure, and suppose it is required to calculate the area of the region $S'$ lying on the surface.

Denote by $S$ the region on the $XOY$-plane, which is the orthogonal projection of $S'$ on that plane. Now pass planes parallel to $YOZ$ and $XOZ$ at common distances $\Delta x$ and $\Delta y$ respectively. As in § 217, these planes form truncated prisms (as $PB$) bounded at the top by a portion (as $PQ$) of the given surface whose projection on the $XOY$-plane is a rectangle of area $\Delta x\Delta y$ (as $AB$), which rectangle also forms the lower base of the prism, the coördinates of $P$ being $(x, y, z)$.

Now consider the plane tangent to the surface $KL$ at $P$. Evidently the same rectangle $AB$ is the projection on the $XOY$-plane of that portion of the tangent plane ($PR$) which is intercepted by the prism $PB$. Assuming $\gamma$ as the angle the tangent plane makes with the $XOY$-plane, we have

$$\text{area } AB = \text{area } PR \cdot \cos\gamma,$$

[The projection of a plane area upon a second plane is equal to the area of the portion projected multiplied by the cosine of the angle between the planes.]

or,
$$\Delta y\Delta x = \text{area } PR \cdot \cos\gamma.$$

But
$$\cos\gamma = \frac{1}{\left[1+\left(\frac{\partial z}{\partial x}\right)^2+\left(\frac{\partial z}{\partial y}\right)^2\right]^{\frac{1}{2}}};$$

[Cosine of angle between tangent plane, (72), p. 266, and $XOY$-plane found by method given in Solid Analytic Geometry.]

hence
$$\Delta y\Delta x = \frac{\text{area } PR}{\left[1+\left(\frac{\partial z}{\partial x}\right)^2+\left(\frac{\partial z}{\partial y}\right)^2\right]^{\frac{1}{2}}},$$

or,
$$\text{area } PR = \left[1+\left(\frac{\partial z}{\partial x}\right)^2+\left(\frac{\partial z}{\partial y}\right)^2\right]^{\frac{1}{2}}\Delta y\Delta x,$$

which we take as the element of area of the region $S'$. We then define the area of the region $S'$ as

$$\lim_{\substack{\Delta y = 0\\ \Delta x = 0}} \sum\sum\left[1+\left(\frac{\partial z}{\partial x}\right)^2+\left(\frac{\partial z}{\partial y}\right)^2\right]^{\frac{1}{2}}\Delta y\Delta x,$$

the summation extending over the region $S$, as in § 217. Denoting by $A$ the area of the region $S'$, we have

$$(B) \qquad A = \iint\limits_S \left[1+\left(\frac{\partial z}{\partial x}\right)^2+\left(\frac{\partial z}{\partial y}\right)^2\right]^{\frac{1}{2}} dy dx,$$

the limits of integration depending *on the projection on the XOY-plane of the region whose area we wish to calculate.* Thus for (**B**) we choose our limits from the boundary curve or curves of the region $S$ in the $XOY$-plane precisely as we have been doing in the previous four sections.

If it is more convenient to project the required area on the $XOZ$-plane, use the formula

$$(C)\qquad A = \iint_S \left[1 + \left(\frac{\partial y}{\partial x}\right)^2 + \left(\frac{\partial y}{\partial z}\right)^2\right]^{\frac{1}{2}} dz dx,$$

where the limits are found from the boundary of the region $S$, which is now the projection of the required area on the $XOZ$-plane.

Similarly, we may use

$$(D)\qquad A = \iint_S \left[1 + \left(\frac{\partial x}{\partial y}\right)^2 + \left(\frac{\partial x}{\partial z}\right)^2\right]^{\frac{1}{2}} dz dy,$$

the limits being found by projecting the required area on the $YOZ$-plane.

In some problems it is required to find the area of a portion of one surface intercepted by a second surface. In such cases the partial derivatives required for substitution in the formula should be found from the equation of the surface whose partial area is wanted.

Since the limits are found by projecting the required area on one of the coördinate planes, it should be remembered that

*To find the projection of the area required on the $XOY$-plane, eliminate $z$ between the equations of the surfaces whose intersections form the boundary of the area.*

*Similarly, we eliminate $y$ to find the projection on the $XOZ$-plane, and $x$ to find it on the $YOZ$-plane.*

This area of a surface gives a further illustration of *integration of a function over a given area.* Thus in **(B)**, p. 414, we integrate the function

$$\left[1+\left(\frac{\partial z}{\partial x}\right)^2+\left(\frac{\partial z}{\partial y}\right)^2\right]^{\frac{1}{2}}$$

over the projection on the $XOY$-plane of the required curvilinear surface.

ILLUSTRATIVE EXAMPLE 1. Find the area of the surface of sphere $x^2 + y^2 + z^2 = r^2$ by double integration.

**Solution.** Let $ABC$ in the figure be one eighth of the surface of the sphere. Here

$$\frac{\partial z}{\partial x} = -\frac{x}{z}, \qquad \frac{\partial z}{\partial y} = -\frac{y}{z},$$

and

$$1 + \left(\frac{\partial z}{\partial x}\right)^2 + \left(\frac{\partial z}{\partial y}\right)^2 = 1 + \frac{x^2}{z^2} + \frac{y^2}{z^2} = \frac{x^2 + y^2 + z^2}{z^2} = \frac{r^2}{r^2 - x^2 - y^2}.$$

The projection of the area required on the $XOY$-plane is $AOB$, a region bounded by $x = 0$, $(OB)$; $y = 0$, $(OA)$; $x^2 + y^2 = r^2$, $(BA)$.

Integrating first with respect to $y$, we sum up all the elements along a strip (as $DEFG$) which is projected on the $XOY$-plane in a strip also (as $MNFG$); that is, our $y$-limits are zero and $MF\left(=\sqrt{r^2 - x^2}\right)$. Then integrating with respect to $x$ sums up all such strips composing the surface $ABC$; that is, our $x$-limits are zero and $OA$ $(=r)$. Substituting in $(B)$, we get

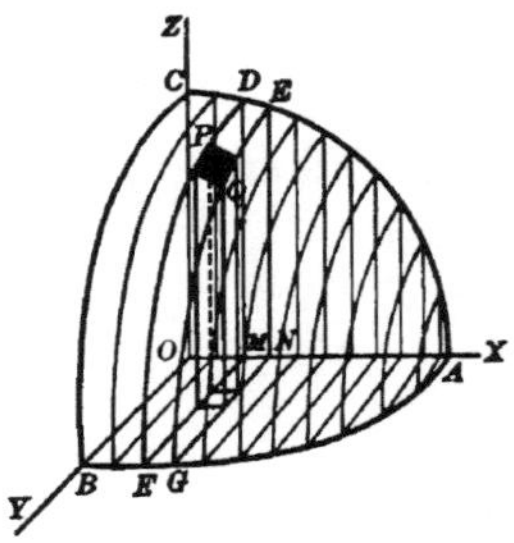


$$\frac{A}{8} = \int_0^r \int_0^{\sqrt{r^2 - x^2}} \frac{r\,dy\,dx}{\sqrt{r^2 - x^2 - y^2}}$$

$$= \frac{\pi r^2}{2},$$

or, $\quad A = 4\pi r^2$. *Ans.*

ILLUSTRATIVE EXAMPLE 2. The center of a sphere of radius $r$ is on the surface of a right cylinder, the radius of whose base is $\frac{r}{2}$. Find the surface of the cylinder intercepted by the sphere.

**Solution.** Taking the origin at the center of the sphere, an element of the cylinder for the $z$-axis, and a diameter of a right section of the cylinder for the $x$-axis, the equation of the sphere is $x^2 + y^2 + z^2 = r^2$, and of the cylinder $x^2 + y^2 = rx$. $ODAPB$ is evidently one fourth of the cylindrical surface required. Since this area projects into the semicircular arc $ODA$ on the $XOY$-plane, there is no region $S$ from which to determine our limits in this plane; hence we will project our area on, say, the $XOZ$-plane. Then the region $S$ over which we integrate is $OACB$, which is bounded by $z = 0$, $(OA)$; $x = 0$, $(OB)$; $z^2 + rx = r^2$, $(ACB)$; the last equation being found by eliminating $y$ between the equations of the two surfaces. Integrating first with respect to $z$ means that we sum up all the elements in a vertical strip (as $PD$), the $z$-limits being zero and $\sqrt{r^2 - rx}$. Then on integrating with respect to $x$ we sum up all such strips, the $x$-limits being zero and $r$.

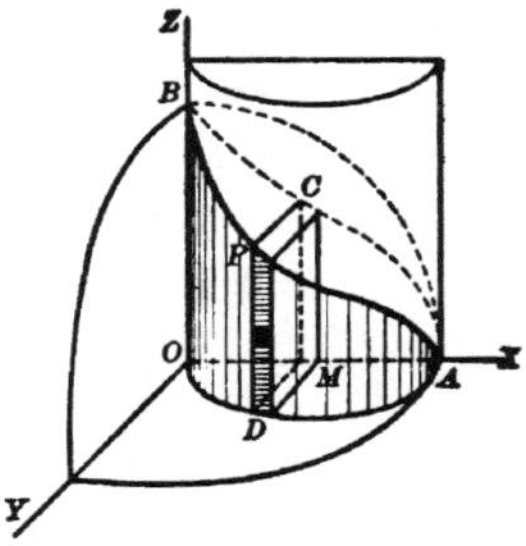


Since the required surface lies on the cylinder, the partial derivatives required for formula $(C)$, p. 415, must be found from the equation of the cylinder.

Hence
$$\frac{\partial y}{\partial x} = \frac{r - 2x}{2y}, \qquad \frac{\partial y}{\partial z} = 0,$$

Substituting in $(C)$, p. 415,

$$\frac{A}{4} = \int_0^r \int_0^{\sqrt{r^2 - rx}} \left[1 + \left(\frac{r - 2x}{2y}\right)^2\right]^{\frac{1}{2}} dz\,dx.$$

Substituting the value of $y$ in terms of $x$ from the equation of the cylinder,

$$A = 2r \int_0^r \int_0^{\sqrt{r^2 - rx}} \frac{dz\,dx}{\sqrt{rx - x^2}} = 2r \int_0^r \frac{\sqrt{r^2 - rx}}{\sqrt{rx - x^2}} dx = 2r \int_0^r \sqrt{\frac{r}{x}}\, dx = 4r^2.$$

### EXAMPLES

1. In the preceding example find the surface of the sphere intercepted by the cylinder.

$$Ans.\quad 4r\int_0^r\int_0^{\sqrt{rx-x^2}}\frac{dydx}{\sqrt{r^2-x^2-y^2}}=2(\pi-2)r^2.$$

2. The axes of two equal right circular cylinders, $r$ being the radius of their bases, intersect at right angles. Find the surface of one intercepted by the other.

HINT. Take $x^2+z^2=r^2$ and $x^2+y^2=r^2$ as equations of cylinders.

$$Ans.\quad 8r\int_0^r\int_0^{\sqrt{r^2-x^2}}\frac{dydx}{\sqrt{r^2-x^2}}=8r^2.$$

3. Find by integration the area of that portion of the surface of the sphere $x^2+y^2+z^2=100$ which lies between the parallel planes $x=-8$ and $x=6$.

4. Find the surface of the cylinder $x^2+y^2=r^2$ included between the plane $z=mx$ and the $XOY$-plane. *Ans.* $4r^2m$.

5. Find the surface of the cylinder $z^2+(x\cos\alpha+y\sin\alpha)^2=r^2$ which is situated in the positive compartment of coördinates.

HINT. The axis of this cylinder is the line $z=0$, $x\cos\alpha+y\sin\alpha=0$; and the radius of base is $r$.

$$Ans.\quad \frac{r^2}{\sin\alpha\cos\alpha}.$$

6. Find the area of that part of the plane $\frac{x}{a}+\frac{y}{b}+\frac{z}{c}=1$ which is intercepted by the coördinate planes. *Ans.* $\frac{1}{2}\sqrt{b^2c^2+c^2a^2+a^2b^2}$.

7. Find the area of the surface of the paraboloid $y^2+z^2=4ax$ intercepted by the parabolic cylinder $y^2=ax$ and the plane $x=3a$. *Ans.* $\frac{56}{3}\pi a^2$.

8. In the preceding example find the area of the surface of the cylinder intercepted by the paraboloid and plane.

$$Ans.\quad (13\sqrt{13}-1)\frac{a^2}{\sqrt{3}}.$$

9. Find the area of that portion of the surface of the cylinder $y^{\frac{2}{3}}+z^{\frac{2}{3}}=a^{\frac{2}{3}}$ bounded by a curve whose projection on the $XY$-plane is $x^{\frac{2}{3}}+y^{\frac{2}{3}}=a^{\frac{2}{3}}$. *Ans.* $\frac{12}{5}a^2$.

10. Find the area of that portion of the sphere $x^2+y^2+z^2=2ay$ cut out by one nappe of the cone $x^2+z^2=y^2$. *Ans.* $2\pi a^2$.

**227. Volumes found by triple integration.** In many cases the volume of a solid bounded by surfaces whose equations are given may be calculated by means of three successive integrations, the process being merely an extension of the methods employed in the preceding sections of this chapter.

Suppose the solid in question be divided by planes parallel to the coördinate planes into rectangular parallelepipeds having the dimensions $\Delta z$, $\Delta y$, $\Delta x$. The volume of one of these parallelepipeds is

$$\Delta z\cdot\Delta y\cdot\Delta x,$$

and we choose it as the element of volume.

Now sum up all such elements within the region $R$ bounded by the given surfaces by first summing up all the elements in a column parallel to one of the coördinate axes; then sum up all such columns in a slice parallel to one of the coördinate planes containing that axis, and finally sum up all such slices within the region in question. The volume $V$ of the solid will then be the limit of this triple sum as $\Delta z$, $\Delta y$, $\Delta x$ each approaches zero as a limit. That is,

$$V = \underset{\substack{\Delta x = 0\\ \Delta y = 0\\ \Delta z = 0}}{\text{limit}} \sum_{R}\sum\sum \Delta z \Delta y \Delta x,$$

the summations being extended over the entire region $R$ bounded by the given surfaces. Or, what amounts to the same thing,

$$V = \iiint_R dz dy dx,$$

the limits of integration depending on the equations of the bounding surfaces.

Thus, by extension of the principle of § 218, p. 401, we speak of volume as the result of integrating the function $f(x, y, z) = 1$ *throughout a given region.* More generally, many problems require the integration of a *variable* function of $x$, $y$, and $z$ throughout a given region, this being expressed by the notation

$$\iiint_R f(x, y, z)\, dz dy dx,$$

which is, of course, the limit of a triple sum analogous to the double sums we have already discussed. The method of evaluating this triple integral is precisely analogous to that already explained for double integrals in § 218, p. 401.

ILLUSTRATIVE EXAMPLE 1. Find the volume of that portion of the ellipsoid

$$\frac{x^2}{a^2} + \frac{y^2}{b^2} + \frac{z^2}{c^2} = 1$$

which lies in the first octant.

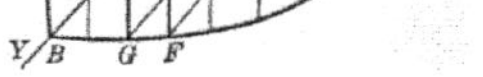


**Solution.** Let $O - ABC$ be that portion of the ellipsoid whose volume is required, the equations of the bounding surfaces being

(1) $\frac{x^2}{a^2} + \frac{y^2}{b^2} + \frac{z^2}{c^2} = 1$, $(ABC)$,

(2) $z = 0$, $(OAB)$,

(3) $y = 0$, $(OAC)$,

(4) $x = 0$, $(OBC)$.

$PQ$ is an element, being one of the rectangular parallelepipeds with dimensions $\Delta z$, $\Delta y$, $\Delta x$ into which the planes parallel to the coördinate planes have divided the region.

Integrating first with respect to $z$, we sum up all such elements in a column (as $RS$), the $z$-limits being zero [from (2)] and $TR = c\sqrt{1 - \frac{x^2}{a^2} - \frac{y^2}{b^2}}$ [from (1) by solving for $z$].

Integrating next with respect to $y$, we sum up all such columns in a slice (as $DEMNGF$), the $y$-limits being zero [from (3)] and $MG = b\sqrt{1 - \frac{x^2}{a^2}}$ [from equation of the curve $AGB$, namely $\frac{x^2}{a^2} + \frac{y^2}{b^2} = 1$, by solving for $y$].

Lastly, integrating with respect to $x$, we sum up all such slices within the entire region $O - ABC$, the $x$-limits being zero [from (4)] and $OA = a$.

Hence
$$V = \int_0^a \int_0^{b\sqrt{1-\frac{x^2}{a^2}}} \int_0^{c\sqrt{1-\frac{x^2}{a^2}-\frac{y^2}{b^2}}} dz dy dx$$
$$= c\int_0^a \int_0^{b\sqrt{1-\frac{x^2}{a^2}}} \left(1 - \frac{x^2}{a^2} - \frac{y^2}{b^2}\right)^{\frac{1}{2}} dy dx$$
$$= \frac{\pi cb}{4a^2}\int_0^a (a^2 - x^2)\, dx = \frac{\pi abc}{6}.$$

Therefore the volume of the entire ellipsoid is $\frac{4\pi abc}{3}$.

ILLUSTRATIVE EXAMPLE 2. Find the volume of the solid contained between the paraboloid of

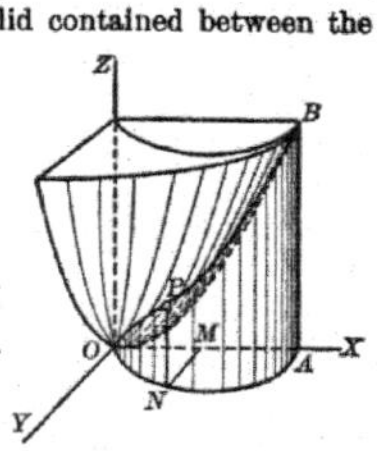


revolution $x^2 + y^2 = az$,

the cylinder $x^2 + y^2 = 2ax$,

and the plane $z = 0$.

**Solution.** The $z$-limits are zero and $NP$ $(= \frac{x^2 + y^2}{a}$, found by solving equation of paraboloid for $z$).

The $y$-limits are zero and $MN$ $(= \sqrt{2ax - x^2}$, found by solving equation of cylinder for $y$).

The $x$-limits are zero and $OA$ $(= 2a)$.

The above limits are for the solid $ONAB$, one half of the solid whose volume is required.

Hence
$$\frac{V}{2} = \int_0^{2a} \int_0^{\sqrt{2ax - x^2}} \int_0^{\frac{x^2+y^2}{a}} dz dy dx = \frac{3\pi a^3}{4}.$$

Therefore
$$V = \frac{3\pi a^3}{2}.$$

## EXAMPLES

1. Find the volume of the sphere $x^2 + y^2 + z^2 = r^2$ by triple integration.

*Ans.* $\frac{4\pi r^3}{3}$.

2. Find the volume of one of the wedges cut from the cylinder $x^2 + y^2 = r^2$ by the planes $z = 0$ and $z = mx$.

*Ans.* $2\int_0^r \int_0^{\sqrt{r^2 - x^2}} \int_0^{mx} dz dy dx = \frac{2r^3m}{3}$.

3. Find the volume of a right elliptic cylinder whose axis coincides with the $x$-axis and whose altitude $= 2a$, the equation of the base being $c^2y^2 + b^2z^2 = b^2c^2$.

*Ans.* $8\int_0^a \int_0^b \int_0^{\frac{c}{b}\sqrt{b^2 - y^2}} dz dy dx = 2\pi abc$.

**4.** Find the entire volume bounded by the surface $\left(\frac{x}{a}\right)^{\frac{1}{2}} + \left(\frac{y}{b}\right)^{\frac{1}{2}} + \left(\frac{z}{c}\right)^{\frac{1}{2}} = 1$, and the coördinate planes. *Ans.* $\frac{abc}{90}$.

**5.** Find the entire volume bounded by the surface $x^{\frac{2}{3}} + y^{\frac{2}{3}} + z^{\frac{2}{3}} = a^{\frac{2}{3}}$. *Ans.* $\frac{4\pi a^3}{35}$.

**6.** Find by triple integration the volume of the solid bounded by the planes $x = a$, $y = b$, $z = mx$ and the coördinate planes $XOY$ and $XOZ$. *Ans.* $\frac{1}{2}mba^2$.

**7.** The center of a sphere of radius $r$ is on the surface of a right circular cylinder the radius of whose basis is $\frac{r}{2}$. Find the volume of the portion of the cylinder intercepted by the sphere. *Ans.* $\frac{2}{3}(\pi - \frac{4}{3})r^3$.

**8.** Find the volume bounded by the hyperbolic paraboloid $cz = xy$, the $XOY$-plane, and the planes $x = a_1$, $x = a_2$, $y = b_1$, $y = b_2$. *Ans.* $\frac{(a_2^2 - a_1^2)(b_2^2 - b_1^2)}{4c}$.

**9.** Find the volume common to the two cylinders $x^2 + y^2 = r^2$ and $x^2 + z^2 = r^2$. *Ans.* $\frac{16 r^3}{3}$.

**10.** Find the volume of the tetrahedron bounded by the coördinate planes and the plane $\frac{x}{a} + \frac{y}{b} + \frac{z}{c} = 1$. *Ans.* $\frac{1}{6}abc$.

**11.** Find the volume bounded by the paraboloid $x^2 + y^2 - z = 1$ and the $XY$-plane. *Ans.* $\frac{\pi}{2}$.

**12.** Find the volume common to the paraboloid $y^2 + z^2 = 4ax$ and the cylinder $x^2 + y^2 = 2ax$. *Ans.* $2\pi a^3 + \frac{16}{3}a^3$.

**13.** Find the volume included between the paraboloid $y^2 + z^2 = 4ax$, the parabolic cylinder $y^2 = ax$, and the plane $x = 3a$. *Ans.* $(6\pi + 9\sqrt{3})a^3$.

**14.** Find the entire volume within the surface $x^{\frac{1}{3}} + y^{\frac{1}{3}} + z^{\frac{1}{3}} = a^{\frac{1}{3}}$.

**15.** Compute the volume of a cylindrical column standing on the area common to the two parabolas $x = y^2$, $y = x^2$ as base and cut off by the surface $z = 12 + y - x^2$.

**16.** Find the volume bounded by the surfaces $y^2 = x + 1$, $y^2 = -x + 1$, $z = -2$, $z = x + 4$.

**17.** Find the volume bounded by $z = x^2 + 2y^2$, $x + y = 1$, and the coördinate planes.

**18.** Given a right circular cylinder of altitude $a$ and radius of base $r$. Through a diameter of the upper base pass two planes which touch the lower base on opposite sides. Find the volume of the cylinder included between the two planes. *Ans.* $(\pi - \frac{4}{3})ar^2$.

# CHAPTER XXX

## ORDINARY DIFFERENTIAL EQUATIONS*

**228. Differential equations. Order and degree.** A differential equation is an equation involving derivatives or differentials. Differential equations have been frequently employed in this book, the following being examples:

(1) $x^2\frac{d^2y}{dx^2}+x\frac{dy}{dx}+y=0.$ Ex. 1, p. 151

(2) $\left(3a\frac{dy}{dx}+2\right)\left(\frac{d^2y}{dx^2}\right)^2=\left(a\frac{dy}{dx}+1\right)\frac{dy}{dx}.$

(3) $\tan\psi\frac{d\rho}{d\theta}=\rho.$ (*A*), p. 84

(4) $\frac{d^2y}{dx^2}=12(2x-1).$ Ex. 1, p. 101

(5) $dy=\frac{b^2x}{a^2y}dx.$ Ex. 2, p. 138

(6) $d\rho=-\frac{a^2\sin 2\theta}{\rho}d\theta.$ Ex. 3, p. 138

(7) $d^2y=(20x^3-12x)dx^2.$ Ex. 1, p. 139

(8) $x\frac{\partial u}{\partial x}+y\frac{\partial u}{\partial y}=5u.$ Ex. 7, p. 194

(9) $x\frac{\partial^2 u}{\partial x^2}+y\frac{\partial^2 u}{\partial x\partial y}=2\frac{\partial u}{\partial x}.$ Ex. 8, p. 204

(10) $\frac{\partial^3 u}{\partial x\partial y\partial z}=(1+3xyz+x^2y^2z^2)u.$ Ex. 7, p. 204

In fact, all of Chapter XI in the Differential Calculus and all of Chapter XXIII in the Integral Calculus treats of differential equations.

An *ordinary differential equation* involves only one independent variable. The first seven of the above examples are ordinary differential equations.

* A few types only of differential equations are treated in this chapter, namely, such as the student is likely to encounter in elementary work in Mechanics and Physics.

A ***partial differential equation*** involves more than one independent variable, as (8), (9), (10).

In this chapter we shall deal with ordinary differential equations only.

The *order* of a differential equation is that of the highest derivative (or differential) in it. Thus (3), (5), (6), (8) are of the *first order;* (1), (4), (7) are of the *second order;* and (2), (10) are of the *third order.*

The *degree* of a differential equation which is algebraic in the derivatives (or differentials) is the power of the highest derivative (or differential) in it when the equation is free from radicals and fractions. Thus all the above are examples of differential equations of the *first degree* except (2), which is of the *second degree.*

**229. Solutions of differential equations. Constants of integration.** A *solution* or *integral* of a differential equation is a relation between the variables involved by which the equation is identically satisfied. Thus

$$(A) \qquad y = c_1 \sin x$$

is a solution of the differential equation

$$(B) \qquad \frac{d^2y}{dx^2} + y = 0.$$

For, differentiating (*A*),

$$(C) \qquad \frac{d^2y}{dx^2} = -c_1 \sin x.$$

Now, if we substitute (*A*) and (*C*) in (*B*), we get

$$-c_1 \sin x + c_1 \sin x = 0,$$

showing that (*A*) satisfies (*B*) identically. Here $c_1$ is an arbitrary constant. In the same manner

$$(D) \qquad y = c_2 \cos x$$

may be shown to be a solution of (*B*) for any value of $c_2$. The relation

$$(E) \qquad y = c_1 \sin x + c_2 \cos x$$

is a still more general solution of (*B*). In fact, by giving particular values to $c_1$ and $c_2$ it is seen that the solution (*E*) includes the solutions (*A*) and (*D*).

The arbitrary constants $c_1$ and $c_2$ appearing in these solutions are called *constants of integration.* A solution such as (*E*), which contains a number of arbitrary essential constants equal to the order of the equation (in this case two), is called the ***general solution*** or the ***complete integral.**** Solutions obtained therefrom by giving particular values to the constants are called ***particular solutions*** or ***particular integrals.***

The solution of a differential equation is considered as having been effected when it has been reduced to an expression involving integrals, whether the actual integrations can be effected or not.

**230. Verification of the solutions of differential equations.** Before taking up the problem of solving differential equations it is best to further familiarize the student with what is meant by the solution of a differential equation by verifying a number of given solutions.

ILLUSTRATIVE EXAMPLE 1. Show that

$$y = c_1 x \cos \log x + c_2 x \sin \log x + x \log x \tag{1}$$

is a solution of the differential equation

$$x^2 \frac{d^2y}{dx^2} - x\frac{dy}{dx} + 2y = x \log x. \tag{2}$$

**Solution.** Differentiating (1), we get

$$\frac{dy}{dx} = (c_2 - c_1) \sin \log x + (c_2 + c_1) \cos \log x + \log x + 1. \tag{3}$$

$$\frac{d^2y}{dx^2} = -(c_2 + c_1)\frac{\sin \log x}{x} + (c_2 - c_1)\frac{\cos \log x}{x} + \frac{1}{x}. \tag{4}$$

Substituting (1), (3), (4) in (2), we find that the equation is identically satisfied.

## EXAMPLES

Verify the following solutions of the corresponding differential equations:

| *Differential equations* | *Solutions* |
|---|---|
| 1. $\left(\frac{dy}{dx}\right)^2 - \frac{dy}{dx} - x\frac{dy}{dx} + y = 0.$ | $y = cx + c - c^2.$ |
| 2. $y\left(\frac{dy}{dx}\right)^2 + 2x\frac{dy}{dx} - y = 0.$ | $y^2 = 2cx + c^2.$ |
| 3. $xy\left[1 - \left(\frac{dy}{dx}\right)^2\right] = (x^2 - y^2 - a^2)\frac{dy}{dx}.$ | $y^2 - cx^2 + \frac{a^2c}{1+c} = 0.$ |
| 4. $\frac{d^3y}{dx^3} + \frac{3}{x}\frac{d^2y}{dx^2} = 0.$ | $y = c_1 x + \frac{c_2}{x} + c_3.$ |
| 5. $\frac{d^2y}{dx^2} - 2k\frac{dy}{dx} + k^2y = e^x.$ | $y = (c_1 + c_2 x)e^{kx} + \frac{e^x}{(k-1)^2}.$ |

* It is shown in works on Differential Equations that the general solution has $n$ arbitrary constants when the differential equation is of the $n$th order.

| *Differential equations* | *Solutions* |
|---|---|
| 6. $\frac{d^4y}{dx^4}-4\frac{d^3y}{dx^3}+6\frac{d^2y}{dx^2}-4\frac{dy}{dx}+y=0.$ | $y=(c_1+c_2x+c_3x^2+c_4x^3)e^x.$ |
| 7. $x^2\frac{d^2y}{dx^2}-5x\frac{dy}{dx}+5y=\frac{1}{x}.$ | $4y=\frac{1}{3x}+c_1x^5+c_2x.$ |
| 8. $x\frac{dy}{dx}-y+x\sqrt{x^2-y^2}=0.$ | $\text{arc sin}\frac{y}{x}=c-x.$ |
| 9. $\frac{dy}{dx}+y\cos x=\frac{\sin 2x}{2}.$ | $y=\sin x-1+ce^{-\sin x}.$ |
| 10. $(1-x^2)\frac{d^2y}{dx^2}-x\frac{dy}{dx}-a^2y=0.$ | $y=c_1e^{a\,\text{arc sin}\,x}+c_2e^{-a\,\text{arc sin}\,x}.$ |
| 11. $\frac{d^2y}{dx^2}+\frac{2}{x}\frac{dy}{dx}=0.$ | $y=\frac{c_1}{x}+c_2.$ |

**231. Differential equations of the first order and of the first degree.** Such an equation may be brought into the form $Mdx+Ndy=0$, in which $M$ and $N$ are functions of $x$ and $y$. Differential equations coming under this head may be divided into the following types:

**Type I. Variables separable.** When the terms of a differential equation can be so arranged that it takes on the form

$$(A)\qquad f(x)\,dx+F(y)\,dy=0,$$

where $f(x)$ is a function of $x$ alone and $F(y)$ is a function of $y$ alone, the process is called *separation of the variables*, and the solution is obtained by direct integration. Thus integrating $(A)$, we get the general solution

$$(B)\qquad \int f(x)\,dx+\int F(y)\,dy=c,$$

where $c$ is an arbitrary constant.

Equations which are not given in the simple form $(A)$ may often be brought into that form by means of the following **rule for separating the variables.**

FIRST STEP. *Clear of fractions, and if the equation involves derivatives, multiply through by the differential of the independent variable.*

SECOND STEP. *Collect all the terms containing the same differential into a single term. If, then, the equation takes on the form*

$$XYdx+X'Y'dy=0,$$

*where $X$, $X'$ are functions of $x$ alone, and $Y$, $Y'$ are functions of $y$ alone, it may be brought to the form $(A)$ by dividing through by $X'Y$.*

THIRD STEP. *Integrate each part separately, as in $(B)$.*

ILLUSTRATIVE EXAMPLE 1. Solve the equation

$$\frac{dy}{dx} = \frac{1+y^2}{(1+x^2)xy}.$$

**Solution.** *First step.* $(1+x^2)xydy = (1+y^2)dx.$

*Second step.* $(1+y^2)dx - x(1+x^2)ydy = 0.$

To separate the variables we now divide by $x(1+x^2)(1+y^2)$, giving

$$\frac{dx}{x(1+x^2)} - \frac{ydy}{1+y^2} = 0.$$

*Third step.* $$\int\frac{dx}{x(1+x^2)} - \int\frac{ydy}{1+y^2} = C,$$

$$\int\frac{dx}{x} - \int\frac{xdx}{1+x^2} - \int\frac{ydy}{1+y^2} = C,$$ p. 329

$$\log x - \frac{1}{2}\log(1+x^2) - \frac{1}{2}\log(1+y^2) = C,$$

$$\log(1+x^2)(1+y^2) = 2\log x - 2C.$$

This result may be written in more compact form if we replace $-2C$ by $\log c$, i.e. we simply give a new form to the arbitrary constant. Our solution then becomes

$$\log(1+x^2)(1+y^2) = \log x^2 + \log c,$$

$$\log(1+x^2)(1+y^2) = \log cx^2,$$

$$(1+x^2)(1+y^2) = cx^2. \quad \textit{Ans.}$$

ILLUSTRATIVE EXAMPLE 2. Solve the equation

$$a\left(x\frac{dy}{dx} + 2y\right) = xy\frac{dy}{dx}.$$

**Solution.** *First step.* $axdy + 2aydx = xydy.$

*Second step.* $2aydx + x(a-y)dy = 0.$

To separate the variables we divide by $xy$,

$$\frac{2adx}{x} + \frac{(a-y)dy}{y} = 0.$$

*Third step.* $$2a\int\frac{dx}{x} + a\int\frac{dy}{y} - \int dy = C,$$

$$2a\log x + a\log y - y = C,$$

$$a\log x^2y = C + y,$$

$$\log_e x^2y = \frac{C}{a} + \frac{y}{a}.$$

By passing from logarithms to exponentials this result may be written in the form

$$x^2y = e^{\frac{C}{a}+\frac{y}{a}},$$

or,

$$x^2y = e^{\frac{C}{a}}\cdot e^{\frac{y}{a}}.$$

Denoting the constant $e^{\frac{C}{a}}$ by $c$, we get our solution in the form

$$x^2y = ce^{\frac{y}{a}}. \quad \textit{Ans.}$$

**EXAMPLES**

| *Differential equations* | *Solutions* |
|---|---|
| 1. $ydx - xdy = 0.$ | $y = cx.$ |
| 2. $(1+y)\,dx - (1-x)\,dy = 0.$ | $(1+y)(1-x) = c.$ |
| 3. $(1+x)\,ydx + (1-y)\,xdy = 0.$ | $\log xy + x - y = c.$ |
| 4. $(x^2 - a^2)\,dy - ydx = 0.$ | $y^{2a} = c\dfrac{x-a}{x+a}.$ |
| 5. $(x^2 - yx^2)\dfrac{dy}{dx} + y^2 + xy^2 = 0.$ | $\dfrac{x+y}{xy} + \log\dfrac{y}{x} = c.$ |
| 6. $u^2dv + (v-a)\,du = 0.$ | $v - a = ce^{\frac{1}{u}}.$ |
| 7. $\dfrac{du}{dv} = \dfrac{1+u^2}{1+v^2}.$ | $u = \dfrac{v+c}{1-cv}.$ |
| 8. $(1+s^2)\,dt - t^{\frac{1}{2}}ds = 0.$ | $2t^{\frac{1}{2}} - \text{arc}\tan s = c.$ |
| 9. $d\rho + \rho\tan\theta d\theta = 0.$ | $\rho = c\cos\theta.$ |
| 10. $\sin\theta\cos\phi d\theta - \cos\theta\sin\phi d\phi = 0.$ | $\cos\phi = c\cos\theta.$ |
| 11. $\sec^2\theta\tan\phi d\theta + \sec^2\phi\tan\theta d\phi = 0.$ | $\tan\theta\tan\phi = c.$ |
| 12. $\sec^2\theta\tan\phi d\phi + \sec^2\phi\tan\theta d\theta = 0.$ | $\sin^2\theta + \sin^2\phi = c.$ |
| 13. $xydx - (a+x)(b+y)\,dy = 0.$ | $x - y = c + \log(a+x)^a y^b.$ |
| 14. $(1+x^2)\,dy - \sqrt{1-y^2}\,dx = 0.$ | $\text{arc}\sin y - \text{arc}\tan x = c.$ |
| 15. $\sqrt{1-x^2}\,dy + \sqrt{1-y^2}\,dx = 0.$ | $y\sqrt{1-x^2} + x\sqrt{1-y^2} = c.$ |
| 16. $3e^x\tan ydx + (1-e^x)\sec^2 ydy = 0.$ | $\tan y = c(1-e^x)^3.$ |
| 17. $2x^2ydy = (1+x^2)\,dx.$ | $y^2 = -\dfrac{1}{x} + x + c.$ |
| 18. $(x - y^2x)\,dx + (y - x^2y)\,dy = 0.$ | $x^2 + y^2 = x^2y^2 + c.$ |
| 19. $(x^2y + x)\,dy + (xy^2 - y)\,dx = 0.$ | $xy + \log\dfrac{y}{x} = c.$ |

**Type II. Homogeneous equations.** The differential equation

$$Mdx + Ndy = 0$$

is said to be homogeneous when $M$ and $N$ are homogeneous functions of $x$ and $y$ of the same degree.* Such differential equations may be solved by making the substitution

$$y = vx.$$

This will give a differential equation in $v$ and $x$ in which the variables are separable, and hence we may follow the rule on p. 424.

* A function of $x$ and $y$ is said to be *homogeneous* in the variables if the result of replacing $x$ and $y$ by $\lambda x$ and $\lambda y$ ($\lambda$ being arbitrary) reduces to the original function multiplied by some power of $\lambda$. This power of $\lambda$ is called the *degree* of the original function.

ILLUSTRATIVE EXAMPLE 1. Solve the equation

$$y^2 + x^2\frac{dy}{dx} = xy\frac{dy}{dx}.$$

**Solution.** $$y^2dx + (x^2 - xy)\,dy = 0.$$

Since this is a homogeneous differential equation, we transform it by means of the substitution $y = vx$. Hence $dy = vdx + xdv$,
and our equation becomes

$$v^2x^2dx + (x^2 - vx^2)(vdx + xdv) = 0,$$
$$x^2vdx + x^3(1 - v)\,dv = 0.$$

To separate the variables divide by $vx^3$. This gives

$$\frac{dx}{x} + \frac{(1-v)\,dv}{v} = 0,$$
$$\int\frac{dx}{x} + \int\frac{dv}{v} - \int dv = C,$$
$$\log x + \log v - v = C,$$
$$\log_e vx = C + v,$$
$$vx = e^{C+v} = e^C \cdot e^v,$$
$$vx = ce^v.$$

But $v = \frac{y}{x}$. Hence the solution is $y = ce^{\frac{y}{x}}$. *Ans.*

**EXAMPLES**

| *Differential equations* | *Solutions* |
|---|---|
| 1. $(x + y)\,dx + xdy = 0.$ | $x^2 + 2xy = c.$ |
| 2. $(x + y)\,dx + (y - x)\,dy = 0.$ | $\log(x^2 + y^2)^{\frac{1}{2}} - \arctan\frac{y}{x} = c.$ |
| 3. $xdy - ydx = \sqrt{x^2 + y^2}\,dx.$ | $1 + 2cy - c^2x^2 = 0.$ |
| 4. $(8y + 10x)\,dx + (5y + 7x)\,dy = 0.$ | $(x + y)^2(2x + y)^3 = c.$ |
| 5. $xy^2dy = (x^3 + y^3)\,dx.$ | $y^3 = 3x^3\log cx.$ |
| 6. $(x^2 - 2y^2)\,dx + 2xydy = 0.$ | $y^2 = -x^2\log cx.$ |
| 7. $(x^2 + y^2)\,dx = 2xydy.$ | $y^2 = x^2 + cx.$ |
| 8. $(2\sqrt{st} - s)\,dt + tds = 0.$ | $te^{\sqrt{\frac{s}{t}}} = c.$ |
| 9. $(t - s)\,dt + tds = 0.$ | $te^{\frac{s}{t}} = c.$ |
| 10. $x\cos\frac{y}{x}\cdot\frac{dy}{dx} = y\cos\frac{y}{x} - x.$ | $xe^{\sin\frac{y}{x}} = c.$ |
| 11. $x\cos\frac{y}{x}(ydx + xdy) = y\sin\frac{y}{x}(xdy - ydx).$ | $xy\cos\frac{y}{x} = c.$ |

**Type III. Linear equations.** A differential equation is said to be *linear* if the equation is of the first degree in the dependent variable (usually $y$) and its derivatives (or differentials). The linear differential equation of the first order is of the form

$$(A) \qquad \frac{dy}{dx} + Py = Q,$$

where $P$, $Q$ are functions of $x$ alone, or constants.

To integrate $(A)$, let

$$(B) \qquad y = uz,$$

where $z$ is a new variable and $u$ is a function of $x$ to be determined. Differentiating $(B)$,

$$(C) \qquad \frac{dy}{dx} = u\frac{dz}{dx} + z\frac{du}{dx}.$$

Substituting $(C)$ and $(B)$ in $(A)$, we get

$$u\frac{dz}{dx} + z\frac{du}{dx} + Puz = Q, \text{ or,}$$

$$(D) \qquad u\frac{dz}{dx} + \left(\frac{du}{dx} + Pu\right)z = Q.$$

Now let us determine, if possible, the function $u$ such that the term in $z$ shall drop out. This means that the coefficient of $z$ must vanish; that is,

$$\frac{du}{dx} + Pu = 0.$$

Then $$\frac{du}{u} = -Pdx,$$

and $$\log_e u = -\int Pdx + C, \text{ giving}$$

$$(E) \qquad u = c_1 e^{-\int Pdx}.$$

Equation $(D)$ then becomes

$$u\frac{dz}{dx} = Q.$$

To find $z$ from the last equation, substitute in it the value of $u$ from $(E)$ and integrate. This gives

$$c_1 e^{-\int Pdx}\frac{dz}{dx} = Q,$$

$$c_1 dz = Qe^{\int Pdx}dx,$$

$$(F) \qquad c_1 z = \int Qe^{\int Pdx}dx + C.$$

The solution of $(A)$ is then found by substituting the values of $u$ and $z$ from $(E)$ and $(F)$ in $(B)$. This gives

$$(G) \qquad y = e^{-\int Pdx}\left(\int Qe^{\int Pdx}dx + C\right).$$

The proof of the correctness of $(G)$ is immediately established by substitution in $(A)$. In solving examples coming under this head the student is advised to find the solution by following the method illustrated above, rather than by using $(G)$ as a formula.

ILLUSTRATIVE EXAMPLE 1. Solve the equation

$$\frac{dy}{dx} - \frac{2y}{x+1} = (x+1)^{\frac{5}{2}}. \tag{1}$$

**Solution.** This is evidently in the linear form $(A)$, where

$$P = -\frac{2}{x+1} \text{ and } Q = (x+1)^{\frac{5}{2}}.$$

Let $y = uz$; then $\frac{dy}{dx} = u\frac{dz}{dx} + z\frac{du}{dx}$. Substituting in the given equation (1), we get

$$u\frac{dz}{dx} + z\frac{du}{dx} - \frac{2uz}{1+x} = (x+1)^{\frac{5}{2}}, \text{ or,}$$

$$u\frac{dz}{dx} + \left(\frac{du}{dx} - \frac{2u}{1+x}\right)z = (x+1)^{\frac{5}{2}}. \tag{2}$$

Now to determine $u$ we place the coefficient of $z$ equal to zero. This gives

$$\frac{du}{dx} - \frac{2u}{1+x} = 0,$$

$$\frac{du}{u} = \frac{2\,dx}{1+x},$$

$$\log_e u = 2\log(1+x),$$

$$u = e^{\log(1+x)^2} = (1+x)^2.* \tag{3}$$

Equation (2) now becomes, since the term in $z$ drops out,

$$u\frac{dz}{dx} = (x+1)^{\frac{5}{2}}.$$

Replacing $u$ by its value from (3),

$$\frac{dz}{dx} = (x+1)^{\frac{1}{2}},$$

$$dz = (x+1)^{\frac{1}{2}}\,dx,$$

$$z = \frac{2(x+1)^{\frac{3}{2}}}{3} + C. \tag{4}$$

Substituting (4) and (3) in $y = uz$, we get the solution

$$y = \frac{2(x+1)^{\frac{7}{2}}}{3} + C(x+1)^2. \quad \textit{Ans.}$$

* Since $\log_e u = \log_e e^{\log(1+x)^2} = \log(1+x)^2 \cdot \log_e e = \log(1+x)^2$, it follows that $u = (1+x)^2$. For the sake of simplicity we have assumed the particular value zero for the constant of integration.

**EXAMPLES**

| *Differential equations* | *Solutions* |
|---|---|
| 1. $\dfrac{dy}{dx} - \dfrac{2y}{x+1} = (x+1)^3.$ | $2y = (x+1)^4 + c(x+1)^2.$ |
| 2. $\dfrac{dy}{dx} - \dfrac{ay}{x} = \dfrac{x+1}{x}.$ | $y = cx^a + \dfrac{x}{1-a} - \dfrac{1}{a}.$ |
| 3. $x(1-x^2)\,dy + (2x^2-1)\,y\,dx = ax^3dx.$ | $y = ax + cx\sqrt{1-x^2}.$ |
| 4. $dy - \dfrac{xy\,dx}{1+x^2} = \dfrac{a\,dx}{1+x^2}.$ | $y = ax + c(1+x^2)^{\frac{1}{2}}.$ |
| 5. $\dfrac{ds}{dt}\cos t + s\sin t = 1.$ | $s = \sin t + c\cos t.$ |
| 6. $\dfrac{ds}{dt} + s\cos t = \frac{1}{2}\sin 2t.$ | $s = \sin t - 1 + ce^{-\sin t}.$ |
| 7. $\dfrac{dy}{dx} - \dfrac{n}{x}y = e^x x^n.$ | $y = x^n(e^x + c).$ |
| 8. $\dfrac{dy}{dx} + \dfrac{n}{x}y = \dfrac{a}{x^n}.$ | $x^n y = ax + c.$ |
| 9. $\dfrac{dy}{dx} + y = \dfrac{1}{e^x}.$ | $e^x y = x + c.$ |
| 10. $\dfrac{dy}{dx} + \dfrac{1-2x}{x^2}y = 1.$ | $y = x^2(1 + ce^{\frac{1}{x}}).$ |

**Type IV. Equations reducible to the linear form.** Some equations that are not linear can be reduced to the linear form by means of a suitable transformation. One type of such equations is

$$(A) \qquad \frac{dy}{dx} + Py = Qy^n,$$

where $P$, $Q$ are functions of $x$ alone, or constants. Equation $(A)$ may be reduced to the linear form $(A)$, Type III, by means of the substitution $z = y^{-n+1}$. Such a reduction, however, is not necessary if we employ the same method for finding the solution as that given under Type III, p. 427. Let us illustrate this by means of an example.

Illustrative Example 1. Solve the equation

$$(1) \qquad \frac{dy}{dx} + \frac{y}{x} = a\log x \cdot y^2.$$

**Solution.** This is evidently in the form $(A)$, where

$$P = \frac{1}{x},\ Q = a\log x,\ n = 2.$$

Let $y = uz$; then

$$\frac{dy}{dx} = u\frac{dz}{dx} + z\frac{du}{dx}.$$

Substituting in (1), we get

$$u\frac{dz}{dx} + z\frac{du}{dx} + \frac{uz}{x} = a\log x \cdot u^2z^2,$$

$$(2) \qquad u\frac{dz}{dx} + \left(\frac{du}{dx} + \frac{u}{x}\right)z = a\log x \cdot u^2z^2.$$

Now to determine $u$ we place the coefficient of $z$ equal to zero. This gives

$$\frac{du}{dx} + \frac{u}{x} = 0,$$

$$\frac{du}{u} = -\frac{dx}{x},$$

$$\log u = -\log x = \log\frac{1}{x},$$

$$u = \frac{1}{x}. \tag{3}$$

Since the term in $z$ drops out, equation (2) now becomes

$$u\frac{dz}{dx} = a\log x \cdot u^2z^2,$$

$$\frac{dz}{dx} = a\log x \cdot uz^2.$$

Replacing $u$ by its value from (3),

$$\frac{dz}{dx} = a\log x \cdot \frac{z^2}{x},$$

$$\frac{dz}{z^2} = a\log x \cdot \frac{dx}{x},$$

$$-\frac{1}{z} = \frac{a(\log x)^2}{2} + C,$$

$$z = -\frac{2}{a(\log x)^2 + 2C}. \tag{4}$$

Substituting (4) and (3) in $y = uz$, we get the solution

$$y = -\frac{1}{x}\cdot\frac{2}{a(\log x)^2 + 2C},$$

or,

$$xy\left[a(\log x)^2 + 2C\right] + 2 = 0. \quad \textit{Ans.}$$

**EXAMPLES**

| *Differential equations* | *Solutions* |
|---|---|
| 1. $\frac{dy}{dx} + xy = x^3y^3.$ | $y^{-2} = x^2 + 1 + ce^{x^2}.$ |
| 2. $(1-x^2)\frac{dy}{dx} - xy = axy^2.$ | $y = \left(c\sqrt{1-x^2} - a\right)^{-1}.$ |
| 3. $3y^2\frac{dy}{dx} - ay^3 = x + 1.$ | $y^3 = ce^{ax} - \frac{x+1}{a} - \frac{1}{a^2}.$ |
| 4. $\frac{dy}{dx}(x^2y^3 + xy) = 1.$ | $x\left[(2-y^2)e^{\frac{y^2}{2}} + c\right] = e^{\frac{y^2}{2}}.$ |
| 5. $(y\log x - 1)\,ydx = xdy.$ | $y = (cx + \log x + 1)^{-1}.$ |
| 6. $y - \cos x\frac{dy}{dx} = y^2\cos x(1 - \sin x).$ | $y = \frac{\tan x + \sec x}{\sin x + c}.$ |
| 7. $x\frac{dy}{dx} + y = y^2\log x.$ | $y^{-1} = \log x + 1 + cx.$ |

**232. Differential equations of the $n$th order and of the first degree.** Under this head we will consider four types which are of importance in elementary work. They are special cases of *linear differential equations*, which we defined on p. 427.

**Type I. The linear differential equation**

$$(A) \qquad \frac{d^n y}{dx^n} + p_1 \frac{d^{n-1} y}{dx^{n-1}} + p_2 \frac{d^{n-2} y}{dx^{n-2}} + \cdots + p_n y = 0,$$

*in which the coefficients* $p_1, p_2, \cdots, p_n$ *are constants.*

The substitution of $e^{rx}$ for $y$ in the first member gives

$$(r^n + p_1 r^{n-1} + p_2 r^{n-2} + \cdots + p_n) e^{rx}.$$

This expression vanishes for all values of $r$ which satisfy the equation

$$(B) \qquad r^n + p_1 r^{n-1} + p_2 r^{n-2} + \cdots + p_n = 0;$$

and therefore for each of these values of $r$, $e^{rx}$ is a solution of $(A)$. Equation $(B)$ is called the *auxiliary equation* of $(A)$. We observe that the coefficients are the same in both, the exponents in $(B)$ corresponding to the order of the derivatives in $(A)$, and $y$ in $(A)$ being replaced by 1. Let the roots of the auxiliary equation $(B)$ be $r_1, r_2, \cdots, r_n$; then

$$(C) \qquad e^{r_1 x}, \quad e^{r_2 x}, \quad \cdots, \quad e^{r_n x}$$

are solutions of $(A)$. Moreover, if each one of the solutions $(C)$ be multiplied by an arbitrary constant, the products

$$(D) \qquad c_1 e^{r_1 x}, \quad c_2 e^{r_2 x}, \quad \cdots, \quad c_n e^{r_n x}$$

are also found to be solutions.* And the sum of the solutions $(D)$, namely,

$$(E) \qquad y = c_1 e^{r_1 x} + c_2 e^{r_2 x} + \cdots + c_n e^{r_n x},$$

may, by substitution, be shown to be a solution of $(A)$. Solution $(E)$ contains $n$ arbitrary constants and is the *general solution* (if the roots are all different), while $(C)$ are *particular solutions.*

Case I. *When the auxiliary equation has imaginary roots.* Since imaginary roots occur in pairs, let one pair of such roots be

$$r_1 = a + bi, \qquad r_2 = a - bi. \qquad i = \sqrt{-1}$$

* Substituting $c_1 e^{r_1 x}$ for $y$ in $(A)$, the left-hand member becomes

$$(r_1^n + p_1 r_1^{n-1} + p_2 r_1^{n-2} + \cdots + p_n) c_1 e^{r_1 x}.$$

But this vanishes since $r_1$ is a root of $(B)$; hence $c_1 e^{r_1 x}$ is a solution of $(A)$. Similarly for the other roots.

The corresponding solution is

$$\begin{aligned} y &= c_1e^{(a+bi)x} + c_2e^{(a-bi)x} \\ &= e^{ax}(c_1e^{ibx} + c_2e^{-ibx}) \\ &= e^{ax}\{c_1(\cos bx + i\sin bx) + c_2(\cos bx - i\sin bx)\}^* \\ &= e^{ax}\{(c_1+c_2)\cos bx + i(c_1-c_2)\sin bx\}, \end{aligned}$$

or,

$$y = e^{ax}(A\cos bx + B\sin bx),$$

where $A$ and $B$ are arbitrary constants.

Case II. *When the auxiliary equation has multiple roots.* Consider the linear differential equation of the third order

$$(F) \qquad \frac{d^3y}{dx^3} + p_1\frac{d^2y}{dx^2} + p_2\frac{dy}{dx} + p_3y = 0,$$

where $p_1, p_2, p_3$ are constants. The corresponding auxiliary equation is

$$(G) \qquad r^3 + p_1r^2 + p_2r + p_3 = 0.$$

If $r_1$ is a root of $(G)$, we have shown that $e^{r_1x}$ is a solution of $(F)$. We will now show that if $r_1$ is a double root of $(G)$, then $xe^{r_1x}$ is also a solution of $(F)$. Replacing $y$ in the left-hand member of $(F)$ by $xe^{r_1x}$, we get

$$(H) \qquad xe^{r_1x}(r_1^3 + p_1r_1^2 + p_2r_1 + p_3) + e^{r_1x}(3r_1^2 + 2p_1r_1 + p_2).$$

But since $r_1$ is a double root of $(G)$,

$$r_1^3 + p_1r_1^2 + p_2r_1 + p_3 = 0,$$

and

$$3r_1^2 + 2p_1r_1 + p_2 = 0. \qquad \text{By § 69, p. 88}$$

* Replacing $x$ by $ibx$ in Example 1, p. 232, gives

$$e^{ibx} = 1 + ibx - \frac{b^2x^2}{\underline{|2}} - \frac{ib^3x^3}{\underline{|3}} + \frac{b^4x^4}{\underline{|4}} + \frac{ib^5x^5}{\underline{|5}} - \cdots, \text{ or,}$$

$$(1) \qquad e^{ibx} = 1 - \frac{b^2x^2}{\underline{|2}} + \frac{b^4x^4}{\underline{|4}} - \cdots + i\left(bx - \frac{b^3x^3}{\underline{|3}} + \frac{b^5x^5}{\underline{|5}} - \cdots\right);$$

and replacing $x$ by $-ibx$ gives

$$e^{-ibx} = 1 - ibx - \frac{b^2x^2}{\underline{|2}} + \frac{ib^3x^3}{\underline{|3}} + \frac{b^4x^4}{\underline{|4}} - \frac{ib^5x^5}{\underline{|5}} - \cdots, \text{ or,}$$

$$(2) \qquad e^{-ibx} = 1 - \frac{b^2x^2}{\underline{|2}} + \frac{b^4x^4}{\underline{|4}} - \cdots - i\left(bx - \frac{b^3x^3}{\underline{|3}} + \frac{b^5x^5}{\underline{|5}} - \cdots\right).$$

But, replacing $x$ by $bx$ in $(A)$, $(B)$, p. 231, we get

$$(3) \qquad \cos bx = 1 - \frac{b^2x^2}{\underline{|2}} + \frac{b^4x^4}{\underline{|4}} - \cdots.$$

$$(4) \qquad \sin bx = bx - \frac{b^3x^3}{\underline{|3}} + \frac{b^5x^5}{\underline{|5}} - \cdots.$$

Hence (1) and (2) become

$$e^{ibx} = \cos bx + i\sin bx, \qquad e^{-ibx} = \cos bx - i\sin bx.$$

Hence $(H)$ vanishes, and $xe^{r_1x}$ is a solution of $(F)$. Corresponding to the double root $r$, we then have the two solutions

$$c_1e^{r_1x}, \qquad c_2xe^{r_1x}.$$

More generally, if $r_1$ is a multiple root of the auxiliary equation $(B)$, p. 432, occurring $s$ times, then we may at once write down $s$ distinct solutions of the differential equation $(A)$, p. 432, namely,

$$c_1e^{r_1x}, \quad c_2xe^{r_1x}, \quad c_3x^2e^{r_1x}, \quad \cdots, \quad c_sx^{s-1}e^{r_1x}.$$

In case $a+bi$ and $a-bi$ are each multiple roots of the auxiliary equation, occurring $s$ times, it follows that we may write down $2s$ distinct solutions of the differential equation, namely,

$$c_1e^{ax}\cos bx, \quad c_2xe^{ax}\cos bx, \quad c_3x^2e^{ax}\cos bx, \quad \cdots, \quad c_sx^{s-1}e^{ax}\cos bx;$$
$$c_1'e^{ax}\sin bx, \quad c_2'xe^{ax}\sin bx, \quad c_3'x^2e^{ax}\sin bx, \quad \cdots, \quad c_s'x^{s-1}e^{ax}\sin bx.$$

Our results may now be summed up in the following **rule for solving differential equations of the type**

$$\frac{d^ny}{dx^n}+p_1\frac{d^{n-1}y}{dx^{n-1}}+p_2\frac{d^{n-2}y}{dx^{n-2}}+\cdots+p_ny=0,$$

*where $p_1, p_2, \cdots, p_n$ are constants.*

FIRST STEP. *Write down the corresponding auxiliary equation*

$$r^n+p_1r^{n-1}+p_2r^{n-2}+\cdots+p_n=0.$$

SECOND STEP. *Solve completely the auxiliary equation.*

THIRD STEP. *From the roots of the auxiliary equation write down the corresponding particular solutions of the differential equation as follows:*

| AUXILIARY EQUATION | | DIFFERENTIAL EQUATION |
|---|---|---|
| (a) *Each distinct real root* $r_1$ | *gives* | *a particular solution* $e^{r_1x}$. |
| (b) *Each distinct pair of imaginary roots* $a \pm bi$ | *gives* | *two particular solutions* $e^{ax}\cos bx$, $e^{ax}\sin bx$. |
| (c) *A multiple root occurring* $s$ *times* | *gives* | $s$ *particular solutions obtained by multiplying the particular solutions* (a) *or* (b) *by* $1, x, x^2, \cdots, x^{s-1}$. |

FOURTH STEP. *Multiply each of the* $n$* *independent solutions by an arbitrary constant and add the results. This gives the complete solution.*

* A check on the accuracy of the work is found in the fact that the first three steps must give $n$ independent solutions.

ILLUSTRATIVE EXAMPLE 1. Solve $\frac{d^3y}{dx^3} - 3\frac{d^2y}{dx^2} + 4y = 0.$

**Solution.** Follow above rule.

*First step.* $r^3 - 3r^2 + 4 = 0$, auxiliary equation.

*Second step.* Solving, the roots are $-1, 2, 2$.

*Third step.* (a) The root $-1$ gives the solution $e^{-x}$.

(b) The double root 2 gives the two solutions $e^{2x}$, $xe^{2x}$.

*Fourth step.* General solution is

$$y = c_1e^{-x} + c_2e^{2x} + c_3xe^{2x}. \quad \textit{Ans.}$$

ILLUSTRATIVE EXAMPLE 2. Solve $\frac{d^4y}{dx^4} - 4\frac{d^3y}{dx^3} + 10\frac{d^2y}{dx^2} - 12\frac{dy}{dx} + 5y = 0.$

**Solution.** Follow above rule.

*First step.* $r^4 - 4r^3 + 10r^2 - 12r + 5 = 0$, auxiliary equation.

*Second step.* Solving, the roots are $1, 1, 1 \pm 2i$.

*Third step.* (b) The pair of imaginary roots $1 \pm 2i$ gives the two solutions $e^x \cos 2x$, $e^x \sin 2x\,(a = 1, b = 2)$.

(c) The double root 1 gives the two solutions $e^x$, $xe^x$.

*Fourth step.* General solution is

$$y = c_1e^x + c_2xe^x + c_3e^x\cos 2x + c_4e^x \sin 2x,$$

or,

$$y = (c_1 + c_2x + c_3\cos 2x + c_4 \sin 2x)e^x. \quad \textit{Ans.}$$

## EXAMPLES

| *Differential equations* | *General solutions* |
|---|---|
| 1. $\frac{d^2y}{dx^2} = 9y.$ | $y = c_1e^{3x} + c_2e^{-3x}.$ |
| 2. $\frac{d^2y}{dx^2} + y = 0.$ | $y = c_1 \sin x + c_2 \cos x.$ |
| 3. $\frac{d^2y}{dx^2} + 12y = 7\frac{dy}{dx}.$ | $y = c_1e^{3x} + c_2e^{4x}.$ |
| 4. $\frac{d^2y}{dx^2} - 4\frac{dy}{dx} + 4y = 0.$ | $y = (c_1 + c_2x)e^{2x}.$ |
| 5. $\frac{d^3y}{dx^3} - 4\frac{dy}{dx} = 0.$ | $y = c_1 + c_2e^{2x} + c_3e^{-2x}.$ |
| 6. $\frac{d^4y}{dx^4} + 2\frac{d^2y}{dx^2} - 8y = 0.$ | $y = c_1e^{x\sqrt{2}} + c_2e^{-x\sqrt{2}} + c_3 \sin 2x + c_4 \cos 2x.$ |
| 7. $\frac{d^3s}{dt^3} - \frac{d^2s}{dt^2} - 6\frac{ds}{dt} = 0.$ | $s = c_1e^{3t} + c_2e^{-2t} + c_3.$ |
| 8. $\frac{d^4\rho}{d\theta^4} - 12\frac{d^2\rho}{d\theta^2} + 27\rho = 0.$ | $\rho = c_1e^{3\theta} + c_2e^{-3\theta} + c_3e^{\theta\sqrt{3}} + c_4e^{-\theta\sqrt{3}}.$ |
| 9. $\frac{d^2u}{dv^2} - 6\frac{du}{dv} + 13u = 0.$ | $u = (c_1 \sin 2v + c_2 \cos 2v)e^{3v}.$ |
| 10. $\frac{d^4y}{dx^4} + 2n^2\frac{d^2y}{dx^2} + n^4y = 0.$ | $y = (c_1 + c_2x)\cos nx + (c_3 + c_4x)\sin nx.$ |
| 11. $\frac{d^3s}{dt^3} = s.$ | $s = c_1e^t + e^{-\frac{t}{2}}\left(c_2 \sin\frac{t\sqrt{3}}{2} + c_3 \cos\frac{t\sqrt{3}}{2}\right).$ |

| *Differential equations* | *General solutions* |
|---|---|
| 12. $\frac{d^3s}{dt^3} - 7\frac{ds}{dt} + 6s = 0.$ | $s = c_1e^{2t} + c_2e^{t} + c_3e^{-3t}.$ |
| 13. $\frac{d^4y}{dx^4} - 3\frac{d^3y}{dx^3} + 3\frac{d^2y}{dx^2} - \frac{dy}{dx} = 0.$ | $y = c_1 + (c_2 + c_3x + c_4x^2)e^{x}.$ |
| 14. $\frac{d^2y}{dx^2} + 3\frac{dy}{dx} - 10y = 0.$ | $y = c_1e^{2x} + c_2e^{-5x}.$ |
| 15. $\frac{d^2y}{dx^2} + 2\frac{dy}{dx} + 10y = 0.$ | $y = e^{-x}(c_1\cos 3x + c_2\sin 3x).$ |
| 16. $2\frac{d^3y}{dx^3} - 3\frac{d^2y}{dx^2} + 2\frac{dy}{dx} + 2y = 0.$ | $y = c_1e^{-\frac{1}{2}x} + e^{x}(c_2\cos x + c_3\sin x)$ |

**Type II. The linear differential equation**

$$(I)\qquad \frac{d^ny}{dx^n} + p_1\frac{d^{n-1}y}{dx^{n-1}} + p_2\frac{d^{n-2}y}{dx^{n-2}} + \cdots + p_ny = X,$$

*where $X$ is a function of $x$ alone, or constant, and $p_1, p_2, \cdots, p_n$ are constants.*

When $X = 0$, $(I)$ reduces to $(A)$, Type I, p. 432,

$$(J)\qquad \frac{d^ny}{dx^n} + p_1\frac{d^{n-1}y}{dx^{n-1}} + p_2\frac{d^{n-2}y}{dx^{n-2}} + \cdots + p_ny = 0.$$

The complete solution of $(J)$ is called the *complementary function* of $(I)$.

Let $u$ be the complete solution of $(J)$, i.e. the complementary function of $(I)$, and $v$ any particular solution of $(I)$. Then

$$\frac{d^nv}{dx^n} + p_1\frac{d^{n-1}v}{dx^{n-1}} + p_2\frac{d^{n-2}v}{dx^{n-2}} + \cdots + p_nv = X,$$

and
$$\frac{d^nu}{dx^n} + p_1\frac{d^{n-1}u}{dx^{n-1}} + p_2\frac{d^{n-2}u}{dx^{n-2}} + \cdots + p_nu = 0.$$

Adding, we get

$$\frac{d^n}{dx^n}(u+v) + p_1\frac{d^{n-1}}{dx^{n-1}}(u+v) + p_2\frac{d^{n-2}}{dx^{n-2}}(u+v) + \cdots + p_n(u+v) = X,$$

showing that $u + v$ is a solution* of $(I)$.

To find a particular solution $v$ is a problem of considerable difficulty except in special cases. For the problems given in this book we may use the following **rule for solving differential equations of Type II.**

FIRST STEP. *Replace the right-hand member of the given equation $(I)$ by zero and solve by the rule on p. 434. This gives as a solution the complementary function of $(I)$, namely,*

$$y = u.$$

* In works on differential equations it is shown that $u + v$ is the complete solution.

SECOND STEP. *Differentiate successively the given equation (I) and obtain, either directly or by elimination, a differential equation of a higher order of Type I.*

THIRD STEP. *Solving this new equation by the rule on p. 434, we get its complete solution*

$$y = u + v,$$

*where the part u is the complementary function of (I) already found in the first step,* and v is the sum of the additional terms found.*

FOURTH STEP. *To find the values of the constants of integration in the particular solution v, substitute*

$$y = v$$

*and its derivatives in the given equation (I). In the resulting identity equate the coefficients of like terms, solve for the constants of integration, substitute their values back in*

$$y = u + v,$$

*giving the complete solution of (I).*

This method will now be illustrated by means of examples.

NOTE. The solution of the auxiliary equation of the new derived differential equation is facilitated by observing that the left-hand member of that equation is exactly divisible by the left-hand member of the auxiliary equation used in finding the complementary function.

ILLUSTRATIVE EXAMPLE 1. Solve

$$(K) \qquad \frac{d^2y}{dx^2} + \frac{dy}{dx} - 2y = ae^{-2x}.$$

**Solution.** *First step.* Replacing the right-hand member by zero,

$$(L) \qquad \frac{d^2y}{dx^2} + \frac{dy}{dx} - 2y = 0.$$

Applying the rule on p. 434, we get as the complete solution of (L)

$$(M) \qquad y = c_1e^x + c_2e^{-2x} = u.$$

*Second step.* Differentiating (K) gives

$$(N) \qquad \frac{d^3y}{dx^3} + \frac{d^2y}{dx^2} - 2\frac{dy}{dx} = -2ae^{-2x}.$$

Multiplying (K) by 2 and adding the result to (N), we get

$$(O) \qquad \frac{d^3y}{dx^3} + 3\frac{d^2y}{dx^2} - 4y = 0,$$

a differential equation of Type I.

*Third step.* Solving by the rule on p. 434, we get the complete solution of (O) to be

$$y = c_1e^x + c_2e^{-2x} + c_3xe^{-2x},$$

or, from (M),

$$y = u + c_3xe^{-2x} = u + v.$$

* From the method of derivation it is obvious that every solution of the original equation must also be a solution of the derived equation.

*Fourth step.* We now determine $c_3$ so that $c_3xe^{-2x}$ shall be a particular solution $v$ of $(K)$.

Substituting $y = c_3xe^{-2x}$, $\frac{dy}{dx} = c_3e^{-2x}(1-2x)$, $\frac{d^2y}{dx^2} = c_3e^{-2x}(4x-4)$ in $(K)$, we get

$$-3c_3e^{-2x} = ae^{-2x}.$$

$$\therefore \; -3c_3 = a, \quad \text{or,} \quad c_3 = -\tfrac{1}{3}a.$$

Hence a particular solution of $(K)$ is

$$v = -\tfrac{1}{3}axe^{-2x},$$

and the complete solution is

$$y = u + v = c_1e^x + c_2e^{-2x} - \tfrac{1}{3}axe^{-2x}.$$

ILLUSTRATIVE EXAMPLE 2. Solve

$$(P) \qquad \frac{d^2y}{dx^2} + n^2y = \cos ax.$$

**Solution.** *First step.* Solving

$$(Q) \qquad \frac{d^2y}{dx^2} + n^2y = 0,$$

we get the complementary function

$$(R) \qquad y = c_1 \sin nx + c_2 \cos nx = u.$$

*Second step.* Differentiating $(P)$ twice, we get

$$(S) \qquad \frac{d^4y}{dx^4} + n^2\frac{d^2y}{dx^2} = -a^2 \cos ax.$$

Multiplying $(P)$ by $a^2$ and adding the result to $(S)$ gives

$$(T) \qquad \frac{d^4y}{dx^4} + (n^2 + a^2)\frac{d^2y}{dx^2} + a^2n^2y = 0.$$

*Third step.* The complete solution of $(T)$ is

$$y = c_1 \sin nx + c_2 \cos nx + c_3 \sin ax + c_4 \cos ax,$$

or,

$$y = u + c_3 \sin ax + c_4 \cos ax = u + v.$$

*Fourth step.* Let us now determine $c_3$ and $c_4$ so that $c_3 \sin ax + c_4 \cos ax$ shall be a particular solution $v$ of $(P)$.

Substituting

$y = c_3 \sin ax + c_4 \cos ax$, $\frac{dy}{dx} = c_3a \cos ax - c_4a \sin ax$, $\frac{d^2y}{dx^2} = -c_3a^2 \sin ax - c_4a^2 \cos ax$

in $(P)$, we get

$$(n^2c_4 - a^2c_4) \cos ax + (n^2c_3 - a^2c_3) \sin ax = \cos ax.$$

Equating the coefficients of like terms in this identity, we get

$$n^2c_4 - a^2c_4 = 1 \quad \text{and} \quad n^2c_3 - a^2c_3 = 0,$$

or,

$$c_4 = \frac{1}{n^2 - a^2} \quad \text{and} \quad c_3 = 0.$$

Hence a particular solution of $(P)$ is

$$v = \frac{\cos ax}{n^2 - a^2},$$

and the complete solution is

$$y = u + v = c_1 \sin nx + c_2 \cos nx + \frac{\cos ax}{n^2 - a^2}.$$

**EXAMPLES**

| *Differential equations* | *Complete solutions* |
|---|---|
| 1. $\frac{d^2y}{dx^2} - 7\frac{dy}{dx} + 12y = x.$ | $y = c_1e^{3x} + c_2e^{4x} + \frac{12x+7}{144}.$ |
| 2. $\frac{d^4y}{dx^4} - 2\frac{d^3y}{dx^3} + 2\frac{d^2y}{dx^2} - 2\frac{dy}{dx} + y = a.$ | $y = c_1 \sin x + c_2 \cos x + (c_3 + c_4x)e^x + a.$ |
| 3. $\frac{d^2s}{dt^2} - a^2s = t + 1.$ | $s = c_1e^{at} + c_2e^{-at} - \frac{t+1}{a^2}.$ |
| 4. $\frac{d^3\rho}{d\theta^3} - 2\frac{d^2\rho}{d\theta^2} + \frac{d\rho}{d\theta} = e^\theta.$ | $\rho = \left(c_1 + c_2\theta + \frac{\theta^2}{2}\right)e^\theta + c_3.$ |
| 5. $\frac{d^4y}{dx^4} - a^4y = x^3.$ | $y = c_1e^{ax} + c_2e^{-ax} + c_3 \sin ax + c_4 \cos ax - \frac{x^3}{a^4}.$ |
| 6. $\frac{d^2s}{dx^2} + a^2s = \cos ax.$ | $s = c_1 \sin ax + c_2 \cos ax + \frac{x \sin ax}{2a}.$ |
| 7. $\frac{d^2s}{dt^2} - 2a\frac{ds}{dt} + a^2s = e^t.$ | $s = (c_1 + c_2t)e^{at} + \frac{e^t}{(a-1)^2}.$ |
| 8. $\frac{d^2y}{dx^2} + 2\frac{dy}{dx} + y = e^{2x}.$ | $y = e^{-x}(c_1x + c_2) + \frac{1}{9}e^{2x}.$ |
| 9. $\frac{d^2y}{dx^2} - y = 5x + 2.$ | $y = c_1e^x + c_2e^{-x} - 5x - 2.$ |
| 10. $\frac{d^2y}{dx^2} - 4\frac{dy}{dx} + 3y = x.$ | $y = c_1e^x + c_2e^{3x} + \frac{1}{3}x + \frac{4}{9}.$ |
| 11. $\frac{d^2y}{dx^2} - 5\frac{dy}{dx} + 6y = e^{nx}.$ | $y = c_1e^{2x} + c_2e^{3x} + \frac{e^{nx}}{n^2 - 5n + 6}.$ |
| 12. $\frac{d^2y}{dx^2} - 3\frac{dy}{dx} + 2y = xe^{nx}.$ | $y = c_1e^x + c_2e^{2x} + \frac{xe^{nx}}{n^2 - 3n + 2} - \frac{(2n-3)e^{nx}}{(n^2 - 3n + 2)^2}.$ |
| 13. $\frac{d^2s}{dt^2} - 9\frac{ds}{dt} + 20s = t^2e^{3t}$ | $s = c_1e^{4t} + c_2e^{5t} + \frac{2t^2 + 6t + 7}{4}e^{3t}.$ |
| 14. $\frac{d^2s}{dt^2} + 4s = t \sin^2 t.$ | $s = \left(c_1 - \frac{t^2}{16}\right)\sin 2t + \left(c_2 - \frac{t}{32}\right)\cos 2t + \frac{t}{8}.$ |

**Type III.** $$\frac{d^ny}{dx^n} = X,$$

where $X$ is a function of $x$ alone, or constant.

To solve this type of differential equations we have the following rule from Chapter XXIX, p. 398:

*Integrate n times successively. Each integration will introduce one arbitrary constant.*

ILLUSTRATIVE EXAMPLE 1. Solve $\frac{d^3y}{dx^3} = xe^x$.

**Solution.** Integrating the first time, $\frac{d^2y}{dx^2} = \int xe^x dx,$

or, $$\frac{d^2y}{dx^2} = xe^x - e^x + C_1. \qquad \text{By } (A), \text{ p. } 347$$

Integrating the second time,

$$\frac{dy}{dx} = \int xe^x \cdot dx - \int e^x dx + \int C_1 dx,$$

$$\frac{dy}{dx} = xe^x - 2e^x + C_1 x + C_2.$$

Integrating the third time,

$$y = \int xe^x dx - \int 2e^x dx + \int C_1 x dx + \int C_2 dx$$

$$= xe^x - 3e^x + \frac{C_1 x^2}{2} + C_2 x + C_3,$$

or, $$y = xe^x - 3e^x + c_1 x^2 + c_2 x + c_3. \quad \textit{Ans.}$$

**Type IV.** $$\frac{d^2y}{dx^2} = Y,$$

where $Y$ is a function of $y$ alone.

The rule for integrating this type is as follows:

FIRST STEP. *Multiply the left-hand member by the factor*

$$2\frac{dy}{dx}dx,$$

*and the right-hand member by the equivalent factor*

$$2\,dy,$$

*and integrate. The integral of the left-hand member will be* *

$$\left(\frac{dy}{dx}\right)^2.$$

SECOND STEP. *Extract the square root of both members, separate the variables, and integrate again.*†

ILLUSTRATIVE EXAMPLE 1. Solve $\frac{d^2y}{dx^2} + a^2y = 0$.

**Solution.** Here $\frac{d^2y}{dx^2} = -a^2y$, and hence is of Type IV.

*First step.* Multiplying the left-hand member by $2\frac{dy}{dx}dx$ and the right-hand member by $2\,dy$, we get

$$2\frac{dy}{dx}\frac{d^2y}{dx^2}dx = -2a^2y\,dy.$$

Integrating, $$\left(\frac{dy}{dx}\right)^2 = -a^2y^2 + C_1.$$

* Since $d\left(\frac{dy}{dx}\right)^2 = 2\frac{dy}{dx}\frac{d^2y}{dx^2}dx$.

† Each integration introduces an arbitrary constant.

*Second step.* $$\frac{dy}{dx} = \sqrt{C_1 - a^2y^2},$$

taking the positive sign of the radical. Separating the variables, we get

$$\frac{dy}{\sqrt{C_1 - a^2y^2}} = dx.$$

Integrating, $$\frac{1}{a}\text{arc sin}\frac{ay}{\sqrt{C_1}} = x + C_2,$$

or, $$\text{arc sin}\frac{ay}{\sqrt{C_1}} = ax + aC_2.$$

This is the same as $$\frac{ay}{\sqrt{C_1}} = \sin(ax + aC_2)$$

$$= \sin ax \cos aC_2 + \cos ax \sin aC_2, \qquad \text{31, p. 2}$$

or, $$y = \frac{\sqrt{C_1}}{a}\cos aC_2 \cdot \sin ax + \frac{\sqrt{C_1}}{a}\sin aC_2 \cdot \cos ax$$

$$= c_1 \sin ax + c_2 \cos ax. \quad \textit{Ans.}$$

## EXAMPLES

| *Differential equations* | *Solutions* |
|---|---|
| 1. $\frac{d^3y}{dx^3} = x^2 - 2\cos x.$ | $y = \frac{x^5}{60} + 2\sin x + c_1x^2 + c_2x + c_3.$ |
| 2. $v\frac{d^3u}{dv^3} = 2.$ | $u = v^2 \log v + c_1v^2 + c_2v + c_3.$ |
| 3. $\frac{d^3\rho}{d\theta^3} = \sin^3\theta.$ | $\rho = -\frac{\cos^3\theta}{27} + \frac{7\cos\theta}{9} + c_1\theta^2 + c_2\theta + c_3.$ |
| 4. $\frac{d^2s}{dt^2} = f\sin nt.$ | $s = -\frac{f}{n^2}\sin nt + c_1t + c_2.$ |
| 5. $\frac{d^2s}{dt^2} = g.$ | $s = \frac{1}{2}gt^2 + c_1t + c_2.$ |
| 6. $\frac{d^ny}{dx^n} = x^m.$ | $y = \frac{\lfloor m\, x^{m+n}}{\lfloor m+n} + c_1x^{n-1} + \cdots + c_{n-1}x + c_n.$ |
| 7. $\frac{d^2y}{dx^2} = a^2y.$ | $ax = \log\left(y + \sqrt{y^2 + c_1}\right) + c_2,$ or, $y = c_1e^{ax} + c_2e^{-ax}.$ |
| 8. $\frac{d^2s}{dt^2} = \frac{1}{\sqrt{as}}.$ | $3t = 2a^{\frac{1}{2}}(s^{\frac{1}{2}} - 2c_1)(s^{\frac{1}{2}} + c_1)^{\frac{1}{2}} + c_2.$ |
| 9. $\frac{d^2y}{dt^2} = \frac{a}{y^3}.$ | $(c_1t + c_2)^2 + a = c_1y^2.$ |
| 10. $\frac{d^2x}{dt^2} = e^{nx}.$ | $t\sqrt{2n} = c_1 \log\frac{\sqrt{c_1^2e^{nx} + 1} - 1}{\sqrt{c_1^2e^{nx} + 1} + 1} + c_2.$ |
| 11. $\frac{d^2y}{dx^2} = x^2\sin x.$ | $y = c_1 + c_2x + (6 - x^2)\sin x - 4x\cos x.$ |

12. $\frac{d^2s}{dt^2} = -\frac{k}{s^2}$. Find $t$, having given that $\frac{ds}{dt} = 0$ and $s = a$, when $t = 0$.

$$\textit{Ans.}\quad t = \sqrt{\frac{a}{2k}}\left\{\frac{a}{2}\left(\text{arc vers}\frac{2s}{a} - \pi\right) - \sqrt{as - s^2}\right\}.$$

## MISCELLANEOUS EXAMPLES

Solve the following differential equations:

1. $\frac{d^4y}{dx^4} - a^2\frac{d^2y}{dx^2} = 0.$

2. $\frac{d^2y}{dx^2} = \frac{a}{x}.$

3. $\frac{dy}{dx} + \frac{2y}{x} = 3x^2y^{\frac{4}{3}}.$

4. $3\frac{dy}{dx} + \frac{2y}{x+1} = \frac{x^3}{y^2}.$

5. $(4y + 3x)\frac{dy}{dx} + y = 2x.$

6. $2\frac{d^2x}{dt^2} + 5\frac{dx}{dt} - 12x = 0.$

7. $\frac{d^2y}{dx^2} - 5\frac{dy}{dx} + 6y = e^{4x}.$

8. $\frac{d^2y}{dx^2} + \frac{dy}{dx} = e^x.$

9. $\frac{dy}{dx} + y = e^{-x}.$

10. $\frac{dy}{dx} + \frac{1-2x}{x^2}y = 1.$

11. $y - x\frac{dy}{dx} = a\left(y^2 + \frac{dy}{dx}\right).$

12. $x^2y\,dx - (x^3 + y^3)\,dy = 0.$

13. $\frac{d^3y}{dx^3} - 3\frac{d^2y}{dx^2} + 4y = 0.$

14. $\frac{dy}{dx} + y\tan x = 1.$

# CHAPTER XXXI

## INTEGRAPH. APPROXIMATE INTEGRATION. TABLE OF INTEGRALS

**233. Mechanical integration.** We have seen that the determination of the area bounded by a curve $C$ whose equation is

$$y = f(x)$$

and the evaluation of the definite integral

$$\int f(x)\,dx$$

are equivalent problems.

Hitherto we have regarded the relation between the variables $x$ and $y$ as given by analytical formulas and have applied analytic methods in obtaining the integrals required. If, however, the relation between the variables is given, not analytically, but, as frequently is the case in physical investigations, graphically, i.e. by a curve,* the analytic method is inapplicable unless the exact or approximate equation of the curve can be obtained. It is, however, possible to determine the area bounded by a curve, whether we know its equation or not, by means of mechanical devices. We shall consider the construction, theory, and use of two such devices, namely, the Integraph, invented by Abdank-Abakanowicz,† and the Polar Planimeter. Before proceeding with the discussion of the Integraph it is necessary to take up the study of *integral curves.*

**234. Integral curves.** If $F(x)$ and $f(x)$ are two functions so related that

$$(A) \qquad \frac{d}{dx}F(x) = f(x),$$

then the curve

$$(B) \qquad y = F(x)$$

is called an *integral curve* of the curve

$$(C) \qquad y = f(x).‡$$

* For instance, the record made by a registering thermometer, a steam-engine indicator, or by certain testing machines.

† See *Les Intégraphes; la courbe intégrale et ses applications*, by Abdank-Abakanowicz, Paris, 1889.

‡ This curve is sometimes called the *original curve.*

The name *integral curve* is due to the fact that from (C) it is seen that the same relation between the functions may be expressed as follows:

$$(D) \qquad \int_0^x f(x)\,dx = F(x). \qquad F(0) = 0$$

Let us draw an original curve and a corresponding integral curve in such a way as easily to compare their corresponding points.

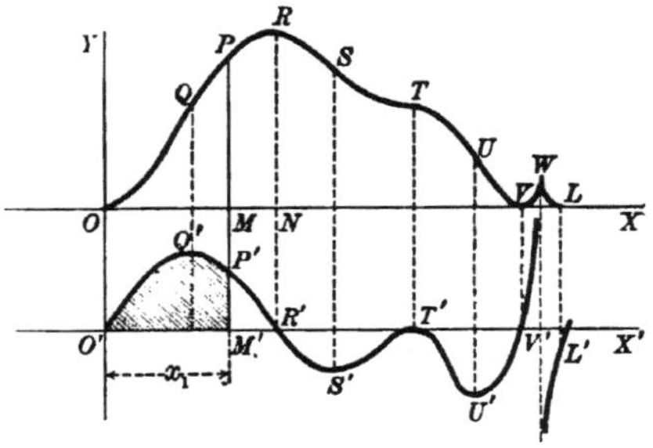


To find an expression for the shaded portion ($O'M'P'$) of the area under the original curve we substitute in (A), p. 365, giving

$$\text{area } O'M'P' = \int_0^{x_1} f(x)\,dx.$$

But from (D) this becomes

$$\text{area } O'M'P' = \int_0^{x_1} f(x)\,dx = [F(x)]_{x=x_1} = F(x_1) = MP.^*$$

**Theorem.** *For the same abscissa $x_1$ the number giving the length of the ordinate of the integral curve (B) is the same as the number that gives the area between the original curve, the axes, and the ordinate corresponding to this abscissa.*

The student should also observe that

(a) For the same abscissa $x_1$ the number giving the slope of the integral curve is the same as the number giving the length of the corresponding ordinate of the original curve [from (C)]. Hence (C) is sometimes called the *curve of slopes* of (B). In the figure we see that at points $O$, $R$, $T$, $V$, where the integral curve is parallel to $OX$, the corresponding points $O'$, $R'$, $T'$, $V'$ on the original curve have zero ordinates, and corresponding to the point $W$ the original curve is discontinuous.

* When $x_1 = O'R'$, the positive area $O'M'R'P'$ is represented by the maximum ordinate $NR$. To the right of $R'$ the area is below the axis of $X$ and therefore negative; consequently the ordinates of the integral curve, which represent the algebraic sum of the areas inclosed, will decrease in passing from $R'$ to $T'$.

The most general integral curve is of the form

$$y = F(x) + C,$$

in which case the *difference* of the ordinates for $x = 0$ and $x = x_1$ gives the area under the original curve. In the integral curve drawn $C = F(0) = 0$, i.e. the general integral curve is obtained if this integral curve be displaced the distance $C$ parallel to $OY$.

(b) Corresponding to points of inflection $Q$, $S$, $U$ on the integral curve we have maximum or minimum ordinates to the original curve.

For example, since $\frac{d}{dx}\left(\frac{x^3}{9}\right)=\frac{x^2}{3}$,

it follows that

(E) $$y=\frac{x^3}{9}$$

is an integral curve of the parabola

(F) $$y=\frac{x^2}{3}.$$

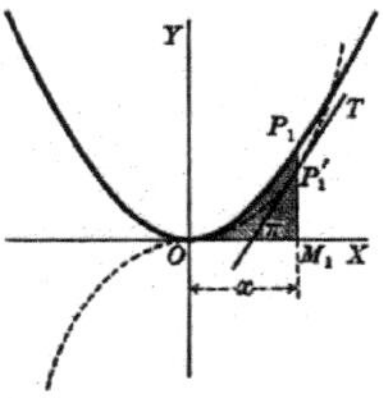


Since from (F) area $OM_1P_1=\int_0^{x_1}\frac{x^2}{3}dx=\frac{x_1^3}{9}$,

and from (E) $M_1P_1'=\frac{x_1^3}{9}$,

it is seen that $\frac{x_1^3}{9}$ indicates the number of linear units in the ordinate $M_1P_1'$, and also the number of units of area in the shaded area $OM_1P_1$.

Also, since from (E) $\frac{dy}{dx}=\frac{x^2}{3}$, or, $\tan\tau=\frac{x_1^2}{3}$,

and from (F) $M_1P_1=\frac{x_1^2}{3}$,

it is seen that the same number $\frac{x_1^2}{3}$ indicates the length of ordinate $M_1P_1$ and the slope of the tangent at $P_1'$.

Evidently the origin is a point of inflection of the integral curve and a point with minimum ordinate on the original curve.

**235. The integraph.** The theory of this instrument is exceedingly simple and depends on the relation between the given curve and a corresponding integral curve.

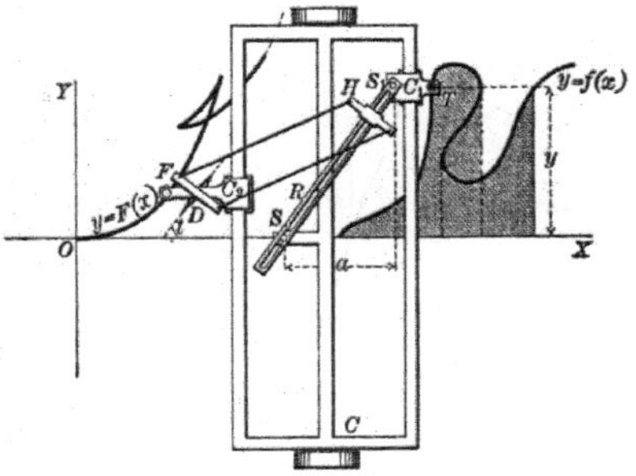


The instrument is constructed as follows: A rectangular carriage $C$ moves on rollers over the plane in a direction parallel to the axis of $X$ of the curve

$$y=f(x).$$

Two sides of the carriage are parallel to the axis of $X$; the other two, of course, are perpendicular to it. Along one of these perpendicular sides moves a small carriage $C_1$ bearing the tracing point $T$, and along the other a small carriage $C_2$ bearing a frame $F$ which can revolve about an axis perpendicular to the surface, and which carries the sharp-edged disk $D$, to the plane of which it is perpendicular. A stud $S_1$ is fixed in the carriage $C_1$ so as to be at the same distance from the axis of $X$ as is the tracing point $T$. A second stud $S_2$ is set in a crossbar of the main carriage $C$ so as to be on the axis of $X$. A split ruler $R$ joins these two studs and slides upon them. A crosshead $H$ slides upon this ruler and is joined to the frame $F$ by a parallelogram.

The essential part of the instrument consists of the sharp-edged disk $D$, which moves under pressure over a smooth plane surface (paper). This disk will not slide, and hence as it rolls must always move along a path the tangent to which at every point is the trace of the plane of the disk. If now this disk is caused to move, it is evident from the figure that the construction of the machine insures that the plane of the disk $D$ shall be parallel to the ruler $R$. But if $a$ is the distance between the ordinates through the studs $S_1$, $S_2$, and $\tau$ is the angle made by $R$ (and therefore also plane of disk) with the axis of $X$, we have

(A) $$\tan\tau = \frac{y}{a};$$

and if $$y' = F(x')$$

is the curve traced by the point of contact of the disk, we have

(B) $$\tan\tau = \frac{dy'}{dx}.^*$$

Comparing (A) and (B), $$\frac{dy'}{dx} = \frac{y}{a}, \text{ or,}$$

(C) $$y' = \frac{1}{a}\int y dx = \frac{1}{a}\int f(x)\,dx = F(x').^\dagger$$

That is (dropping the primes), the curve

$$y = F(x)$$

is an *integral curve* of the curve

(D) $$y = \frac{1}{a}f(x).$$

The factor $\frac{1}{a}$ evidently fixes merely the *scale* to which the integral curve is drawn, and does not affect its *form*.

A pencil or pen is attached to the carriage $C_2$ in order to draw the curve $y = F(x)$. Displacing the disk $D$ before tracing the original curve is equivalent to changing the constant of integration.

**236. Polar planimeter.** This is an instrument for measuring areas mechanically. Before describing the machine we shall take up the theory on which it is based.

**237. Calculation of the area swept over by a moving line of constant length.**

Consider the area $ABQB'A'PA$ swept over by the line $AB$ of constant length $l$. Let $PQ$ and $P'Q'$ be consecutive positions of the line, $d\theta$ = angle $POP'$ = change in

* Since $x = x' + d$, where $d$ = width of machine, and therefore $\frac{dy'}{dx'} = \frac{dy'}{dx}\cdot\frac{dx}{dx'} = \frac{dy'}{dx}$.

† It is assumed that the instrument is so constructed that the abscissas of any two corresponding points of the two curves differ only by a constant; hence $x$ is a function of $x'$.

direction of $PQ$, and $ds$ = circular arc described about $O$ by the middle point $R$ of the line. Using differentials, we have

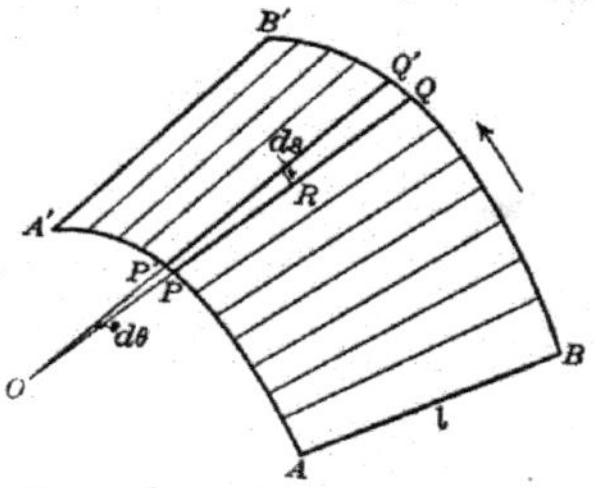


$$\text{area of } OQQ' = \tfrac{1}{2}\overline{OQ}^2 d\theta,^*$$

$$\text{area of } OPP' = \tfrac{1}{2}\overline{OP}^2 d\theta.$$

$$\therefore \text{area of } PQQ'P' = \tfrac{1}{2}\overline{OQ}^2 d\theta - \tfrac{1}{2}\overline{OP}^2 d\theta$$
$$= \tfrac{1}{2}(OQ+OP)(OQ-OP)d\theta$$
$$= OR \cdot PQ d\theta$$
$$= l \cdot OR d\theta = l ds.$$

Summing up all such elements,

$$(A) \qquad \text{area } ABQB'A'PA = \int l ds = l\int ds = ls,$$

where $s$ = displacement of the center of the line in a direction always perpendicular to the line.† To find $s$, let the line be replaced by a rod having a small wheel at the center $R$, the rod being the axis of the wheel. Now as the rod is moved horizontally over the surface (paper), the wheel will, in general, both slide and rotate. Evidently

$$s = \text{distance it rolls}$$
$$= \text{circumference of wheel} \times \text{number of revolutions}.$$

$$(B) \qquad \therefore s = 2\pi r n,$$

where $r$ = radius of wheel, and $n$ = number of revolutions.

Substituting $(B)$ in $(A)$, we get

$$(C) \qquad \text{area swept over} = 2\pi r l n.$$

So far we have tacitly assumed that the areas were swept over always in the same direction. It is easy to see, however, that the results hold true without any such restriction, provided areas are taken as positive or negative according as they are swept over towards the side of the line on which $ds$ is taken positive, or the reverse. Choose signs as indicated in the figure. If the line $AB$ returns finally to its original position, $A$ and $B$ having described closed curves, it is evident that the formula above will give (taking account of signs) the excess of the area inclosed by the path of $A$ over that inclosed by the path of $B$.

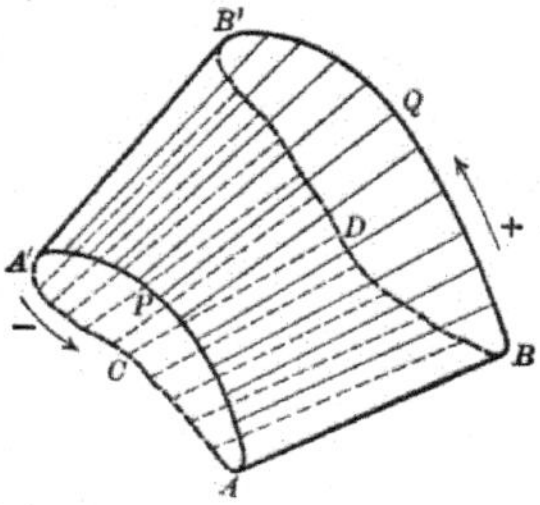


For

$$\text{positive area} = ABQB'A'PA = ABDB'A'PA + \text{closed curve } BQB'DB,$$
$$\text{negative area} = B'A'CABDB' = ABDB'A'PA + \text{closed curve } APA'CA.$$

Finding the difference, we have

$$\text{net area} = \text{closed curve } BQB'DB - \text{closed curve } APA'CA.$$

* Area of circular sector $= \frac{1}{2}$ radius × arc $= \frac{1}{2}OQ \cdot OQ d\theta = \frac{1}{2}\overline{OQ}^2 d\theta$.

† It should be observed that $s$ will not be the length of the path described by the center $R$ unless $AA'$ and $BB'$ are the arcs of circles with the center at $O$.

Now if the area of one of these closed curves (as $APA'CA$) is zero, that is, $A$ keeps to the same path both going and returning, *the area swept over by the line will equal the area of the closed curve* $BQB'DB$.

A simple and widely used type of polar planimeter was invented by Amsler, of Schaffhausen, in 1854. This consists essentially of two bars $OA$ and $AB$, freely jointed at $A$, $OA$ rotating about a fixed point $O$ and $AB$ being the axis of a wheel situated at its center $R$, and having a tracing point at $B$. Now if the tracing point completely describes the closed curve, $A$ will oscillate to and fro along an arc of a circle (as $CD$), describing a contour of zero area. Hence the area swept over by the bar $AB$ exactly equals the area of the closed curve, and is given by the formula

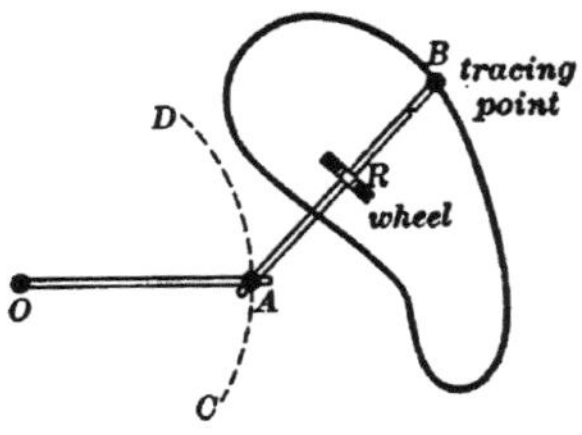


(*D*) area of closed curve $= 2\pi rln$,

where $l$ = length of bar $AB$,

$r$ = radius of wheel,

$n$ = number of revolutions indicated on the wheel after the tracing point has made one complete circuit of the curve.

**238. Approximate integration.** Since the value of a definite integral is a measure of the area under a curve, it follows that the accurate measurement of such an area will give the exact value of a definite integral, and an approximate measurement of this area will give an approximate value of the integral. We will now explain two approximate rules for measuring areas.

**239. Trapezoidal rule.** Instead of inscribing rectangles within the area, as was done in § 204, p. 361, it is evident that we shall get a much closer approximation to the area by inscribing trapezoids. Thus divide the interval from $x = a$ to $x = b$ into $n$ equal parts and denote each part by $\Delta x$. Then, the area of a trapezoid being one half the product of the sum of the parallel sides multiplied by the altitude, we get

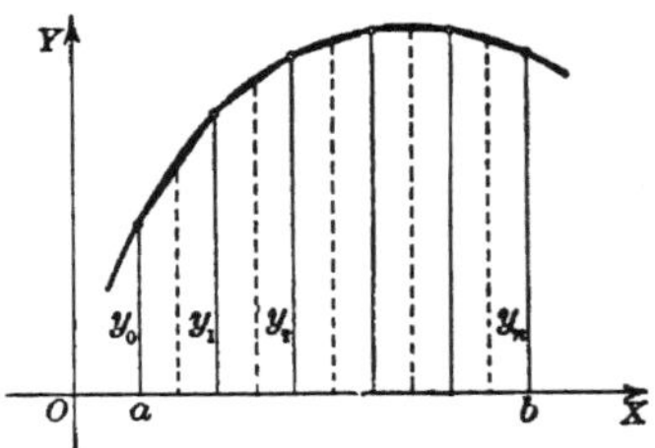


$$\tfrac{1}{2}(y_0 + y_1)\Delta x = \text{area of first trapezoid,}$$

$$\tfrac{1}{2}(y_1 + y_2)\Delta x = \text{area of second trapezoid,}$$

$$\cdot \quad \cdot \quad \cdot \quad \cdot \quad \cdot \quad \cdot$$

$$\tfrac{1}{2}(y_{n-1} - y_n)\Delta x = \text{area of } n\text{th trapezoid.}$$

Adding, we get

$$\tfrac{1}{2}(y_0 + 2y_1 + 2y_2 + \cdots + 2y_{n-1} + y_n)\Delta x = \text{area of trapezoids.}$$

Hence **trapezoidal rule** is

$$(A)\qquad \textbf{area} = (\tfrac{1}{2}y_0 + y_1 + y_2 + \cdots + y_{n-1} + \tfrac{1}{2}y_n)\,\Delta x.$$

It is clear that the greater the number of intervals (i.e. the smaller $\Delta x$ is) the closer will the sum of the areas of the trapezoids approach the area under the curve.

ILLUSTRATIVE EXAMPLE 1. Calculate $\int_1^{12} x^2dx$ by the trapezoidal rule, dividing $x = 1$ to $x = 12$ into eleven intervals.

**Solution.** Here $\frac{b-a}{n} = \frac{12-1}{11} = 1 = \Delta x$. The area in question is under the curve $y = x^2$. Substituting the abscissas $x = 1, 2, 3, \cdots, 12$ in this equation, we get the ordinates $y = 1, 4, 9, \cdots, 144$. Hence, from $(A)$,

$$\text{area} = (\tfrac{1}{2} + 4 + 9 + 16 + 25 + 36 + 49 + 64 + 81 + 100 + 121 + \tfrac{1}{2}\cdot 144)\cdot 1 = 577\tfrac{1}{2}.$$

By integration $\int_1^{12} x^2dx = \left[\frac{x^3}{3}\right]_1^{12} = 575\tfrac{2}{3}$. Hence, in this example, the trapezoidal rule is in error by less than one third of 1%.

**240. Simpson's rule (parabolic rule).** Instead of drawing straight lines (chords) between the points of a curve and forming trapezoids, we can get a still closer approximation to the area by connecting the points with arcs of parabolas and summing up the areas under these arcs. A parabola with a vertical axis may be passed through any three points on a curve, and a series of such arcs will fit the curve more closely than the broken line of chords. We now divide the interval from $x=a=OM_0$ to $x=b=OM_n$ into an *even* number ($=n$) of parts, each equal to $\Delta x$. Through each successive set of three points $P_0$, $P_1$, $P_2$; $P_2$, $P_3$, $P_4$; etc., are drawn arcs of parabolas with vertical axes. From the figure

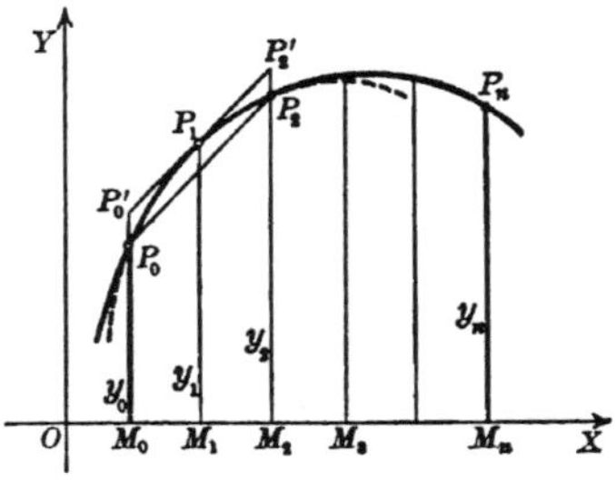


area of parabolic strip $M_0P_0P_1P_2M_2$ = area of trapezoid $M_0P_0P_2M_2$

+ area of parabolic segment $P_0P_1P_2$.

But the area of the trapezoid $M_0P_0P_2M_2 = \frac{1}{2}(y_0 + y_2)\,2\,\Delta x$

$= (y_0 + y_2)\,\Delta x,$

and the area of the parabolic segment $P_0P_1P_2$

= two thirds of the circumscribing parallelogram $P_0P_0'P_2'P_2$

$= \frac{2}{3}[y_1 - \frac{1}{2}(y_0 + y_2)]\,2\,\Delta x = \frac{2}{3}(2\,y_1 - y_0 - y_2)\Delta x.$

Hence area of first parabolic strip $M_0P_0P_1P_2M_2$

$$= (y_0 + y_2)\Delta x + \tfrac{2}{3}(2y_1 - y_0 - y_2)\Delta x$$

$$= \frac{\Delta x}{3}(y_0 + 4y_1 + y_2).$$

Similarly, second strip $= \dfrac{\Delta x}{3}(y_2 + 4y_3 + y_4)$,

third strip $= \dfrac{\Delta x}{3}(y_4 + 4y_5 + y_6)$,

. . . . .

$n$th strip $= \dfrac{\Delta x}{3}(y_{n-2} + 4y_{n-1} + y_n)$.

Adding, we get

$$\frac{\Delta x}{3}(y_0 + 4y_1 + 2y_2 + 4y_3 + 2y_4 + \cdots + 2y_{n-2} + 4y_{n-1} + y_n)$$

as the sum of these areas. Hence **Simpson's rule** is ($n$ being even)

$$(B) \qquad \textbf{area} = \frac{\Delta x}{3}(y_0 + 4y_1 + 2y_2 + 4y_3 + 2y_4 + \cdots + y_n).$$

As in the case of the trapezoidal rule, the greater the number of parts into which $M_0M_n$ is divided, the closer will the result be to the area under the curve.

Illustrative Example 1. Calculate $\int_0^{10} x^3dx$ by Simpson's rule, taking ten intervals.

**Solution.** Here $\dfrac{b-a}{n} = \dfrac{10-0}{10} = 1 = \Delta x$. The area in question is under the curve $y = x^3$. Substituting the abscissas $x = 0, 1, 2, \cdots, 10$ in $y = x^3$, we get the ordinates $y = 0, 1, 8, 27, \cdots, 1000$. Hence, from ($B$),

$$\text{area} = \tfrac{1}{3}(0 + 4 + 16 + 108 + 128 + 500 + 432 + 1372 + 1024 + 2916 + 1000) = 2500.$$

By integration, $\int_0^{10} x^3dx = \left[\dfrac{x^4}{4}\right]_0^{10} = 2500$, so that in this example Simpson's rule happens to give an exact result.

### EXAMPLES

1. Calculate the integral in Illustrative Example 1 (above) by the trapezoidal rule, taking ten intervals. *Ans.* 2525.

2. Calculate $\int_1^5 \dfrac{dx}{x}$ by both rules when $n = 12$. *Ans.* Trap. 1.6182; Simp. 1.6098.

3. Evaluate $\int_1^{11} x^3dx$ by both rules when $n = 10$. *Ans.* Trap. 3690; Simp. 3660.

4. Calculate $\int_1^{10} \log_{10} xdx$ by both rules when $n = 10$. *Ans.* Trap. 6.0656; Simp. 6.0896.

5. Evaluate $\int_0^2 \dfrac{dx}{1+x^2}$ by both rules when $n = 6$. *Ans.* Trap. 1.0885; Simp. 1.0906.

6. Calculate $\int_0^{60°} \sin x dx$ by both rules for ten-degree intervals.

7. Evaluate $\int_2^6 x^3 dx$ by both rules for $n = 12$.

8. Find the error in the evaluation of $\int_0^{10} x^4 dx$ by Simpson's rule when $n = 10$.

9. Evaluate $\int_0^1 e^x dx$ by Simpson's rule when $n = 10$.

**241. Integrals for reference.** Following is a table of integrals for reference. In going over the subject of Integral Calculus for the first time, the student is advised to use this table sparingly, if at all. As soon as the derivation of these integrals is thoroughly understood, the table may be properly used for saving time and labor in the solution of practical problems.

SOME ELEMENTARY FORMS

1. $\int (du \pm dv \pm dw \pm \cdots) = \int du \pm \int dv \pm \int dw \pm \cdots.$

2. $\int a dv = a \int dv.$

3. $\int df(x) = \int f'(x) dx = f(x) + C.$

4. $\int x^n dx = \frac{x^{n+1}}{n+1} + C, n \neq -1.$

5. $\int \frac{dx}{x} = \log x + C.$

FORMS CONTAINING INTEGRAL POWERS OF $a + bx$

6. $\int \frac{dx}{a+bx} = \frac{1}{b} \log (a + bx) + C.$

7. $\int (a + bx)^n dx = \frac{(a+bx)^{n+1}}{b(n+1)} + C, n \neq -1.$

8. $\int F(x, a + bx) dx.$ Try one of the substitutions, $z = a + bx$, $xz = a + bx$.

9. $\int \frac{x dx}{a+bx} = \frac{1}{b^2} [a + bx - a \log (a + bx)] + C.$

10. $\int \frac{x^2 dx}{a+bx} = \frac{1}{b^3} [\frac{1}{2} (a + bx)^2 - 2a(a + bx) + a^2 \log (a + bx)] + C.$

11. $\int \frac{dx}{x(a+bx)} = -\frac{1}{a} \log \frac{a+bx}{x} + C.$

12. $\int \frac{dx}{x^2(a+bx)} = -\frac{1}{ax} + \frac{b}{a^2} \log \frac{a+bx}{x} + C.$

13. $\int \frac{x dx}{(a+bx)^2} = \frac{1}{b^2} \left[ \log (a + bx) + \frac{a}{a+bx} \right] + C.$

14. $\int \frac{x^2 dx}{(a+bx)^2} = \frac{1}{b^3} \left[ a + bx - 2a \log (a + bx) - \frac{a^2}{a+bx} \right] + C.$

15. $\int \frac{dx}{x(a+bx)^2} = \frac{1}{a(a+bx)} - \frac{1}{a^2} \log \frac{a+bx}{x} + C.$

16. $\int \frac{x dx}{(a+bx)^3} = \frac{1}{b^2} \left[ -\frac{1}{a+bx} + \frac{a}{2(a+bx)^2} \right] + C.$

**FORMS CONTAINING $a^2+x^2$, $a^2-x^2$, $a+bx^n$, $a+bx^2$**

17. $\int \frac{dx}{a^2+x^2} = \frac{1}{a}\tan^{-1}\frac{x}{a} + C; \quad \int \frac{dx}{1+x^2} = \tan^{-1}x + C.$

18. $\int \frac{dx}{a^2-x^2} = \frac{1}{2a}\log\frac{a+x}{a-x} + C; \quad \int \frac{dx}{x^2-a^2} = \frac{1}{2a}\log\frac{x-a}{x+a} + C.$

19. $\int \frac{dx}{a+bx^2} = \frac{1}{\sqrt{ab}}\tan^{-1}x\sqrt{\frac{b}{a}} + C.$

20. $\int \frac{dx}{a^2-b^2x^2} = \frac{1}{2ab}\log\frac{a+bx}{a-bx} + C.$

21. $\int x^m(a+bx^n)^p dx$

$$= \frac{x^{m-n+1}(a+bx^n)^{p+1}}{b(np+m+1)} - \frac{a(m-n+1)}{b(np+m+1)}\int x^{m-n}(a+bx^n)^p dx.$$

22. $\int x^m(a+bx^n)^p dx = \frac{x^{m+1}(a+bx^n)^p}{np+m+1} + \frac{anp}{np+m+1}\int x^m(a+bx^n)^{p-1}dx.$

23. $\int \frac{dx}{x^m(a+bx^n)^p}$

$$= -\frac{1}{(m-1)ax^{m-1}(a+bx^n)^{p-1}} - \frac{(m-n+np-1)b}{(m-1)a}\int \frac{dx}{x^{m-n}(a+bx^n)^p}.$$

24. $\int \frac{dx}{x^m(a+bx^n)^p}$

$$= \frac{1}{an(p-1)x^{m-1}(a+bx^n)^{p-1}} + \frac{m-n+np-1}{an(p-1)}\int \frac{dx}{x^m(a+bx^n)^{p-1}}.$$

25. $\int \frac{(a+bx^n)^p dx}{x^m} = -\frac{(a+bx^n)^{p+1}}{a(m-1)x^{m-1}} - \frac{b(m-n-np-1)}{a(m-1)}\int \frac{(a+bx^n)^p dx}{x^{m-n}}.$

26. $\int \frac{(a+bx^n)^p dx}{x^m} = \frac{(a+bx^n)^p}{(np-m+1)x^{m-1}} + \frac{anp}{np-m+1}\int \frac{(a+bx^n)^{p-1}dx}{x^m}.$

27. $\int \frac{x^m dx}{(a+bx^n)^p} = \frac{x^{m-n+1}}{b(m-np+1)(a+bx^n)^{p-1}} - \frac{a(m-n+1)}{b(m-np+1)}\int \frac{x^{m-n}dx}{(a+bx^n)^p}.$

28. $\int \frac{x^m dx}{(a+bx^n)^p} = \frac{x^{m+1}}{an(p-1)(a+bx^n)^{p-1}} - \frac{m+n-np+1}{an(p-1)}\int \frac{x^m dx}{(a+bx^n)^{p-1}}.$

29. $\int \frac{dx}{(a^2+x^2)^n} = \frac{1}{2(n-1)a^2}\left[\frac{x}{(a^2+x^2)^{n-1}} + (2n-3)\int \frac{dx}{(a^2+x^2)^{n-1}}\right].$

30. $\int \frac{dx}{(a+bx^2)^n} = \frac{1}{2(n-1)a}\left[\frac{x}{(a+bx^2)^{n-1}} + (2n-3)\int \frac{dx}{(a+bx^2)^{n-1}}\right].$

31. $\int \frac{xdx}{(a+bx^2)^n} = \frac{1}{2}\int \frac{dz}{(a+bz)^n}$, where $z = x^2$.

32. $\int \frac{x^2dx}{(a+bx^2)^n} = \frac{-x}{2b(n-1)(a+bx^2)^{n-1}} + \frac{1}{2b(n-1)}\int \frac{dx}{(a+bx^2)^{n-1}}.$

33. $\int \frac{dx}{x(a+bx^n)} = \frac{1}{an}\log\frac{x^n}{a+bx^n} + C.$

34. $\int \frac{dx}{x^2(a+bx^2)^n} = \frac{1}{a}\int \frac{dx}{x^2(a+bx^2)^{n-1}} - \frac{b}{a}\int \frac{dx}{(a+bx^2)^n}.$

35. $\int \frac{xdx}{a+bx^2} = \frac{1}{2b}\log\left(x^2+\frac{a}{b}\right) + C.$

36. $\int \frac{x^2dx}{a+bx^2} = \frac{x}{b} - \frac{a}{b}\int \frac{dx}{a+bx^2}.$

37. $\int \frac{dx}{x(a+bx^2)} = \frac{1}{2a}\log \frac{x^2}{a+bx^2} + C.$

38. $\int \frac{dx}{x^2(a+bx^2)} = -\frac{1}{ax} - \frac{b}{a}\int \frac{dx}{a+bx^2}.$

39. $\int \frac{dx}{(a+bx^2)^2} = \frac{x}{2a(a+bx^2)} + \frac{1}{2a}\int \frac{dx}{a+bx^2}.$

### Forms containing $\sqrt{a+bx}$

40. $\int x\sqrt{a+bx}\,dx = -\frac{2(2a-3bx)\sqrt{(a+bx)^3}}{15b^2} + C.$

41. $\int x^2\sqrt{a+bx}\,dx = \frac{2(8a^2-12abx+15b^2x^2)\sqrt{(a+bx)^3}}{105b^3} + C.$

42. $\int \frac{xdx}{\sqrt{a+bx}} = -\frac{2(2a-bx)}{3b^2}\sqrt{a+bx} + C.$

43. $\int \frac{x^2dx}{\sqrt{a+bx}} = \frac{2(8a^2-4abx+3b^2x^2)}{15b^3}\sqrt{a+bx} + C.$

44. $\int \frac{dx}{x\sqrt{a+bx}} = \frac{1}{\sqrt{a}}\log \frac{\sqrt{a+bx}-\sqrt{a}}{\sqrt{a+bx}+\sqrt{a}} + C$, for $a > 0$.

45. $\int \frac{dx}{x\sqrt{a+bx}} = \frac{2}{\sqrt{-a}}\tan^{-1}\sqrt{\frac{a+bx}{-a}} + C$, for $a < 0$.

46. $\int \frac{dx}{x^2\sqrt{a+bx}} = \frac{-\sqrt{a+bx}}{ax} - \frac{b}{2a}\int \frac{dx}{x\sqrt{a+bx}}.$

47. $\int \frac{\sqrt{a+bx}\,dx}{x} = 2\sqrt{a+bx} + a\int \frac{dx}{x\sqrt{a+bx}}.$

### Forms containing $\sqrt{x^2+a^2}$

48. $\int (x^2+a^2)^{\frac{1}{2}}dx = \frac{x}{2}\sqrt{x^2+a^2} + \frac{a^2}{2}\log\left(x+\sqrt{x^2+a^2}\right) + C.$

49. $\int (x^2+a^2)^{\frac{3}{2}}dx = \frac{x}{8}(2x^2+5a^2)\sqrt{x^2+a^2} + \frac{3a^4}{8}\log\left(x+\sqrt{x^2+a^2}\right) + C.$

50. $\int (x^2+a^2)^{\frac{n}{2}}dx = \frac{x(x^2+a^2)^{\frac{n}{2}}}{n+1} + \frac{na^2}{n+1}\int (x^2+a^2)^{\frac{n}{2}-1}dx.$

51. $\int x(x^2+a^2)^{\frac{n}{2}}dx = \frac{(x^2+a^2)^{\frac{n+2}{2}}}{n+2} + C.$

52. $\int x^2(x^2+a^2)^{\frac{1}{2}}dx = \frac{x}{8}(2x^2+a^2)\sqrt{x^2+a^2} - \frac{a^4}{8}\log\left(x+\sqrt{x^2+a^2}\right) + C.$

53. $\int \frac{dx}{(x^2+a^2)^{\frac{1}{2}}} = \log\left(x+\sqrt{x^2+a^2}\right) + C.$

54. $\int \frac{dx}{(x^2+a^2)^{\frac{3}{2}}} = \frac{x}{a^2\sqrt{x^2+a^2}} + C.$

55. $\int \frac{x\,dx}{(x^2+a^2)^{\frac{1}{2}}} = \sqrt{x^2+a^2} + C.$

56. $\int \frac{x^2dx}{(x^2+a^2)^{\frac{1}{2}}} = \frac{x}{2}\sqrt{x^2+a^2} - \frac{a^2}{2}\log\left(x+\sqrt{x^2+a^2}\right) + C.$

57. $\int \frac{x^2dx}{(x^2+a^2)^{\frac{3}{2}}} = -\frac{x}{\sqrt{x^2+a^2}} + \log\left(x+\sqrt{x^2+a^2}\right) + C.$

58. $\int \frac{dx}{x(x^2+a^2)^{\frac{1}{2}}} = \frac{1}{a}\log\frac{x}{a+\sqrt{x^2+a^2}} + C.$

59. $\int \frac{dx}{x^2(x^2+a^2)^{\frac{1}{2}}} = -\frac{\sqrt{x^2+a^2}}{a^2x} + C.$

60. $\int \frac{dx}{x^3(x^2+a^2)^{\frac{1}{2}}} = -\frac{\sqrt{x^2+a^2}}{2\,a^2x^2} + \frac{1}{2\,a^3}\log\frac{a+\sqrt{x^2+a^2}}{x} + C.$

61. $\int \frac{(x^2+a^2)^{\frac{1}{2}}dx}{x} = \sqrt{a^2+x^2} - a\log\frac{a+\sqrt{a^2+x^2}}{x} + C.$

62. $\int \frac{(x^2+a^2)^{\frac{1}{2}}dx}{x^2} = -\frac{\sqrt{x^2+a^2}}{x} + \log\left(x+\sqrt{x^2+a^2}\right) + C.$

FORMS CONTAINING $\sqrt{x^2-a^2}$

63. $\int (x^2-a^2)^{\frac{1}{2}}dx = \frac{x}{2}\sqrt{x^2-a^2} - \frac{a^2}{2}\log\left(x+\sqrt{x^2-a^2}\right) + C.$

64. $\int (x^2-a^2)^{\frac{3}{2}}dx = \frac{x}{8}(2\,x^2-5\,a^2)\sqrt{x^2-a^2} + \frac{3\,a^4}{8}\log\left(x+\sqrt{x^2-a^2}\right) + C.$

65. $\int (x^2-a^2)^{\frac{n}{2}}dx = \frac{x(x^2-a^2)^{\frac{n}{2}}}{n+1} - \frac{na^2}{n+1}\int (x^2+a^2)^{\frac{n}{2}-1}dx.$

66. $\int x(x^2-a^2)^{\frac{n}{2}}dx = \frac{(x^2-a^2)^{\frac{n+2}{2}}}{n+2} + C.$

67. $\int x^2(x^2-a^2)^{\frac{1}{2}}dx = \frac{x}{8}(2\,x^2-a^2)\sqrt{x^2-a^2} - \frac{a^4}{8}\log\left(x+\sqrt{x^2-a^2}\right) + C.$

68. $\int \frac{dx}{(x^2-a^2)^{\frac{1}{2}}} = \log\left(x+\sqrt{x^2-a^2}\right) + C.$

69. $\int \frac{dx}{(x^2-a^2)^{\frac{3}{2}}} = -\frac{x}{a^2\sqrt{x^2-a^2}} + C.$

70. $\int \frac{x\,dx}{(x^2-a^2)^{\frac{1}{2}}} = \sqrt{x^2-a^2} + C.$

71. $\int \frac{x^2dx}{(x^2-a^2)^{\frac{1}{2}}} = \frac{x}{2}\sqrt{x^2-a^2} + \frac{a^2}{2}\log\left(x+\sqrt{x^2-a^2}\right) + C.$

72. $\int \frac{x^2dx}{(x^2-a^2)^{\frac{3}{2}}} = -\frac{x}{\sqrt{x^2-a^2}} + \log\left(x+\sqrt{x^2-a^2}\right) + C.$

73. $\int \frac{dx}{x(x^2-a^2)^{\frac{1}{2}}} = \frac{1}{a}\sec^{-1}\frac{x}{a} + C;\ \int \frac{dx}{x\sqrt{x^2-1}} = \sec^{-1}x + C.$

74. $\int \frac{dx}{x^2(x^2-a^2)^{\frac{1}{2}}} = \frac{\sqrt{x^2-a^2}}{a^2x} + C.$

75. $\int \frac{dx}{x^3(x^2-a^2)^{\frac{1}{2}}} = \frac{\sqrt{x^2-a^2}}{2a^2x^2} + \frac{1}{2a^3}\sec^{-1}\frac{x}{a} + C.$

76. $\int \frac{(x^2-a^2)^{\frac{1}{2}}dx}{x} = \sqrt{x^2-a^2} - a\cos^{-1}\frac{a}{x} + C.$

77. $\int \frac{(x^2-a^2)^{\frac{1}{2}}dx}{x^2} = -\frac{\sqrt{x^2-a^2}}{x} + \log(x+\sqrt{x^2-a^2}) + C.$

### Forms containing $\sqrt{a^2-x^2}$

78. $\int (a^2-x^2)^{\frac{1}{2}}dx = \frac{x}{2}\sqrt{a^2-x^2} + \frac{a^2}{2}\sin^{-1}\frac{x}{a} + C.$

79. $\int (a^2-x^2)^{\frac{3}{2}}dx = \frac{x}{8}(5a^2-2x^2)\sqrt{a^2-x^2} + \frac{3a^4}{8}\sin^{-1}\frac{x}{a} + C.$

80. $\int (a^2-x^2)^{\frac{n}{2}}dx = \frac{x(a^2-x^2)^{\frac{n}{2}}}{n+1} + \frac{a^2n}{n+1}\int (a^2-x^2)^{\frac{n}{2}-1}dx.$

81. $\int x(a^2-x^2)^{\frac{n}{2}}dx = -\frac{(a^2-x^2)^{\frac{n+2}{2}}}{n+2} + C.$

82. $\int x^2(a^2-x^2)^{\frac{1}{2}}dx = \frac{x}{8}(2x^2-a^2)\sqrt{a^2-x^2} + \frac{a^4}{8}\sin^{-1}\frac{x}{a} + C.$

83. $\int \frac{dx}{(a^2-x^2)^{\frac{1}{2}}} = \sin^{-1}\frac{x}{a}; \int \frac{dx}{\sqrt{1-x^2}} = \sin^{-1}x.$

84. $\int \frac{dx}{(a^2-x^2)^{\frac{3}{2}}} = \frac{x}{a^2\sqrt{a^2-x^2}} + C.$

85. $\int \frac{xdx}{(a^2-x^2)^{\frac{1}{2}}} = -\sqrt{a^2-x^2} + C.$

86. $\int \frac{x^2dx}{(a^2-x^2)^{\frac{1}{2}}} = -\frac{x}{2}\sqrt{a^2-x^2} + \frac{a^2}{2}\sin^{-1}\frac{x}{a} + C.$

87. $\int \frac{x^2dx}{(a^2-x^2)^{\frac{3}{2}}} = \frac{x}{\sqrt{a^2-x^2}} - \sin^{-1}\frac{x}{a} + C.$

88. $\int \frac{x^m dx}{(a^2-x^2)^{\frac{1}{2}}} = -\frac{x^{m-1}}{m}\sqrt{a^2-x^2} + \frac{(m-1)a^2}{m}\int \frac{x^{m-2}}{(a^2-x^2)^{\frac{1}{2}}}dx.$

89. $\int \frac{dx}{x(a^2-x^2)^{\frac{1}{2}}} = \frac{1}{a}\log\frac{x}{a+\sqrt{a^2-x^2}} + C.$

90. $\int \frac{dx}{x^2(a^2-x^2)^{\frac{1}{2}}} = -\frac{\sqrt{a^2-x^2}}{a^2x} + C.$

91. $\int \frac{dx}{x^3(a^2-x^2)^{\frac{1}{2}}} = -\frac{\sqrt{a^2-x^2}}{2a^2x^2} + \frac{1}{2a^3}\log\frac{x}{a+\sqrt{a^2-x^2}} + C.$

92. $\int \frac{(a^2 - x^2)^{\frac{1}{2}}}{x} dx = \sqrt{a^2 - x^2} - a \log \frac{a + \sqrt{a^2 - x^2}}{x} + C.$

93. $\int \frac{(a^2 - x^2)^{\frac{1}{2}}}{x^2} dx = - \frac{\sqrt{a^2 - x^2}}{x} - \sin^{-1} \frac{x}{a} + C.$

FORMS CONTAINING $\sqrt{2ax - x^2}$, $\sqrt{2ax + x^2}$

94. $\int \sqrt{2\,ax - x^2}\, dx = \frac{x - a}{2} \sqrt{2\,ax - x^2} + \frac{a^2}{2} \operatorname{vers}^{-1} \frac{x}{a} + C.$

95. $\int \frac{dx}{\sqrt{2\,ax - x^2}} = \operatorname{vers}^{-1} \frac{x}{a}; \int \frac{dx}{\sqrt{2\,x - x^2}} = \operatorname{vers}^{-1} x + C.$

96. $\int x^m \sqrt{2\,ax - x^2}\, dx = - \frac{x^{m-1} (2\,ax - x^2)^{\frac{3}{2}}}{m + 2} + \frac{(2\,m + 1)\,a}{m + 2} \int x^{m-1} \sqrt{2\,ax - x^2}\, dx.$

97. $\int \frac{dx}{x^m \sqrt{2\,ax - x^2}} = - \frac{\sqrt{2\,ax - x^2}}{(2\,m - 1)\,ax^m} + \frac{m - 1}{(2\,m - 1)\,a} \int \frac{dx}{x^{m-1} \sqrt{2\,ax - x^2}}.$

98. $\int \frac{x^m dx}{\sqrt{2\,ax - x^2}} = - \frac{x^{m-1} \sqrt{2\,ax - x^2}}{m} + \frac{(2\,m - 1)\,a}{m} \int \frac{x^{m-1} dx}{\sqrt{2\,ax - x^2}}.$

99. $\int \frac{\sqrt{2\,ax - x^2}}{x^m} dx = - \frac{(2\,ax - x^2)^{\frac{3}{2}}}{(2\,m - 3)\,ax^m} + \frac{m - 3}{(2\,m - 3)\,a} \int \frac{\sqrt{2\,ax - x^2}}{x^{m-1}} dx.$

100. $\int x \sqrt{2\,ax - x^2}\, dx = - \frac{3\,a^2 + ax - 2\,x^2}{6} \sqrt{2\,ax - x^2} + \frac{a^3}{2} \operatorname{vers}^{-1} \frac{x}{a}.$

101. $\int \frac{dx}{x \sqrt{2\,ax - x^2}} = - \frac{\sqrt{2\,ax - x^2}}{ax} + C.$

102. $\int \frac{x\,dx}{\sqrt{2\,ax - x^2}} = - \sqrt{2\,ax - x^2} + a \operatorname{vers}^{-1} \frac{x}{a} + C.$

103. $\int \frac{x^2 dx}{\sqrt{2\,ax - x^2}} = - \frac{x + 3\,a}{2} \sqrt{2\,ax - x^2} + \frac{3}{2} a^2 \operatorname{vers}^{-1} \frac{x}{a} + C.$

104. $\int \frac{\sqrt{2\,ax - x^2}}{x} dx = \sqrt{2\,ax - x^2} + a \operatorname{vers}^{-1} \frac{x}{a} + C.$

105. $\int \frac{\sqrt{2\,ax - x^2}}{x^2} dx = - \frac{2 \sqrt{2\,ax - x^2}}{x} - \operatorname{vers}^{-1} \frac{x}{a} + C.$

106. $\int \frac{\sqrt{2\,ax - x^2}}{x^3} dx = - \frac{(2\,ax - x^2)^{\frac{3}{2}}}{3\,ax^3} + C.$

107. $\int \frac{dx}{(2\,ax - x^2)^{\frac{3}{2}}} = \frac{x - a}{a^2 \sqrt{2\,ax - x^2}} + C.$

108. $\int \frac{x\,dx}{(2\,ax - x^2)^{\frac{3}{2}}} = \frac{x}{a \sqrt{2\,ax - x^2}} + C.$

109. $\int F(x, \sqrt{2\,ax - x^2})\, dx = \int F(z + a, \sqrt{a^2 - z^2})\, dz$, where $z = x - a$.

110. $\int \frac{dx}{\sqrt{2\,ax + x^2}} = \log (x + a + \sqrt{2\,ax + x^2}) + C.$

111. $\int F(x, \sqrt{2\,ax + x^2})\, dx = \int F(z - a, \sqrt{z^2 - a^2})\, dz$, where $z = x + a$.

### Forms containing $a + bx \pm cx^2$

112. $\int \frac{dx}{a + bx + cx^2} = \frac{2}{\sqrt{4ac - b^2}} \tan^{-1} \frac{2cx + b}{\sqrt{4ac - b^2}} + C$, when $b^2 < 4ac$.

113. $\int \frac{dx}{a + bx + cx^2} = \frac{1}{\sqrt{b^2 - 4ac}} \log \frac{2cx + b - \sqrt{b^2 - 4ac}}{2cx + b + \sqrt{b^2 - 4ac}} + C$, when $b^2 > 4ac$.

114. $\int \frac{dx}{a + bx - cx^2} = \frac{1}{\sqrt{b^2 + 4ac}} \log \frac{\sqrt{b^2 + 4ac} + 2cx - b}{\sqrt{b^2 + 4ac} - 2cx + b} + C.$

115. $\int \frac{dx}{\sqrt{a + bx + cx^2}} = \frac{1}{\sqrt{c}} \log (2cx + b + 2\sqrt{c}\sqrt{a + bx + cx^2}) + C.$

116. $\int \sqrt{a + bx + cx^2}\,dx$

$$= \frac{2cx + b}{4c}\sqrt{a + bx + cx^2} - \frac{b^2 - 4ac}{8c^{\frac{3}{2}}} \log (2cx + b + 2\sqrt{c}\sqrt{a + bx + cx^2}) + C.$$

117. $\int \frac{dx}{\sqrt{a + bx - cx^2}} = \frac{1}{\sqrt{c}} \sin^{-1} \frac{2cx - b}{\sqrt{b^2 + 4ac}} + C.$

118. $\int \sqrt{a + bx - cx^2}\,dx = \frac{2cx - b}{4c}\sqrt{a + bx - cx^2} + \frac{b^2 + 4ac}{8c^{\frac{3}{2}}} \sin^{-1} \frac{2cx - b}{\sqrt{b^2 + 4ac}} + C.$

119. $\int \frac{xdx}{\sqrt{a + bx + cx^2}} = \frac{\sqrt{a + bx + cx^2}}{c} - \frac{b}{2c^{\frac{3}{2}}} \log (2cx + b + 2\sqrt{c}\sqrt{a + bx + cx^2}) + C.$

120. $\int \frac{xdx}{\sqrt{a + bx - cx^2}} = -\frac{\sqrt{a + bx - cx^2}}{c} + \frac{b}{2c^{\frac{3}{2}}} \sin^{-1} \frac{2cx - b}{\sqrt{b^2 + 4ac}} + C.$

### Other Algebraic Forms

121. $\int \sqrt{\frac{a + x}{b + x}}\,dx = \sqrt{(a + x)(b + x)} + (a - b) \log (\sqrt{a + x} + \sqrt{b + x}) + C.$

122. $\int \sqrt{\frac{a - x}{b + x}}\,dx = \sqrt{(a - x)(b + x)} + (a + b) \sin^{-1} \sqrt{\frac{x + b}{a + b}} + C.$

123. $\int \sqrt{\frac{a + x}{b - x}}\,dx = -\sqrt{(a + x)(b - x)} - (a + b) \sin^{-1} \sqrt{\frac{b - x}{a + b}} + C.$

124. $\int \sqrt{\frac{1 + x}{1 - x}}\,dx = -\sqrt{1 - x^2} + \sin^{-1} x + C.$

125. $\int \frac{dx}{\sqrt{(x - \alpha)(\beta - x)}} = 2 \sin^{-1} \sqrt{\frac{x - \alpha}{\beta - \alpha}} + C.$

### Exponential and Trigonometric Forms

126. $\int a^x dx = \frac{a^x}{\log a} + C.$

127. $\int e^x dx = e^x + C.$

128. $\int e^{ax} dx = \frac{e^{ax}}{a} + C.$

129. $\int \sin x dx = -\cos x + C.$

130. $\int \cos x dx = \sin x + C.$

131. $\int \tan x dx = \log \sec x = -\log \cos x + C.$

132. $\int \cot x dx = \log \sin x + C.$

133. $\int \sec x dx = \int \frac{dx}{\cos x} = \log (\sec x + \tan x) = \log \tan \left(\frac{\pi}{4} + \frac{x}{2}\right) + C.$

134. $\int \operatorname{cosec} x dx = \int \frac{dx}{\sin x} = \log(\operatorname{cosec} x - \cot x) = \log \tan \frac{x}{2} + C.$

135. $\int \sec^2 x dx = \tan x + C.$

136. $\int \operatorname{cosec}^2 x dx = -\cot x + C.$

137. $\int \sec x \tan x dx = \sec x + C.$

138. $\int \operatorname{cosec} x \cot x dx = -\operatorname{cosec} x + C.$

139. $\int \sin^2 x dx = \frac{x}{2} - \frac{1}{4}\sin 2x + C.$

140. $\int \cos^2 x dx = \frac{x}{2} + \frac{1}{4}\sin 2x + C.$

141. $\int \sin^n x dx = -\frac{\sin^{n-1} x \cos x}{n} + \frac{n-1}{n}\int \sin^{n-2} x dx.$

142. $\int \cos^n x dx = \frac{\cos^{n-1} x \sin x}{n} + \frac{n-1}{n}\int \cos^{n-2} x dx.$

143. $\int \frac{dx}{\sin^n x} = -\frac{1}{n-1}\frac{\cos x}{\sin^{n-1} x} + \frac{n-2}{n-1}\int \frac{dx}{\sin^{n-2} x}.$

144. $\int \frac{dx}{\cos^n x} = \frac{1}{n-1}\frac{\sin x}{\cos^{n-1} x} + \frac{n-2}{n-1}\int \frac{dx}{\cos^{n-2} x}.$

145. $\int \cos^m x \sin^n x dx = \frac{\cos^{m-1} x \sin^{n+1} x}{m+n} + \frac{m-1}{m+n}\int \cos^{m-2} x \sin^n x dx.$

146. $\int \cos^m x \sin^n x dx = -\frac{\sin^{n-1} x \cos^{m+1} x}{m+n} + \frac{n-1}{m+n}\int \cos^m x \sin^{n-2} x dx.$

147. $\int \frac{dx}{\sin^m x \cos^n x} = \frac{1}{n-1} \cdot \frac{1}{\sin^{m-1} x \cos^{n-1} x} + \frac{m+n-2}{n-1}\int \frac{dx}{\sin^m x \cos^{n-2} x}.$

148. $\int \frac{dx}{\sin^m x \cos^n x} = -\frac{1}{m-1} \cdot \frac{1}{\sin^{m-1} x \cos^{n-1} x} + \frac{m+n-2}{m-1}\int \frac{dx}{\sin^{m-2} x \cos^n x}.$

149. $\int \frac{\cos^m x dx}{\sin^n x} = -\frac{\cos^{m+1} x}{(n-1)\sin^{n-1} x} - \frac{m-n+2}{n-1}\int \frac{\cos^m x dx}{\sin^{n-2} x}.$

150. $\int \frac{\cos^m x dx}{\sin^n x} = \frac{\cos^{m-1} x}{(m-n)\sin^{n-1} x} + \frac{m-1}{m-n}\int \frac{\cos^{m-2} x dx}{\sin^n x}.$

151. $\int \sin x \cos^n x dx = -\frac{\cos^{n+1} x}{n+1} + C.$

152. $\int \sin^n x \cos x dx = \frac{\sin^{n+1} x}{n+1} + C.$

153. $\int \tan^n x dx = \frac{\tan^{n-1} x}{n-1} - \int \tan^{n-2} x dx + C.$

154. $\int \cot^n x dx = -\frac{\cot^{n-1} x}{n-1} - \int \cot^{n-2} x dx + C.$

155. $\int \sin mx \sin nx dx = -\frac{\sin(m+n)x}{2(m+n)} + \frac{\sin(m-n)x}{2(m-n)} + C.$

156. $\int \cos mx \cos nx dx = \frac{\sin(m+n)x}{2(m+n)} + \frac{\sin(m-n)x}{2(m-n)} + C.$

157. $\int \sin mx \cos nx dx = -\frac{\cos(m+n)x}{2(m+n)} - \frac{\cos(m-n)x}{2(m-n)} + C.$

158. $\int \frac{dx}{a + b\cos x} = \frac{2}{\sqrt{a^2-b^2}}\tan^{-1}\left(\sqrt{\frac{a-b}{a+b}}\tan\frac{x}{2}\right) + C$, when $a > b$.

159. $\int \frac{dx}{a + b\cos x} = \frac{1}{\sqrt{b^2 - a^2}} \log \frac{\sqrt{b-a}\tan\frac{x}{2} + \sqrt{b+a}}{\sqrt{b-a}\tan\frac{x}{2} - \sqrt{b+a}} + C$, when $a < b$.

160. $\int \frac{dx}{a + b\sin x} = \frac{2}{\sqrt{a^2 - b^2}} \tan^{-1} \frac{a\tan\frac{x}{2} + b}{\sqrt{a^2 - b^2}} + C$, when $a > b$.

161. $\int \frac{dx}{a + b\sin x} = \frac{1}{\sqrt{b^2 - a^2}} \log \frac{a\tan\frac{x}{2} + b - \sqrt{b^2 - a^2}}{a\tan\frac{x}{2} + b + \sqrt{b^2 - a^2}} + C$, when $a < b$.

162. $\int \frac{dx}{a^2\cos^2 x + b^2\sin^2 x} = \frac{1}{ab} \tan^{-1}\left(\frac{b\tan x}{a}\right) + C$.

163. $\int e^{ax}\sin nx dx = \frac{e^{ax}(a\sin nx - n\cos nx)}{a^2 + n^2} + C$; $\int e^{x}\sin x dx = \frac{e^{x}(\sin x - \cos x)}{2} + C$.

164. $\int e^{ax}\cos nx dx = \frac{e^{ax}(n\sin nx + a\cos nx)}{a^2 + n^2} + C$; $\int e^{x}\cos x dx = \frac{e^{x}(\sin x + \cos x)}{2} + C$.

165. $\int xe^{ax}dx = \frac{e^{ax}}{a^2}(ax - 1) + C$.

166. $\int x^n e^{ax}dx = \frac{x^n e^{ax}}{a} - \frac{n}{a}\int x^{n-1}e^{ax}dx$.

167. $\int a^{mx}x^n dx = \frac{a^{mx}x^n}{m\log a} - \frac{n}{m\log a}\int a^{mx}x^{n-1}dx$.

168. $\int \frac{a^x dx}{x^m} = -\frac{a^x}{(m-1)x^{m-1}} + \frac{\log a}{m-1}\int \frac{a^x dx}{x^{m-1}}$.

169. $\int e^{ax}\cos^n x dx = \frac{e^{ax}\cos^{n-1}x\,(a\cos x + n\sin x)}{a^2 + n^2} + \frac{n(n-1)}{a^2 + n^2}\int e^{ax}\cos^{n-2}x dx$.

170. $\int x^m \cos ax dx = \frac{x^{m-1}}{a^2}(ax\sin ax + m\cos ax) - \frac{m(m-1)}{a^2}\int x^{m-2}\cos ax dx$.

## Logarithmic Forms

171. $\int \log x dx = x\log x - x + C$.

172. $\int \frac{dx}{\log x} = \log(\log x) + \log x + \frac{1}{2^2}\log^2 x + \cdots$.

173. $\int \frac{dx}{x\log x} = \log(\log x) + C$.

174. $\int x^n \log x dx = x^{n+1}\left[\frac{\log x}{n+1} - \frac{1}{(n+1)^2}\right] + C$.

175. $\int e^{ax}\log x dx = \frac{e^{ax}\log x}{a} - \frac{1}{a}\int \frac{e^{ax}}{x}dx$.

176. $\int x^m \log^n x dx = \frac{x^{m+1}}{m+1}\log^n x - \frac{n}{m+1}\int x^m \log^{n-1} x dx$.

177. $\int \frac{x^m dx}{\log^n x} = -\frac{x^{m+1}}{(n-1)\log^{n-1} x} + \frac{m+1}{n-1}\int \frac{x^m dx}{\log^{n-1} x}$.

# INDEX

(The numbers refer to pages.)

Spotted something that isn't quite right?

Let us know and we'll fix it for future reprints. And as a thank you for your diligence, we'll send you a free Sothis book of your choice.

publisher@sothis.press

This page intentionally left blank

Made in United States
Orlando, FL
04 November 2026

90206684R00282